ENVIRONMENT, AGRICULTURE
AND
POVERTY

About the Editors

Ali Mohammad, M.A., PG. D. is former Chairman, Department of Geography, Aligarh Muslim University, Aligarh. His field of specialization is agricultural geography, food security and nutritional problems in rural India. Earlier he has worked as a Senior Research fellow in Jawaharlal Nehru University; as an Associate Professor at the Centre of West Asian Studies, Aligarh Muslim University, Aligarh; Associate Professor in Egerton University, Kenya; and Professor in Asmara University East Africa, Eritrea, Dr. Ali Mohammad has to his credit numerous research papers and more than one dozen books on agriculture, food security and nutritional deficiency.

Abdul Munir, Senior Reader, Department of Geography, Aligarh Muslim University, Aligarh, U.P.

S.H. Siddiqui, Senior Reader, Department of Geography, Aligarh Muslim University, Aligarh, U.P.

Rocket Ibrahim, Director, Jamia Millia Islamia, (Central University), Jamia Nagar, New Delhi.

ENVIRONMENT, AGRICULTURE AND POVERTY

VOLUME ONE

Editors
Ali Mohammad
Abdul Munir
Shamsul Haque Siddiqui
Rocket Ibrahim

CONCEPT PUBLISHING COMPANY, NEW DELHI-110059

ISBN

978-81-8069-270-3 (Set)
978-81-8069-286-4 (Vol. II)
978-81-8069-285-7 (Vol. I)
978-81-8069-287-1 (Vol. III)

Edition 2016

Published and Printed by

Concept Since 1974

Concept Publishing Company Pvt. Ltd.
A/15-16, Commercial Block, Mohan Garden
New Delhi-110059 (India)
T: 25351460, 25351794, *F*: 091-11-25357109
E: publishing@conceptpub.com, W: www.conceptpub.com
Editorial Office: H-13, Bali Nagar, New Delhi-110 015, India.

Cataloging in Publication Data--Courtesy: D.K. Agencies (P) Ltd. <docinfo@dkagencies.com>

Environment, agriculture and poverty / editors, Ali Mohammad ... [et al.].
3 v. cm.
Papers presented at a conference.
Includes bibliographical references.
Includes index.
ISBN 9788180692703

1. Agriculture—Environmental aspects—India—Congresses. 2. Agriculture—Environmental aspects—Asia—Congresses. 3. Poverty—Government policy—India—Congresses. 4. Poverty—Government policy—Asia—Congresses. I. Ali Mohammad, 1946-

DDC 630.954 22

INTRODUCTION

The future of humanity is at critical juncture as the population continues to grow at a high rate in most of the regions surrounding South Asia, which is characterized as world's most deficit region with high level of poverty. The looming shortage of basic needs has compelled the man to irrational exploitation of his immediate surrounding, *i.e.*, natural resources available which has adversely affected the environment. In regional context increasing greenhouse gas emission, ozone depletion and acid rain, loss of bio-diversity, deforestation and loss of top soil and fertility, shortage of water, food and fuel indicate how the natural systems are being pushed even closer to their limits of exploitation.

In order to fulfil the basic needs and to have a hunger free society, we should harness the available resources and the public distribution system in a judicious manner.

In this country and other surrounding countries agriculture is still a main source of survival but highly affected by environmental phenomenon because of the poor infrastructure and poor socio-economic base. With the increase of population and other demographic characteristics we want to raise more food, raw material and for other purpose such as food security and changing demand. Therefore, we have to produce very much more per hectare because we have very little more land, which we can bring under cultivation. Therefore, multiple cropping and intensive input based agriculture is going to be surely a characteristic of the future. This agricultural system require intensive use of chemical fertilizers and irrigation that leads to ecological imbalances. What are going to be the long-term effects of some of these changes which we try to introduce? The high yielding varieties have made a great impact which has aroused the enthusiasm of the farmers but it needs high amount of modern inputs. Probably under these conditions there will be more outbreaks of pests and diseases in the severe form than in the past. What will be the long range effects of some of these practices?

In spite of substantial increase in global food production, FAO has estimated that about 790 million people in the developing countries and about 34 million people in developed and transition economies were suffering from under-nourishment during 1995-1997.

Approximately, 75 per cent of these food deficit population is located in Asia, the Pacific and sub-Sahara region. The largest number of food insecure people is in India (about 200 million).

Agriculture alone cannot solve the problem of hunger and poverty unless and until other supporting factors are not applied. The agricultural production should reach to the weaker sections of the society. Moreover, efforts should be made to generate employment in rural areas. This will improve the purchasing power of the rural masses, particularly the landless labours and artisans.

The regional dimension of environmental degradation is also very dangerous for its inhabitants throughout the world. Change of the environment in one region of country affected others and the problems had to be tackled on a global or regional scale. The issue of environmental diversity with diversified pattern of resource endowments should be understood according to the need of the people living in the region.

An integrated approach is needed to have a sustainable development of all the resources available in the diversified environment, so that it should be viable for man in anchoring the poverty.

About 65 per cent of the livelihood of the people is dependent on agriculture in India by providing employment but poverty has become a way of life for those totally dependent on agriculture because of the small and marginal size of landholding. It is also observed that per hectare productivity has increased two and half times after independence and degradation of environment, especially the soil nutrients depletion has become a major issue for the sustainable agriculture. Input-output ratio is being disturbed in intensive agricultural pockets and it is observed that the model of Punjab agricultural development is not favourable to continue. It shows that environment should be protected in a situation where India is characterised as the home of the largest number of poorest population which is not at all desirable. Therefore, poverty alleviation programme, schemes and planning will affect the environment. But at the cost of development, environment should be protected or not, poverty will prevail at mass level in India.

ALI MOHAMMAD

CONTENTS

Volume One

ENVIRONMENT

VOLUME TWO

AGRICULTURE

VOLUME THREE

POVERTY

CONTRIBUTORS

A. Taheri Tizro
College of Agriculture,
Razi University,
Kermanshah, Iran.

A.K. Srivastava
Haryana Institute of Public Administration,
Sector 18, Gurgaon,
Haryana.

Abdul Munir
Senior Reader,
Department of Geography,
Aligarh Muslim University,
Aligarh.

Abdul Rauf Shah
Research Associate,
Geography Unit,
Center of Central Asian Studies,
University of Kashmir,
Srinagar (J&K).

Agha M. A. Siddiqui
Department of Architecture and Planning,
University of Roorkee,
Roorkee-247667.

Ahmad Ali Bozorgzad
Islamic Azad University,
Najafabad, Esfahan,
Iran.

Ahmad Mujtaba Siddiqui
Research Scholar,
Department of Geography,
Aligarh Muslim University,
Aligarh.

Ali Mohammad,
Professor, Department of Geography,
Aligarh Muslim University,
Aligarh (U.P.)

Alpana Chandra Scott Lecturer,
I.T. Degree College,
Lucknow.

Archana Gupta
Department of Grography,
Allahabad University,
Allahabad.

B.R. Kareriya
Research Scholar,
Department of Geography,
Banaras Hindu Univeısity,
Varanasi.

B.R.K. Sinha
Reader, Department of Geography,
Visva Bharti University,
Shantiniketan-731235.

Balak Ram
Central Arid Zone Research Institute,
Jodhpur-342003.

Beik Mohammadi
Department of Geography,
University of Isfahan,
Iran

D.D. Vishwakarma
Assistant Professor,
Department of Geography,
Government P.G. College,
Chhindwara, (M.P.).

D.K. Mishra
Department of Grography,
Delhi School of Economics,
University of Delhi,
Delhi.

G.M. Mir
Reader, Geography Unit,
Centre of Central Asian Studies,
University of Kashmir,
Srinagar (J&K).

G.S. Chauhan
B.R. Ambedkar College,
(University of Delhi),
Geeta Colony,
Delhi-110 003.

Gheesa Lal
Central Arid Zone Research Institute,
Jodhpur-342003.

H.N. Mishra
Department of Geography,
Allahabad University,
Allahabad.

J.S. Chauhan
Central Arid Zone Research Institute,
Jodhpur-342003.

K.V. Sundaram
Founder Chairman,
Bhoovigyan Vikas Foundation,
641, Sector-A,
Pocket-C, Vasant Kunj,
New Delhi.

Kumar Nikhil
Scientist,
Environmental Management Group,
Central Mining Research Institute,
Dhanbad, Bihar.

M. Ahmad
Scientist,
Environmental Management Group,
Central Mining Research Institute,
Dhanbad (Bihar).

M.P. Tripathi
Department of Geography,
Tribhuwan University,
P.N. Campus, Pokhra,
Nepal

Mehdi Momeni
Najafabad Islamic Azad University,
Najafabad, Iran.

Mohammad Jafar Zomorrodian
Department of Geography,
Ferdowsi University Mashhad,
Mashhad, Iran.

Mohammad Shafi
(Prof. Emeritus) (Padam Shri)
Department of Geography,
Aligarh Muslim University,
Aligarh.

Mohd. Taufique
Research Scholar,
Department of Geography,
Aligarh Muslim University,
Aligarh.

Mumtaj Ahmed
Research Scholar,
Department of Geography,
Aligarh Muslim University,
Aligarh.

Nizamuddin Khan
Senior Lecturer,
Department of Geography,
Aligarh Muslim University,
Aligarh.

R.B. Singh
Department of Geography,
Delhi School of Economics,
University of Delhi,
Delhi-110007.

R.S. Dixit
Reader-Head,
Post-Graduate Department of Geography,
Isabella P.G. College,
Lucknow University,
Lucknow, U.P.

R.S. Sangwan
Department of Geography,
University College (M.D.U.),
Rohtak, Haryana.

Rais Ahmad
Reader,
Department of Agricultural Economics and Business Management,
Aligarh Muslim University,
Aligarh (U.P.).

Rajanish Kumar
G.B. Pant Social Science Institute,
Jhusi-Allahabad.

Rocket Ibrahim
Professor,
Department of Geography,
Aligarh Muslim University,
Aligarh (U.P.)

S. Hasan Ansari
Reader,
Department of Geography,
M.D. University,
Rohtak (Haryana).

S.B. Singh
Head,
Department of Geography,
Banaras Hindu University,
Varanasi.

S.W.A. Ashraf
Professor,
Department of Geography,
Aligarh Muslim University,
Aligarh (U.P.)

Saifuddin
Remote Sensing Application Centre,
7-B, Old Shibli Road,
Aligarh Muslim University,
Aligarh.

Sayed Hassan Hossaini Abari
Department of Geography,
Esfahan University,
Esfahan, Iran.

Shamsul Haque Siddiqui
Senior Reader,
Department of Geography,
Aligarh Muslim University,
Aligarh.

Soorya Lal Amatya
Department of Geography,
University of Kathmandu,
Kathmandu, Nepal.

Sumanta Chatterjee
Research Scholar,
Department of Geography,
Visva Bharti University,
Shantiniketan-731235.

Tapeshwar Singh
Reader,
Department of Geography,
M.M.H. College (C.C.S. University),
Ghaziabad (U.P.).

Yasir Saeed Hanafi
Research Scholar,
Department of Geography,
Aligarh Muslim University, Aligarh.

ENVIRONMENT

1

KEY NOTE ADDRESS

MOHAMMAD SHAFI *(Professor Emeritus)*

The theme of the conference is Environment, Agriculture and Poverty. An analysis of the theme will show that agriculture stands at the centre-piece in the theme. Seventy per cent of the population is dependent on agriculture directly or indirectly and agriculture constitutes the single most important occupation of the people. It provides not only occupation to the people, but also food, feed, fibre, fuel and shapes the life of people in diverse ways.

What is the essential criteria for the development of agriculture. Should we exploit the resources pertaining to agriculture to the maximum extent possible ? or land-use should take care of ecological balance also. The answer would be that agriculture should be sustainable. In other words the fertility of the soil should be maintained and should not lead to exhaustion, the cropping pattern should conform to the prevailing environment, and productivity of the land should ensure a reasonable yield per hectare of the crops.

Indian yields of crops are one of the lowest in the world. Except sugarcane, the yields of all other crops are 1/3 or 1/4 of the advanced countries of the world. Even in some cases they do not compare well with some of the developing countries, low yield leads to low income, low income leads to ill health and poverty and poverty leads to low yield. The Indian farmer is caught in a vicious circle. So the need of the hour is soil and water conservation and management revolution. We had green revolution which has increased the foodgrain production substantially, but in order to attain sustainability in agriculture, we need the second revolution of soil and water conservation.

China has less cultivated land than India but its foodgrain production is three or four times than that of India. Its land fertility is amazing as she makes extensive use of nightsoil and compost to make the land fertile and prevents land degradation.

There are substantial areas in India that are affected by soil erosion or land degradation. There is urgent need that all such areas be identified with the help of remote sensing and field surveys, where geographer can play a notable role.

Another area where geographers can make their contribution is in the integrated watershed management in the catchment of flood prone rivers.

In India large quantities of biomass are produced by the plant and animal systems. It is estimated that about 1000 million tonnes of organic wastes in the form of crop residues and another 500 million tonnes of cattle dung and animal droppings are available annually in the country. These materials contain about 6 million tonnes of nitrogen, 2.5 million tonnes of phosphates, and 5 million tonnes of potassium.

If these plant nutrients are recycled, it will *(a)* improve the agricultural economy, and *(b)* the crop land would become much better.

Again the rural compost, which can be prepared from the rural wastes, would be about 60 million tonnes and the urban wastes would contribute about 20 million tonnes of compost.

It may be noted that in rural areas 80 per cent of the needs is met from fuel-wood and cowdung. If the arrangement for the energy needs of the rural people is not met before the material is used as manure, there will be a crisis. The answer is biogas on a massive scale. The Sulabh International Social Service is doing excellent work in this connection.

Even if sustainable agriculture is attained and the environment is taken care of, the million dollar question remains: the alleviation of poverty which is closely linked with the growth of population. India has made commendable effort in reducing the growth rate of population from 2.4 per cent in 1981-1991 to 1.6 per cent per annum in 2000 A.D. but even with this rate of growth in 2020, 240 million tonnes of foodgrains would be needed at the current rate of nutrition (half kg of foodgrains per day).

The thesis earlier advocated was to increase production and when supply is more than demand, prices would come down and poverty will be gradually reduced. The theory was in vogue for quite some time. Now the concept is changed. Reduce poverty and production will consequently increase. Even the World Bank has veered round this view. How to reduce poverty and bring about social transformation? The Government of India promulgated umpteen schemes towards the alleviation of rural poverty. For example 1969-74 two schemes were floated: (1) Crash Scheme for Rural Employment; (2) Pilot Intensive Rural Employment Programme. In 1980 came the National Rural Employment Programme. In 1983 a new scheme was introduced: Rural Landless Employment Guarantee Programme 1989—Jawahar Rozgar Yojna (which combined NREP and RLEGP) and finally IRDP, where the objective was announced to identify the poorest of the poor and they will be provided assets instead of giving cash to carry on some business. All these schemes have not made any significant dent in poverty reduction. The IRDP on paper is a good theoretical proposition but could not be put in practice in the real spirit.

Poverty alleviation needs a three pronged attack:

1. Block survey of the resources through village Panchayat leading to identification of growth poles and growth centres.
2. Establishment of small scale industries based on local and if need be on imported resources, and supplying the needed technology wherever necessary, to the co-operatives looking after the industries.
3. Development of infrastructure, road linkages to urban centres for the sale of goods and providing employment to the rural people and increasing their purchasing power.

The main focus of employment will be the landless labourers, small and marginal farmers, share croppers and artisans. It is through the Panchayats that education of each block be ensured, and women be given more power and as far as possible be made earning members. If the purchasing power of the people is not ensured, reduction of poverty and social transformation will be a mirage. If employment on the above lines is generated in rural areas it will stern the tide of migration of people from the rural to the urban areas and the quality of life in the cities will be improved. The Geographers can play a vital role in this national task.

2

NATURAL ENVIRONMENT, CONSERVATION, MANAGEMENT AND POLICIES

MEHDI MOMENI

INTRODUCTION

The earth enjoys its own specific physical and chemical features, obtained over billions of years. The essential conditions which have made life possible on it are included in the unisphere framework. This is a thin layer of water and soil making up a relatively small portion of the earth's shell and a rather short altitude of the air, comprising vital gases of high importance the correlation of which with soil have provided the earth's complex and delicate evolution. This correlation has caused the destruction of every natural system on earth to leave long-lasting large reactions.

Despite his immense progress in science and technology, man today has proved incompetent in the management of his environment. The major interrelated factors of his failure are : Rapid population growth, Natural Sources Overuse and Exhaustion and Environmental Pollution.

RAPID POPULATION GROWTH

From the earliest time, when man made his first appearance on the earth until about 8000 B.C., the population of the earth did not exceed 1,50,00,000 in all.

This amounted to 200 million about the advent of Jesus Christ.

By 1825, when Malthus introduced his famous treatise on population explosion, it reached one billion. Then on, thanks to technological advances, industrialization and modern medicine, population growth speeded up with a higher rate than before. Within the span of a century it doubled and from 1925 to 1976—a half century—it doubled again to 4 billions, and by 1990 it increased to 5.3 billions (Kendy, 1994, p. 37).

Based on an estimate made by the World Bank, from 1993 on, increase rate in population has been 100 million a year. Thus, the world population, according to this estimate rises to 6.5 by 2000 and will rise to 8.7 by 2030 (Deuchland, 1994).

Population growth is mostly a developing country's problem, the growth rate is 0.22 per cent in European countries and 3 per cent annually in African countries (Kifter, 1999, p. 37). This problem arises from different fertility level in various countries, which has declined sharply in industrial countries (Abedin, Dor Koosh, 1999).

Population growth implies increase in air, water and soil pollution, destruction of natural sources and, in general, decline in the quality of life standards.

One of the immediate outcomes of incrêase in population is the unwise urban expansion and growth which have led to extensive ecological issues, especially after the Industrial Revolution.

Urban Life and its Ecological Outcomes

Urban life, once a merit for city dwellers, is presently a demerit and menace to man's life and his environment. Cities are the origin of civilizations, learning, culture, economy and man's innovations. But urban life (city-dwelling) enjoyed a rapid growth in the twentieth century.

In the mid-twentieth century, the population of merely two cities exceeded or amounted to 10 million people, while at the present time there are 45 metropolitan cities, which with their suburbs, have inhabited 5 to 10 million people each (George, 1993, p. 80).

The third world cities cope with painful issues: They suffer epidemics and also vast migration of villages.

Despite the cities, failure to produce productive jobs, the migration keeps going on. This phenomenon, known as "over-urbanization", is an indication of the growth of the cities without development and growth in economy.

Urban growth, today, is a process of "Globalization" and indeed, for man to keep pace with living in the framework of big cities, what a big price he has to pay! Actually, how can we plan for such mammoth "inhabitats"?

Along with the expansion of cities, natural gifts and values, as the bed and respiratory system of the environment, are subjected to destruction daily, and man is left alone, exhausted, depressed and deprived of natural attractions (Momeni, 1998, p. 79).

The matchless features of the urban ecosystem are its anthropological sides, comprising all aspects of population—cultural, social, psychological, economic and political. Therefore, the major goals of urban and environmental planning in modern world must pivot:

(a) Preservation of Nature *versus* Constant Exploitation.
(b) Preservation and promotion of life standards.

NATURAL SOURCES, OVERUSE AND EXHAUSTION

The effects of progressional population growth and per capita consumption of natural resources, raised the exploitation of resources five-fold. The consumption rates of most metals and ores, since 1930 have exceeded their total consumption in past ages. A book, entitled "Natural Resources in Future America", published in the mid 1960s, estimated that by the year 2000, compared with the sixties, food production will rise three-fold, energy five-fold, metal alloys production three-fold and wood consumption three-fold, too. The latest researches and controls have revealed that these estimates are reliable and we should expect more misuse and unwise employment of natural resources in the new century (Hagett, 1995, p. 344). A hint to statistical facts can clarify the actualities of the ecosystems.

Water

World food production depends, to a large extent, on the possibility of increase in water by rain, rivers and subterranean sources. But both, the rivers and the underground water sources are presently suffering a shortage. The volume and combination of the earth's waters are as follows: 97 per cent of the sea water is salty and the remainder which amounts to 42 million cubic kilometres is fresh. 77 per cent of the fresh water has frozen in glaciers — Polar or non-polar; 22.4 per cent is imprisoned as subterranean water; 35 per cent forms lakes and pools;

0.04 per cent is in the atmosphere and the rest - 0.01 per cent runs in the rivers. Therefore, considering the population growth and the inevitable water consumption in urban sections, in agricultural and industrial sectors, against the limited available fresh water, the seriousness of the issue man faces becomes obvious.

Arable Pieces of Land

Arable land is restricted and the total agricultural pieces of land on earth amounts to 32 million square kilometres. Supposing that every 4,000 square metres of land can feed just an individual, the available agricultural areas on the earth can supply food for 8 billion people (the political map, 1990). But this may increase, provided technology succeeds in raising soil fertility.

Because of the lack of technology and the capital needed to preserve the soil, based on some estimates, nearly 20 per cent of the upper layers of the soil has been ruined and consequently food production has universally decreased.

Wheat Production

A U.N. report, published in 1996, indicates that wheat production has declined since 1986. In the present conditions, an area of about 240 million hectares has been devoted to the cultivation of wheat (Pur fikoiei, 1996, p. 117) on the earth, while 840 million people are living below the poverty line. According to a research made by the FAO, about 500 million children are underfed and are suffering from irrevocable mental and physical handicaps.

Forty years ago, Jouzneh De Castro the destinguished geographer and the FAO's ex-President, announced that "Man, with the advancement of technology will have no difficulty providing food for billions of people in future".

But why the nutrition experts' calculations proved wrong and that man has already arrived in the twenty first century and still fails to provide enough food is an issue which can have numerous solutions. No doubt, there are many political, social, economic, geographical and ecological factors involved in this respect.

Environmental pollution, on the one hand, and the climatic upheavals—and outcome of the nursery conditions, as some scholars believe—have brought about detrimental impacts on global agriculture, notably crops. One the other hand, and imbalance in world food consumption has aggravated the situation.

Destruction of Forests

Water, soil, air and vegetation, the most valuable natural assets of the earth have always been victims of man's expansionism, menacing global life, in general, and man's in specific.

Employing wood to furnish heat and lighting, the removal of trees to provide clearings for farming and raw-materials for pulp (paper) causes the destruction of wood-lands. For instance, about 11.3 million hectares of equatorial rainy jungles are annually ruined (Hunt, 1997).

Trees absorb carbon dioxide (CO_2) and emit oxygen into the atmosphere instead; 23 per cent of the world CO_2 problem is the result of the destruction of the forests (Fotooreh Chian, 1997).

Uncontrolled Fishing

The marine ecology consists of a variety of elements and is susceptible and vulnerable to foreign substances which are intentionally emptied in the sea or enter it.

The constant ruin of the sea-bed which is an outcome of incorrect pattern of fishing, the destruction of the sea animals living in the Coral rocks, the encroachment of the sea on the land, the ruin of the places where fish is bred, accumulation of poisonous materials in the sea, ... all are good examples of encroachment on the sea, which are improperly considered as development.

In some parts of the world fishing does not comply with standard norms; for instance, the number of the fishermen is too high while the amount of the fish in the area is limited. The annual global catch has long been 87 million a year (Parfit, 1995).

ENVIRONMENTAL POLLUTION

Today, man's activities, from the viewpoint of size compete with those of nature. Accordingly, what in the past aroused local ecological crises, causes global issues now. Furthermore, some ecological problems, such as increase in carbon dioxide, once unimportant, are now threatening basic ecological cycles on the earth. Air pollution, ozone layer impairment, increase in nursery gases, water resources contamination, soil pollution, the production of poisonous refuse and the like are all considered evil constituents of man's environment today.

Air Pollution

The air, as the most basic need of all living-being is a natural combination of nitrogen azote (78 per cent), oxygen (21 per cent), argon (9 per cent) and carbon dioxide (.03 per cent) (Bisser, 1994, p. 48). The rest is made up of small percentages of rare gases such as : xenon, hydrogen and helium. Ecology can naturally withstand restricted contaminations by self-refinement. Technological advance has imbalanced ecology. The sources which pollute the air fall in two classes: natural and artificial.

Notable among the natural sources poluting the air are the active volcanoes which pour, gases such as SO_2 and CO_2 into the atmosphere. Sandstorms, duststorms and thermal springs which raise sulphuric gases into the air are pollutants whose restricted activities are considered a part of the natural cycle.

Artificial sources which contaminate the air are of 4 types: Industries, trade and home industries, motor vehicles and miscellaneous industries.

Industries

Industrial revolution is mostly related to the new sources of energy and no industry comes into being without it. Industries are the means of man's advance in life, but they make pollution and depending on the production they have and the fossil fuel they use add smoke, CO_2, SO_2 and so forth to the air. The industrial refuse of the 24 major industrial countries amounted to 90 million tonnes in 1990 (Hunt, 1997). Home industries and trade units, too, produce smoke, CO_2 and SO_2 gases and have their own shares in air pollution.

Motor Vehicles

Motor vehicles produce carbon monoxide, carbon dioxide, azotic oxides, lead and hydrocarbons and thus pollute the air. Based on the available statistics, at present, nearly 600 million cars run in the world and indication of the importance of car industry in man's daily life (the annual global car output is 60 million per year).

Effects of Pollution

The factors which affect man's activities negatively have increased in number and are hazardous to his future. Every year, about 145 million tonnes of sulphur dioxide, 250 million tonnes of dust, 70 billions cubic metres of gas, almost 1 million tonnes of the compounds of lead, tens of thousands of fluorides and chlorides and immense amounts of benzpyrene are emptied in the air many of which are poisonous and carcinogenic.

According to WHO statistics, the world has tested 60,000 kinds of poisonous substances to which 3000 new ones are annually added (Global problems, 1978).

Mortality rate and its tight correlation with smoke pollution is another dangerous result of the above mentioned gases.

Not only do these gases affect man directly, but they also widely affect ecology, notable among which are: increase in the earth's temperature

— Acid rains
— Ozone layer impairment

Increase in the Earth's Temperature

Ever since the Industrial Revolution began and coal consumption went up, the unfavourable effects of this fossil fuel have shown themselves. This increase in consumption has raised carbon dioxide content four-fold, the nursery impact of which has augmented temperature, in the past two centuries, about 10 degrees Fahrenheit (Solivam, 1977).

The earth's temperature data are preserved in Godard Institute, a part of the NASA. Based on these data, the mean temperature for 1995 is 0.3 in excess to that of the 1992. In the past century, the earth's mean temperature increase was between 0.3 to 0.6 centigrade. Such a rise has been unprecedented since the end of the last Ice Age about 1000 years ago (Vital, 1996).

Ozone Layer Impairment

Air pollution by SCFC has impaired the ozone layer of the atmosphere. The five per cent reduction in the ozone content of the air has raised ultraviolet reception on the earth from 7.5 to 15 per cent, which results in skin cancer rise (Eaglemann, 1981).

Decrease in the thickness of the ozone layer can raise ultraviolet radiation of the sun, and generally cause precocious diseases and delayed diseases. Erythma (sunburn) and photokratitis are included in the first group and different kinds of skin cancer, cataract and the weakening of the immunological system of the body are in the delayed type.

Water and Soil Pollution

Most countries have economic problems providing fresh water. Every year, about 32 cubic kilometres of industrial refuse enter subterranean water sources.

The oceans have also become polluted and 10 million tonnes of crude oil and chemical refuse enter them annually (Science and Vie 1991).

Marine ecosystems, unlike ones, contain micro-organisms. Most of the garbage and refuse, dumped in the rivers, finally enter the seas and contaminate them. Therefore, it is believed that 77 per cent of the sea pollution is caused by land refuse.

Some heavy metals such as copper, magnesium, zinc and iridium which are extracted through industrial processes enter the soil in different ways. There, they change form into solid solutions and are absorbed by plants, the products of which are very dangerous.

FIGHTING CONTAMINATIONS (CULTURAL REACTIONS)

Preservation and improvement of ecology is the removal of the faults and detriments of technology. This necessitates providing not only safe and clean water, air, plants, animals, birds and the like but also decreasing sound pollution and minimizing the waste and refuse and lowering the expenditures.

The expenses that unsafe environments have and cause short stagnation in economy are comparable, to some extent, with economic criteria. Obtaining a favourable level in cleanliness raises welfare expenses considerably and affects GNP (Gross National Products). These expenses are shown in the following formula.

$$W_c = P_c + A_c + D_c$$

W_c = Ecological pollution of destruction expenses.

P_c = Pollution prevention expenses.
A_c = Expenses to avoid pollution impairments.
D_c = Pollution damages.

Therefore, it is in the capacity of the ecology sector to create a kind of technology to minimize not only the refuse but also the pollution expenses.

Thoman and Corbin studied the interaction between man, culture and nature and in the assessment of economic activities and their relations with the impacts of the interactions between the environment and economic activities he believed in three domains : Natural-cultural probability domain, economically practical domain and economic reality domain (Tavalla-ei 1997, p. 24).

Increase in industries, the application of advanced technology and the grave global ecological issues have all led the lovers of the earth, in all countries, to form numerous associations and organizations such as :

International Institute of Environment, Environmental Protection Agency and Green Peace.

Green Peace group is the most important group that supports environmental conservation and has 5 million members in 24 countries. In the past 25 years, it has had increasing social and cultural activities all aimed at fighting ecological hazard.

It has also had its direct and indirect presence in many international organizations and conventions.

CONCLUSIONS AND SUGGESTIONS

Man's economic activities have had unfavourable ecological effects on the earth. Exploitation and application of natural resources depend on urbanity level, industrialisation and intensification of economic activities. Industrial countries are now responsible for 80 per cent of ecological issues: USA 30 per cent; western Europe 25 per cent; former Soviet Union and Eastern Europe 25 per cent and the developing countries 13 per cent only (Global problems, 1978).

Ecological issues, today, have got global or planetary dimensions and it is in the capacity of the powerful institutions to see to them. These issues are so enormous and far-reaching that they involve not only the destiny of the whole planet but also the social and ecological complexities all of which are grounded in the very dominant patterns of production. The crises will escalate provided the issues of production, distribution, technology and development be fundamentally settled, at universal level.

Today's actual ecological issues are environmental pollution of air, water, soil and sound; insanitary living condition in huts; lack of access to safe drinking water; sub-standard sanitation, urban crime increase, violence, traffic... all of which are the direct results of economic inequalities. The gap between the poor and the rich had doubled in the past four decades, and so is the situation between countries and social classes (Khoshnevis, 1997, p. 5).

Environmental preservation is creating a harmony between man and the environment he lives in which necessitates balanced action and interaction on both parts. It also means making stable conditions for society to meet its needs moderately and logically, without depriving the coming generations of natural resources and being fully aware of its responsibilities. To achieve these objectives I suggest :

1. Preservation,
2. Recycling,
3. Application of renewable resources,
4. Repair and improvement,
5. Population control, and
6. Adaptation to Nature.

To achieve the above objectives, man is in need of a spiritual and cultural development. This could be the acquisition of a new outlook and culture based on the friendship among nations and men's living in harmony with nature, that can provide solutions to increasing international crises. And at the threshold of the twenty-first century, the establishment of new political and cultural institutions — local or international — to maintain inter-nations dialogues that can help to plan and cooperate for a better and promising world.

REFERENCES

Abedin Dorkosh (1994), *Urban Economy*, University Publication Centre.

Biser, Artur (1994), *Earth* (translated by Abass Jafary) Tehran-Gitashenasi, p. 48.

Chongnon, S.A. (1973), *Atmospheric Alteration from Man Made Changes*, University of Victoria, p. 134.

Deuchland, No. 5, October 1994.

Docastore, J. (1968), *Geopolitic of Hunger* (translated by Monir, J.), University of Tehran.

Eaglemann, R.J. (1981), *Focus on the Ozone Layer,* Environ Conservation, p. 148.

Economist, 15 Feb., 1992.

Fatorechian, Saied (1997), *Hamshahry Newspaper*, No. 1147.

Global Problems and the Future of Mankind, Progress Publishers, 1978.

Goddard Institute, for space studies.

Haggett Piter (1995), *Geography: A Modern Synthesis* (translated by Shahpor Godarzy) Samt, p. 344.

Holdgato, M. (1982). *The World Environment.*

Hunt John (1991), The grip tightens, *Financial Times*, p. VI.

Hunt John (1997), Environment rising level of concern, *Financial Times*, p. VII.

Intergovernmental panel on change (IPC).

George, Piere (1993), *Geography of Population* (translated by Dr. Siros Sahamy, Mashhed, Nica), p. 80.

Khoshnevis, Mohammad Hasan (1997), *Hamshahry Newspaper*, No. 1090.

Ghesas, Mohammad (1989), *World in the Twentieth Century,* p. 56.

Kendy, Pole (1994), *Twenty First Century* (translated Abass Mokhber), Tehran, p. 37.

Kifter, Netben (1999), *Population* (translated by Mohammad Taghy Sabzei), Afarinesh, N. 144.

Mankind at the turning point, 1974.

Mazumdar, S. Schimmel, H. Higgins (1982), Relation of daily mortality to air pollution, *Environ Health*, p. 213.

Mahmody, Ali (1987), *Economy of Sources*, University of Tehran, p. 11.

Momeni, Mehdi (1998), *Human Geography*, Mohtasham, p. 79.

Nori, Jafar (1989), *The Influence of Industrial Development in Kimiya*, No. 7.

Omidpanah, Parviz (1986), *Ecology*, Esfahan, pp. 11-18.

Parfit Michael (1995), *National Geography*, Nov. 1995.

Purfikoi Ali (1996), Geopolitic, Mashhad, Vaghefy, p. 117.

Polad Dej Mohammad (1987), *Industrial Placement*, Tehran, p. 171.

Rice Robert (1990), Along nightmare in litigation, *Financial Times*, p. VIII.

Riordn, T.O. (1976), *Environment*, London, Pion, p. 228.

Science and Vie, Nov. 1991.

Salivan Walter (1977), *Scientist Fear heavy use of coal may bring adverse shift in climate*, New York.

Tavakoly, Mohammd Bagher (2000), *Research Design* (UVR) in Esfahan, p. 4.

Tavallay Simin (1997), *Geographical Economy*, Tehran, p. 24.

The Political Map of the World, Progress Publishers, 1990.

Vital Signs Lester, R., Brown *et al.* (1996), World Watch Institute.

Ward Barbara (1995), 5 *Ideas That Changed The World.*

3

WATER USE AND SOCIAL ORGANISATION IN THE DESERT BORDER OF IRAN

The Zāyandeh-Rūd River

S.H. HOSSAINI ABARI

The following contribution is an attempt to open some Iranian past theories and managing systems concerning the justified and elaborate managing rules due to the scarcity of water resources and the critical needs for water in the desert bordering areas of Iran (*i.e.* the Zāyandeh-Rūd River of Esfahān).

The Zāyandeh-Rūd River

The Zāyandeh-Rūd, the most honourable constant river in the inner plateau of Iran, rises in the Kuhrang mountains in the Bakhtiāry state, some 360 kms sounth-west of Esfahan.[1] It flows north-east for 45 kms through the Bakhtiāry country, watering a narrow strip on its banks, and then, passing through the Bakhtiāry pastural areas in Chahār Mohāl Va Bakthyāry and Fridan among a number of gorges, enters the Faridan district. There in its way to the entrance of the Zāyandeh-Rūd reservoir lake dam, joins by a number of tributaries and sub-tributaries, amongst them some (Dimeh, Khorbeh-Pelasgān and Naleshkanān) which mostly drain the western and north-western areas of the region, are the main. The Zāyandeh-Rūd from its lake dam to the Gāvkhūni Marsh along with a distance of about 240 km in its way, does not join any noticeable tributary, but using its water in irrigation, urban usages and industries.

The total discharge of the Zāyandeh-Rūd originally has been an average of 960 million m^3,[2] but during the four past decades its water has increased through two diverting tunnels, namely Kubrang first and Kuh Rang second[3], and the third tunnel with an annual average discharge of 250 million m^3. That finally calculates the amounts of water to be 1650 million m^3, is to be used in the irrigating an area of 9500 ha. in its down stream.[4]

General Situation of the Zāyandeh-Rūd Region

A description of the climatic conditions in Iran but mainly in its central plateau is essential for understanding the need for water and its management system.[5]

In the desert border of the mouth of the Zāyandeh Rūd region the amounts of precipitation received are between 50 and 120 mm per annum (national average 250-300 mm at the source area of Zāyandeh Rūd, in the highest peaks of Zardkuh 4450 m height and almost about 50 mm in the mouth of the Zāyandeh Rūd (the Gāvkhūni Marsh) 1440 m height,[6] and they are very heterogeneously distributed. In summer, the temperatures are over 40ºC. in the shade (in Gāvkhūni area) and in winter they are occasionally 32°C. at a max[7], in Zardkuh, large differences in night time and day time temperatures are typical of such diversified climatic conditions in high ranges up to down areas of Zāyandeh-Rūd Region.[8]

Though in this region, agriculture without irrigation specially in the second half of the area (about up to 2/3rd of its area) is possible only by irrigation in the cold season's cereals (*paizeh*) and spring cereals (*bahareh*).[9] In the irrigated area of the Zāyandeh Rūd Valley, desert border beings can be defined as in which life is virtually impossible without human efforts to utilise and do a careful management on the run offs of the Zāyandeh Rūd. Therefore, from the early history of the Esfahan City (the main metropol in Zāyandeh Rūd Region) some elaborate systems of water management has been always done on it, of which there, a well known regulation has been in action from the A.H. 923 (A.D. 1524) the Safavid times, popularly known as Shaikh Bāhai's Tūmār. This system remained in force until A.H. 1315 Shamsi (A.D. 1936), when it underwent alternation, owing to the fact of government orders, due to which the cultivation of cotton was subtituted in the lands watered by the Zāyandeh Rūd for the cultivation of rice.[10]

The details of this regulation are found in an interesting document, popularly known as Shaikh Bāhai's Tūmār, to be found in the general

Finance Office of Esfahan. The precise of the origin of Tūmār is not surely known. What is certain is that the sharing of the water of this river according to historical documents goes back to about 1700 years ago, in the early times of Sassnid Dynasty (the Ardeshir of Bābak), about A.D. 200.[11]

A Brief of the Managing System of Tūmār

Among the Persian documents a tūmār has been a long role on which some important regulations or arrangements were written. So that it would remain using for the long times, mostly on the paper or leather, the Tūmār of Zāyandeh Rūd water irrigation system was written on a cow leather (skin), in a number of copies. One to be kept in the sealing office for irrigation of the province, one as the guideline for the use of mirāb, and some for other uses.

The general copy of the Tūmār of Shaikh Bāhai is not the original one, but in recent years, it has been rewritten on paper in 27 pages (A. 4). The words and figures of the Tūmār have been in the traditional Persian. But now it is transferred into the recent Persian words and figures.

The following words are written at the beginning page of the Tūmār of Shaikh Bāhai :

"He is God most high, His Power is Mighty"[12]

"The sealing of the late honourable prince, his resting place be in the paradise, God make light his tomb, the Shah Tahmasb, he commanded that, since some differences had arisen in the arrangements of the water of the holy river of Zāyandeh-Rūd of Esfahan, so the component authorities of the powerful government should appoint a number of the trustworthy and aged men and the exalted and honourable Mostoufies to establish under the signature of the headmen of the watering boluks, which re-regulate and share the water of the Zāyandeh-Rūd, honestly and to the best of their knowledge, the share and lot of each village and hamlet in each boluk, according to its capacity and need, and enter these in the registers under guarantee, so that, that regulation should be put in to execution. And since the boluk of Jey is in the middle of the boluks watered by Zāyandeh-Rūd, as it has been a tradition from the past times that one of the reliable and trustworthy head men of that boluk should be charged with the responsibility of the office of Mirab; and the same procedure as it was followed in the former days must be followed at the present time. It is hereby established that the honourable officials and government respresentatives in the village and headmen and aged of the above mentioned boluks and the Mirabs, Mobashers, Mādisālārs and subordinate officials of the river, in obedience with this high command, having established, according to the instructions in the Tūmār, the water right of every place, according to its share shall not act its share contrary to this and having avoided being punished and called to account by the high authorities of the state shall honour the agreement. Written in the month of Rajab-ol-Morajjab A.H. 923" (20th July-18th August 1517).

The main interest of the Tūmār lies in the description it gives of the regulation of the Zāyandeh-Rūd, but it has also a subsidiary geographical interest, for in the Tūmār it can be realised certain geographical views (physical and human), *i.e.* the detail of the name and the water shares of each village, hamlet and the bolūk all through the main irrigation basin of Zāyandeh-Rūd from the Lenjān to the Gāvkhūni, an area of about 160 km, in length and about 65,000 ha.[13] in 7 bolūks namely as : Lenjān, Alenjān, Mārbin, Jey Va Barzrūd, Kerārj, Baraān and Rūdashtain as Table 3.1 shows :

Table 3.1 : Shows the bolūks and their shares[14]

Boluks	*The original shares*	*Secondary shares*	*Tertiary shares*
1. Lenjān	6	113.5	357
2. Alenjān	4	47.5	315
3. Mārbin	4	29	282
4. Jey & Barzrūd	6	27	674
5. Kerārj	3	12	287
6. Baraān	4	14	840
7. Rūdashtain	6	22	243
Total	33	275	3098

The regulation was allotted under three different rules, Āzād, Mokhtass and Moshtarek, so that having been divided according to location. With regard to the need and extent of the cultivation of each bolūk, the division of each bolūk was not subject to any variation, and being based on the sources of the river, even in years of drought or scarcity of water it required no alternation. Therefore, in the different conditions of availability or scarcity of water, the effect on the watering of the bolūks was the same.

Table 3.1 shows some geographical facts related to the water management of Zāyandeh Rūd river, as follows :

1. The final bolūks of Baraān and Rūdashtain were allowed solely to cultivate the winter cereals, due to the reasons : *(a)* In summer the runoff of the river is scarce, and *(b)* The evaporation in deserty Gāvkūni is abundant (more than 3500 mm annually). Therefore, the length of distance and high evaporation would loose the laying small amount of water of the river. So it was better to give those-two bolūks as the Mokhtass shared of the river only in two periods. The first was, when they ploughed the lands for wheat crops seeded and the second, was for the final irrigation of them.
2. The Bolūk of Kerārj, only could use the natural drains of the river in the hot seasons, therefore, that bolūk should divert its shares to the upper bolūks of Alenjān and alenjan for the rice cultivation. And this did not bring any harm for its main garden croppings in the summer.
3. In the summer season the water of the Zāyandeh-Rūd was not allowed to pass the Esfahan City. Hence the Tanāvobi or Rotational regulation worked only for the four bolūks of the west Esfahan only. But the Tanāvobi rule, in the normal years worked for each two of these bolūks by parallel. In the years by the scarce of water, the regulating mentioned carefully.

The Regulation of the Tūmār[15]

As was mentioned earlier, the use of Zāyandeh-Rūd runoff was allotted under three different rules, namely as : 1. Āzād, 2. Moshtared and 3. Mokhtass.

1. Āzād, in Persian literatures, Āzād means (a thing which is) free, and in Zāyandeh-Rūd irrigation system it means that, in the cold seasons when there is not a forced need for the irrigation of crops as well as the source of the river being increased due to the precipitation, therefore, in all over the

region each of the cultivated crops may use the water whatever it needs (during 195 days a year). But from the beginning of the month of Āzar (the last month of autumn) until the 15th day of Khordād (75 days past the No Vrūz)* the water was not to be under regulation. So that, during 195 days each bolūk could make good any provision deficiencies in the irrigation of its crops. The surplus, after the wheat crops had been irrigated, flowed to the Gāvkhūni. From the second month of spring until the 15th of Khordād (5th of June) if the water was abundant, it was sufficient for the irrigation of wheat crops of all the bolūks, but if it was scarce, from the 15th of Ordibehesht (5th of May) onwards, it did not reach the bolūks of Rūdashtain and Barāān, for which, special provisions were made (as it come later on).

2. Mokhtass (being sole or to use solely), Mokhtass prevailed at two seasons of the year. The first was from the 15th of Khordād to the 28th of the same month, during the 15 days, when the wheat in the other bolūks having been irrigated, the water was reserved for the final irrigation of the wheat crops of the bolūks of Rūdashtain and Barāān; and the second period was from the 15th Ābān (6th of Nov.) until the end of the month (21st of Nov. to 5th Dec.), during which the water of the river was reserved for the cultivation of those two bolūks. In each of these periods of fifteen days it was reserved for full days for the sole use of Rūdashtain and then the bolūk of Barāān also for five days shared it. During the first of these two periods of fifteen days, a little water from seven or eight places was also allowed to flow into the irrigation canals (Mādis) to Alenjān to be given to the newly planted rice.
3. Mushtarek (the joint rights among the irrigation bolūks watering by the Zāyandeh-Rūd) from the 15th of Tir (22nd of June) to the late Ābān (21st of Nov.) the water was shared in two ways. Firstly, in six periods of fifteen days and secondly in one period of eight days the water from the Kalleh Bridge (the start point of watering under the regulation of the Tūmār in West Esfahan) was shared among four bolūks in four keshiks (the area watered by the river is divided into six keshiks. Four in the west of Esfahan and two in the east of it),

* The yearly months of the Tumar all are calculated as being 30 days each.

Table 3.2 : The Watering Regulation of Zayandeh-Rud[16] according to the Tumar of Shaikh Bāhai

Bolūks	*Khordād*	*Tir*		*Mordad*		*Shahrivar*		*Mehr*		*Aban*		
Days	*16th upto 30th*	*1st upto 18th*	*19th upto 30th*	*1th upto 15th*	*16th upto 30th*	*1st upto 15th*	*16th upto 30th*	*1st upto 18th*	*19th upto 28th*	*29th upto 8th*	*9th upto 15th*	*16th upto 30th*
	May-June	*June-July*		*June-Sept.*		*June-Sept.*		*Sept.-October*		*Sept.-Nov.*		
Lenjān & Alenjan	—	18 days	—	9 days	9 days	9 days	— 9 days	11 days	10 days	— —	—	—
Mārbin & Jey va Barzrūd	—	—	12 days	— 6 days	— 6 days	— 6 days	6 days	7 days	—	10 days seeding	—	—
Kerārj & Barāān	—	—	—	—	—	— —	— —	— —	—	—	7 days seeding	—
Rūdasht	15 days for seedings	—	—	—	—	— —	— —	— —	—	—	—	15 days seeding

of Lenjān, Alenjān, Mārbin and Jey va Barzrūd, by turns according to Table 3.2.

Table 3.2 shows the irrigation calendar of the Zāyandeh Rūd in hot seasons.

From the 1st of Tir (21st of June) Lenjān had the right to the water to wherever it reached for three days, and Alenjān up to the Falāvarjān Bridge (fifteen kms. west of Esfahan) four days, separately, Lenjān had the right to the water for two days, and Mārbin, Jey va Barzrūd up to the Mārnān Bridge (the entrance of Esfahan) for six days. During the period of eighteen days it was shared from the 1st of Mehr (23rd of Sept.) to the end of that month.

For eleven days in Lenjān and Alenjān and Mārbin, in three bolūks, and for seven days by Jey va Barzrūd in one bolūk, so that the summer crops should be irrigated.

From the 1st of Mehr (23rd of Sept.) to the 15th of Ābān which was the beginning of the second period when the water was appropriated for Rūdashtain, the water was also shared by seven bolūks as follows :

Lenjān and Alenjān ten days, Mārbin, Jey va Barzrūd ten days, Kerārj three days and Barāān four days, the sharing of the water in rotation was known as Tanāvobi.

The regulation of the water by the Mokhtass, Moshtarek and Tanāvobi principles was in accordance with the exigencies of the cultivation and the need for water of the crops sown in each bolūk. Thus the water was appropriated for Rūdashtain and Barāān, at the time when crops were in great need, and would be completely destroyed if they did not have water, whereas the wheat of the Moshtarek lands had no need of water at that time, for during the Āzād period, the wheat crops of the Moshtarek bolūks had, by the 15th of Khordād (5th of June) been irrigated four times and, therefore, required no further water and suffered no harm. In the Mokhtass lands, on the other hand, in years when water was scarce, also from the 15th of Ordibehesht (5th of May) onwards the water was cut off from their wheat crops after they had been irrigated once or twice, and hence if the last water were not to reach the wheat crops of these bolūks, they would naturally be destroyed. Similarly in the Moshtarek bolūks, when it was the turn of the bolūks of Alenjān and Lenjān to be irrigated, Mārbin, Jey va Barzrūd did not go without water for more than three periods of nine days, during which time they did not require water.

The Tanāvobi principle which prevailed in the Moshtarek bolūks was also in accordance with their needs, because during the time that the seeds were taking root in Lenjān and Alenjān, which was from the 15th of Ordibehesht to the end of the month, and until the end of Āzād period (15th of Khordād), the ground was inundated and therefore able to do without water for six or seven days.

On the 1st of Tir (21st of June) when the transplantation of rice began, the seeds cropped in the lands belonging to Rūdashtain, having been flooded, and then lands of Lenjān and Alenjān were in their turn flooded for two or four times, and the rice in Lenjān and Alenjān was transplanted. While the water was being thus distributed in rotation, the periods during which Lenjān had no water did not exceed during four or six days, but Alenjān was without water for eleven days when the water belonged to Mārbin, Jey va Barzrūd and during that time careful attention had to be paid to the rice cultivation of Alenjān.

The above regulation of the waters of the Zāyandeh-Rūd was according to special rights of the system, and was divided according to three further principles: by bolūks, by streams (mādis) and by villages or hamlets.

1. *By bolūks,* the water allotted to the bolūks or a bolūk, was due to the boundaries which were limited by a keshik (guarding), and was regulated according to shares allotted to the stream, and which of those were divided according to shares allotted to the streams within its boundaries.
2. *By the streams*—Each village or hamlet (mazra'e) received the water according to its rightful share by means of the stable systematic regulation of the irrigation channels which were blocked by special stone and plaster, namely "Lat" up to a certain height and width, so that, in any condition of runoff, the amount of waterflow was according to its share, either in scarcity or plentiful of water in the river.
3. *By Village or Hamlets*—The water allotted to some bolūks such as Mārbin, Jey va Barzrūd etc., was regulated by Taq, *i.e.* one day or one night's water, and each taq referred to a certain portion of the village lands (Arāzi or Sahrā) and was divided into Fenjan, Peng or Piāleh (Iranian water clock), with reference to the lands of the taq, so that finally each cultivator in the bolūks knew that days of the week and for how many hours and how many minutes each Jarib (a farming

unit equal to 1000 sq. m) of cultivation, which he possessed, had water and he should cultivate his land accordingly.

The Social System of the Zāyandeh-Rūd Management

The services related to the Zāyandeh-Rūd river were both social and physical, as the control of the regulation and the sharing its water, the protection and maintenance of the irrigating establishments, the election of Mirab and his subordinates etc.

Since the Zāyandeh-Rūd is being the sole constant source of the irrigating water in the desert bordering area of the biggest city of the inter plateau of Iran, it had been always needed a careful control on water use. Therefore, the state governments traditionally consignmented all its managing affairs to the local people. It is the reason that during the long about 2000 years of its history of irrigation, the kings and local governors rarely came in direct touch with its managing system. There are only a few data about the past kings and the dynasties relating with the Zāyandeh-Rūd. There are some indications as follows :

1. The first time was at about 1780 years ago (228-238 A.D.), that the Ardeshir of Babak (the founder of Sasanid Dynasty) commanded that "since some disorders appeared in the regulation of the Zāyandeh-Rūd waters, an engineer, named Mehrin-e-Vardan should go and rearrange it."
2. The next time was in the time of Saldugid Which their court was in Esfahān (475-600 A.H.), (1000-1125 A.D.). They decided to increase the water of Zāyandeh-Rūd through rud trough the diversion of some waters from the Karun segment area (the Kuhrang mountains), but they did not start it.[17]
3. The third time was, as it is popularly known, the Shah Tahmasb command and the establishment of the regulation 923 A.H. known as The Shaikh Bāhai's Tūmār (923 A.H. or 1517 A.D.).[18]
4. In the fourth time with the aim of increasing the demand of the river, Shah Abbas the Great (1001 A.H.) through an open canal, in the Kuh-Rang area, started the work, but after his death, it was stopped.[19]
5. The last was in 1327 to 1332 A.H. (1948-1953 A.D.), that the first tunnel of Kuh-Rang was digged and it increased some

250 million m^3 of water to the main Zāyandeh-Rūd, and it followed to dig the tunnel No. 2 and the tunnel No. 3, of which the last is being under way. The estimate has been made that, after the tunnel No. 3 is done, the total demand of the water will be increased at about 1650 million m^3 of which 80 per cent (950 million m^3) is for irrigation, and the rests for urban use and industries.[20]

As it is mentioned earlier, the management of Zāyandeh-Rūd water and its services at the time of the traditional regulation (the Tūmār of Shaikh Bāhai) up to 1347 A.H. or 1968 was entirely consigned to the people, so that all the system was democratic and the governments or the state governors rarely had a direct hand over it.[21]

Table 3.3 : The Original Shares of the Bolūks and their Representatives[22]

Bolūks	*Original shares*	*Representatives*
Lenjān	6	6
Alenjān	4	6
Mārbin	4	4
Jey va Barzrūd	6	6
Kerārj	3	3
Barāān	4	4
Rūdashtain	6	6
Total	33	33

Hence, the Zāyandeh-Rūd in all its managing affair, through at least about four hundred years, was ruling as follows :

1. Each of the irrigating bolūks appointed a number of representatives among thier landowners, so that, the total of them although the irrigated area of the river calculated at 33 ones, as it is shown in Table 3.3.
2. Before the exact time of the need for the regulated use of the water (in the Nov. Ruz times), the above 33 representatives, as it was commended in the Tūmār of Shaikh Bāhai, were to appoint a capable, confident, and trustworthy man as well as

being expert on the water regulation of the river, as the Mirāb, from the bolūk of Jey as the Mirāb, to take charge of all the managing system of the river for a period of 165 days of a year (the regulated times of the water, from 5th Khordād to the end of Ābān). They also had a right to fail him if he was not able best to do his duty.

3. The Mirāb himself appointed a number of 6 persons, as his assistants (Keshiks) to look after the watering and sharing of the bolūks (one for each bolūk).
4. The Keshiks also then, appointed about 12 or 13 men as the Mādi-Sālārs (the chief of mādis) to manage the water in the area of each of the bolūks.
   Though the mirāb himself had the right to appoint his own assistants, the 33 named representatives should have a good recommendation upon them. Another mention also should be made that the representative of the assistants of mirāb should be as it was for himself. Then they might be reappointed by the mirābs for several years.
   The salary of Mirāb and his assistants came according to the water share in each bolūk (by the production not more than 1/100th of total raw harvest) of each cultivator.
5. Each of the hamlets had to appoint one of the capable cultivators to manage the sharings in the cultivated lands, namely as the Sarjūg. At the mean time, the Sarjūghs were also to choose and send some traditionally fixed number of the local cultivators for guarding the mādis in upper bolūks when their own turn of water came, and who has to send the needed manpower for other irrigation services, but the Sarjūghs did not ask any wage.[23]

All the payments, and services for the river and irrigational affairs in the Zāyandeh-Rūd irrigating area was based upon the one's sharing of water. And it seemed that it was the main reason due to which, during all the Esfahān history of the Zāyandeh-Rūd irrigation times, the above elaborate system of its water managing rule, has been growing its development and peacefulness in the area. But during the government rule on the river, since 1347 (1968), there has not been any trust on having an assured water for the cultivated lands. Due to which, it is the first time in the history of Esfahan and its river, that the Zāyandeh-Rūd has been dried for about 10 months (1379 A.H. or

2000 A.D.) and it may be increased for the months coming in 1380 (2001).

Unfortunately they cannot believe that in the growing countries it is obviously proved that "the government managing, specially on the economic affairs is suffering some chronic illness, and in Zāyandeh-Rūd every one can see with his own eyes.

Is the Tūmār an Original One and the Work of the Famous Shaikh Bāhai ?

As it was mentioned earlier, the following words has been written : at the beginning page of the Tūmār "Written in the sealing office of the late Shah Tahmasb". This is clearly that the Shah Tahmasb did not ascend the throne in the year 923 A.H. but he did in the year, about years after that time. The Tūmār also is not have been the work of the well known Shaikh Bāhai. He himself was born in 953 A.H., about 30 years later.[24]

Moreover, there are several documents and books referring the oldest regulation of the Zāyandeh-Rūd irrigating waters, *i.e.* Ebne Houghal, in 375 A.H. says that "The Zāyandeh-Rūd water is the most regulated ones in the world, of which not any drops run loosely. In each of the months of the year nine days it should flow to the rurality of Rūdasht."[25]

The Ebn-e-Rosteh also in 290 A.H. (920 A.D.) announced that, "In the Zāyandeh-Rūd irrigation region, each of every plant has its own share of the water".[26]

In the details of the Tūmār also several points is referring to the past regulations and sharings of the water.

There are more indications in the Tūmār itself that which proves the regulation not being the work of Shaikh Bāhai, but because he was well known learned man, learned in mathematics, it was attributed to him. In the some way that many old buildings are, at the present time atributed to the Shah Abbas, and the architecture of Shaikh Bāhai.

Moreover, in Iran there are so many works which popularly attributed to him, and he himself did not referred to them in his written materials.

There remains one thing to say, that, so the Shaikh has been the most well known learned man and authority in his own time, may have revised on the regulation of the water of the Zāyandeh-Rūd and mentioned some shares to the Court Gardens.[27]

The Tūmār itself also refers on its own priority in various ways.

Conclusion

In the desert bordering Esfahan, as being the most important economic and cultural region in the central plateau of Iran, from the early past, the people and the governments thought that, it would be a critical task of using the scarce amount of water of the Zāyandeh-Rūd, in an elaborate and justified way. So that, it should be helping the inhabitants, both in the town and villages, due to which, the settlements are not depopulated, the landscapes do not become waste lands, and the desertification does not increase. Hence, the careful use of water in the long history of Esfahan became the main order of life. Therefore, the managing affairs of the river in all its aspects was consigned to the local people through the dynasties, and the local governments. They also came to believe that, the local people of course, are the best managers of their own resources. And especially it has proved, in the Zāyandeh-Rūd river and its management during the past long history of Iran.

In comparison only can be said that due to this type of management even in the droughts also the river has not been dried for more than one and a half month, but in the time of writing these words, due to the ill management of the government staff, the river has been dried for more than ten months, and the people are afraid that it may be followed for the months to come. The main suggestion of this chapter is that "in Iran as it is one of the bases of the human culture of cultivation the best way of developing the rural resources is that, enhance the fruitful experiences of our own ancestors and join them with the modern knowledge of time.

Key Words

Bolūk : A traditional administrative area, in the Zāyandeh-Rūd region at present some of the past bolūks are diverted to a township (the Lejan, Alenjān and Mārbin), some have remained as rural areas, without any town or city (The Barāān and Kerārj), some are as a governary (The Rūdasht), and the bolūk of Jey va Barzrūd is covered by the extension of the big city of Esfahan.

Keshik	:	The area watered by the Zāyandeh-Rūd was divided into six keshiks (the Lenjān-Alenjān-Mārbin and Jey va Barzrūd-Barāān and Rūdasht). The keshik of Barāān charged the Mādis of Kerārj also.
Mādi	:	A mother canal through which several streams are now diverted.
Mādi Sālār	:	The man who was appointed by the Mirāb, charging a mādi of a bolūk. For each madi there was one mādi sālār.
Mobasher	:	A man who was appointed by the local cultivators of each village or hamlet, charging watering in the lands. His other name was Sarjūg.
Mirāb	:	The official incharge of the Zāyandeh-Rūd. His qualifications explained before.
Mostoufi	:	The state account.
Tag	:	One day or one night, mainly in the watering affairs.
Nov-Rūz	:	In Iran, the official year begins from the first day of the month of Farvardin, which is the first day of spring and the end of a long cold winter. The Nov-Rūz is the greazz

NOTES

1. Hossaini Abari, S.H. 2000, Zāyandeh-Rūd from the source to its month, Golhā, Esfahān, pp. 4-7.
2. Shafaghi, Sirus, *The Geography of Esfahān City*, Esfahān University, 1975, p. 46.
3. Mahmudiān, Mohammad, 1965, *The Zāyandeh Rūd of Esfahān*, University of Esfahān, p. 18.
4. The General Office of Water, Esfahān, The Archive.
5. Amini, Saiwush, 1999, Water Use and Forms of Social Organisation... in Iran, Institute for Scientific Co-operation, Tubingen, pp. 88-98 and Andereas. H. Luck, *Water Management Planning in Arid Regions...* 1999, Institute for Scientific Co-operation, *ibid.*, p. 68.
6. The Ministry of Energy, 1975, The Map 1/250000, Tehran, Archive.
7. Hossaini Abari, 2000, *ibid.*, p. 12.
8. Hossaini Abari, S.H., *Human Geography of Zāyandeh Rūd Region*, Ph.D. Thesis (A.M.U.), Aligarh, India, p. 18.
9. Hossaini Abari, S.H., *Reconstruction of an Old Dam*, Esfahān University, Faculty of Letters and Humanities, *Seasonal Magazine*, Vols. 10 & 11, p. 35.

10. Al-Esfahāni Hamzeh, Ebn-e-Hassan, *The Beuties of Esfahān, The History of Prophets and Kings*, 290 A.H., p. 35, Leiden.
11. According to the Field Investigations.
12. Samiee, Kāzem, 1928, (The Copywriting of the Tūmār-e-Haikh Bāhai), Rāh Nejāt, Esfahān.
13. Hossaini Abari, S.H., 2000, *ibid.*, p. 6.
14. According to the Tūmār of Shaikh Bāhai, *ibid.*
15. *Ibid.*
16. After Hossaini Abari 2000, *ibid.*
17. Al Esfahāni, *ibid.*, p. 35.
18. Hossaini Abari, S.H., 2000, *ibid.*
19. Charden, Jan, The Travel Account, (16th century), Translated by Mahmūd Abbasi, Tehran, 1959.
20. Puri Attari, Ali, 1985, The Management of Zāyandeh-Rūd Water, Department of Geography, Esfahān University, B.A. Desertation, p. 151.
21. According to the Tūmār of Shaikh Bāhai, *ibid.*
22. *Ibid.*
23. The Field Investigations.
24. Shaikh Bāhai, Banuoddin, "Kushkul" 1999, translated by : Azizollāh Kāsseb, Tehran, p. 12.
25. Ebn-e-Hougal, 370 A.H. (1670 A.D.), The Travel Account, translated by Jafar-Shoār, Tehran, 1966, p. 325.
26. Ebn-e-Rosteh, (Al-Alagh-on-Nafisah), Written in 740 A.H. (1066 A.D.), translated by Egbāl, Abbas, 1939, (The General History of Iran), Tehran, p. 346.
27. Nafissi, Saeed, The Conditions of Shaikh Bāhai, Tehran.

APPENDIX 3.1

The copy of pages one and two of the Tūmar of Shaikh Bāhāi pages no. One and Two.

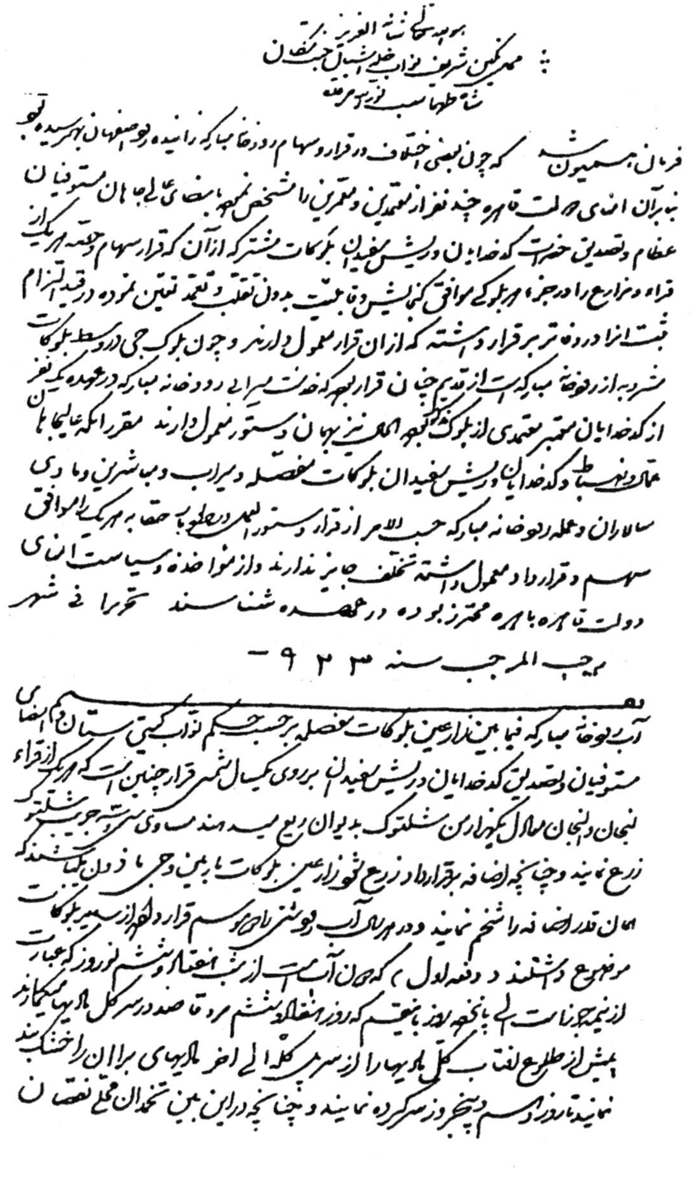

فرمان همایون شد که چون بعضی اختلاف در قرار و سهام رودخانه مبارکه زاینده رود صفاهان بهم رسیده بود
بنابر آن امنای دولت قاهره چند نفر از معتمدین و معتبرین را مشخص نموده باتفاق عالیجاهان مستوفیان
عظام و تصدیق حضرات کدخدایان و ریش سفیدان بلوکات مشترکه از آن که قرار سهام جمیعه هر یک از
قراء و مزارع را در جزء هر بلوکی موافق گنجایش و قابلیت بدون تعلب و تقلب تعیین نموده در قید التزام
ثبت انداز در دفاتر برقرار داشته که از آن قرار معمول دارند و چون بلوک [illegible] در وسط بلوکات
شروع به از رودخانه مبارکه است از قدیم چنان قرار بود که خدمت میرابی رودخانه مبارکه در عهده یک نفر
از کدخدایان معتبر معتمدی از بلوک [illegible] الحال نیز بهمان دستور معمول دارند مقرر آنکه عالیجاهان
عمال و [illegible] و کدخدایان و ریش سفیدان بلوکات مفصله و میراب و مباشرین و مادی
سالاران و عمله رودخانه مبارکه حسب الامر از قرار دستور العمل [illegible] حقابه هر یک را موافق
سهم و قرار داد معمول داشته تخلف جایز ندارند و از مواخذه و سیاست امنای
دولت قاهره باهره محترز بوده در عهده شناسند تحریرا فی شهر
رجب المرجب سنه ۹۲۳

آب رودخانه مبارکه فیما بین مزارعین بلوکات مفصله بر حسب حکم نواب گیتی ستان و امضای
مستوفیان و تصدیق کدخدایان و ریش سفیدان بر روی یکسال شمسی قرار چنین است که هر یک از قراء
[illegible]
[illegible]

نقصان ماشش برسد روز ششم و هفتم گوشه آبی به ... رسند و ونش ... که خاک آب میکشد
بدستور ونش اول خشک بند و سرگرده نمایند و مالی فدی چون در شهر جاری است چهار
سنگ آب بدهند و مالی نیا صرم جی چون بعضی از ... از آن شرب میشود
و معمول است قرار چنان شده که یکروز قبل از ونش روشنی ... آب داغ بگذارد
بدست کدخدا و مالی سالار و مردم ... بسپارد که تا آخر ونش از آن قرار معمول دارند
و تخلف ننمایند و برای آنها در پنج روز سرگرده شرکت میداشتند —

در ایام شمسی

از بابت آنچه مشترک است فیما بین بلوکات چون هنگام و فور آب و سیلاب است چندان احتیاج ندارند متعرض نشده

از اول ... ابتدای حمل ... تا نهایت نیمه ... آخر ... که موسم زمستان است
هنگام و فور آب و سیلاب و احتیاجی ندارند

حمل ثور جوزا الی نیمه ... قوس جدی دلو حوت

از بابت آنچه مختص هر بلوک است بعضی مشترک و برخی مختص هر بلوکی سوای سهام غیر معین

سهم ... سهمی پنج یوم

سهم بلوک رودشتین که از سایر بلوکات ممنوع شده و حقابه او را در دو موسم قرار داده اند بشرح ذیل

دفعه اول از ... از روز چهار و ششم نهایت نیمه ثور و از بشرح فوق
دفعه ثانی که خشک آب است از نیمه عقرب الی آخر بشرح فوق

سهم سایر بلوکات که بعضی مشترک و برخی مختص است سوای سهام غیر معین

ایامیکه موسم بذرکاری است چنانکه آب وفور دارد و چنانکه قلت دارد ونش میشود از انجام ...

4

REMOTE SENSING AND GIS BASED APPRAISAL OF WASTELANDS FOR SUSTAINABLE DEVELOPMENT

A Study of Barmer District (Rajasthan), India

BALAK RAM, GHEESA LAL And J.S. CHAUHAN

INTRODUCTION

Uneconomic use and over-exploitation of land and water resources by fast growing human population over the last three decades has caused land degradation and intensification of desertification process. In north-western arid zone of India which witnesses more fragile ecosystem, such environmental degradation has posed a serious problem. In order to restore and save these limited and precious natural assets from further degradation and meet the increasing food, fodder and fuel requirements, the proper identification, mapping and evaluation of wastelands is a pre-requisite. The recently developed remote sensing and G1S technologies and availability of better resolution data, have revolutionised the mapping of wastelands and other natural resources in more accurate and faster way. In the present study the wasteland mapping of Barmer the westernmost district of arid Rajasthan has been done on 1:50,000 scale through visual interpretation of I RS-IC LISS-III geocoded FCC supported with field survey. The wastelands, base, administrative and watershed layers have been digitized and digital (vector) data base has been created under G1S environment. About 33.87 per cent area of the district has been mapped under seven different wasteland categories of whom desertic sands, degraded pastures/grazing land and land with or without scrub are dominant. The spatial distribution, characteristics and present status of different wastelands are discussed. Measures and technologies are also suggested for rehabilitation and development of different wasteland categories in order to save them from further degradation and put them into appropriate and sustainable production systems.

STUDY AREA

Barmer district of Rajasthan with a total geographical area of 28,387 sq km extends from 70° 05' 06" to 72° 50' 55" E longitudes and from 24° 38' 52" to 26° 30' 41" N latitudes (Fig. 4.1). The district is bounded by Jaisalmer and Jodhpur districts towards north; Jalor district towards south; Jodhpur and Pali districts towards north-east and Tharparkar district of Pakistan towards west (Joseph,1962). As per 1991 census the district has a population of 14,35,222 (7,59,077 males and 6,76,145 females) with a density of 51 persons/sq km. (Anon.,1991). The district constitutes 8.29 per cent area and 3.26 per cent population of Rajasthan State. The livestock population of the district as per 1997 livestock census is 41,76,661 (Anon., 1996).

ENVIRONMENT

Climate

The climate of the district is extremely arid with a mean aridity index of 85.5 and standard deviation of 8.7 mm. Extremes of temperatures, erratic and fitful rainfall, high dust storms during summer and frequent droughts mark climatic characters. May and June are the hottest months recording mean maximum temperatures of 41.9°C and 40.3°C respectively which may often cross 45°C. During December and January the mean minimum temperatures are lowest being in the order of 10° to 12°C with lowest mean monthly temperatures recorded being 9.2°C and 7.5°C respectively. Occasionally for short period the minimum temperatures reach below freezing point and frost conditions prevail. The mean relative humidity during morning hours is 59 per cent and that of evening hours 35 per cent. In August these values attain peak to an order of 81 and 59 per cent respectively. Monsoonal rainfall generally occur from first week of July to last week of September. The mean monthly wind speeds over the region vary between 5.3 and 14.2 kmph. The mean annual rainfall of the district (1901-1998 average) is 268.35 mm which varies from (NW)

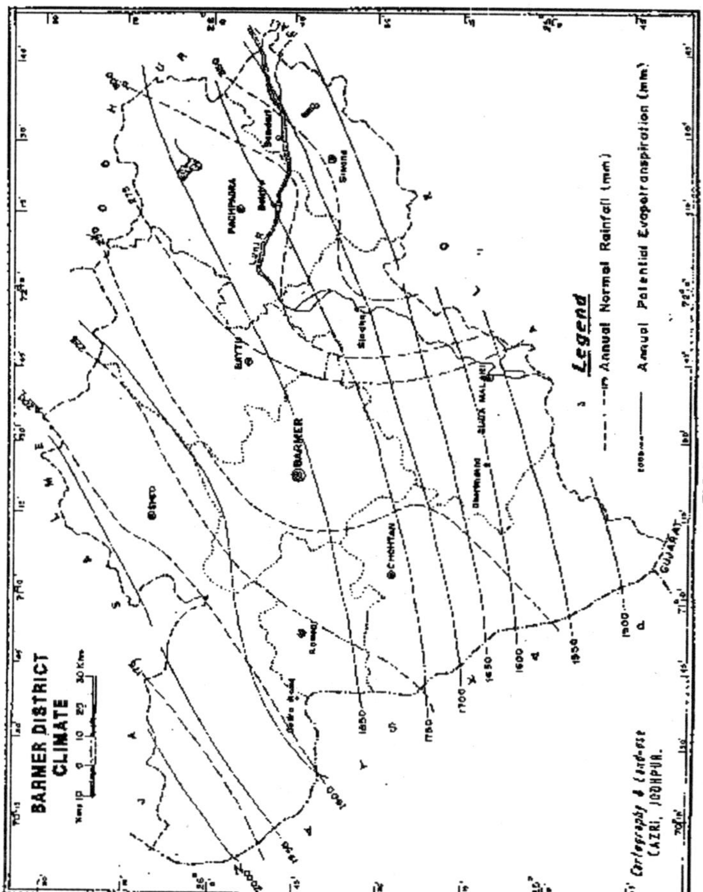

Fig. 4.1 : Barmer District Climate

209.7 mm at Sheo to (SE) 345.8 mm at Siwana occurring in 9.4 to 15.2 rainy days. The coefficient of variation (C.V.) of annual rainfall vary from 55.2 per cent at Sheo to 77.0 per cent at Chohtan. The potential evapo-transpiration (PE) across the district varies between 1500 to 2000 mm/year (Chatterji *et al.*).

Physiography and Drainage

Apart from an offshoot of the Aravalli hills in the east, the area is a vast sand covered tract of hummocks and high sand dunes with sub-stratum of gneiss, hornblende and quartz which here and there rise up through the sands. 'Chhappan ka-pahar' of this region attains a maximum height of 975 m above m.s.l. West of Luni river the region present a picture of vast sandy plain dotted with sand dunes. In the extreme north and west this sandy plain is broken by a series of longitudinal and parabolic dunes upto 4 km in length and 110 m in height. The general scope of the district is towards south. The height from mean sea level decreases from 315 m in the north to 11 m in south in Rann of Kachchh.The ephemeral Luni is the only river of this region which has its outlet in Rann of Kachchh. Others are Sukri, Mithri, Sukri -2 and Lik nadi (dried). The districts have few saline depressions like Pachpadra salt basin. Thob-ka-Rann, Sanwarla-ka-Rann and that of Vavtalai, Redana and Punaberi.

Soils

Soils of the district are dominantly coarse textured, covered with sand sheeting, hummocks and dunes. These have large variability with respect to the depth of lime concretionary layer and sand hummockiness and in the height and density of dunes on the surface. Medium fine textured alluvium soils occur along the Luni river.

Vegetation

The flora of the district is that of a hot desert region. Most of the flowering plants are shrubs and wild grasses. The grasses do not, however, survive for more than a few months after rains. Important tree species of the region are : *Prosopis cineraria* (Khejri), *Tecomella undulata* (Rohira), *Capparis decidua* (Kair), *Calligonum polygonoides* (Phog), *Calotropis procera* (Akra), *Salvadora oleoides* (Jal), *Ziziphus nummularia* (Bordi), *Acacia senegal* (Kumut), *Prosopis juliflora* (Vilayati-babool), *Euphorbia caducifolia* (Danda thor), *Grewia tenax* (Gangani) and *Azadirachta indica* (Neem). Dominant shrubs and grass species are: *Leptadenia pyrotechnica* (Khimp), *Aerva pseudotomentosa* (Bui), *Lycium barbarum* (Murali), *Crotolaria burhia* (Siniya), *Tephrosia purpurea* (Dhamasa), *Demostachya bipinnata* (Dab), *Cenchrus biflorus* (Bhurut), *Cenchrus setigerus* (Dhaman), *Lasiurus sindicus* (Sewan), *Aristida funiculata* (Lampra), *Panicum turgidum* (Murut), *Eleusine compressa* (Tantia), *Dichanthium annulatum* (Karad), *Cymbopogon jwarancusa* (Burgas), *Sporobolus marginatus* (Kharia-ghas) and *Suaeda fruticosa* (Kala lana).

Land-Use and Agriculture

Monocropped rainfed agriculture with fallowing system is most predominant land-use of the district constituting 52.28 per cent of the total geographical area and 68.48 per cent of the total agricultural land. Groundwater irrigated farming is concentrated along the younger alluvial plain of Luni, Siwana intermontane basin and interdunal area in Guda Malani tehsil. Such irrigated and often double cropped area constitute 2.2 per cent of the total district area and 4.13 per cent of the total agricultural land. Permanent pastures constitute 7.3 per cent. As per 1997-98 land records data forest, hills, barren and uncultivable land and land put to non-agricultural uses constitute 0.92, 1.92, 2.70 and 2.57 per cent area respectively. The area under current fallow, 2.5 years fallow and culturable waste constitute 8.93, 12.94 and 8.24 per cent respectively. Net sown area is 54.48 per cent. The intensity of cultivation decreases from east to west with the declining trend of rainfall. Out of the total cropped area of the district, bajra and guar constitute 51 and 23 per cent respectively. Kharif pulses *viz.* moth and moong contribute 17.9 per cent, sesamum 0.1 per cent, rape and mustard 1.25 per cent and wheat 0.9 per cent area respectively. Other crops are cumin, isabgol, castorseed, jowar, taramira, chillies, onion and fodder crops. Rabi crops like taramira and mustard are also produced on conserved moisture. The total foodgrain production during 1995-96 was 1,05,663 tonnes, oilseeds 20,184 tonnes and condiments and spices 6998 tonnes.

METHODOLOGY

The methodology of wastelands classification system, interpretation of satellite data and preparation of wasteland maps, is same as adopted by NRSA (1986). Similarly, the methodology for preparation of wastelands digital (vector) data base in GIS environment has also been followed as per manual published by NRSA (1997, 1998). The wasteland mapping of the district covering 63 toposheets has been done through visual interpretation of IRS-IC-LISS-III geocoded FCC of November and December, 1997. Boundaries of different wasteland categories were verified during the field survey. The same were finalised after considering the tehsil-wise land use data of 1997-98, earlier landuse maps prepared by CAZRI in 1989 and 1990 and discussions held with local people (Anon.)1990 and Balak Ram *et al.*, 1989).

Toposheet-wise base, administrative,wasteland and watershed layers were prepared and digitized. Thereafter, using ARC/INFO GIS workstation and following the procedure laid down in the manual of NRSA for preparation of wastelands digital data base, their editing, labelling, projection, transformation, edge matching and overlaying processes were completed. Statistical output was then generated. After their verification the desired corrections were incorporated and final toposheet-wise and watershed-wise statistics were generated. Thereafter using DMP 126(R) Plotter the final maps were generated on sunscreen transparent film in B/W and on plotter paper in colour. The complete output has also been transformed in 650 MB CD.

Spatial Distribution and Characteristics of Different Categories of Wastelands

The statistics generated through digital (vector) data base created in GIS reveal that out of the total geographical area of Barmer district (28,387 sq km) 33.87 per cent or 9614.41 sq km has been identified and mapped under seven wasteland categories (Balak Ram and Gheesa Lal, 1999). Sands-desertic has emerged as most dominant wasteland category constituting 21.13 per cent of the total district area and 62.39 per cent of the total wastelands. This is followed by degraded pastures/ grazing land 6.20 per cent, land with or without scrub 4.15 per cent, barren rocky/stony waste/sheet rock area 1.03 per cent, land affected by salinity/alkalinity 0.90 per cent, under-utilised/degraded notified forest 0.45 per cent, and mining/industrial waste 0.01 per cent respectively (Table 4.1 and Fig. 4.2). The spatial distribution and characteristics of different categories of wastelands are discussed below.

Table 4.1 : Wastelands Statistics of Barmer District, Rajasthan

Sl. No.	*Category*	*Area (ha)*	*Percentage to total district area*	*Percentage to total wastelands*
1.	Land with or without scrub	117940.96	4.15	12.27
2.	Land affected by salinity/ alkalinity	25418.66	0.90	2.64
3.	Underutilized/degraded notified forest land	12697.20	0.45	1.32
4.	Degraded pastures/grazing land	176035.38	6.20	18.31
5.	Sands-Desertic	599848.17	21.13	62.39
6.	Mining/industrial wastelands	370.49	0.01	0.04
7.	Barren rocky/stony waste/ sheet rock area	29130.52	1.03	3.03
	Total	961441.38	33.87	100.00

Total District Area : 28,38,700 ha.

Land with or without Scrub

Rocky/gravelly pediments localy known as 'Magra' and shallow gravelly uplands have been classified under this category. These are highly degraded and denuded land with exposed gravels and rock exposures. Low lying areas and rills are supported by scrub vegetative *viz. Capparis decidua* (Kair), *Calotropis procera* (Akra), *Salvadora oleoides* (Pilu-Jal), *Leptadenia pyrotechnica* (Khemp), *Eleusine compressa* (Tantia) and *Cenchrus biflorus* (Bhurut), etc. An area of 1,17,940.96 ha which is 4.15 per cent of the total geographical area of the district and 12.27 per cent of the total wastelands, has been mapped as lands with or without scrub. They are chiefly occur in eastern Sheo, Barmer, Pachpadra and Siwana tehsils. In Sheo tehsil such wastelands occur around Jhinkali, Rajra, Sheo, Jhanpali, Mungeria, Bhiyar, Dharvi Khurd, Nimbla, Thumbi, Agoria, Mokhab, and Giral villages. In Barmer tehsil these lands are concentrated between Baisala, Sihani, Patrasar, Radwa and Barmer town. In

Fig. 4.2 : Barmer District Wastelands

Pachpadra tehsil lands of this category occur around Nakora hills; between Patodi and Thob; between Mohanpura, Nyuri, Deriya, Nagona and Mandli and between Korna and Gangawas. In Siwana intermontane basin such lands are found around the Chhappan ka Pahar, Lohiya ka Pahar and Dantala, Bhimgoda, Nimle, Kerli, Koliyasar, Siwana, Devra, Mokalsar and Samdari hills. Moderate to severe water erosion is the dominant hazard. Such lands are also identified along the Luni river between Sindhari and Dandali and near Sarnu, Sanpa, Kamthai, Santara, Nausar and Thareri villages. These wastelands need control of water erosion, reduction in biotic pressure and rehabilitation.

Land Affected by Salinity/Alkalinity

An area of 25,418.66 ha has been mapped as salt affected land which constitutes 0.90 per cent of the total district area and 2.64 per cent of the total wastelands. Most of these areas are natural saline depression locally called as 'Rann'. Secondary salinized lands due to application of saline/sodic groundwater occur in the irrigated areas of Siwana and Balotra but their size is unmappable. Soils are coarse and fine textured moderately deep to very deep with high salinity, devoid of vegetation cover, low infiltration, high runoff potential and very high TSS (more than 3000 ppm) in groundwater. Vegetation can only be seen around fringe areas and in patches where sands have been deposited. Pachpadra, Thob and Sanwarla are important saline depressions. Other salt affected areas occurring in the district are near Patodi and Richholi in Pachpadra tehsil; Samuja and Bamsin in Siwana tehsil; Jaidu, Chuli, Sonri, Redana and Botiya in Barmer tehsil; Nausar in Baytu tehsil; Piparli, Jhakarda and Kottilani in Guda Malani tehsil and north western fringe of Rann of Kachchh between Bhalgaon and Dasauriya, near Gira, Pindaran, Vavtalai, Puniaberi and Sonji villages in the southern part of Chohtan tehsil.

Underutilised/Degraded Notified Forest Land

The forest of this region are mixed xeromorphic thorn forest. These cannot be called as true forest owing to poor density, crown cover, ground cover and quantity of forest produce. These forest lands generally occur over wastelands *viz.* rocky/stony hilts, saline depressions and high sand dunes. Out of the total underutilised/

degraded notified forest area 90.33 per cent is concentrated over two large and parallel Siwana hills and rest 9.67 per cent over Nakora hills in Pachpadra tehsil. The lower granitic Siwana hills run from Kundal in the west to Kathari in the east The upper rhyolitic hills called "Chhappan ka Pahar" run from Siner and Gura Nal in the west to Mokalsar in the east. Nakora hills forest area is occurring from Sinli in the west to Mewanagar and Jasol in the east. *Anogeissus pendula* (Dhokra) community is dominant on Siwana hills. In their lower slopes *Maytenus emarginatus* (Kankhera), *Commiphora wightii wightii* (Guggal). and *Moringa concanesis* are found. In Nakora hills forest area *Acacia senegal* (Kumat), *Euphorbia caducifolia* (Danda-Thor) and *Grewia tenax* (Gangari) tree species are dominant. Among prevailing grass community are: *Erenopogon faveolatus, Dactyloctenium, Cymbopogon jwarancusa* (Burgas), and *Aristida adscensionis* (Lompra) are important. Steep slopes, high runoff potential, very poor groundwater potential, highly degraded vegetation cover, very shallow and gravelly miscellaneous soils and insignificant infiltration are major characteristics of these areas. In spite of all these limitations, technologies are there to develop such lands to a certain extent.

Degraded Pastures/Grazing Land

This is the most important land-use system of arid region. Every village has some village common grazing land locally known as gochar, agor, oran, bir, etc. Altogether an area of 1,76,035.38 ha has been mapped under this category which constitutes 6.20 per cent of the total district area and 18.31 per cent of the total wastelands. Such lands are distributed throughout the region and hence occur on all types of habitats with different vegetative cover.

Pastures occurring on sand dunes and sandy undulating interdunal plain are dominated by *Prosopis cineraria* (Khejri), *Aerva pseudotomentosa* (Bui), *Crotolaria burhia* (Senia), *Calotropis procera* (Akra), *Cerodendrum phlomoides* (Arni), *Calligonum polygonoides* (Phog), *Acacia jacquemontii* (Bawli), *Lycium barbarum* (Murali), *Leptadenia pyrotechnica* (Khemp), and *Maytenus emarginata* (Kankhera). Dominant grass species are *Panicum turgidum* (Murat), *Cymbopogon jwarancusa* (Burgas), *Cenchrus biflorus* (Bhurut) and *Lasiurus sindicus* (Sewan). Pastures on younger and older alluvial plains comprise mainly *Prosopis cineraria, Capparis decidua* (Kair),

Zizyphus nummularia (Bordi), *Leptadenia pyrotechnica, Salvadora oleoides* (Jal), *Mimosa hamata* (Jhujni) and *Calotropis procera. Prosopis juliflora* and *Tamarix articulata* (Jhau) occur mainly in younger alluvial plains. Important grasses are *Cenchrus biflorus, Tephrosia purpurea* (Dhamasa) and *Eleusine compressa* (Tantia). On rocky/gravelly pediments and buried pediments *Capparis decidua, Ziziphus nummularia, Leptadenia pyrotechnica* and *Calotropis procera* are dominant tree species. Saline areas particularly have *Prosopis juliflora, Suaeda fruticosa* (Kala lana), *Capparis decidua* and *Sporobolus marginatus* (Kharia ghas).

The spatial distribution of pastures/grazing lands primarily govern by the intensity of agricultural land; occurrence of marginal lands and wastelands, density of population and aridity conditions. Therefore the size of pastures and their density increases from east to west with the decline in annual rainfall and density of population and increasing trend of aridity. Below 200 mm annual rainfall zone animal husbandry is predominant while agriculture is done only in patches. Between 200-300 mm rainfall zone the distribution of pastures is moderate and confined to marginal lands and wastelands. In the areas getting more than 300 mm annual rainfall are of the smaller in size. So far the regional distribution is concerned the pastures are mainly concentrated in north western and western part of the district comprising Sheo and Chohtan tehsils. Sheo tehsil has nearly 18 per cent of its total area under permanent pastures which constitutes about 57 per cent of the total area under this category in the district. This is followed by Chohtan tehsil where major concentration is towards north and along Pakistan border. At several places forest department has done fodder and fuelwood plantation and pasture development work with *Acacia tortilis* (Israeli babool). It is evident that most of the pastures are in highly degraded condition mainly due to heavy biotic pressure, wind and water erosion and salinity hazard.

Sands-Desertic

This is the most dominant category of wasteland covering 5,99,848.17ha area of Barmer district. It constitutes 21.13 per cent of district area and 62.39 per cent of total area under wastelands. The soils of sand dunes are fine sand to loamy fine sand in texture, loose to semi-compact and very deep. Wind erosion and deposition and shifting of active sand dunes are major hazards of such a terrain. Most of the stabilized dunes have their upper flank active.The vegetative composition over sand dunes comprises *Clerodendrum phimoides* (Arni), *Calotropis procera* (Akra), *Calligonum polygonoides* (Phog) and *Aerva pseudotomentosa* (Bui). Leeward side of the dunes are generally supported with *Acacia senegal* (Kumut). Other species are *Acacia jacquemontii, Lycium barbarum, Leptadenia pyrotechnica, Maytenus emarginatus* and *Haloxylon salicornicum.* Among grass species *Cenchrus biflorus, Panicum turgidum* (Murut), *Lasiurus sindicus* and *Eleusine compressa* are important.

In general, sandy waste areas occur intensively and dominantly in north-western, western and southern parts of the district, though such wastelands occur almost in everywhere. West of Harsani in Sheo tehsil sandy waste are associated with scrub. Important places with sizeable area of sand dunes are Dudaberi, Sokhra, Nand, Sonri, Nimla, Suwala, Balasar and Kotra. Most hazardous sandy waste are located near Jaisindhar, Rohidi, Sundra, Girab and Gadra road. In Chohtan tehsil sandy wastes are concentrated along Pakistan border right from Bakhasar in the south to Ramjan-ki-Gaphan towards north. Large area under this category occurs near Talsar, Mithran, Nawatala, Dedusar, Mate-ka-tala, Chohtan, Dhanau and Konra (80 m). Near Chohtan shifting dunes are most problematic and need proper stabilization. In Barmer tehsil sandy wastes are most dominant near Sonri, Gehun, Dhanora, Bhadres, Para, Sihani, Ramsar, Ranigaon, Shivkar, and east of Baytu. High sand dunes occur mainly in Chohtan and Guda Malani tehsils. In Guda Malani these lands are concentrated near Jaraji-ka-Mandir, Bhukan, Bhalisar, Band, Khudala and Mangta. Sand dune near Chhotu goes upto 94 m high, near Bhimthal 84 m and near Khariberi 80 metres. Sandy wastes in Pachpadra tehsil are found along Lik river bed, Dudwa, Nawtala, Sajiyali, windward side of Nakora hills, Borawas, Akali Bagundi and towards south of Tapra and Jagsa. In Siwana tehsil sandy wastes are found in south-western part near Kankhi, Indrana, Siner, Than Mata Hinglaj and along prevailing windward side of hills. During high dust storms such lands become more active and causes loss to sheep and goats and disruption of road and rail traffic. The State Forest Department has done sand dune stabilization work at several places.

Mining/Industrial Wastelands

Unscientific mining of gypsum, lignite, bentonite, fuller's earth and building stones has also created wasteland in this region. Gypsum in the district occurring at Kawas, Uttarlai, Shivkar, Gagariya, Chhitar ka par, Rateu and Agoriya. Bentonite is occurring in patches in Akli, Thumbli, Harbecha, Sonri, Bisala, Bhadres and Mahabar. Selicious earth occur in Akli bentonite mines. Mining of lignite is very much in progress at Akli and Giral by the Rajasthan State Mineral Development Corporation (RSMDC). Mining of stones for building and road material is done wherever it is possible. The mining of gypsum cause the denudation of vegetative cover, loss to topography, depletion of water resources, decline in soil fertility, surface crusting and soil erosion. The lignite is occurring at 20 to 60 m deep hence the wastelands are created by dumping highly acidic excavated material in reverse order *i.e.* sand, calcium carbonate, shale clay and lignite waste. Looking into all the mapping limitation an area of 370.49 ha has been delineated under this category. Exploration of lignite and gypsum has better prospects in Barmer district in the years to come. Hence there is great need to develop technology to treat these mine spoil wastelands and rehabilitate them from the very beginning to save the environment degradation. The discharge of toxic effluents from dyeing and printing industries at Balotra and Jasol has severely degraded multiple cropped land and also polluted the groundwater.

Barren Rocky/Stony Waste/Sheet Rock Area

Barren rocky/stony waste/sheet rock area constitute 29,130.52 ha or 1.03 per cent of the total district area and 3.03 per cent of the total area under wastelands. Excepting the part of hills covered by forest, all the hills occurring in the district has been classified under this category. These hills which are mainly of rhyolite, granite and sandstone formation give barren rocky/stony outlook. Vegetation can only be seen in their joints, fractures, and valley regions. Steep slope, high runoff, very poor groundwater potential, active water erosion and very shallow gravelly soils in patches are major characteristics of such wastelands.These are mainly concentrated within Siwana intermontane basin between Kankhi, Indrana, Meli and Mokalsar; around Barmer town between Baisala, Sihani, Patrasar and Radwa villlages. Besides these lands are also occur near Chohtan, Dhorimanna, Redana, Setrau, Bhachhbar, Mangta, Kotara, Thob, Mandli, Nagona and Korna villages. At places forest department has carried out afforestation work.

RECOMMENDATIONS FOR WASTELAND RECLAMATION AND MANAGEMENT

Land with or without Scrub

Degraded rocky/stony and gravelly upland and shallow hard pan soil areas with very scanty and poor distribution of unpalatable shrubs and grasses as well as very low production potential, are recommended for development of silvipastures. The first and foremost requirement to establish such production system is to undertake improved soil and water conservation measures to minimise soil loss, spread the rainwater evenly and allow it to penetrate at lower depth and proper harvesting of rainwater. This includes the construction of contour furrow and trenching at an interval of 3.5 m on moderately sloping land. Vegetative barriers may also be practised as per topographic needs. On steep sloping land, terracing should be practised. Suitable rain water harvesting techniques *viz.* half-moon terraces, ridge and furrow and microcatchment should be undertaken. Suitable grass species like *Cenchrus ciliaris, Dichanthium annulatum, Sehima nervosam* and *Cenchrus setigerus* should be grown on furrows (Singh and Kar, 1997).

For ensured and continuous supply of quality fodder suitable tree and shrub species *viz. Prosopis cineraria, Acacia tortilis, Acacia senegal, Colophospermum mopane, Hardwickia binata, Cordia gharaf, Grewia tenax* and *Ziziphus spinacristii* should be raised. Their seedlings should be planted at 20 m row to row distance and 5 m tree to tree spacing in contour furrows. In hard pan soil areas such trees can be grown by breaking hard concretionary strata by crow-bar.

Reclamation and Management of Salt Affected Land

Natural salt affected lands

In order to increase the production potential of these lands plant species should be selected according to the salinity level. For soils having EC (dSm^{-1}) 8-16 *Acacia leucocephala, Dichrostachya nutans* and *Acacia nilotica* can be grown. For areas having EC more than 16 the suitable tree/shrub species are *Tamarix aphylla* (Jhau), *Salvadora* sp., *Haloxylon salicornicum, Salsola baryosma* (Jerio Lana), *Suaeda fruticosa* and *Atriplex* sp. Suitable salt tolerant grasses are *Sporobolus marginata* (Khari ghas), *Dichanthium annulatum, Chloris virgata* and

Cenchrus setigerus. Grasses are grown through pelleted seeds or rooted slips on ridges of 0.8 to 1.0 m height created east-west direction of a regular interval of 5-6 m apart in between the tree rows.

Reclamation of Sands deteriorated due to irrigation with high RSC water

The reclamation measures include the application of gypsum @ 100 per cent of soil requirement. Application of gypsum @ 14 tons/ha for wheat and 24 tons/ha for mustard, cumin and isabgol is recommended to reclaim such lands. This is to be repeated after 3-5 years depending upon the quality of groundwater used for irrigation. In situation where the quality of irrigation water is very poor, the lands can be reclaimed for producing rainfed kharif crops. In first year, gypsum @ 50 per cent of soil requirement should be applied in soil before rains.

Gap Filling and Regeneration of Underutilized Degraded Notified Forest Land

In Barmer district degraded forest areas are mostly occur on hills and their foothill zone. Due to high runoff and moderate to severe water erosion, these areas require rehabilitation to check erosion of skeletal soil. In this case soil conservation measures like contour furrowing at 8-10 m interval with 20-30 cm deep furrow, 60 cm wide contour trenching, and bunding are required to be developed for improving the soil moisture status. Tree species like *Acacia senegal*, *Anogeissus pendula*, *Grewia tenax* and *Maytenus emarginata* can be grown to enrich the forest area by transplanting poly bags raised seedlings. Ground flora may be enriched by *Sehima nervosum*, *Cenchrus setigerus* and *Cymbopogon jwarancusa*. The afforestation and gap filling activities should be carried out at proper time *i.e.* within first fortnight of the occurrence of first monsoon good soaking shower. Initial watering and seedling mortality should be timely monitored and quick replacement be done. After planting a saucer shaped depression around the plant having 15 cm deep and 1 m diameter in plain areas and a crescent shaped ridge of 15 cm high may be provided on sloppy areas to collect and store rain water.

Re-seeding and Development of Degraded Pasture/Grazing Lands

The main objective to develop such common lands is to provide a palatable vegetation cover on these lands and regeneration of the degraded vegetation cover by providing suitable soil and moisture conservation measure and the protection. First of all barbed wire fencing or ditch digging and bund formation should be carried out. Then contour furrows, 60 cm wide and 25 cm deep are constructed in grasslands at a distance of 10-15 m apart, across the slope of the land. In another technique of inter-row water harvesting (IRWH) system 30 cm wide ditch alternate with 70 cm wide raised beds are made (Gupta,1998). The grasses/legumes/shrubs are grown on the edge of the ditch. For planting trees in the grazing/oran lands piting-discing techniques should be followed.The area in question for rehabilitation should be divided into four to six compartments and one part may be taken up each year.The re-seeding should be done with high yielding species like *C.ciliaris, C. setigerus, D. annuiatum, P. antidotale* and *Lasiurus sindicus.* Transplantation of top feed species like *A. tortilis, A.nilotica, Prosopis cineraria, Salvadora oleoides, A. senegal, Dichrostachys nutans, Ziziphus nummularia, Calligonum polygonoides* and *Capparis decidua.* Forage cutting should be preferred over grazing (Yadav, *et al.*, 1997).

Stabilisation and Management of Sand Dunes

Shifting or active sand dunes

The technology for stabilisation of shifting or active sand dunes requires erection of strand barbed wire fencing and Stablishment of chess board pattern micro wind breaks with locally available brush wood or undershrubs and shrubs like *Crotolaria burhia, Aerva persica, Leptadenia pyrotechnica, Clerodendrum phlomoides, Ziziphus nummularia, Calotropis procera, Tephrosia purpurea, T. falciformis,* etc. at right angle to the prevailing wind direction and transplantation of disease resistant and fast growing tree/shrub species like *Acacia tortilis, Acacia nubica, Cercidium floridum* and *Calligonum polygonoides.* In between the tree species, the rooted slips of locally available indigenous perennial grasses like *Panicum antidotale, P. turgidum, Lasiurus sindicus* and *Cenchrus ciliaris* may be transplanted.

Stabilized sand dunes

Due to intense biotic activities the top, crest and upper part of the flanks of almost all the parabolic dunes are covered with loose sand. During monsoon period, the live chess board system may be made in the first year by putting finger to thumb thick cuttings of 40 to 60 cm long pieces in the sand with local available shrubs. In subsequent years, these may be filled with Bawli, Israeli babool, Phog, etc. The dunes have big clumps of Gramna (*Panicum antidotale*), Murut (*P. turgidum*), and Sewan (*Lasiurus sindicus*) grasses. The root slips of these grasses may be used in covering the loose sandy crests and tops of the dunes in the first year of stabilisation. Seeds of the above species may also be dibbled just after first shower.

The ground flora can be enriched by growing the adopted grasses like *Lasiurus sindicus*, *Cenchrus ciliaris*, *Panicum antidotale* and *Citrullus colocynthis*. For semi-stabilized/stabilized sand dunes, some species like *Acacia salicina*, *Hardwickia binata*, *Albizia lebbek*. *Azadirachta indica*, *Maytenus emarginata*, *Balanites roxburghii* and *Tecomella undulata* are recommended. Thus a blend of suitable plant species, planting technique and after care, can develop the area for silvipasture and also control wind erosion.

Rehabilitation and Development of Mined Spoil Areas

Gypsum mined spoil areas

The rehabilitation of gypsum mined wasteland can be achieved through an optimum combination of rain water harvesting, soil profile modification and appropriate plant species (Anon.,1998) in the following manner.

Land shaping : The surface of the disposed gypsum mine spoils shaped into terraces and slopes for providing a catchment for harvesting precipitation and a terrace for transplanting saplings. The rainwater harvesting techniques are microcatchment, half moon terraces and ridge and furrow.

Soil profile modification : To modify soil profile planting should be done in auger holes of 15 cm diameter and 1 m depth. The auger holes be filled up with necessary growing medium consisting of a mixture of fine sand, farmyard manure and top soil in equal proportions.

Plantation : Plantation of a mixture of top feed tree, shrub and grass species *viz. Acacia nubica, Acacia tortilis, Azadirachta indica,*

Cercidium floridum, Dichrostachys nutans, Prosopis cineraria, Salvadora oleoides, Salvadora persica and *Tamarix aphylla* is to be done.

Lignite mined spoil areas

The first requirement for reclamation is land shaping and construction of bench terraces or half moon terraces as per requirement for rainwater harvesting. This is followed by covering of surface with 45 to 60 cm layer of fine sand and murram to neutralize H_2SO_4, soil profile modification by digging pits and filling them with fine sand, pond silt or clay and FYM and plantation of suitable tree, shrub and grass species.

Land affected by industrial effluents

The reclamation measures require proper drainage system to carry effluents, physical and chemical treatment of effluents at source level *i.e.* within the industrial unit and biological treatment at Common Effluents Treatment Plant (CETP).

Rehabilitation and Development of Barren Rocky/Stony Waste/ Sheet Rock Area

Adoption of soil conservation measures, contour furrowing, contour trenching and contour terracing is the first requirement to rehabilitate such vulnerable areas. The suitable species for plantation are *Acacia senegal, Anogeissus pendula* (for Siwana region only), *Grewia tenax, Acacia leucophloea, Maytenus emarginatus, Commiphora wightii* and *Euphorbia caducifolia.* The ground flora can be enriched by growing grass species like *Sehima nervosum, Cenchrus setigerus* and *Cymbopogon jwarancusa* between tree/shrub rows. A brick wall (1 m height) may be raised around the treated area. Instead, a live hedge of *Acacia senegal* and/or *Euphorbia caducifolia* can also be developed. At suitable sites 'tanka' or other water harvesting structures should be constructed to store water for the establishment of plantation.

CONCLUSIONS

Through visual interpretation of IRS IC LISS-III FCC supported with field survey the wastelands in Barmer district are mapped on 1:50,000 scale and digital data base has been created under GIS environment. Watersheds are delineated only in Luni basin area in eastern part while there is no drainage in other parts of the district. About 33.87 per cent area of the district and digital data base has been created under GIS environment. Watersheds are delineated only in Luni basin area in eastern part while there is no drainage in other parts of the district. About 33.87 per cent area of the district has been mapped under different wasteland categories. Sands-Desertic has emerged as most dominant wasteland category occupying 21.13 per cent district area and occurring dominantly in north western, western and southern part of the region. Degraded pasture/grazing land constitute 6.20 per cent area occurring throughout the region but more dominantly in Sheo tehsil. Land with or without scrub (4.15 per cent) are concentrated around Barmer, Sheo, Siwana and Thob. Shallow gravelly soil cover, water erosion and lesser and unpalatable vegetative cover are major problems. Notified forest areas (0.45 per cent) occur mainly within Siwana and Nakora hills. These are highly degraded and have poor production potential. Land affected by salinity/alkalinity (0.90 per cent), barren rocky/stony waste/sheet rock area (1.03 per cent) and mining/industrial waste (0.01 per cent) are other wastelands identified in the district. Although more than 50 per cent area of the district is under agriculture (mainly rainfed) but production potential is low. Sand drift, salinity/sodicity and repeated droughts are major problems.

In spite of all natural hazards and physio-climatic constraints the district has enough potentiality for development of silvi-pastures, permanent pastures and forest areas with the application of suitable technologies, involvement of local people and political will. Detailed mapping is required for identification and reclamation of lands deteriorated due to irrigation with high RSC/ polluted ground water.

Acknowledgements

The authors are highly grateful to Dr. Pratap Narain, Director and Dr. I.C. Gupta, Head of Natural Resources and Environment Division, Central Arid Zone Research Institute, Jodhpur for providing all facilities and encouragement during the course of this study.

REFERENCES

Anon.,1990. Report on land use/land cover, Barmer district, Rajasthan, CAZRI, Jodhpur, p. 19.

Anon.,1991. District Census Handbook, Barmer. Directorate of Census Operations. Rajasthan, Jaipur, p. 499.

Anon., 1996. Statistical Abstract Rajasthan, 1996. Directorate of Economic and Statistics, Jaipur, p. 409.

Anon., 1998. Rehabilitation of gypsum mined wastelands in the Indian Desert, Best Management Practice Bulletin-1, CAZRI, Jodhpur, p. 23.

Balak Ram, Sen, A.K. and Chauhan, J.S. 1989. *Present Land Utilisation in Barmer District, Rajasthan*. CAZRI, Jodhpur, p. 29.

Balak Ram and Gheesa Lal, 1999. *Wastelands in Barmer District, Rajasthan*, CAZRI, Jodhpur, p. 35.

Chatterji, P.C. and Joshi, D.C. (Ed.) 1989. *Integrated Natural and Human Resources Appraisal*, Barmer district. CAZRI, Jodhpur, p. 179.

Gupta, J.P. *et al.* 1998. Project Report on Demonstration of technology for biomass production and reclamation of degraded land in arid zone, CAZRI, Jodhpur, p. 19.

Joseph, D.C. 1962. Gazetteer of India, Rajasthan, Barmer, *Rajasthan District Gazetteers*, Jaipur, p. 277.

NRSA, 1986. Manual of procedure for wastelands mapping using remote sensing techniques. NRSA, Hyderabad, p. 52.

NRSA, 1997. Manual of procedure for preparation of wastelands digital data base using remote sensing and GIS techniques. NRSA, Hyderabad, p. 90.

NRSA, 1998. Wasteland manual — 15 steps for preparation of digital data base. NRSA. Hyderabad, p. 24.

Singh, S. and Kar, A. (Ed.) 1997. *Desertification Control—In the arid ecosystem of India for sustainable development*. Agro-Botanical Publishers, Bikaner, p. 393

Yadav, M.S. *et al.* (Ed.) 1997. Silvipastoral systems in arid and semi-arid ecosystems. CAZRI, Jodhpur, p. 472.

5

ENSO IMPACTS ON DROUGHT OCCURRENCE IN INDIA

TAPESHWAR SINGH

INTRODUCTION

Drought is amongst a few devastating natural disasters in India. Although there is no universally acceptable definition of drought, the word is associated with prolonged and abnormal rainfall deficiency. It is a recurrent phenomenon. Occurrence of drought episode upsets the country's food security by seriously hitting agriculture, which still continues to be the mainstay of our farming communities. Indian agriculture is said to be gamble with monsoon. Drought occurs in both high as well as low rainfall area. It is a creeping phenomenon, making an accurate prediction of either its onset or end is a difficult task. The pattern of drought occurrence and its frequency and severity varies regionally. In our country drought episode is considered inevitable and an unavoidable part of '*Karma*' (destiny).

Thus, interannual as well as spatial variations in the summer monsoon rain over India have a profound socio-economic impact. Indian summer monsoon contributes nearly 80 per cent of the annual rainwater potential. Failure of monsoon leads to drought, crop failure and in the more extreme cases, famine. Droughts in successive years or prolonging continuously for more than one or two years have a disastrous effect on food production and the national economy. The cause of drought is a complex question and it needs comprehensive research. Some researchers have attempted to link drought episode with ENSO (EL Nino/Southern Oscillation) phenomena. There is however, a growing consensus among the international scientific community that ENSO phenomena have links with the extreme weather events *i.e.* cyclones, floods, droughts etc. The challenging question before us is whether these droughts over different parts of the world occur randomly or appear simultaneously with some large scale atmospheric and/or oceanic phenomena? Several studies over the past decade or so have shown that inter-annual variation in many parts of the world and ENSO events in the tropics are generally related, for example, drought in Australia, Indonesia, India and parts of Africa and heavy rains and floods in the Pacific Coast of South America (Nicholls 1991[1], Ropelewski and Harpert 1987[2]). Prior to Sikka (1980[3]) many workers—Lockyer and Lockyer (1902[4]), Walker (1923[5], 1924[6]), Bjerknes (1969[7]). Shukla & Misra (1977[8]), Khandekar (1979[9]), and others had also drawn attention on the links between global-scale oscillations in surface pressure and summer monsoon rain over India and surrounding regions. However, Sikka (1980) for the first time, related monsoon droughts over India with the occurrence of ENSO event. Since then several workers have attempted to study the links between Indian monsoon rainfall and ENSO/anti-ENSO events (Angell 1981[10], Rasmusson and Carpenter 1983[11], Mooley and Parthasarathy 1983[12], Parthasarathy and Pant 1985[13], Bhalme *et al.* 1980[14], Webster and Yang 1992[15], Mohanty and Ramesh 1993[16], Chattopadhyay and Bhatia 1996[17], Sikka, 1998[18], Kane 1998[19] and Victor *et al.* 1995[20]). ENSO has been subject to major diagnostic and modelling studies in the past 25 years or so, particularly after the 1982 ENSO event. Although most of the studies suggest some links between ENSO (anti-ENSO) events and deficient summer monsoon rainfall over India. Different workers selected the ENSO years based on differing criteria, for example, Sea Surface Temperature Anomaly (SSTA) over eastern Pacific, Quinn *et al.* (1987[21]), Rasmusson and Carpenter (1982[22]), Southern Oscillation Index (SOI) Harpert and Ropelewski (1992[23]) and several other scientists have used atmospheric and oceanic parameters like disruption of fishery and marine bird life off the coasts of Peru and Equador. In the light of this ambiguity in the definition of ENSO years serious differences prevail among the scientists about the listing of the ENSO years. Moreover, a confusion prevail when ENSO event continues for two years, which year should be identified as ENSO year for its teleconnection study with summer monsoon rainfall. However, for the present study years defined by different workers like Deser and Wallace (1987[24]), Quinn *et al.* (1987), Shukla and Paolino (1983[25]), Rasmusson and Carpenter (1983), Sikka (1998) and others are considered as ENSO year. In this study no attempt has been made to define ENSO years. Before a discussion on the links between ENSO phenomena and drought episodes in India, a brief account of ENSO is felt necessary,The El Nino/Southern Oscillation is one of the most pronounced Oscillations in the tropics. El Nino now more generally refers to as warming of the tropical Pacific basin that occurs roughly every three to seven years in association with a weakening of the trade winds. The flip side of El Nino, La Nina is characterized by stronger than normal trade winds and usually cold sea-surface temperature in the tropical Pacific. Both El Nino and La Nina are accompanied by swings in atmospheric pressure between the eastern and western Pacific known as the Southern Oscillation. These phenomena are collectively referred to as ENSO (El Nino/ Southern Oscillation) McPhaden (1999[26]). The general mechanism underlying ENSO involves large scale ocean-atmosphere interactions and equatorial ocean dynamics (Neelin *et al.* 1998[27]). But each El Nino and La Nina is unique in the combination of its strength, duration and pattern of development. Irregularity in the ENSO cycle can be seen in both instrumental record dating back to the middle of the last century and in proxy data (such as late sediments, coral growth rings and ice cores) going back hundreds or even thousand of years (Rodbell *et al.* 1999[28]). So in principle, it should not be surprising that an unusually strong El Nino occurs every so often (McPhaden, 1999). The two phenomena—EL Nino/Southern Oscillation are atmospheric and oceanic parts of an elegant climate fluctuations (Obasi, 1996[29]). Although it involves only one-fifth of the circumference of the planet but it transforms weather around the globe. According to historical record 600,000 people died in just one region of India from the epic droughts of the 1789-1793 El Nino (Supple, 1999[30]). The evolution of ENSO exhibits conspicuous irregularities from one cycle to another (Wang *et al.* 1999[31]). The ENSO years listed in our study are 21 in number, *i.e.*, 1877, 1884, 1891, 1896, 1899, 1902, 1905, 1911, 1918, 1925, 1928, 1941, 1951, 1965, 1969, 1972, 1977, 1982, 1987, 1992 and 1998. However, we have identified drought years on the basis of long time rainfall data series—annual, seasonal and homogeneous regions of India.

OBJECTIVES

1. To compute drought years for all India rainfall data series during the period 1871-1998 (128 years).

2. To assess the link between ENSO event and drought episodes on different spatial scales.
3. To examine the links between ENSO events on the simultaneous performance of annual, summer monsoon for All India and five homogeneous region's rainfall and to list non-ENSO years but drought as well as non-drought ENSO years. Thus, the approach in our study is to examine the links between ENSO years and All-India drought episodes and summer monsoon season as well as different homogeneous regions. We have also attempted to discuss some peculiar features of ENSO-drought relation.

METHODS

In the present study we have used Parthasarathy *et al.* (1995[32]) data sets. This includes a listing of monthly, seasonal and annual rainfall data series for the period 1871-1994 for All India, 5 homogeneous regions and 29 meteorological sub-divisions. It is prepared on the basis of a fixed and well distributed network of 306 raingauge stations over India, each representing districts of different sizes by proper area-weightage. Rainfall data since 1995 onward are provided by the Hydrology Division of India Meteorological Department (IMD), New Deihi.The contiguous area considered measures 2.88×10^6 sq. km. which represents about 90 per cent of the country's total geographical area. The mean, standard deviation and coefficient of variation are also presented. All-India annual, monsoon season and Homogeneous Region's rainfall normal as well as percentage departure from the normal for all ENSO/Drought years are also computed. Different methods are used by investigators to define meteorological drought such as Ramdas Method (1950[33]), IMD Method (1971[34]), Drought Area Indices Method, Monsoon Deficiency Index (Mooley and Parthasarathy 1982[35], Parthasarathy *et al.,* 1992[36]) and Standard Deviation Methods. In the present study standard deviation method is used to classify the performance of monsoon rains based on the departure from the S.D. (δ). This criteria is used for All-India, all 5 Homogeneous Regions of India as well as summer monsoon rainfall. On the basis of this criteria we have determined different categories of drought on different spatial scales. The drought is considered to have occurred if the deficiency of rainfall is equal to or above 1.0 δ. On the basis of this method we have defined various classes of

Table 5.1: Indian Monsoon and Homogeneous Region Droughts (1871-1998)

(Rainfall Departure in Percentage)

Sl.No.	*Drought Years*	*Standard Deviations (δ)*	*All India*		*North-West India Region*	*West Central India Region*	*Central North-East India Region*	*North-East India Region*	*South-Peninsular India Region*
			Seasonal	*Annual*					
1.	1873	-1.0	-11.15	-13.96	3.00	-10.80	-12.73	-17.85	-15.39
2.	1877*	-2.9	-29.11	-13.15	-61.50	-28.80	-38.68	-3.8	-9.60
3.	1899*	-2.6	-26.17	-25.57	-66.90	-42.92	-1.11	15.64	-34.94
4.	1901	-1.6	-15.29	-10.84	-46.45	-9.09	-18.52	-4.05	-9.40
5.	1904	-1.2	-11.97	-10.61	-36.55	-16.88	6.20	-6.33	-13.72
6.	1905*	-1.6	-15.93	-15.56	-45.29	-21.05	-8.39	9.35	-16.65
7.	1911*	-1.4	-13.59	-11.27	-55.67	-13.36	1.85	-0.23	-17.11
8.	1918*	-2.1	-23.63	-21.30	-64.31	-32.58	-13.00	26.35	-38.61
9.	1920	-1.7	-15.63	-15.46	-21.96	-29.14	-12.47	-5.76	-8.37
10.	1928*	-0.9	-9.89	-4.95	-12.61	-7.79	-16.05	-1.45	-12.41
11.	1941*	-1.4	-14.55	-12.15	-12.69	-26.93	-8.96	0.77	9.28
12.	1951*	-2.0	-13.29	-10.80	-43.16	-12.87	-14.90	4.92	-3.37
13.	1965*	-1.5	-16.78	-20.00	-25.86	-19.38	-21.01	-0.15	-12.30
14.	1966	-1.4	-13.20	-11.03	-16.39	-19.53	-22.38	8.52	-0.36

(Contd.)

Table 5.1 *(Contd.)*

Sl.No.	*Drought Years*	*Standard Deviations (δ)*	*All India*		*North-West India Region*	*West Central India Region*	*Central North-East India Region*	*North-East India Region*	*South-Peninsular India Region*
			Seasonal	*Annual*					
15.	1968	-1.1	-11.47	-11.82	-30.16	-15.85	-12.7	8.47	-2.74
16.	1972*	-2.3	-23.40	-22.18	-36.59	-26.70	-19.30	-14.92	-18.30
17.	1974	-1.2	-12.24	-10.73	-38.73	-18.96	-16.01	19.48	9.46
18.	1979	-1.9	-16.96	-11.74	-3.08	-20.18	-34.10	-5.08	-3.58
19.	1982*	-1.3	-13.73	-10.18	-31.27	-10.59	-10.01	-13.83	-11.97
20.	1985	-1.0	-10.86	-7.62	-30.37	-13.89	1.76	-7.78	-9.65
21.	1986	-1.3	-12.83	-8.80	-30.41	-10.86	-7.76	-10.87	-10.92
22.	1987*	-1.8	-18.20	-11.48	-59.76	-20.85	-11.53	-12.00	-18.30
23.	1991	-1.0	-7.96	-7.80	-27.44	-10.44	-9.55	-7.40	16.38
Mean	—	—	852.40	1090.3	490.00	933.20	1002.40	1419.20	659.40
Per Ann.	—	—	78.20	100.30	89.90	86.30	83.30	68.70	57.00
.STD.	—	—	84.70	103.50	132.40	125.90	112.80	121.30	98.30
Cov.	—	—	9.90	9.30	27.00	13.05	11.02	8.50	14.90

*ENSO events.

Limit of 2.0 δ or 25 per cent deficiency or more is phenomenal drought on All-India basis. Years—1918 and 1972 are approaching the limit marginally.

droughts. The drought is classified as *mild* if the rainfall deficiency is between 1.0 to 1.24 δ, *moderate* between 1.25 δ to 1.49 δ, *severe* between 1.50 to 2.0 δ and *phenomenal* if the deficiency is more than 2.0 δ. Negative departure of seasonal rainfall from the normal by 25 per cent to 49 per cent is categorized as phenomenal drought and if the negative departure from the normal is by 50 per cent or more it is said to be *calamitous*. Droughts which have occurred during the period under study on (*i*) All-India basis, (*ii*) Monsoon season basis, and (*iii*) Homogeneous Regions of India basis have been listed in Tables 5.1 and 5.2.

RESULTS AND DISCUSSION

All-India Drought

In this section we discuss the occurrence of drought on All-India and summer monsoon season basis. Table 5.1 reveals that there have been 23 years when summer monsoon drought has occurred when the deficiency of rain has been 1.0 δ or more. And of these 17 years witnessed moderate drought, 13 years severe and only 4 years phenomenal *viz.* 1877, 1899, 1951 and 1972 and all these years are ENSO years. It is also reflected from this table that the frequency of drought was more during the period 1899 or 1920, it decreased sharply during 1921 to 1960 and again increased during 1961 to 1990. The period 1921 to 1964 (43 years) did not witness a single severe and phenomenal drought despite 3 ENSO events of 1925, 1941 and 1951. We also notice from this table that there are 4 years since 1871 to present, that India as a whole suffered from monsoon rainfall deficiency ranging between 25 per cent to 49 per cent of the normal. There is no year on record on All-India scale when the deficiency was more than 50 per cent. We also notice that during this period at least 19 ENSO events happened. If we compare the drought years over All-India basis with each of the 5 homogeneous regions basis we found that hardly 50 per cent years are common to each other, even for moderate or severe drought events on All-India scale, some of the regions do not experience drought of the same magnitude in many of those years. In terms of mild and above classes it is almost similar but differs significantly in terms of moderate and above and severe and above classes. It shows that there is much variability on interannual basis in the seasonal rainfall on homogeneous regions basis than on All-India basis. The West Central India Region and South Peninsular India Region have much less number of droughts under moderate and above classes than All-India basis. Phenomenal drought over North West India Region occurred in 1987 after 1918 (interval of 69 years), over West Central India Region in 1972 after 1941 (interval of 31 years), over Central North India Region in 1966 after 1877 (interval of 89 years), over North East India Region in 1992 after 1896 (interval of 96 years) and over South Peninsular India Region in 1952 after 1918 (interval of 34 years). These deviations raise questions to investigate the causes of these variations. It must be noted, that the standard deviation as well as normal rainfall deficiency over different regions differ considerably. Thus, despite ENSO events probability of occurrence of phenomenal drought on All-India basis is very low *i.e.* once in 30 years or so and the phenomenal drought of 1972 occurred after a lapse of 54 years. These figures further point to fairly high stability in the India's summer monsoon rainfall. However, there is no uniform cyclic trend of occurrence of drought on All-India and regional scales. Further, we do not find any clustering of drought except on 2 occasions of two successive years—1904 and 1905 and 1965 and 1966 only, and one occasion of 3 continuous drought years—1985, 1986 and 1987.

Drought over Homogeneous Regions

We have shown in Table 5.2 the rainfall deficient years for each Homogeneous Region based on standard deviation from the normal. Percentage departure from the normal is given in Table 5.1.

Table 5.2 : Incidence of Droughts of Different Categories over Homogeneous Regions of India

Homogeneous Regions	*Mild & Above $\geq 1.0\,\delta$*	*Moderate & above $\geq 1.25\,\delta$*	*Severe & above $\geq 1.50\,\delta$*	*Phenomenal $\geq 2.0\delta$*
North West India Region	22	13	8	6
West Central India Region	20	10	10	7
Central North East India Region	20	15	6	3
North East India Region	22	19	11	4
South Peninsular India Region	18	8	4	3

Table 5.2 shows that for the individual homogeneous region 18 to 22 drought years have occurred under mild or above, 8 to 19 under moderate or above, 4 to 11 under severe or above and 3 to 7 under phenomenal drought. It shows that there is a considerable inter-regional variability with regard to different classes of drought. The phenomenal droughts are rare except North West India and West Central India regions where 6 and 7 events respectively have occurred. It is important to note that even an area of high rainfall regime of India such as North East India Region is not immune from such event, may these areas be rarely suffering from drought. It can also be noticed that among the common years of All-India scale severe to higher intensity drought and regional scale drought of the same magnitude, almost all years except 1920, 1974 and 1979 are ENSO years. This confirm an important feature of India's summer monsoon season drought *i.e.* ENSO is linked to the occurrence of severe and phenomenal drought over All-India scale as well as on regional scale in North West India Region and West Central India Region. There are years when drought conditions prevailed over different regions while it did not happen so over a contiguous homogeneous region or on All-India basis simultaneously, even during the ENSO years. It points to the fact that planetary scale feature like ENSO could be one driving force of drought episodes in India but not exclusive. The interaction between planetary and regional scale features may result in chaotic nature of simultaneous occurrence of drought on All-India basis homogeneous region basis. The probability of occurrence of drought over entire areal extent of India simultaneously is almost negligible. Even in the worst drought years there could be meteorological sub-divisions where the rainfall would be normal or even excess. Thus, the scale of deficient or excess rainfall is not uniform on different spatial scales.

Drought Years but not ENSO Events

It is found that of these 23 drought years, 11 years (1873, 1901, 1904, 1920, 1966, 1968, 1974, 1979, 1985, 1986 and 1991) are not associated with ENSO episodes. There is a secular variability in the ENSO-Monsoon relationship. This is a complex question and a challenge to climate science. It is expected that several question relating to ENSO-Monsoon connections on interannual and decadal scale will be better answered through data and modelling studies during CLIVAR and ICRP in the next decade or so (Sikka, 1998). This indicates that the existence of an ENSO episode is not sure indication of impending monsoonal droughts. Thus, the performance Indian monsoon may be associated with many other atmospheric, oceanic and terrestrial conditions. Therefore, ENSO is not a single condition for the drought occurrence in our country.

ENSO Events but not Drought

The years—1880, 1891, 1896, 1902, 1925, 1969, 1983, 1992 and 1998 were ENSO years but were not associated with incidence of drought event in India. During these years monsoon rains were either on positive side of the normal or did not show any significant departure from the normal. For example, despite '*super ENSO*' of 1998 this year India experienced more than 26 per cent rainfall from the normal and monsoon was considered to be excellently good. This poses a challenging question about ENSO-Monsoon relations, which we have to answer.

In the light of the above discussions we find that drought occurrence is frequent but has no cyclic trend. Evolution of ENSO also exhibits conspicuous irregularities from one cycle to another. The existence of an ENSO phenomenon is not a sure indication of impending monsoonal droughts since it has not consistently performed for the prediction of a monsoonal drought. The incidence of various classes of drought might be associated with other oceanic, atmospheric and terrestrial conditions, because atmosphere has a lot of chaotic noise. ENSO is just one factor. The ENSO event is not always sure sign of non-occurrence of excess rainfall. The ENSO years are generally associated with the deficient rainfall. However, all 4 phenomenal drought have links with ENSO events. ENSO events have a significant modulating effect on Indian monsoon. The result of this study by no means is perfect but certainly gives a much better indication of the climatic condition that is likely to prevail following ENSO strike. In the next millennium further detailed investigations on a much larger database of ENSO-drought relations is felt necessary for objectively forecasting such happenings.

Acknowledgements

The author is grateful to the Director General, IMD, New Delhi for providing necessary rainfall data and Dr. D.R. Sikka, Chairman MONTCLIM, ICRP, Department of Science and Technology (Government of India) and Dr. Shunil K. Dash, Principal Scientific Officer, Centre for Atmospheric Sciences, Indian Institute of Technology, New Delhi for constructive comments and suggestions.

NOTES

1. Nicholls, N. (1991). Teleconnection and heath—Teleconnection linking worldwide climate anomalies, (Cambridge University Press), 493-503.
2. Ropelewski, C.F. and Harpert, M.S. (1987). Global and regional scale precipitation patterns associated with the El Nino/Southern Oscillation. *Mon. Weather Rev.*. 115:1606-1626.
3. Sikka, D.R. (1980). Some aspects of large scale fluctuations of summer monsoon rainfall over India in relation to fluctuations in the planetary and regional scale circulation parameters. *Earth and Planetary Sci.*, 89:179-195.
4. Lockyer, N. and Lockyer, W.J.S. (1902). On the similarity of the short-period pressure variation over large areas, *Proc. Roy. Soc.*, London. 73: 457-470.
5. Walker, G.T. (1923). Correlation in seasonal variations of weather. *VIII Mem. India Meteor. Dept*, 24:75-131.
6. Walker, G.T. (1924). Correlation in seasonal variations of weather. *IX Mem. India Meteor. Dept.*. 24; 333-345.
7. Bjerknes, J. (1969). Atmospheric teleconnections from the equatorial Pacific. *Mon. Weath. Rev.*, 97: 163-172.
8. Shukla, J, and Misra, B.M. (1977); Relationship between sea surface temperature and wind speed over the central Arabian Sea and monsoon rainfall over India. *Mon. Weath. Rev.*, 105:993-1002.
9. Khandekar, M.L. (1979). Climate teleconnections from the equatorial Pacific to the Indian monsoon : analysis and implications. *Arch Metero. Geophys. Bioklim.*, A28 :159-168.
10. Angell, J.K. (1981) Comparison of variations in atmospheric quantities with sea surface temperature variations in the equatorial Pacific. *Mon. Weath Rev.*, 109: 230-243.
11. Rasmusson, E.N. and Carpenter, T.H. (1983). The relationship between eastern equatorial Pacific, sea surface temperature and rainfall over India and Sri Lanka. *Mon. Weath Rev.*, 111:517-528.
12. Mooley, D.A. and Parthasarathy, B. (1983). Indian summer monsoon and El Nino. *PAGEOPH*. 121:339-352.
13. Parthasarathy, B. and Pant, G.B. (1985). Some features of a long homogeneous series of summer monsoon rainfall and the Southern Oscillation. *J. Clim.* 5: 369-378.
14. Bhalme, H.N., Sikder, A.B. and Jadhav, S.K. (1990). Relationship between planetary scale waves, Indian monsoon rainfall and ENSO, *Mausam*. 41(2); 279-284.
15. Webster, P.J. and Yang, S. (1992). Monsoon and ENSO: Selectively interactive systems. *Quart. J.R. Met. Soc.*. 118: 877-926.

16. Mohanty, U.C. and Ramesh, K.J. (1993). Characteristics of certain surface meteorological parameters in relation to the interannual variability of Indian summer monsoon. *Proc. Ind. Acad. Sci.,* (Earth and Planet Sci.), 102: 73-88.
17. Chattopadhyay, J. and Bhatia, R. (1996). A re-examination of ENSO/anti-ENSO events and simultaneous performance of the Indian summer monsoon. *Mausam.* 47(1): 59-66.
18. Sikka, D.R. (1998). Monsoonal Droughts in India. In press.
19. Kane, R.P. (1998). El Nino, Southern Oscillation, equatorial eastern Pacific sea surface temperatures and summer monsoon rainfall in India. *Mausam.* 49(1): 103-114.
20. Victor, U.S. *et al.* (1995). El Nino's effects on southwest monsoon rainfall in Andhra Pradesh, India. *Drought Network News*, 7(2): 20-23.
21. Quinn, W.H. *et al.* (1978). Historical trends and statistics of the Southern Oscillation, El Nino and Indonesian droughts. *Fish. Bull.*, 76 : 663-678.
22. Rasmusson, E.M. and Carpenter. T.H. (1982). Variations in tropical sea surface temperature and surface wind fields associated with the Southern Oscillation/ El Nino. *Mon. Weath. Rev.*, 110:354-384.
23. Harpert, M.S. and Ropelewski, C.F. (1992). Surface temperature patterns associated with Southern oscillation. *J. Clim.,* pp. 577-593.
24. Deser, C. and J.M. Wallace. (1987). El Nino events and their relation to the southern oscillation, 1925-(1986). *J. Geophys. Res..* 14:189-196.
25. Shukla, J. and Paolino, D.A. (1983). The southern oscillation and long-range forecasting of the summer monsoonal rainfall over India. *Mon. Weath. Rev.*, 111:1830-1837.
26. McPhaden, M.J. (1999). The Child Prodigy of 1997-1998. *Nature.* 390: 559-562.
27. Neelin, J.D. *et al.* (1998). *J. Geophys. Res.,* 103: 14261-14290.
28. Rodbell, D.T. *et al.* (1999). *Science*, 283: 950-954.
29. Obasi. G.P.O. (1996). Climate, climate change, variability and productivity, Rajiv Gandhi Institute for Contemporary Study, Paper No. 26.
30. Supple, C. (1999). Nature's Vicious Cycle. *National Geographic*, 105: 73-95.
31. Wang, B. *et al.* (1999). Stochastic Dynamics of El Nino-Southem Oscillation. *WANGETAL.* Jan. (1999):5-23.
32. Parthasarathy, B.; Munot, A.M.; and Kothawale, D.R. (1995). Monthly and seasonal rainfall series for All-India, homogeneous regions and meteorological sub-divisions 1871-1994. IMD, Pune.
33. Ramdas, L.A. (1950). Rainfall and agriculture. *Ind. J. Met. Geophys.* 1: 262.
34. I.M.D. (1971). *Rainfall and Droughts in India.* IMD, Pune.
35. Mooley D.A. and Parthasarthy, B. (1982). Fluctuations in the deficiency of the summer monsoon over India and their effect on economy. *Arch. Met. Geophys. Bioklim Ser. B.*, 30: 383-398.
36. Parthasarathy, B.; Rupa Kumar; and Kothawale, D.R. (1992). Indian summer monsoon rainfall indices 1871-1990. *Met. Mag.*, 126: 174-186.

6

RESPONSE OF ECOLOGICAL PARAMETERS ON THE AGROECOLOGY

A Study of Himachal Pradesh, India

YASIR SAEED HANAFI and AHMAD MUJTABA SIDDIQUI

Introduction

This chapter aims to discuss the response of ecological parameters on the agroecology of Himachal Pradesh. It is well known that all the crops can be grown in varieties of environment but responded best only in optimal environmental conditions (*i.e.* gives highest productivity), which we call eco-friendly conditions for that crop and the study of such relationships comes under the discipline of agroecology. In its true sense agroecology is the study of physical characteristics of environment, climate and soils in relation to the optimal agricultural production with least deterioration.

A botanist, Justin Liebig, stated in 1840, that any environmental factor that most closely approached the critical minimum for a species will tend to be a limiting factor for that species. Investigation during the beginning of the twentieth century have set ecology and ecological relationship more and more on scientific factors, a very considerable literature has developed and numerous theories have been advanced since the appearance of Blackman's initial paper in 1905. Blackman's axiom of limiting factor is in reality an elaboration of Liebig's Law of Minimum which in its essence may be as "the yield of any crop always depends on that nutrient constituent which is present in the minimum amount". V.L. Shelford in 1913, expanded Liebig's observations to include the maximal limits of an environmental factor. Thus Shelford Law of Tolerance states that "any value below a critical minimum or a quantity or factor in excess of the critical maximum would exclude

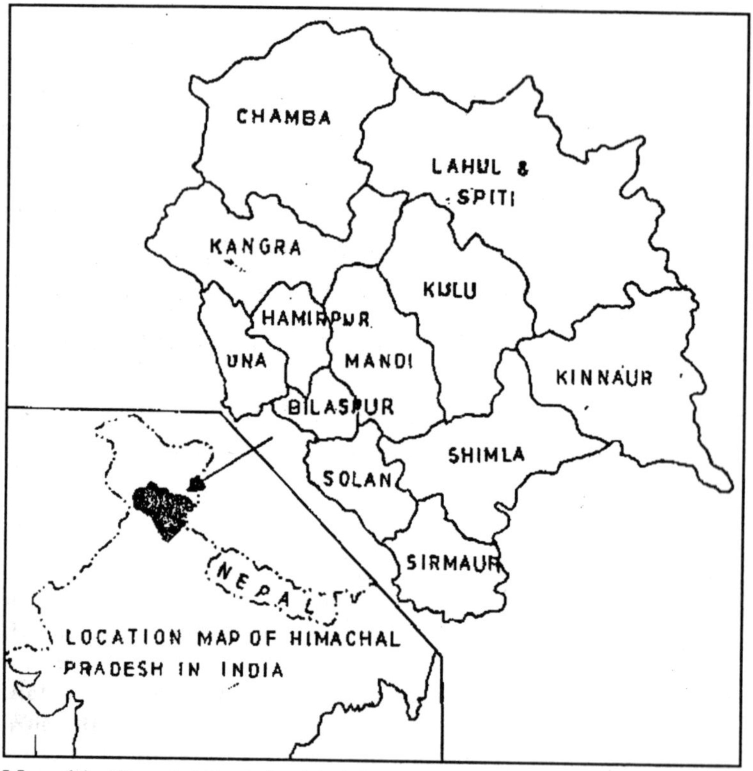

Map. 6.1 : Himachal Pradesh : Administrative Division

certain organisms from environmental areas". Weavers and Clements in 1929, stated that "every plant is a product of the condition under which it grows. The term 'agroecology' for the first time proposed by B.M. Bensin in 1930, to apply for the detailed studies of commercially important crop plants by the use of ecological methods. It is observed that Bensin deal only commercially important crops, so a large number of other crops excluded from the field of agroecology.

THE PROBLEM

Presently, the agroecology of this Himalayan region is threatened by two-fold human interactions. First the quest to increase the area under cultivation and second to get maximum possible agricultural return from a unit area. This in turn has given rise to three main problems of agroecological nature in the region. First, to increase the area under cultivation, clearance of forest areas is badly affecting agro-ecological balance. Consequently, the cultivation is now pushed to forested areas, marginal lands and upslopes without taking into account the suitability of these lands for agriculture. This unscientific and irrational land transformation process has brought about very serious environmental crisis in terms of erosion of top soils and reduction in groundwater recharge in this ecologically sensitive region. Second, to get maximum return from limited agricultural land, heavy doses of chemical fertilizer, insecticides, pesticides and irrigation are being applied without any proper knowledge of actual requirement of the agricultural land. This has resulted in the problems like development of toxic soils, elimination of favourable pests and insects, toxicated water flow in the out drainage river channels, multiplication of pesticide-resistant pests, concentration of harmful salt in the agricultural fields, water logging, salinity and alkalinity etc. Thirdly, good quality agricultural lands are being brought under non-agricultural uses like roads, buildings and water channels.

The process of agricultural resource development has a large and long-term impact on the natural component of a region through the destruction of original ecosystem. The unscientific and irrational agricultural transformation process now operating in this ecologically fragile Himachal region, caused changing pattern of agricultural land-use and cropping which introduced a variety of new conditions to the traditional agroecology of the region.

The ecological balance of flora, fauna and forests is being drastically disturbed by the rapid increase in human population which results in decrease of natural habitats and loss of biodiversity.

BASIC OBJECTIVES

This study tries to fulfil the following objectives :

— To find out the causes of changing cropping pattern. Are these changes towards the ecological adaptation of crops.

— To find out the ecological causes of inequalities of agricultural development in the region.

— To find out the suitability of existing cropping pattern with prevailing ecological conditions.

— To suggest the ecological requirement for optimal growth of various crops.

— To check the agroecological threats from becoming unmanageable problems of the region.
— To identify the deriving forces for deteriorating agro-ecological scenario of the region.
— To suggest the methods to move from maximization of production towards the optimal use of agro-ecosystem,

DATABASE AND METHODOLOGY

Secondary sources of data has been used for this study with an interval of twenty years *i.e.* 1974-75 and 1994-95, because both the point of time are well ahead of the period of Green Revolution in India, and in this period the technology is not showing any drastic change and so the effect of ecology on the agriculture is easily assessed. The climatic data used in this study is published by Regional Meteorological Centre, New Delhi. While the data of cropping and land-use pattern is published by Directorate of Land Records of Himachal Pradesh and Forest Department of Himachal Pradesh. The information on soil types and characteristics gathered from All India Soil and Land-Use Survey of India, New Delhi. The ecological requirement for various crops has been collected through literature survey on these crops.

For this study cropping pattern has been calculated as per cent share of area under each crop from Gross Cropped area, while productivity is simply calculated by dividing the production by area. The changes in cropping pattern and productivity are correlated to know the economic aspects of agriculture on high yield-high spread, high yield-low spread, low yield-high spread and low yield-low spread basis.

The productivity and spread are further correlated with ecological suitability of these crops. This analysis based only on spread of foodgrain crops *i.e.* wheat, rice, maize, barley, pulses and total foodgrains.

ECOLOGICAL PARAMETERS

It is clear from the Liepig's Law of Minimum and Shelford Law of Tolerance, that all the crops have some minimum and maximum requirements of ecological parameters for the sustenance of their life, while below and above of these cardinal points of requirement the crop suffers badly. On the same time somewhere between these cardinal points a situation exists where the crop responded best or optimum. Since so many parameters are involved in the growth of a plant, it is very difficult or almost impossible to find out the cardinal point for one parameter because all the parameters depends on each other. But rainfall, temperature and characteristics of soil are of immense importance and required in different quantities at different stages of their life cycle. A brief account of requirements of ecological parameters for the crops under study are given below:

Ecological Requirement of Wheat Crop

Wheat is mainly a Rabi (winter) season crop in India. It is cultivated from sea level upto even 3000 metres of altitude. It requires about 14°C to 15°C optimum average temperature at the time of ripening. During the growth and flowering stages, excessive high or low temperatures and drought are harmful. Temperature above 25°C during grain filling period tend to depress grain weight. Cloudy weather, with high humidity and low temperature is conducive for rust attack. 25 cm. to 75 cm. of rainfall is favourable for proper growth of this crop. Soils with a clay loam having moderate water holding capacity are ideal for wheat cultivation. Very porous and excessively drained soils should be avoided. Heavy soils with good drainage are suitable for wheat cultivation under dry conditions. Heavy soils with poor drainage are not suitable, as wheat is sensitive to water logging.

Ecological Requirement of Rice Crop

Rice growing seasons vary in different parts of India, depending upon temperature, rainfall and other climatic conditions. It is cultivated from sea level to as high as 3000 metres of altitude. Rice crop needs hot and humid climate. The average temperature required throughout the life period of the crop ranges from 21°C to 37°C. At the time of tillering, the crop requires a higher temperature than for growth. At the time of flowering, the temperature requirement is in the range of 26.5°C to 29.5°C while at the time of ripening the temperature should be between 20°C to 25°C. A well distributed rainfall of 25 cm. to 150 cm. is necessary for the growth of rice crop. Soils having good water retention capacity with good amount of clay and organic matter are ideal for the rice cultivation. Clay or clay loam are most suited for rice cultivation. Rice being a semi-aquatic crop, grows best in submerged conditions. It grows well in soils having a *p*H range between 5.5 to 6.5.

Ecological Requirements of Maize Crop

Maize is a warm weather crop. It grows from sea level up to 3000 metres of altitude. It is a Kharif (monsoon) season crop mainly in northern India. It requires considerable moisture and warmth from germination to flowering. The most suitable temperature for germination is 21°C and for growth 32°C. Extremely high temperature and low humidity during flowering damage the crop, resulting in poor grain formation. Rainfall of 50 cm. to 75 cm. with good drainage is conducive for proper growth. It is very sensitive to stagnant water particularly in early stages of life cycle. Maize is best adapted to well drained sandy loam to silty loam soils. It will not thrive on heavy clays. It can grow successfully in soils whose *p*H ranges from 5.5 to 7.5.

Ecological Requirement of Barley Crop

Barley requires cool weather during early growth and warm and dry weather at maturity. It can be grown from sea level up to 4000 metres of altitude. This crop requires 40 cm. to 90 cm. of rainfall for optimal growth. But it is a drought resistant, so barley suits areas of scanty rainfall too. Similarly the optimal temperature requirement for this crop is 15°C to 18°C and it can give economic returns. Barley thrives best on well drained, fertile deep-loam soils. Being a salt tolerant crop, it can do well even in salt affected soils. It gives higher yields than wheat and other cereal crops on moderately saline-alkaline soils.

Ecological Requirement of Gram Crop

Gram is a winter season crop. It is best suited to the areas having 60 cms. to 90 cm. of rainfall per annum. It is grown from sea level to an altitude of 2000 metres. Severe cold or frost are injurious to it and badly affected at the time of flowering. Excessive rain soon after the sowing or at flowering and fruiting cause heavy loss. The best type of soils for gram is one that is well drained and not too heavy. It is not suited to soils having a *p*H higher than 8.5.

Ecological Requirement of Lentil Crop

Lentil requires cold climate. It can tolerate severe winter and frost to a great extent. It can be grown up to an altitude of 3000 metres. It remain unaffected by rain at any stage of its growth, including flowering and fruiting. The optimum temperature for its growth is 18°C to 30°C. Well drained loam soils are best for lentil cultivation.

Ecological Requirement of Pea Crop

Pea require cold climate. Moderate temperatures are essential throughout the growing season. 22°C of temperature is considered favourable for germination. High temperature are more injurious to pea crop than frost. High humidity is harmful to pea crop as it favours incidence of disease. The optimum temperature suitable for its growth is 13°C to 18°C. Well drained loam soil is considered best for pea cultivation. Optimum soil *p*H is 6.5

Ecological Requirement of Urd Crop

Urd is an important pulse crop require a hot and humid growing season. It can be grown successfully from sea level up to an elevation of 1800 metres. The annual rainfall of 60 cm. to 75 cm. is optimum. Heavy rain during flowering stage are harmful and adversely affect the production. It can be grown on all types of soils ranging from sandy loam to heavy clay, except alkali and saline soils. Loam or slightly heavy soils with neutral *p*H are best suited for Urd cultivation.

CHANGING CROPPING PATTERN

The cropping pattern of the Himachal Pradesh as a whole is on a declining trend from 1974-75 to 1994-95, and the percentage of net cultivated area has declined from 18.56 per cent to 16.70 per cent in this time period. But the variation at district level shows that in six districts, the net cultivated area has decreased while in other six districts it is increased at the cost of pastures and other lands earlier used for grazing of animals.

It is revealed from Table 6.1, that the food crops are predominant in the cropping pattern and occupying 91.27 per cent of total cropped area in 1974-75, while in 1994-95, it declined to spread only over

Table 6.1: Changing Cropping Pattern in Himachal Pradesh (1974-75 to 1994-95)

District	*Wheat*		*Rice*		*Maize*		*Barley*		*Pulses*		*Total Foodgrain*	
	1974-75	*1994-95*	*1974-75*	*1994-95*	*1974-75*	*1994-95*	*1974-75*	*1994-95*	*1974-75*	*1994-95*	*1974-75*	*1994-95*
Bilaspur	36.41	45.05	6.71	3.80	39.19	44.39	0.51	0.83	13.52	2.97	96.36	97.03
Chamba	26.19	30.37	5.07	4.83	41.34	43.46	9.65	4.98	5.31	5.45	93.85	92.21
Hamirpur	33.05	48.82	7.40	4.44	37.42	44.52	0.30	0.28	6.36	0.83	98.23	98.89
Kangra	40.05	42.38	18.61	16.78	23.47	26.34	2.50	1.40	4.79	2.00	89.79	89.04
Kinnaur	14.32	6.45	0.35	0.00	2.85	5.38	22.52	17.20	2.16	9.68	91.29	68.82
Kullu	32.51	36.12	6.84	3.58	24.53	27.43	11.81	5.11	5.82	6.13	93.00	82.28
Lahaul & Spiti	18.18	6.06	0.00	0.00	0.00	0.00	41.61	21.21	6.99	33.33	78.32	66.67
Mandi	38.05	42.06	16.57	13.27	25.50	29.28	3.78	2.74	3.95	2.74	93.88	91.60
Shimla	30.98	24.76	4.85	3.14	19.77	18.38	8.59	4.76	4.31	5.90	82.44	64.19
Sirmour	36.38	37.47	6.63	6.91	32.04	32.48	4.69	3.71	7.09	5.63	89.77	87.34
Solan	30.86	35.61	6.29	5.91	34.22	38.48	2.98	2.58	15.75	6.21	91.10	88.79
Una	36.92	44.71	3.50	2.37	36.65	42.20	0.01	0.00	14.21	2.09	91.31	91.36
Total	34.90	38.73	10.14	8.55	28.85	31.87	4.54	2.72	7.04	3.77	91.27	87.56

Note: All the values are in per cent from gross sown area.

87.56 per cent of total cropped area. If analyzed the foodgrain crops separately, it is found that the percentage share of wheat and maize increases from 34.90 per cent to 38.73 per cent and 28.85 per cent to 31.87 per cent in the same period of time respectively. On the other hand areal spread of rice, barley and pulses has declined from 10.14 per cent to 8.55 per cent, 4.54 per cent to 2.72 per cent and 7.04 per cent to 3.77 per cent respectively.

Wheat is the most dominant crop of the State in terms of areal spread and only three districts shows the declining trend in area under wheat, these are Kinnaur where the share of area under wheat declined from 14.32 per cent to 6.45 per cent, in Lahaul & Spiti the area declined from 18.18 per cent to 6.06 per cent, and in Shimla, it has declined from 30.98 per cent to 24.76 per cent . The causes of decline in Kinnaur and Lahaul & Spiti are attributed to cold desert type of climate while in case of Shimla, it is because of the sprawl of urbanisation. The most significant increase in area under wheat is observed in Hamirpur where wheat alone occupied 48.82 per cent of total cultivated land in 1994-95, while in 1974-75 it occupied only 33.05 per cent. The other wheat dominated districts in terms of area are Bilaspur (36.41 per cent to 45.05 per cent), Kangra (40.05 per cent to 42.38 per cent), Mandi (38.05 per cent to 42.06 per cent) and Una (36.92 per cent to 44.71 per cent).

The second most dominant crop of the State in terms of area is maize. Most of the districts shows an increase in area except in Shimla, where it shows a minor decrease from 19.77 per cent to 18.38 per cent. Kinnaur has an insignificant share in cropping pattern while in Lahaul & Spiti, it is totally absent. Bilaspur, Chamba, Hamirpur and Una are the most significant districts where maize crop occupied more than 40 per cent of total cropped area during 1994-95.

Rice which is one of the most dominant crop of India shows insignificant situation and occupied only 10.14 per cent of area in 1974-75 which further decline to 8.55 per cent in 1994-95. It is evident from Table 6.1, that all the districts shows declining trends of rice in the cropping pattern except in Sirmour, where the area under rice increased from 6.63 per cent to 6.91 per cent. Kangra and Mandi are the most significant districts in terms of area where rice occupied 18.61 per cent and 16.57 per cent in 1974-75 while in 1994-95 again the same districts occupies the highest spread with 16.78 per cent and 13.27 per cent respectively and rest of the districts occupied even below 7 per cent of the total cultivated area.

Barley is an insignificant crop of India but this crop has low water requirement than wheat and gives higher yield on moderately saline-alkaline soils and so a better replacement in the areas where wheat shows uneconomic. In Himachal Pradesh it occupied less than 5 per cent area in 1974-75 which further declined to less than 3 per cent in 1994-95. The districts of Kinnaur and Lahaul & Spiti which showed insignificant in terms of wheat and maize are the most significant under barley crop. Though the pattern is declining but these districts have a considerable share of barley in the total cultivated land. In Kinnaur, barley occupies 17.20 per cent area while Lahaul & Spiti, it occupies 21.21 per cent of area, previously it occupies 22.52 per cent and 41.61 per cent area respectively.

Pulses are an important component of foodgrain crop because of their high nutritive value (protein contents ranges between 17 per cent to 27 per cent), they form an important segment of balanced diet. In the State as a whole the percentage share of area under pulses has declined from 7.04 per cent to 3.77 per cent. Bilaspur, Una and Solan are the districts of great concern as the area under pulses in these districts has declined badly from 13.52 per cent to 2.97 per cent, 14.21 per cent to 2.09 per cent and 15.75 per cent to 6.21 per cent respectively. Again the districts of Kinnaur and Lahaul & Spiti which shows insignificant in the State are found significant here as the area in these districts has increased from 2.16 per cent to 9.68 per cent and 6.99 per cent to 33.33 per cent respectively.

SPATIAL VARIATION OF PRODUCTIVITY

It is revealed from Table 6.2, that an overall productivity of foodgrain in the state is recorded 1662 kg./hect, in 1994-95 as against the 1232 kg./hect. of 1974-75. It is analyzed at district level that the productivity of total foodgrains has increased in almost all the districts of the state except in Bilaspur and Lahaul & Spiti where it declines from 1485 kg./hect. to 1320 kg./hect. and from 1946 kg./hect. to 1454 kg./hect. respectively. The districts showing significant increase in productivity are Chamba, Kullu, Mandi and Sirmour where it increased from 1263 kg./hect. to 1909 kg./hect., 1124 kg./hect. to 1894 kg./hect., 1206 kg./hect. to 1893 kg./hect. and 1295 kg./hect. to 1866 kg./hect. respectively. While Una, Solan, Hamirpur and Kangra falls under the moderate productivity districts during 1994-95 with productivity of 1731kg./hect., 1664 kg./hect., 1583 kg./hect. and 1515 kg./hect. respectively.

Table 6.2: Changing Pattern of Productivity of Foodgrains in Himachal Pradesh (1974-75 to 1994-95)

District	*Wheat*		*Rice*		*Maize*		*Barley*		*Pulses*		*Total Foodgrain*	
	1974-75	*1994-95*	*1974-75*	*1994-95*	*1974-75*	*1994-95*	*1974-75*	*1994-95*	*1974-75*	*1994-95*	*1974-75*	*1994-95*
Bilaspur	1026	1153	1165	826	2086	1594	1379	1400	447	333	1485	1320
Chamba	1013	1354	863	1129	1953	2778	1225	1218	272	371	1263	1909
Hamirpur	1376	1591	1011	1187	1493	1638	1380	1500	327	333	1365	1583
Kangra	1115	1687	979	1392	1608	1432	1294	1167	326	325	1236	1515
Kinnaur	1151	1000	1006	N.A.	1738	2000	2109	1000	160	1111	704	843
Kullu	1209	2165	1943	1428	3275	2130	1294	1900	315	583	1124	1894
Lahaul & Spiti	3245	2000	N.A.	N.A.	N.A.	N.A.	2605	1285	2450	1454	1946	1454
Mandi	1209	1541	943	1445	1749	2902	1294	1066	925	355	1206	1893
Shimla	1151	1307	1345	1272	1390	1917	1421	1500	222	387	1246	1371
Sirmour	1301	1727	861	889	1505	2649	1379	965	425	341	1295	1866
Solan	1182	1868	955	2256	1879	1626	1421	1353	456	292	1049	1664
Una	1174	1872	915	2176	1236	1627	1319	N.A.	1006	266	1211	1731
Total	1123	1625	1048	1392	1875	2022	1140	1296	519	419	1232	1662

Note: The productivity is in kilogram per hectare.
N.A. = Data Not Available.

It is evident from Table 6.2, that the maize crop shows the highest productivity at both points of time which is 1875 kg./hect. and 2022 kg./hect. in the state as a whole. At district level Mandi ranked first with 2902 kg./hect. followed by Chamba with 2778kg./hect. and Sirmour with 2649 kg./hect. are the significant districts in 1994-95. If we analyze the growth in productivity, Sirmour shows 76 per cent, followed by Mandi and Chamba with 66 per cent and 42 per cent growth in productivity respectively. Bilaspur (2086 kg./hect. to 1594 kg/hect.), Kangra (1608 kg./hect. to 1432 kg./hect.), Kullu (3275 kg./hect. to 2130 kg./hect.) and Solan (1879 kg./hect. to 1626 kg./hect.) are the districts where the productivity of maize has declined during the period 1974-75 to 1994-95.

Wheat is the second most dominant crop of the State in terms of productivity. It is noted from Table 6.2, that the productivity of wheat in the State as a whole is recorded as 1625 kg./hect. in 1994-95, while it was only 1123 kg./hect. in 1974-75. A district level variation shows that in all the districts except Kinnaur and Lahaul & Spiti, the productivity of wheat has increased. Kullu, Una and Solan falls under the high productivity districts of the State with average productivity of more than 1850 kg./hect. followed by Sirmour, Kangra, Hamirpur and Mandi with moderate productivity while Bilaspur with 1153 kg./hect. is the lowest productivity district of the State.

Rice which is not a significant crop in cropping pattern of the State showed an increase in productivity from 1048 kg./hect. to 1392 kg./hect. in 1974-75 and 1994-95 of the State as a whole. The variation at district level shows that Bilaspur with 826 kg./hect. is the lowest productivity district in 1994-95 while Solan with 2256 kg./hect. is the highest productivity district of the State followed by Una with 2176 kg./hect. In terms of growth in productivity it is estimated that Una with 138 per cent and Solan with 136 per cent are the most dominant districts of the State in terms of productivity as well as the growth in productivity during 1974-75 and 1994-95. The Kullu district is a matter of great concern as the productivity declined from 1943 kg./hect. to 1428 kg./hect. or a negative growth of '–26.50 per cent' On the other hand the rapid urbanization coupled with shrinking agricultural land is a big problem to sustain the livelihood of growing population.

The productivity of barley, which has a meagre share in the cropping pattern is noted to be 1296 kg./hect. in 1994-95, with the growth of 13.68 per cent over the productivity of 1974-75. Kullu with 1900 kg./hect. followed by Hamirpur and Shimla both with

1500 kg./hect. are the most significant districts in the State during 1994-95. Severe decline in productivity is observed in Kinnaur (from 2109 kg./hect. to 1000 kg./hect.) and in Lahaul & Spiti (from 2605 kg./hect. to 1285 kg./hect.) these districts are facing the cold desert like climate. Not only the productivity but area under barley in these districts also shows a severe decline in the same span of time.

The productivity of pulses in the State as a whole was very poor in 1974-75, *i.e* 519 kg./hect. which further declines to only 419 kg./hect. in 1994-95. Lahaul & Spiti with 2450 kg./hect., Una with 1,006 kg./hect. and Mandi with 925 kg./hect. are the most important districts in terms of productivity of pulses in the State during 1974-75 while rest of the districts could not reach even 500 mark in productivity. The situation is observed worst in 1994-95 when Lahaul & Spiti with 1,454 kg./hect., Kinnaur with 1,111 kg./hect. and Kullu with 583 kg./hect. could be able to cross the 500 mark in productivity while rest of the districts having the productivity even below the 400 kg./hect.

RESPONSE OF ECOLOGICAL PARAMETERS ON AGRICULTURE

Earlier in late sixties and early seventies a mad race for the growth of agriculture led to fast depletion of resources and rapid deterioration of habitat[1]. Ecological imbalance thus created are disproportionately large in areas having fragile ecology. "Now a consensus has emerged that ecology is needed to be saved from misdirected development. In this way regional development is needed to be defined as eco-development which represents the degree of concern for conservation of environment. From the angle of agro-ecological development an understanding of the requirements of ecological parameters such as rainfall, temperature, humidity and characteristics of soils etc. for optimum agricultural return is a must. Any ecological parameter can be limiting in excess or below the critical minimum. No factor affecting plant life is so thoroughly clear as the influence of water (Schimper, 1903). Ecological importance of water to the plant is fundamental and almost surpasses that of light and heat (Warming, 1909). Water is the chief limiting factor in the growth of most crops (Robinson, 1927).

At least a minimum amount of water must be present in the soil for the preservation of plant life. There is also an optimum moisture level on which plant is expected to give maximum responses beyond which growth suffers accordingly. Since so many factors are involved

Table 6.3: District-wise Agro-Ecological Parameters of Himachal Pradesh (1975 and 1995)

	Rainfall (in cm)		*Minimum Temperature (°C)*		*Maximum Temperature (°C)*		*Humidity*	
	1975	*1995*	*1975*	*1995*	*1975*	*1995*	*1975*	*1995*
Bilaspur	165.9	150.2	2.0 to 20.0	N.A.	21.0 to 42.5	N.A.	82	N.A.
Chamba	187.3	186.0	(-) 1.2 to 12.0	N.A.	15.0 to 32.6	N.A.	60	N.A.
Hamirpur	138.6	99.2	N.A.	N.A.	N.A.	N.A.	N.A.	N.A.
Kangra	131.2	229.0	2.4 to 17.9	2.8 to 18.0	16.9 to 35.6	18.5 to 38.6	66	62
Kinnaur	83.9	84.2	N.A.	(-) 9 to 12.4	N.A.	8.3 to 29.8	N.A.	58
Kullu	N.A.	103.8	(-) 5.2 to 15.3	(-) 1.3 to 16.1	20.3 to 38.0	20.7 to 39.4	87	80
Lahaul & Spiti	N.A.	119.3	N.A.	N.A.	N.A.	N.A.	N.A.	N.A.
Mandi	110.7	128.6	0.5 to 18.0	0.6 to 20.2	25.2 to 33.2	21.0 to 39.6	69	80
Shimla	137.4	125.5	(-) 1.1 to 12.7	(-) 1.8 to 14.6	12.9 to 29.6	13.2 to 29.4	54	61
Sirmour	108.8	184.5	N.A.	1.1 to 16.5	N.A.	20.6 to 46.8	N.A.	63
Solan	110.8	136.8	N.A.	11.1	N.A.	35	N.A.	N.A.
Una	117.7	194.8	N.A.	N.A.	N.A.	31.4	N.A.	75

Note: N.A. = Either the data is not available or not accepted.
Humidity is recorded at 08.30 IST.

in growth of a plant, the cardinal points for any ecological parameter are very difficult to know but up to some extent we can correlate the ecological parameters to the agricultural development of a region.

It is evident from Table 6.3, that the ecological parameter discussed earlier and their correlation with Table 6.1 and Table 6.2, that in case of wheat, the rainfall is sufficient enough according to required level and, therefore, the productivity and percentage share of wheat in cropping pattern is increased in all the districts except in Kinnaur and Lahaul & Spiti because temperature is found much lower than the critical minimum and the characteristics of the soil are also not found suitable for the cultivation of wheat in these cold desert like districts. As a result the productivity and percentage share of wheat in cropping pattern has declined. On the other hand, the condition for maize crop is more or less the same as of wheat, the rainfall is more than the sufficient but good drainage balanced it according to requirements. The temperature in some districts like Chamba, Shimla and Kullu is below the critical requirement but the characteristics of soil are favourable for this crop in these districts, so the temperature could not affect to a great extent and overall required parameters for the crop boost the productivity and percentage share to the total cultivated land.

In case of rice, the productivity and share in cropping pattern is not comparable with wheat and maize but the response of ecological parameters existing at two points of time is clearly shown from the fluctuation of productivity at district level. There are wide variation in the rainfall in Kangra, Mandi, Solan and Una in 1974-75 and 1994-95. Productivity of rice also vary accordingly in these districts where not only the rainfall but temperature and humidity requirement are also favourable in these districts for the rice crop. The rainfall in Bilaspur, Hamirpur shows severe decline, so the productivity also decline accordingly. In Kullu the temperature, humidity and soil characteristics at low land are favourable but the productivity of rice has declined significantly because the chief limiting factor *i.e.* rainfall was insufficient in this district during 1994-95.

Barley is a crop grown generally where the wheat gives uneconomic returns, though the ecological condition for barley is suitable but the ignorance towards this minor crop results in lower yield than the other dominant crops, because of its meagre share as compared to the other crops, the proper attention is not given for this crop. This is clear from the district of Kinnaur and Lahaul & Spiti where there is major share under this crop, and the productivity was observed high during 1974-75 while in 1994-95 the share of this crop declines in these districts so the productivity also declined accordingly. In these districts the overall situation is also affected by the introduction of horticultural crops at large scale.

The ecological parameters for the pulses are not suitable in terms of low temperature while most of the pulses required warm climate except pea which require cold climate. Similarly a large number of pulses crops can sustain under low water availability while high rainfall and high humidity affect on their productivity. So on the basis of existing ecological parameters we can say that low productivity and lower share of pulses attributable towards unfavourable ecological parameters.

CONCLUSION

It is concluded from the above discussion that the crops which have favourable or require ecological parameters in the State, are sprawling in the cropping pattern and responded with maximum return. So, it is said that the changing cropping pattern of the State is because of the ecological adaptation of the crops for optimal return. Its condition is very similar to the Darwin's theory of survival for the fittest. It is also concluded from the above discussion that the crops responded according to the prevailing ecological parameters in combination to each other and unwanted excess or insufficiency of any one affect badly on the yield. It is said that the changing cropping pattern of the State is an ecological readjustment of the natural agro-ecosystem as it is evident from the increased productivity mostly in all the crops in spite of area adjustment with the ecological condition.

It is also noted here, that the agro-ecological threats are because of the mad race of maximisation of production by the application of modern techniques and other developmental works accelerated in the Himalayan region like mining and quarrying, construction of roads, commercial problems pertaining to landslides, deforestation, air and water pollution and soil erosion.

It is, therefore, imperative to integrate agricultural development policies with environmental management and planning on ecological principle. The agro-ecological planning in this context, is referred to as scientific and integrated framework for ecologically sustainable and economically viable development of agriculture through ensuring optimum and most rational utilization of environmental resources.

A further research is needed and suggested that cropping pattern should be readjusted in terms of the ecology and a crop should occupy the largest share based on the ecological conditions. Hence it is concluded that land best suitable for any crop should be introduced to achieve the higher return from the land.

NOTE

1. Habitat refers to sum total of all external conditions affecting the development, special response and growth of plants (organism). Since the term habitat was first used by botanists, and especially ecologists, it is best to apply it to the description of physiological conditions influencing the distribution and growth of plants as contrasted with social environment which deals with the influence of a variety of factors other than those concerned with the direct growth requirements determining the distribution of the crop plants.

REFERENCES

Chandel, R.S. and S.S. Thakur (1999). Insect Pests of Hops and their Control in Lahaul Valley of Himachal Pradesh, *Indian Farming,* Vol. 48, No. 3, pp. 27-28.

Colinvaux, P.A. (1973). *Introduction to Ecology,* John Wiley and Sons, Inc., New York.

Darwin, C. (1859). *The Origin of Species by Natural Selection,* Modern Library, New York.

Dhaliwal, G.S. and D.S. Kler (1995). *Principles of Agricultural Ecology,* Himalayan Publishing House, New Delhi.

Dubey, Y.P. and B.P. Kaistha (1998). Production Potential of Lahaul Valley—Constraints and Remedies, *Indian Farming,* Vol. 48, No. 9, p. 21.

Klages, K.H.W. (1942). *Ecological Crop Geography,* The MacMillan Co., New York.

Liebig, J. (1840). *Chemistry in its Application to Agriculture and Physiology,* Taylor and Walton, London.

Park, S. Nobel (1983). *Biophysical Plant Physiology and Ecology,* W.H. Freeman & Co., New York.

Purseglove, J.W. (1968). *Tropical Crops: Dicotyledons,* Longman Group Limited, London.

Robbins, W.W. (1927). *Principles of Plant Growth,* Wiley, New York.

Schimper, A.F.W. (1903). *Plant Geography Upon Physiological Basis,* Clarendon Press, Oxford.

Shelford, V.E. (1907). Preliminary notes on the distribution of the tiger beetles (Cicindela). and its relation to plan succession, *Biol. Bull;* 14; 9-14.

Singh, C. (1983). *Modern Techniques of Raising Field Crops,* Oxford & IBH Publishing Co. Pvt. Ltd. New Delhi.

Singh, R.B. (1998). *Sustainable Development of Mountain Environment in India and Canada,* Oxford & IBH Publishing Co. Pvt. Ltd., New Delhi.

Singh, Tejvir (1980). *Studies in Himalayan Ecology and Development Strategies,* The English Book Store, New Delhi.

Slatyer, R.D. (1973). *Plant Response to Climate Factors,* UNESCO Publication, Paris.

Sood, K.C.; Awasthi, K.R.; and B.C. Sood (1999). Constraints in Rabi Pulses Production in Himachal Pradesh and Quality Seed Production of Gram, *Indian Farming,* Vol. 49, No. 8, pp. 30-32.

Warming, J.E.W. (1909). *Oecology of Plants,* Clarendon Press, Oxford, pp. i-xi. 1-422.

Weaver, J.E. and Clements, F.E. (1929), *Plant Ecology,* Tata McGraw-Hill, New York.

Wilsie, C.P. (1974). *Crop Adaptation and Distribution,* Eurasia Publishing House Pvt. Ltd., New Delhi.

7

IMPACT OF FLOOD, MITIGATION MEASURES AND STRATEGY FOR SUSTAINABLE DEVELOPMENT

A Study of Haryana, India

G.S. CHAUHAN, R.S. SANGWAN and A.K. SRIVASTAVA

INTRODUCTION

Floods are considered the most devastating hazards among all the natural calamities. These hazards poses great impact on the various human activities and economy of a nation. Each and every year flash floods exert a heavy toll of human life and loss of flora and fauna and national property. Flood hazards are the extreme events caused by natural processes and human factors and some of the exceptional cases exceeds the tolerable magnitude, makes adjustment difficult and resultantly in catastrophic loss of huge property. While flood is the cumulative hazard which usually occurs when there is an appreciable increase in rainfall from the average in the catchment area. Although, it is totally a natural phenomena but human factors are equally important in order to bring such hazards. Moreover, when the degree of such natural calamities like floods, is acute causing huge loss to lives and properties, then becomes a hazard.

Meanwhile, floods are the extreme events and beyond control and predictive power of man. But such disasters frequency and degree of damage of loss may be minimised by the large scale efforts made by man. The flash floods which are known as one of the major hazards have proved a very serious problem not only for our country but it has appeared as a global problem in the current years. This problem is more serious and a matter of concern especially for South Asia including India. Presently North India is suffering from this annual problem.

At local level it is generally known as human induced problem. These hazards have great impact on different parts of the region as well as their socio-economic condition and developmental activities up to the maximum extent. Apart from these, the anthropogenic activities and overall physical setting of a landscape is also disturbed and influenced by such natural disasters. Hence, floods hazards are posed a severe threat to the existing fragile ecosystem and ecological sustainability.

The recent furious floods in Haryana (1995) have been considered the worst floods of the century in the history of the State. Such floods scenario never been so baffling in the State so far. These have been created a havoc in whole of the State and it has badly damaged infrastructure; property and economy as well as socio-economic conditions of the people.

THE STUDY AREA

The study area which is located in the northern part of the country consists of 17 districts. This small agrarian State possesses a continental type of climate. The climate of the region is characterized by high temperature in May and June and low temperature in December and January. While normal rainfall and dust storms usually occurred in the south-western parts in the month of June. Moreover, the maximum rainfall is perceived in the months of August and September. Although a little and scanty rainfall is also gained in the winter season by the certain parts especially by the temperate cyclones (Western disturbances).

Geographically the State is located approximately between 27° 30' and 30° 50' north latitudes and between 70° 30' and 77° 34' east longitudes (Map 7.1). It comprises about 44,244 sq. km. area which is hardly more than one per cent (1.34%) of the country. The State with a population of 16.4 million (1,64,63,000) in India, is one of the densely populated State of the north-west India. The density of population was 274 in 1981 and 372 persons per sq km. in 1991, which covers about 1.93 per cent population of the country. It has been estimated by the State government that more than 1.8 million people in 13 cities and towns and about 2,043 villages in at least 15 districts were devastated and shifted to the safer places. About 253

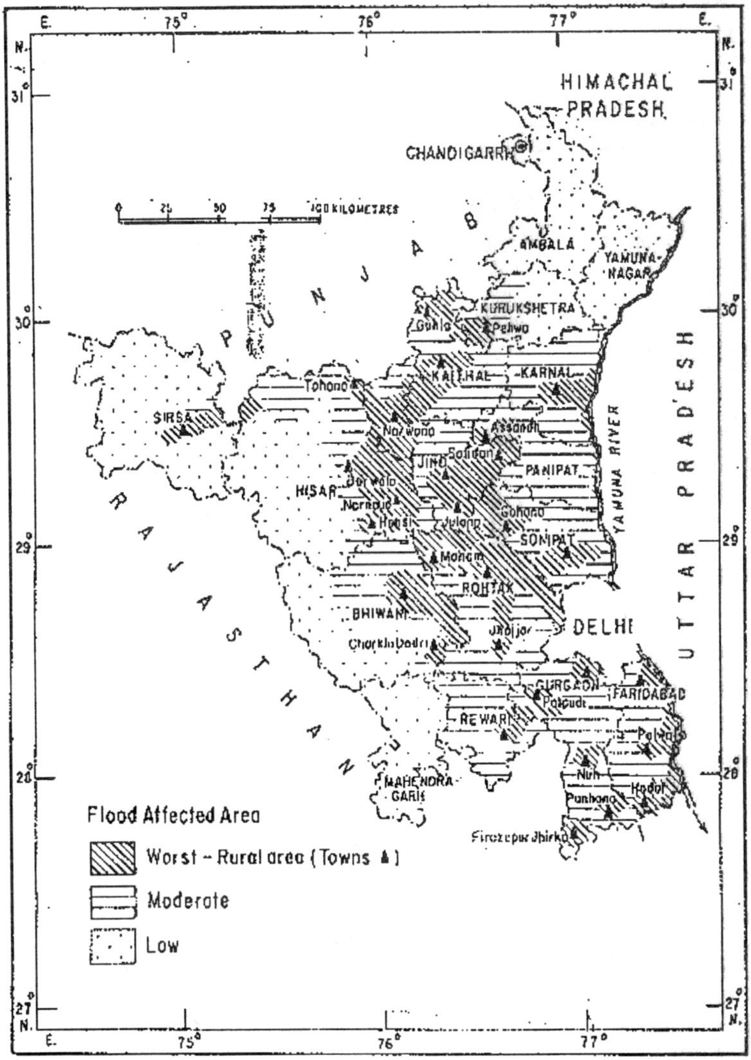

Map 7.1 : Haryana : Flood affected area, 1995

cattle died and as many as fled away in rain water. More than 72,000 houses and at least 1,558 roads in various parts had been extensively damaged whereas traffic on 36 roads had been suspended for several weeks and in some cases several months. In case of National Highway

No. 10 which runs Delhi to Fazilka, a major sector of road between Meham and Rohtak was badly damaged and in this section the traffic had been suspended at least for three months (Map 7.2).

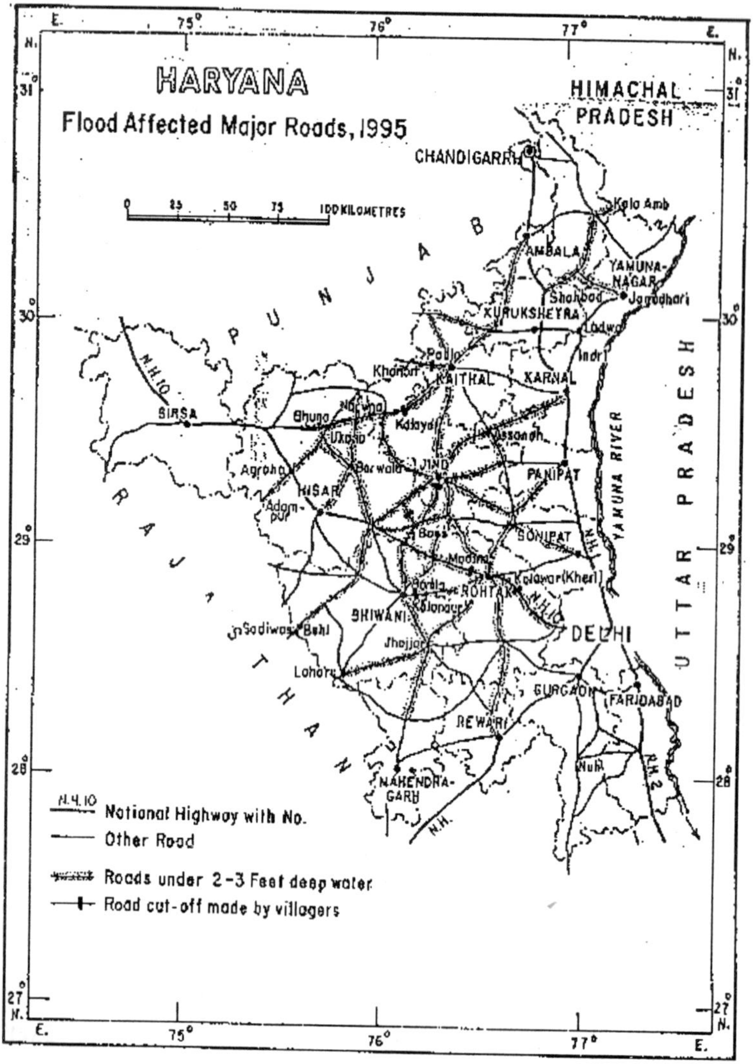

Map 7.2 : Haryana : Flood affected major roads, 1995

The recent fierce floods proved very disasters and destructive to the State economy as well as its agricultural crops too. It has been estimated that about Rs. 20,000 million property was hit and damaged by these floods. The floods figures reveal that the State was badly affected and hampered as a result various standing crops damaged and spoiled. Moreover, the infrastructure was also in the grip of furious floods. In the last hundred years the State never saw such devastated floods.

Ultimately the existing situation was very alarming and critical and it was like a big catastrophe liable for bringing countless sufferings and pain to a sizeable population, both rural and urban. Out of 17 districts, 10 districts *i.e.* Rohtak. Bhiwani, Hisar, Jind, Sirsa. Kaithal, Sonepat, Faridabad, Rewari and Gurgaon were affected.

SPATIAL EXTENT OF HARYANA FLOOD

Almost all districts including their concerned villages were affected and influenced. But Rohtak was the worst affected district followed by Bhiwani district. The broad description of flood as well as spatial extent have been given in Map 7.1.

Rohtak

Although, it has been visibly observed in some of the previous years that this legendary town is almost every year has to become the victim of the natural calamities like flood. But in comparison to earlier catastrophe, the flash floods which occurred recently were more aggressive, furious and beyond the control. Virtually, this central town is considered often more vulnerable to floods. The most dominated and responsible factor for such natural events is cuplike landscape of this populace town and this ankles more easy and heavy inflow of rain water inside the town and simultaneously very difficult of its out flow. That is why, the town was highly damaged and influenced by the incessant rain followed by heavy and powerful floods. Apart from its various colonies and settlements, the adjoining towns were also got the same tastes and experiences. It has been stated that almost all areas of the town were under six to ten feet of deep water while the various major roads leading to adjoining towns were worst affected. The flood water was flowing unruly into the several areas and many of them were totally submerged. Moreover, an almost identically feeling often persists in the 1,300 odd villages affected by floods. About 250 villages (Table 7.1) of the district have worst affected while, about 24 villages have been completely cut off by the State as well as its various approachable centres. The similar situation was also facing the other towns and villages of the district were Meham, Jhajjar and including their affiliated villages. Rural settlements were also in the hit list of furious floods. Evidently, hundred of villages were directly affected.

Bhiwani

The Bhiwani district which is located in the semi-arid tract was perhaps first time in the grip of the floods. The walled city itself including the hundred of villages situated in and around it were also badly affected. Most of the localities noticed were under 6 to 8 feet deep water. About 128 villages of the district were influenced. Another equally hit adjoining town was Charkhi Dadri, followed by its associated villages. Flood situation in Bhiwani, Dadri towns and over 250 villages was very grim and terrible.

Hissar

Like the other towns as well as several villages the Hissar district and its various parts were also in the floods. Hansi, Tohana and some of villages of Hissar sub-division were highly affected. Out of them, Hansi town and it associated villages were totally cut off by the State. In Hansi sub-division as many as 55 villages have been affected and similar fatal situation was facing another town Barwala, 30 km from Hissar. The situation in this small town was highly grim and several settlements and colonies were nearly 6 feet deep water. Apart from these, the adjoining villages were too going through equally critical situation. Furthermore, the Tohana an another town, has also been hit, and entire area was appearing as a mini island and all approaching roads leading to the town were fully waterlogged.

Other Affected Districts

The situation in Jind district has deteriorated as the number of flood affected villages has increased from 100 to 227. Out of the most affected villages about 90 villages have been marooned and have been

Table 7.1: Dimensions of Flood in Haryana

S. No.	*Name of the District*	*Affected village*	*Nos. of total hit villages*	*Nos. of total vil. completely cut off by State & Distt. HQ.*	*Worst Affected Towns/Villages*	*Total loss (in crore) due to crops damage including damage of property*	*Rainfall (in mm) on first day at various stations*	*Name of the Kharif crops destroyed*****	*Agricul. land area under flood water (in acre)*	*Depth of Water (in feet)*****	
										Rural	*Urban*
(1)	*(2)*	*(3)*	*(4)*	*(5)*	*(6)*	*(7)*	*(8)*	*(9)*	*(10)*	*(11)*	*(12)*
1.	Rohtak	1300 odd villages	250	24	Rohtak, Meham, Jhajjar and Jassia, Kahni, Khamar, Rithal, Makrauli Kalan & Khurd, Phogat, Pulangi, Rurki, Mungan, Ason, Mokhra Khas, Azaib, Madina, Bahu-Akbarpur, Balamba, Sunaria, Mokhra, Maina, Bhambheva, Nindana, Bainsi, Gugaheri, Samar Gopalpur and Kailanga.	339* 1500**	86	Cotton, Jowar, Sugarcane, Guar and Oilseeds	2,00,000	3-6	6-10
2.	Hisar	260	225	20	Hansi, Tohana, Barwala, Basskhurd, Akbarpur, Badhahpur and Azmshahpura, Petwar, Badala,	68*	31.7	Cotton, Jowar, Guar, Sugarcane, Oilseeds, Bajra,	85,000	4-7	4-6

(Contd.)

Table 7.1 *(Contd.)*

(1)	*(2)*	*(3)*	*(4)*	*(5)*	*(6)*	*(7)*	*(8)*	*(9)*	*(10)*	*(11)*	*(12)*
					Narnaud, Gangankheri, Majra, Mohala, Badcchappar, Ugalan, Sisai, Kumbha- Thurana, Sisai, Kharbala, Sorkhi, Bhattol, Kharkara, Khanda Kheri, Sirsod, Bichpari, Kheri, Barki and Talwandi Rana			Moong			
3.	Bhiwani	328	250	22	Bhiwani and Dadri, Tosham Towns and Talu, Dhanana, Chang, Tigrana, Doki, Kitlana, Mundhal, Kharak, Bamla, Nindan, Bran, Ghuskani, Gujrani & Bondkalan	200*	—	Bajra, Jowar, Guar, Cotton, Moong and Oilseeds	90,534	3-6	6-8
4.	Jind	227	100	90	Jind, Narwana, Safidon, Julana and Uchana Towns and Pillukhera Alewa, Dhakal, Belarkha, Dumarkha, Hathoo, Amargarh, Mohalkhera, Nagura, Pega, Dhanori, Hatt, Dhamtan, Sahib, Singwal, Chattar, Thua, Zil, Khatkar, Baroda, Kar Sindhu, Sudkan, Lodhar, Barsola, Bhikhewala, Bhahri and Bhagana	65*	—	Paddy, Jowar, Bajra, Cotton, Guar and Oilseeds	60,000	5-7	4-6

(Contd.)

Table 7.1 *(Contd.)*

(1)	(2)	(3)	(4)	(5)	(6)	(7)	(8)	(9)	(10)	(11)	(12)
5.	Kaithal	123	67	23	Kaithal, Kalayat and Gulha towns and Gilana, Songari, Pega Taragarh, Kithana, Rahra, Bhana, Balu-Batta, Pai, Bad-Sikri, Sergarh, Rajaund, Serda, Songal-Majra, Kharak-Pandavan, Kotra and Kheri-lamba	42*	—	Paddy, Arhar, Jowar and Sugarcane	35,000	6-7	2-4
6.	Gurgaon	22	5	2	Gurgaon, Nuh, Nagina, Pataudi, Firozpur-Zirka towns and Inchapuri, Jatola Jorisapka	10*	28.2	Bajra, Arhar, Guar, Jowar and Oilseeds	14,000	3-5	2-3
7.	Rewari	25	12	5	Rewari town and Balawas, Kishangarh Khalilpur, Kumbhawas and Bhurpur	21*	—	Bajra, Jowar, Arhar, Guar and Oilseeds	34,436	3-5	2-4
8.	Sirsa	28	13	4	Sirsa town and Nejadela Musahibwala, Sangar and Kharia	8*	—	Bajra, Cotton Guar, Jowar and oilseeds	18,000	4-6	2-3
9.	Karnal	67	40	12	Karnal, Assand Towns and Syyad, Chapra, Nabiabad, Dabkauli, Nagli, Halmama, Zabti Chapra, Kamalpur, Hansamajra, Chandraon, Chaugaon and Khukni	4*	34.6	Paddy, Jowar and Sugarcane	20,000	4-6	2-3

(Contd.)

Table 7.1 *(Contd.)*

(1)	(2)	(3)	(4)	(5)	(6)	(7)	(8)	(9)	(10)	(11)	(12)
10.	Kuru-kshetra	45	25	7	Pehowa town and Jognakhera, Jhansa, Thol, Nayagaon, Mundakhera	2*	—	Paddy, Jowar Sugarcane and Oilseeds	1,200	4-5	2-3
11.	Faridabad	50	12	15	Faridabad, Hodal, Palwal towns and Thanthri, Satwaghesh, Indira Nagar Hafala, Basantpur, Mohabatpur, Lalpur, Dhadar, Khidwali, Alipur, Aminpur, Bhopani, Dadsia, Sherpur and Nicholi	15*	—	Jowar, Arhar Paddy, Bajra, Guar and Maize	60,000	4-6	2-4
12.	Sonepat	144	98	3	Sonepat, Gohana towns and Rindana, Dhanana, Bhawar, Gadwal, Ahulana, Madiana, Nagar, Kathura, Gorar and Katwal	3*	—	Sugarcane, Jowar, Guar	2,00,000	4-6	3-4

Source : * All figures are estimated by the State government officials and (agriculture department) which are released by the State Government time to time in various newspapers, Magazines and Bulletins.

** Figures are based on public opinion.

*** Based on personal surveys and people's perception.

**** Personal Survey.

ultimately cut off by the State and consequently flood water was flowing continuously throughout the town and adjoining villages destabilizing all short of anthropogenic activities. The worst hit smaller towns of the district were Safidon, Julana, Pilukhera, Aleva, Narwana including Uchana blocks. Out of them, Julana and Narwana blocks were highly affected and the water level recorded at these places were 4 to 6 feet deep.

The Kaithal district has too assumed alarming situation. The rain water freely entered in the town at number of places on huge scale. Most of the residential areas of the town submerged under 2 to 4 feet of flood water. About 123 villages were badly affected and out of these 67 including 23 villages across the Ghaggar river in Gulha sub-division were absolutely marooned and visibly appearing like a small island. The major cause for the flash floods in the Sirsa district and its adjoining villages was the Ghaggar, a seasonal river which was on its full swing in floods hours and its level has relatively increased suddenly resultantly many villages of the surrounding tracts were flooded. The situation in Sonepat district and its concerning towns particularly Gohana sub-division was terrible. Many nearby towns have been submerged. The major factor liable for flood in and around the Gohana was the Yamuna which of course, overflowing subsequently submergence of several villages under flood water. Gohana town remain cut off from the nearby towns upto several days. About 144 villages in the countryside have been affected.

In Gurgaon district the severely hit areas of the district was Mewat region including the district headquarters. There were 22 villages including Nun and Punhana had been equally affected. However, the sub-divisional towns of Nuh, Nagina and Ferozepore Zhirka in Mewat region worst struck followed by a breach in bundh in Bharatpur district (Rajasthan). Flood water entered at least 5 villages located along the Aravalli bundh. Simultaneously, residential areas in Nuh town were also hit. The water in Sahibi rivulet had also flooded the Pataudi block and adjoining villages.

Villages situated in Khadar areas of Faridabad district had also been totally cut off from the rest of the State due to heavy rains and sudden rise in Yamuna water. 15 villages located in the *Khadar* belt with a combined population of more than 18,000 persons were affected. The low lying areas in Palwal and Hathin blocks had been inundated with rain water. Nearly 60,000 acres of land belonging to about 50 villages was under 4 to 6 feet of water. Floods hit 12 villages of the district and around 15 villages of the district (*Khadar* area) had been totally cut off from the rest of the State. Similarly nearly 25 villages of the Rewari district and half a dozen colonies of Rewari town had been badly struck. Most of the low-lying areas (settlements) in and around the town are under 2 to 4 feet deep water which had also damaged thousand of houses. However, in Karnal district the Karnal town and various villages of the district which are close to the Yamuna adjacent areas were highly affected in comparison to the other areas. Although Karnal sub-division was fully in the grip of flood.

Meanwhile, in comparison to the various previously discussed districts, Kurukshetra district was least affected. Although, 25 villages of the district had been noticed to be flood affected.

MAIN CAUSES OF HARYANA FLOOD

Although, there are a number of causes for the occurrence of the flash floods all over the State including its various tracts. But the major factors are following:

1. Constant and excessive rainfall within a short span of time.
2. Defective and poor drainage system including inadequate sewerage system in towns.
3. Overflow in the major rivers, canals and including drains due to the maximum discharge of water in the Yamuna and various tributaries, canals and drains.
4. Breaches made by the local people in the main canals and drains including highways as well as below railway lines.
5. Accumulation of debris and silt in the major rivers, canals, and drainage bed (course).
6. Lack of ring dams (*bandhs*) around the villages and towns (Flow in the major drain of the State No. 8 which ultimately joins Yamuna in Delhi flowing through the main flood-prone areas).

IMPACT OF FLOOD

Meanwhile, almost all the developmental activities were hit on large scale and more apparently the State economy was destabilized. The

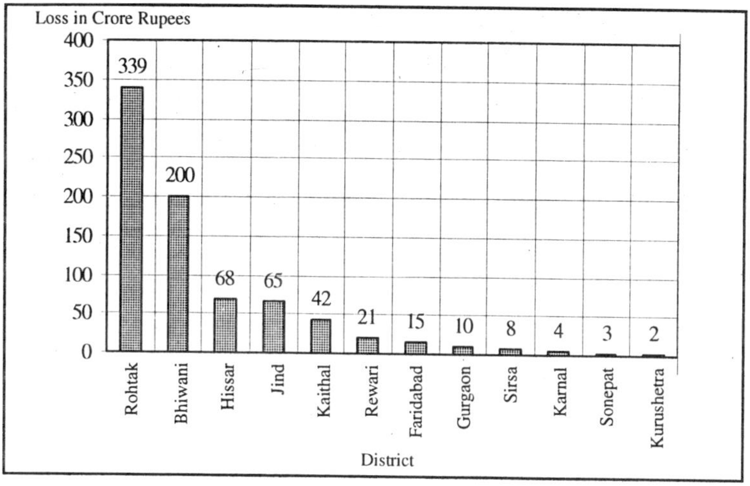

Graph 7.1 : Total Loss of Crops and Property by Haryana's 1995 Flood

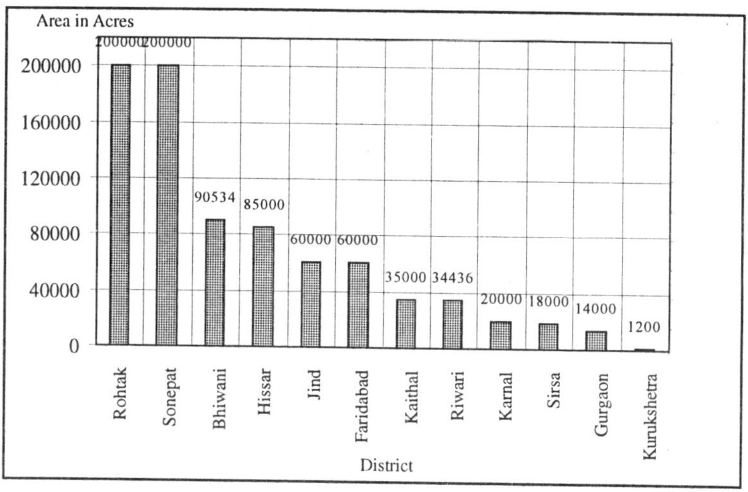

Graph 7.2 : Haryana Agricultural Land Area Submerged under Flood Water

equally other important reason for the more destruction and impact was the occurrence of powerful storm with high velocity followed by torrential rain. On the whole widespread damage was visibly seen in most of the parts and ultimately the socio-economic activities of the

State's people were almost negligible and endangered. All sorts of activities both human and livestock were halted and the flash floods have very intensive impacts on them.

Apart from them, thousand of people either became homeless or as many as were wandering for shelter. Hundred of people died or fled away in flood water. The similar situation was also facing the livestock. Many hundred livestock fled away and numbers of died in the wake of or shortage of fodder. Besides this around 2.2 million acres of agricultural land was converted into waste land and countless trees especially on road, rivers and canals side were uprooted and agricultural crops worth million of rupees were drastically spoiled and destructed. Simultaneously roads and railway lines were hit particularly the roads network of the State. Likewise, thousand of houses either totally collapsed or badly damaged. Even then, the total impact on the various types of activities including different sectors are given below:

Population

It has been visibly and apparently seen that maximum impact of these flash floods were ultimately on huge population. The main sufferers were particularly farmers and rural dwellers of low lying areas and poor people. Thousands of villages of the State were inundated and flood water unruly entered in the villages, towns and cities subsequently millions of people vacated their houses and such people were shifted to the safer places.

The situation was so much critical and grim that it was very hard to supply them even the essential commodities. Simultaneously, the affected population were too facing certain problems especially food, shelter, clothing including clean drinking water. Their houses both *kaccha* and *pucca* were either totally collapsed or damaged. As a result, they have to vacate their houses and compelled to stay under the open sky.

Livestock

Like human population, livestock population was also affected. Hundred of cattle heads fled away in the flood water and as many as died.

Apart from the above mentioned two major sectors, agriculture, socio-economic conditions, and all anthropogenic activities were too equally hit. Moreover, the other worst affected sectors were dominantly road and rail route, power supply and telecommunication. Flash floods had the following impacts:

(i) Millions of people became homeless and looked for shelter for several days.

(ii) Thousands of houses and settlements were collapsed and highly damaged.

(iii) Hundreds of persons fled away on the flood water and equal numbers died.

(iv) Thousands of livestock either fled away or died in lack of fodder shortage.

(v) Almost all kharif crops especially paddy, cotton, bajra sugarcane, jowar and oilseeds were destroyed due to Submergence in water around 3 to 4 feet deep for a long time.

(vi) Most of the agricultural land was under many feet flood water and was not in a condition for further cultivation in certain areas.

(vii) Maximum fertile soil have been eroded on large scale in the flood water especially along the embankments of canals, rivers and drains.

(viii) Due to the heavy and incessant rains water-logging problem had been created in certain pockets of the worst affected areas.

(ix) Most of agricultural land had been converted in saline and alkaline areas and problems of alkalinity and salinity had taken place.

(x) Large quantity of water remain standing at various places for a long period as a result the underground watertable suddenly rose up.

(xi) Due to the overflow in rivers, canals and drains, the seepage problems were visibly observed at many places especially at some of the vulnerable points.

(xii) Thousand hectares of land had been converted into the wasteland followed by barren land.

(xiii) Production of various kharif crops either decreased or gone down.

(xiv) Like highway the other State highways and other

associated roads were submerged in water subsequently failure of traffic services.

(xv) Due to the breaches in railway tracks, the railway services were also cancelled at various rail roads section.

(xvi) The power supply was totally damaged because almost electric poles were uprooted by the speedy winds accompanied by heavy rains.

(xvii) The telecommunication system was also hit on large scale.

(xviii) All socio-economic and commercial activities were severely affected.

On the whole, it has been assessed that the maximum flood impact were clearly observed roads (network) including district roads, Stàte highways and national highways. Due to flooding of roads, traffic services were suspended at various places. The most significant roads leading to Delhi and joining the national capital to Fazilka and other commercial towns. National Highway No. 10 flooded at number of places and suffered extensive damage at several places consequently all commercial as well as economic activities were stopped and suspended up to at least three months. Simultaneously, the various other State highways including link roads also suffered. Moreover, *kaccha* link roads (unmetalled) were badly affected.

Due to the occurrence of floods the rail traffic were too affected on large scale resulting suspension of rail services at so many important places especially Bhiwani, Sonepat and Jind district. Rail traffic on the Panipat-Rohtak Bhiwani section and Jind-Panipat, Jind-Delhi had been suspended resulting in cancellation of some important trains.

Meanwhile, the other towns remained cut off by rail link from the effected towns and cities including the State and national capital. Moreover, several cuts have been made on the railway tracks at many places mainly to drain out the flood water resulting the disruption of rail traffic.

Telephone Services

The flash have also hit the modern and most important tool of means and communication, telephone services. These stated services got fault and most of the telephone sets were dead in several affected towns and cities. The fault may be probably to water seepage into the underground cable ducts. Scores of telephones went out of order at many places. However, due to the high velocity wind followed by torrential rain caused widespread damage phone poles were uprooted by high speed winds at several places leading to disruption of services at a number of places including Bhiwani, Sonepat, Rohtak, Kaithal and Jind districts.

Electricity

Like the other services, electricity was also highly struck and high-velocity winds also uprooted the power and electric poles as a result disrupting the power supply on the various villages, towns and cities. Hundreds of transformers were submerged in the flood water and badly damaged.

PEOPLE'S PERCEPTION

Flash floods are no doubt, the natural calamitics beyond the control of man. It was the common perception of the people that the recent floods was so devastated and destructed such furious floods never seen by the State people in the last 100 years. Meanwhile, it was also the equally important and a general view that man is himself responsible for such catastrophe. In the rural areas several breaches were made by cuts in overflowing and drains to save their localities and consequently almost all areas of State come up in the tight grip of these furious floods. The older people said that man has full hand to bring such natural calamities. If man does not interfere directly, the floods may damage less rather than it did actually. Moreover, there was no ample time to take quick precautionary measures. People could not assess that the rain would continue to a long period otherwise they would have shifted themselves on safer places.

STRATEGY FOR SUSTAINABLE DEVELOPMENT

First of all, the State as well as Central Government should formulate a policy to prevent and mitigate such natural calamities. Concrete steps and an urgent initiative should be taken during such critical juncture. Perhaps the first and most important work of the government in this direction should be to make the drainage system more efficient and successful especially to drain out flood water more frequently. It will be very beneficial in order to reduce the menace of furious floods. Moreover the existing drainage system in these areas should be made wider and deep generally where frequent floods come so that it can be able to carry out more flood water.

New drains should be constructed in the flood-prone areas particularly in low lying tracts which don't have no source to drain out excessive flood water. The other equally important constraint in order to mitigate such natural events is to clean up all the vulnerable canals, drains in the presence of expert engineers right before the occurrence of rainy season (Monsoon Seasons). So that the deposited silts and diversion of the river bed should be removed and due to this process their carrying capacity may be increased and floods damage may be mitigated up to some extent.

Meanwhile, more and more ring dams should be built around all the worst affected villages to save the largely settled population. Similarly in the urban centres cities/towns sewerage followed by disposal system should be strengthened and improved. Likewise, dams should also be constructed especially to check, soil erosion. The other valuable key factor in connection to make reservoirs in the low lying areas especially for flood water. Such flood water may be used further for purposes especially for fish production. Stone studs may also be constructed at some of the vulnerable points to save the agricultural land by erosional process. Onward, to know the real picture survey should be conducted of all affected areas and more stress should be on the various methods and techniques to drain out flood water.

For achieving the same goal, flood-prone areas should be identified and mapped. All affected areas aerial survey should also be made especially to assess the loss of agricultural crops, vegetation cover and property and to know the exact flow of existing flood water. During the flood, houses along the canals, rivers and drains sides strict vigil by army/police personnel should be taken to face any eventuality and to prevent the local people to make cuts in them. Although, the embankments, dams and drainage channel provide total protection against flood. The another useful method for the mitigation is the flood plain zoning. It is considered the most effective and reliable method for floods prevention. But before identification of flood plain zoning, there is an urgent need for the better understanding of behaviour of rivers, that can help to prevent occurrence of floods. Although construction of dams including reservoirs is also necessary to mention here that the level of National Highways, State Highways including local roads, existing level should be raised just above the level of settlements and adjoining areas. Massive afforestation programmes should be launched particularly along the banks of canals, rivers and drains to strengthen their embankments. Similar, practices should be carried out along the ring bandhs in local rural areas to check the soil erosion.

Apart from all these stated efforts described in the strategy, the other significant and important point is warning system which should be installed and improved. By the application of advance warning system which is now possible in the modern age with satellite based advance disaster warning systems. By adopting various methods and strategies, it may be possible to save the precious lives of millions and would be able to mitigate the flood damage.

CONCLUSION

It has been observed that almost all parts of the mentioned affected regions especially Rohtak and Bhiwani, thousands of villages were in the hit list of flash floods. Millions of people were wandering for shelter. It has also been the people perception that the local people are liable for the floods by making cuts and breaches not only in the canals and drains but also in the various roads. Almost all socio-economic and commercial activities National/State highways and local roads were badly hampered on large scale subsequently, the State economy was fully a destabilised.

The high speed winds which further uprooted countless trees and electric poles. Moreover, the relief operations the totally failure and secondly traffic system was also disrupted. Similarly, all the kharif crops fell down in flood water and ultimately destroyed..

It is necessary to mention here that within a short span of time, the whole area got around 900 mm within three days of continuous rain. Meanwhile the average annual range of the State rainfall is about 550 to 750 mm. It was even more than the average rainfall.

For the mitigation of flood hazards, there is an urgent need first of all to improve the drainage system followed by disposal system. Secondly, dams, reservoirs and ring bandhs should be built in those areas which are vulnerable to floods. New modern techniques should be applied for advance warning which is possible now through the satellite and remote sensing services.

REFERENCES

Aggarwal, A. (ed.) (1991). *Floods, Flood Plains and Environmental Myths*, CSE, New Delhi.

Ansari, A.A. (1987). The Kosi : A Study in River Regime, Unpublished Ph.D. Thesis, Department of Geography, University of Delhi.

Beven, K. and Carling, P. (1989). *Floods: Hydrological Sedimentological and Geomorphological Implications*, John Wiley, New York.

Brammer, H. (1990). "Floods in Bangladesh", *The Geographical Journal*, Vol. 156, pp. 12-22.

Das, K.N. (1968). "Westword shift in the course of the Kosi". *NGJI*. Vol. XIV, Part. I.

Desai, C.G. (1982). *The Kosi River: Its Morphology and Mechanics in Retrospect and Prospect*, C.W.C., New Delhi.

Dhanju, M.S. (1984). Floods plains mapping of Gangetic basin using Landsat Imagery, *COSPAR, Contribution of Space Observation to Water Resources Management*, V.V. Salomanson and P.D. Bhavsar, pp. 215-218.

Ganga Flood Control Commission, (1993). *A Comprehensive Plan of Flood Control for the Kosi Sub-basin*, Report, Vol. I. Government of Bihar, Patna.

8

CHALLENGE OF LAND DEGRADATION

An Indian Experience of Conservation

TAPESHWAR SINGH

INTRODUCTION

Soil is a natural gift bestowed to human kind. It is basic to life. It is the primary means of food production directly supporting the livelihood of most rural population and indirectly of everyone. It is an essential component of terrestrial ecosystem, sustaining their primary producers (all living vegetables and decomposers, micro-organisms, herbivores, carnivores) while providing major sinks for heat energy, nutrients, water and gases. Soil in physical term is merely the unconsolidated material on the earth's surface yet is the hub for a host of life-supporting processes and is testimony to the marvels of natural environment. In fact, all forms of life on the land system derive their nutrition either directly from the soil or through other system controlled by soil. The soil combined with water determines the potential of an area for development.

Commenting on the significance of soil the great thinker Aristotle described soil as the stomach of the plant. Even now over 90 per cent of the world's food comes from the soil and less than 10 per cent comes from both inland water and the ocean. Being finite and highly heterogeneous for different uses it is under competing demands for different purposes to meet the requirements for food, fodder, fibre, fuel, mineral, settlements etc. for the ever increasing population. However, the per capita availability of land in India is declining. For example, per capita availability of land in India was 0.89 ha in 1950 which declined to 0.38 ha in 1990 and is likely to further decline to 0.33 ha in 2000 AD and shall only be 0.19 ha by 2035 AD. The position of per capita availability of agricultural land is also poor as it declined to 0.18 ha in 1990 from 0.48 ha in 1950 and likely to further decline to 0.088 ha by 2035. Thus, with increasing population growth and imposing increasing demand for land it is imperative to protect the quality and improve the productive potential of available land resource and to save this scarce and vital resource for the future generation from all degradation processes *i.e.,* soil and water erosion, water-logging, salinity, alkalinity, *gullies and ravines,* shifting cultivation, soil nutrient mining etc. through appropriate conservation measures.

Despite soil being the prerequisite for the whole life support system its quality is progressively declining. Crop production is becoming more difficult and more expensive as fertility of soil, its water holding capacity and depth decline. Wastelands are increasingly becoming evident in vulnerable areas. This sort of change in soil nutrient status is bound to limit a highly productive and sustainable system. Land degradation is an on-going process without interruption. Land degradation, soil erosion and floods are interrelated environmental problems and adversely affect the functioning of the whole life support system. The emerging crises of degrading resource base and sustainability is related to over increasing population pressure competing for the finite and unrenewable land resources. The per capita of cultivable land declining means that there will be no sufficient land available for agricultural expansion. The scientific community is faced with the challenge of not only increasing productivity per unit area by rationalising land-use but also preserving the resource base for the posterity (Sehgal, 1995). Thus, soil is an endangered asset because its wasteful consumption goes on unabated. No other human-induced process affecting agriculture with the possible exception of climatic change holds the potential for such far-reaching impact as soil degradation.

SOIL DEGRADATION

Soil degradation is defined as a decrease in soil quality as measured by changes in soil properties and processes, and the consequent decline in productivity in terms of production now and in the foreseeable future. Soil erosion is one of the main processes of degradation and consists of physical detachment of soil particles by wind and water and their transport elsewhere in the landscape, to rivers and water storages or to the sea. Land degradation is a composite term describing the aggregate diminution of the productive potential of the land, including its major uses (rainfed arable, irrigated, rangeland, forestry), its farming systems (*e.g.* smallholder subsistence) and its value as an economic resource. Food and Agricultural Organization (FAO, 1977) has defined soil (land) degradation as 'a process which lowers the current and/or the potential capability of soil to produce (quantitatively and/or qualitatively) goods or services'. Soil or land degradation then affects the 'flow' or renewable characteristics of soils by reducing the (potential) biotic productivity of the 'resource. Closely related issues are the environmental effects in the form of excessive siltation of streams and other water bodies (reservoirs, rivers, lakes, tanks etc.) and numerous socio-economic effects that stem directly from the problems of productivity decline and environmental deterioration. Thus, a land is said to be 'degraded' when its productivity declines. In other words land degradation connotes loss of land productivity, qualitatively or quantitatively through a number of processes. The degradation of land may be either by natural agencies or its over-exploitation by man. Of course, change in soils over geologic times is a natural process but the acceleration in degradation in recent years is caused mainly by increasing anthropogenic intervention in the form of cropping, grazing, rotation system and management.

Soil degradation occurs through erosion of top soil, loss of organic matter, depletion of nutrients through repeated harvests, waterlogging and salinization, acidification, compaction, pollution of groundwater and both decline and rise of groundwater tables. However, degradation or denudation of land is one of the major causes of ecological imbalance in the country. It poses great challenge to crop productivity in a number of ways simultaneously.

Process of Soil Degradation

According to T.O. Riorden (1995) broadly, there are six processes of soil degradation which are as follows :

1. *Water erosion :* Splash, sheet and gully erosion, as well as mass movements such as landslides.
2. *Wind erosion :* The removal and deposition of soil by wind.
3. *Excess of salts :* Processes of the accumulation of salt in the soil solution (salinization) and of the increase of

exchangeable sodium on the cation exchange of soil colloids (sodication and alkalinization).

4. *Chemical degradation :* A variety of processes related to leaching of bases and essential nutrients and the build-up of toxic elements. *p*H related problems such as aluminium toxicity and P-fixation also included.
5. *Physical degradation :* An adverse changes in properties such as porosity, permeability, bulk density and structural stability, often related to a decrease in infiltration capacity and plantwater deficiency.
6. *Biological degradation :* Increase in rate of mineralization of humus without replenishment of organic matter.

The magnitude and extent of land degradation varies with land-use system. A number of factors such as soil type, topographic factors, slope, steepness, length and shape, rainfall, and vegetation cover etc. also affect the rate of soil erosion and degradation.

Magnitude of the Global Problem

With the dangerous anthropogenic progressive interference in the ecosystem the land resource is gradually becoming scarce in quantity and degraded in quality. As already mentioned earlier the magnitude of different kinds of land degradation and depletion in India is extensive. The magnitude of land degradation worldwide has been recently estimated using the standardized methodology—the Global Assessment of Soil Degradation (GLASOD).

More recent information suggests that the estimates of the extent of land degradation and its effects on productivity as reported in GLASOD have been exaggerated in a number of cases. At the same time, however, the effects of land degradation are often masked by the compensating effects of improvements in agricultural technology or increasing applications of plant-nutrients so that yields may have been increasing even on degrading land. Also in some cases, farmers and rural communities in some areas of Africa and Asia have demonstrated that land which GLASOD classifies as unrestorable can in fact be rehabilitated.

Soil erosion is by far the most widespread cause of degradation with water erosion being the principal agent. Nearly one billion hectares of arable land in developing countries are estimated to be degraded and the productivity is being moderately or severely affected. Some 9 million ha worldwide, of which 5 million ha are in Africa, have had their original biotic factors fully destroyed and reached the point where rehabilitation is probably uneconomic (Alexandratos, 1995). There is widespread evidence of erosion resulting in losses greatly in excess of 50 tonnes of soil per hectare per year—losses that may be five or more times the natural rate of soil formation. Accelerated water erosion is causing extensive damage to the soil resources in all the continents. Worldwide, about 25,000 million tonnes of soil are being washed away each year, ending up in river and finally the ocean (FAO, 1992). In India, 25 per cent of the total geographical area is subjected to serious water erosion. Agriculture and associated activities cause an estimated loss of 3,334 million tonnes of soil annually, of which about 1,572 M.T. find their way to the ocean, wind is another notorious soil degrading agent. According to a FAO study on soil degradation, 22.4 per cent of Africa, north of equator and 33.5 per cent of west Asia are under severe threat of degradation caused by wind erosion. In India too, the arid regions of Gujarat, Rajasthan and Haryana are exposed to extensive wind erosion.

Magnitude of Soil Degradation in India

Over the periods of last two and half decades a number of studies have helped to identify and analyse the magnitude of different kinds of soil degradation. Still, there are controversies in the estimation of total degraded land in the country. Even estimates made by different departments of the Government of India and other organisations such as National Remote Sensing Agency (NRSA) also differ considerably. Total degraded land of different kinds is now estimated officially to be in the vicinity of 175 Mha (Table 8.1). However, some other estimates put the figure at 163 Mha (see Brandon, 1994 for these alternative estimates) and 130 Mha. The total geographical area of the country is 329 Mha of which 141.15 Mha is subjected to water and wind erosion and about 34 Mha (33.72%) is affected by special land degradation problems. Therefore, the unaffected land area is only 47 per cent of the total geographical area. The average annual loss of nutrients from top soil lies in the range of 5.37 to 8.40 million tonnes (Government India, 2000).

Table 8.1: Distribution of Estimated Areas Under Different Soil Erosion, Land Degradation and Land Utilisation Problems (Estimated)

(Million Hectares)

Sl. No.	*Affected Area/ Problem Area*	*With 1976-77 Land Utilisation Statistics & Reports till 1980-81*	*With 1981-82 Land Utilisation Statistics & Reports till 1984-85*
(1)	*(2)*	*(3)*	*(4)*
1.	Geographical Area	329.00	329.00
2.	Area subject to water and wind erosion	150.00	141.25
3.	Area degraded through special problems		
	3.1. Water logged area	6.00	8.51
	3.2. Alkali soil	2.50	3.58
	3.3 Acidic soil	—	4.50
	3.4 Saline soil including coastal sandy areas	5.50	5.50
	3.5 Ravines and gullies	3.97	3.97
	3.6 Area subject to shifting cultivation	4.36	4.91
	3.7 Riverine and Torrents	2.73	2.73
	Total 3.1 to 3.7	25.06	33.70

(Contd.)

Table 8.1 *(Contd.)*

(1)	*(2)*	*(3)*	*(4)*
4.	Total Problem area (2+3)	175.06	174.95
	4.1 Annual Average loss of nutrients from land is worth (million tonnes)	5.37 to 8.40	5.37 to 8.40
5.	Average annual loss of production for not developing ravines is worth (million tonnes)	3	3
6.	Average annual rate of encroachment of table lands by ravines	0.008	0.008
7.	Average area annually subject to damages through shifting cultivation	1	1
8.	Total flood prone areas	40	40
9.	Total drought prone areas	260	260

Source : Government of India (2000) Water and Related Statistics, Central Water Commission, p. 305.

Land Degradation Estimates

Table 8.2 gives the estimates of the nature and magnitude of wasteland in different States of the country (G.O.I., 1998). However, according to the remote sensing survey conducted by the NRSA during 1986-93 covering 241 critically affected districts of 21 States and union territories, the estimates of wasteland stands at 3,56,470.11 square kilometres *i.e.* only 17.49 per cent of the total geographical area covered. The very high percentage of wasteland area in Jammu and Kashmir (48.32%) and Himachal Pradesh (37.72%) are due to snow cover and degraded forest. Nagaland (54.08%), Assam (57.02%), Manipur (37.72%) are due to shifting cultivation and in Rajasthan (25.18%) due to sandy area. Among all the States covered in the project Punjab has the minimum (4.31%) and Assam has the maximum (57.02%) of area under category of wasteland. When one considers that India's total geographical area is in the neighbourhood of 329 Mha, the magnitude of this degradation and its significance as a problem of natural resource management becomes amply clear. It is perhaps, the single most important natural resource management problem in the country today.

We can recall that area affected by water and wind erosion stands at 141.25 Mha (about 43 per cent of the total geographical area). While faulty farming practices have accelerated the erosion process on undulating topography, deforestation, salinization and associated waterlogging arise mainly due to the faulty irrigation management. When water use is excessive, the groundwater table rises and brings with it dissolved salts from the lower layers through capillary action and leads to salinizations.

Various forms of water erosion — sheet erosion, rill erosion and gully erosion, not only render agricultural land less suitable for crop production but also cause flooding of vast areas and silting of reservoirs as a result of downstream movements of eroded materials from the head-water regions. The life and capacity of many water reservoirs have been reducing at an alarming rate as would be seen from Table 8.3.

Soil Nutrient Mining

Soil Nutrient Mining is another important soil resource degradation process brought into focus by the on-going debate on environmental

Table 8.2: State-wise Estimated Total Special Problem Areas, Areas Treated and Balance yet to be Treated

(Lakh Hectares)

Sl. No.	*Name of the State*	*Total Problem Area due to*			*% of Area Treated till 1996-97 (Prov.)*	*% of Area yet to be Treated*
		Soil Erosion	*Land Degradation*	*Total*		
1.	Andhra Pradesh	115.02	7.29	123.31	12.97	87.03
2.	Arunachal Pradesh	24.44	2.10	26.54	1.54	98.46
3.	Assam	22.17	7.82	29.99	7.84	92.16
4.	Bihar	42.60	22.92	65.52	24.62	75.38
5.	Gujarat	99.46	26.40	125.86	33.67	76.33
6.	Haryana	15.91	25.71	41.62	19.77	80.23
7.	Himachal Pradesh	19.14	—	19.14	19.64	80.36
8.	J & K	8.83	0.10	8.93	35.95	64.05
9.	Karnataka	109.89	4.14	114.03	32.77	67.23
10.	Kerala	17.57	1.78	19.35	25.32	74.68
11.	Madhya Pradesh	196.10	11.07	207.17	25.43	74.57
12.	Maharashtra	191.81	6.65	198.46	55.76	44.24
13.	Meghalaya	8.37	2.65	11.02	12.25	87.75
14.	Orissa	45.78	32.25	78.03	11.21	88.79
15.	Punjab	10.07	22.23	32.30	33.62	66.38
16.	Rajasthan	167.30	174.92	342.22	5.85	94.15
17.	Tamil Nadu	36.40	1.82	38.22	51.43	48.59
18.	Uttar Pradesh	71.10	60.05	131.15	33.2	66.10
19.	West Bengal	10.33	32.70	43.03	11.43	88.57
20.	Others	22.19	12.95	35.14		
	Total	1234.48	455.55 +14.65*	1690.03 +14.65	23.81	76.19
Grand Total		1234.48	470.20	1704.68	19.08	80.02

Source: Govt. of India (1998) Water and Related Statistics, Central Water Commission, p. 307.

* Sandy areas appear not reported along Coastal Saline Soil.

Table 8.3 : Siltation of Selected Indian Reservoirs

Sl. No.	*Reservoir*	*Assumed Rate*	*Observed Rate*	*Expected Life as percentage of Designed Life*
		(Acre-feet per annum)		
1.	Bhakhara	23,000	23,475	48
2.	Maithon	684	5,980	11
3.	Mayurakshi	538	2,080	27
4.	Nizam Sagar	530	8,725	6
5.	Panchet	1,982	9,533	21
6.	Ramganga	1,089	4,366	25
7.	Tungabhadra	9, 796	41,058	24

Source: Government of India, 1994, Central Water Commission, Ministry of Water Resources, New Delhi.

accounting. The shortening of fellow and prolonged crop occupance without adequate technological responses to replace the soil nutrient taken out by crops with or without organic or mineral fertilizer inputs, leguminous crops, nitrogen-fixing algae and so on, is responsible for reducing the nutrient status of soils and the actual and potential crop yields. Consequently, it threatens very sustainability of the agricultural production itself. We cannot forget that the carrying capacity of the soil is finite and degradation is not easily reversible in the present social and economic conditions.

Indiscriminate use of agrochemicals and overuse of irrigation water in the process of agricultural intensification have rendered millions of hectares of fertile soil unfit for cultivation. For example, in Punjab cropping intensity is 176 per cent and rice-wheat system is extensively followed, the yield level of each crop around 3 tonnes per hectare is resulting in a removal of 450 kg. per hectare nutrients annually, which is far in excess of the 171 kilograms per hectare of plant nutrients being added to the soil. In fact, negative balances are common features in intensive cultivation (Shekhon and Puri, 1986).

A recent study conducted by agricultural-scientist Devinder Sharma (*Hindustan Times,* 3 Oct. 1996) reveals that continued water-logging, imbalances in watertable, erosion, nitrate and arsenic pollution etc. had reduced the actual acreage of irrigated land by 60 million hectares. In a shocking revelation this study indicates that organic matter in the soil of Punjab has declined to 0.2 per cent as a result of excessive use of fertilizers. This figure compares very badly against the existing ratio of 2 per cent in China.

LAND DEGRADATION IMPACT ASSESSMENT

In India, due to various limitations in land degradation assessment, it is extremely difficult to assess the actual extent of loss resulting from land degradation. However, the Centre for Science and Environment in its magazine—*Down to Earth* (September, 1996), brought to the fore that our country is losing Rs. 34,000 crore in the form of environmental degradation annually. This staggering estimation is alarming enough to cause concern not only to the environmentalists and policy-makers but also to the whole society which has some stakes as well as sympathy with the generations to come.

Global Scenario

Land degradation is one of the major alarming global concerns too as revealed by the world statistics on land degradation (Table 8.4).

Table 8.4 : World Statistics on Land Degradation

(Area in Million Hectares)

1.	Total Geographical Area		13,500
2.	Area unfit for Sustaining Plant Growth		10,222
3.	Total Cultivated Area		3,278
	(a) High Productivity	(13%)	426
	(b) Medium Productivity	(27%)	885
	(c) Low Productivity	(60%)	1,967
4.	Estimated worldwide rate of loss of land per year		5-7
5.	Projected worldwide rate of loss of land per year by the end of 20th Century (FAO, 1981)		10

Source: Singh, D.R. (1996).

If land degradation process continues unabated at the present alarming rate, around one-third of the world arable land would become degraded and productivity will be reduced by one-fifth. Since 1945, 1.2 billion hectares, an area roughly equivalent to the size of India and China combined together, have been eroded at least to the point where their original biotic functions are impaired.

According to another estimate (Abrol, 1990) up to 24 billion tonnes of sediment has already reached the ocean during the last few decades. Further, the productivity of about 20 Mha per year has been reduced to almost zero.

Some scientists have expressed concern that changes in terrestrial insolation and a globally warmer situation will intensify desertification. Desertification in turn, if not quickly mastered could degrade large areas in the tropics. We can foresee the magnitude of the impending problem by the fact that the area of cropland already subjected to desertification is about 30 per cent of the world's land surface. If we are to survive we cannot be reckless and ignorant to our scarce natural resources.

Consequences of Soil Degradation

Since environmental equilibrium is the prerequisite for the survival of human civilization any aberration in it is bound to deter its survival. Land degradation leads to the deterioration of the overall productive potential of the soil. In turn, soil degradation will result in decrease of the productive land ultimately forcing the farmers to encroach upon unsuitable and vulnerable lands—wasteland, slopes and other marginal areas, and the 'Village Commons'. Thus, per capita productive land and soil quality rapidly diminish and the cycle of soil degradation and the pauperization of peasantry continues even faster. As soil degrades its infiltration and water holding capacity worsens. As a result, water loss in run-off may increase from 20-50 per cent of the total rainfall. Soil erosion impact of fertility losses is very much more than the removal of a topsoil slice. Land degradation also causes siltation of the water storages and courses, and flash floods thereby increasing land pressure and further degradation of environment. If conservation and regeneration of this primary resource are not adequate, it would have disastrous ecological consequences. Hence, we must conserve our soil resources through modern soil conservation practices and techniques for better life.

SOIL CONSERVATION MEASURES

The response of society to land degradation is soil conservation, defined as any set of measures that controls or prevents soil erosion or maintains soil fertility (Stocking, 1995). Water related degradation

Table 8.5: Soil degradation by type and cause (classified as moderately to excessively affected)

(million ha)

	Water erosion	*Wind erosion*	*Chemical degradation*	*Physical degradation*	*Total (Million ha)*
Regions					
Africa	170	98	36	17	321
Asia	315	90	41	6	452
South America	77	16	44	1	138
North and Central America	90	37	7	5	139
Europe	93	39	18	8	158
Australasia	3	–	1	2	6
Total	748	280	147	39	1214
Major Causes	(%)	(%)	(%)	(%)	
Deforestation	43	8	26	2	384
Overgrazing	29	60	6	16	398
Mismanagement of arable land	24	16	58	80	339
Other	4	16	10	2	93
Total	100	100	100	100	1214

Source : Adapted from ISRIC/UNEP (1991).

comprises major part of the degraded land (Table 8.5). Further, water related degradation extends to a number of different agro-climatic regions. Hence, different kinds of land degradation can be designated as 'water-related degradation'. The most important ones among them are—water-caused erosion resulting in gullied or ravinous lands or just sinking away of agricultural land as in several cases in the hillsides and water-logging and salinity (Chopra, 1996) Michael Stocking (1995) observes that:

> "*water conservation,* especially in drier zones, is very closely related in such a way that many techniques of soil conservation

achieve short-term benefit through increasing plant-available water rather than through saving soil. The objective of soil and water conservation is to achieve consistent and lasting production from land while keeping soil loss at or below the soil's rate of renewal. Since sub-soil formation rates are in the region of 0.5-1 tonne/ha/year—which is equivalent to natural rates of soil erosion—soil conservation strategies should not allow greater soil losses than those occurring under a complete natural vegetation. Because any disturbance of the soil causes an increase in soil loss, we have a dilemma : all agricultural land and land-use systems are theoretically unsustainable. In practice, therefore, 'tolerable' soil losses are taken to be either some arbitrary figure which is known to be achievable in practice or the level at which soil fertility can be maintained with inputs of technology at a medium term, say up to 30 years. The latter approach to setting target levels of allowable soil loss in a conservation programme is gaining ground. It recognises that topsoil formation rates, can easily reach ten times formation rates, it acknowledges that farmers utilise their soil 'capital' of nutrients at certain times, restoring them later by fallowing and it allows for the role of the land manager in compensating for nutrient and organic matter losses by specific interventions such as green manuring, cover crops, fertilizer and so on. Nevertheless, a nagging doubt remains. The implication of the current state of knowledge on erosion and soil formation rates is that overall soil depth will decline, even if the quality of the topsoil may remain."

However, he has outlined some examples of modern techniques of soil and water conservation (Table 8.6) that have certainly some advantages over the past practices.

It goes without saying that soil and water conservation is nothing new to India. Ancient Indian farmers used to utilise water harvesting and crop and land husbandry for raising crops. *The Rig Veda, Puranas, Arthashastra* and many other ancient inscriptions speak of soil and water conservation measures. Worldwide, Homer, Virgil and Plato have also made passing references to environmental problems and need for their conservation. Nevertheless, it is only recently that soil conservation has been recognised as a necessary strategy for restoring degraded land and protecting and improving crop productions. In response to the difficulties of applying high cost solutions in resource

Table 8.6 : Examples of Modern Techniques of Soil and Water Conservation

Technique	*Description*	*Resource Implication*
Tillage Practices		
Strip Tillage	Conditioning soil along narrow strips in or adjacent to seed rows, leaving rest of soil undisturbed.	Special equipment possible crusting/infiltration problems in unfilled parts.
Basin listing or tied ridging	Formation of contour bunds (earth banks) with constructed 'ties' or banks between bounds. Listing done by machinery and tied ridging usually by hand.	Need for dedicated machinery and/or considerable extra labour. Danger of water-logging.
Conservation tillage	Technique of light harrowing and retention of crop residues at surface.	Special machinery. Crop residues not available for other uses *e.g.* forage, fuel.
Minimum/Zero tillage	Use of herbicides, then direct drilling into residues. Very little disturbance to the soil.	Expensive chemical and danger of pollution and soil compaction, saving on conventional saving cost.
Land Formation Techniques		
Contour bunds	Earth banks up to 2 m wide across slope to form barrier to run and break slope into shorter segments.	Additional labour and/or equipment.
	Varieties narrow of wide-based; coutour or graded.	Varieties have different costs, *e.g.* narrow-based loses about 14% of plantable area.
Terraces	Earth embankment and major reformation of surface. Three main types: diversion, retention and bench terrace.	Large labour and equipment requirement.Continual maintenance.
Terracettes	Small constructions usually to harvest water. Known by many local names, *e.g.* fish-scale terraces in China for rows of perennial crops; eyebrow terraces for tree planting in semi-arid zones.	Labour and continuing maintenance requirement.
Stabilization Structure		
Gabions	Stone and rock-filled bolsters to protect vulnerable surfaces, *e.g.* bridges, culverts.	High cost, transport of stone, continual maintenance.
Gully Control Dams	Usually constructed of brushwood across a gully.	Materials, labour and maintenance, often fail in heavy storms.

Source: Stocking, Michael (1995).

poor environments, an understanding of indigenous responses to land degradation has developed. Small farms are the characteristics of the developing countries where low cost biological methods of soil quality improvements are appropriate. Large scale estates and commercial productions are the farm features of the developed countries where machinery and conservation structures are the obvious first choice of land managers. Difficulties may crop up where an attempt is made to impose a conservation strategy developed in one country with a different physical, agro-climatic and socio-economic conditions on a quite another situation. While dealing with question of land degradation we are confronted with a number of questions that cannot be properly answered. The deterioration of production environment due to growing human and livestock population is exacerbating erosion and causing nutrient loss and productivity of soil. Ecosystem adapts naturally to the resource utilisation to ensure that the quality of these resources are not threatened and to enable socio-economic development in a sustainable manner. For resource poor farmers in developing countries like ours, the issue may be starker. It may rest between a choice of starvation now or starvation in the future. In fact, if there is a tussle between maintaining current livelihood and looking after conservation works, livelihood will win. Despite these conditions, in future, all development should be sustainable and dictated by the preservations of the tottering land resources and ecological balance. However, in the conservation process the following questions appear to be reckoning:

(i) Is it possible in the present socio-economic conditions to prevent land or resource degradation completely ?

(ii) Is it the limitation inherited in the natural process of degradation which, however, permit limited option of mitigation of the problems ?

(iii) Is it possible with the present population growth rate and more and more demand for food to desist from further agricultural intensification ?

(iv) Are we capable of regenerating the degraded soil or/ and carrying it back to the degraded surface ?

(v) Can we afford to keep land fallow for regaining nutrient or change the land-use which may substitute the spectacular gains of farming ?

(vi) Why should farmers erode their soil while many know fully well that their practices jeopardize their future well-being ?

The land degradation dates back even before man occupied the earth. However, recently, man interference with it has exacerbated degradation. Although, man has acquired enough power to control the nature but nature is also becoming more and more furious and menacing. For the environment manager inevitably some explanatory issues can be addressed and some cannot. Land degradation and soil erosion have been known for centuries but people, have also adjusted their lives and practices to accommodate a declining resource base caused by these human-induced accelerated processes. Erosion is difficult to see; it is insidious and incremental. However, soil conservation can be oriented towards the three main concepts as given hereunder.

Watershed Management

Currently, the concept of development on watershed basis has been one of the major thrust planning areas in the country. This natural unit is evolved through interaction of rainwater with landmass and typically comprises forest area, arable and non-arable land including wasteland and degraded land. Development programmes envisaged under its purview include almost every activity concerning land, water and biomass. The entire strategy framework is holistic since it combines the technologies on soil and water conservation with appropriate agronomic practices against the backdrop of alternative land-uses. It can produce a distinct change in the condition of soil health. The main moto of the watershed management is the efficient biomass production in the areal unit through protection, conservation and improvement of land and water resources. This technique has been proved to be very effective means of in *situ* and in-field conservation of rainwater, controlling soil erosion and deforestation, and increasing groundwater recharge. This approach is thus, aimed at to enhance the productive capacity of land and its carrying capacity by stressing on the integration and development of onsite resource utilization and management practices. It is the integration of various activities necessary to develop the watershed into one composite whole, which can implement and reinvigorate the existing natural resource system and maintain the ecological balance.

Ecological Farming

Despite unparallel achievement on food front conventional farming system has unleashed many agroecological problems including land degradation. The alarming land degradation has led the planners and scientists to find out an ecologically sound crop husbandry practices *i.e.* ecological or organic farming. This farming practice basically depends on the principles of ecology. This system aims at enhancing the agricultural productivity per unit of land and water without endangering the production environment. It is considered to be capable of fulfilling twin objectives of achieving food security and ecological sustainability. It is ecologically sound, economically viable and socially just and humane. This system is devoid of chemical inputs. Under this system the biological potential of the soil and the underground water resources are conserved and protected from natural and human-induced degradation or depletion by adopting suitable cropping models besides natural or biological means of pests, weeds and disease management by which both the soil life and beneficial interaction are also stimulated and sustained so that the system achieve self regulation and stability as well as capacity to produce agricultural outputs at levels that are profitable, enduring over time and consistent with the carrying capacity of the managed ecosystem. It will prove most effective alternative system that can prevent soil nutrient mining and improve growing sustainability crisis.

Integrated Nutrient Management

On account of continued decline of soil nutrient with the runoff water and decreasing application of organic fertilizers to the crops the productivity status of soil is continuously becoming precarious. These are deficient not only in major nutrients but also in secondary nutrients like sulphur and trace elements like zinc and iron. Fertilizer use is confined to major crops and certain specific areas. This situation is unlikely to be changed in near future. Therefore, we have to resort to integrated nutrient management with particular emphasis on ecological farming.

A great deal of work done on conservation in different parts of our country indicates that the benefits of resource conservation practices can be harvested and productivity can be enhanced to a considerable extent on a sustainable basis (Sharma *et al.,* 1994). Conservation practices can be grouped into four broad categories as follows :

1. Rainwater management
2. Soil conservation practices
3. Biological methods of resource conservation
4. Indore method: integrated approach for water management.

Rainwater Management

The measures for conserving rainwater *in situ* and its safe disposal includes various types of land configurations, namely, ridge and furrow system, graded furrows, raised and sunken beds, broad bed and furrows etc. These measures take care of drainage in growing crops, *in situ* percolation of rainwater and thereby conservation of nutrients and soil (Sharma, 1991). The experimental results of these techniques have shown tremendous success in soil and water conservation process.

Soil Conservation Practices

Some land treatments and water disposal systems have been found to control soil erosion from black clay soils. Broaded and furrows, graded furrows and broad based bunds, diversion bunds, loose boulder structures and bench terracing are the commonly recommended treatments. Contour bunding has been used as soil conservation measures in many places, but such conservation structures have been reported to adversely affect the crop yield (Kampen, 1974). Some promising soil conservation measures are briefly discussed below :

Reclamation and Control of Gullies

Masonry structures are costly and inapt in soil conservation. Gabion structures to reclaim gullies and stabilise waterways, and growing of grasses in the waterways along natural depressions have been found more effective. It is a semi-flexible structure of loose boulders packed into prefabricated galvanised iron wire netting and serves as a drop structure. These structures are porous, flexible and allow the runoff water to pass through while retaining the silt upstream.

Bunds

Bunds of various kinds have been found very effective to check erosion losses. In this system graded bunds are constructed across the moderate slope to control soil erosion. They are spaced at suitable vertical intervals so as to divide the total catchment area into small parts and runoff water from each part is carried away along these bunds without causing diversion of runoff water coming off a hillock into cropped fields, saves the land from degradation.

Stabilisation of Washes

Generally, washes formed in the cultivated fields develop into gullies if not controlled at the initial stage. Stabilisation of washes can be achieved by leaving that strip of land fallow where gullies are about to form. In due course grass will grow, otherwise, plant it. If washes are in advance stage of erosion and gully is already formed it may be necessary to provide drop structures.

Provision of Drainage between Waterways

Safe disposal of runoff water of the areas between waterways, is essential. For this open drainage channels on the grade are to be constructed on suitable interval from ridge to grassed waterways to save the land from erosion.

Appropriate Plant Cover

Appropriate plant cover is essential during rainy season for efficient utilisation of land and water resources and for minimising soil erosion. Under harsh conditions a vegetation cover which can provide organic matter, helps in weathering rocks and then stabilise the soil, is the ideal choice. Such characteristics are possessed by legumes and grasses which are drought tolerant and secrete acids to help in weathering rocks. It has been proved beneficial in Bhati Mines area which is at abysmal land degradation (*The Times of India*, Oct. 22, 1996).

Conservation Tillage

Conservation tillage is, perhaps, the most recent of the soil conservation practices and is one which holds considerable promise in developed countries and in those countries of the developing world with the potential of mechanization. According to Crosson *et al.* (1992) conservation tillage practices differ from conventional tillage in three ways : *(i)* they rely on some instrument other than moldboard plough to prepare the land for planting; *(ii)* they leave enough residue from the previous crop on the soil surface to significantly reduce erosion; and *(iii)* they rely more on herbicides and less on mechanical cultivation to control weeds. In the United States nearly one-third of the harvested cropland is under this tillage system. The costs are comparatively low in comparison to conventional farming system.

Biological Methods of Resource Conservation

A considerable amount of soil profile-stored moisture is lost during post-rainy season and dryspell during rainy season also consequently affecting the soil productivity and exposing the soil to degradation. Soil mulching, straw mulching, vertical mulch, plastic mulching etc. are thought to be some of the effective biological practices for soil and moisture conservations.

Indore Method: Integrated Approach for Water Management

The Indore Method of water conservation in and around the fields have been developed (Shroff *et al.,* 1992) at Indore. This method has been found useful not only conserving rainwater but also in draining out excess water and conserving soil erosion. In this method small bunds (25 to 35 cm high) are made around the field. This technique facilitates keeping rainwater at a place where it falls, thereby, increasing infiltration deep into the soil profile and save the field from the soil washes and gullies formations. It presents a holistic approach keeping in view the environmental degradation and ecological farming.

In addition to the measures outlined above, there are some low cost methods highly suited for small and poor farmers. They are contour or terraced farming, mulching crop residue and weed biomass management, strip cropping, minimum tillage off-season tillage, land levelling, land-use according to capacity and vegetative barrier etc. These measures are more effective and economic in conserving, maintaining as well as preventing soil's properties and quality reduction.

However, soil and water conservation in India has been carried out on a limited scale in the form of piecemeal policies. Bali (1988) observed that India has taken 30 years to treat only 3 million hectares of land. In early phases these programmes were confined to the improvement of agricultural lands and contour bunding was the main conservation activity undertaken. Some other programmes (NREP, DPAP, DDP etc.) have also had some components of wasteland development and land and water conservation. Two programmes with an explicitly ecological approach centred on watershed as the unit of management have been implemented as generally sponsored schemes. These are : *(i)* Soil conservation in the catchment of flood prone river valley projects started in the Third Plan, and *(ii)* Integrated Watershed Management in the catchment of flood prone rivers launched in the Sixth Plan. Government initiated approach has a well spelt out technology, backed by the research being carried out in different organisations such as I.C.A.R. and are expected to yield good results.

CONCLUSION

This chapter provides a sound basis to conclude that conservation of soil, water and nutrient can be effected by employing the technique suitable for specific area and location to protect both quality and productivity of land. By over-exploitation of our production environment we are heading towards a future which appears to be bleak for the generations to come. Therefore, if we are to survive and lead a reasonably decent life we can't afford to be reckless and rapacious to our environment, we must devise suitable country and area specific strategies for attaining agricultural sustainability without endangering the resource base. Such strategies must address themselves to various socio-economic and environmental conditions and objectives. It is felt that the quality of land can be protected as well as upgraded through widespread application of environment-farmers-friendly soil protection measures to restore and improve the ecological conditions in the degraded ecosystems. Thus, the quality of land could be considerably improved through effective land resource management.

REFERENCES

Abrol, I.P. (1990). *Caring for Our Soil Resources.* Proc. International Sym. on "Water Erosion Sedimentation and Resource Conservation". Oct. 9-13, 1990, CWCRIT, Dehradun, pp. 1-6.

Alexandratos, Nikos (1995). *World Agriculture : Towards 2010*, A FAO Publication, John Wiley and Sons, Chichester, England, p. 355.

Bali, J.S. (1988). A Critical Appraisal of Past and Present Policies and Strategies of Waste-Land Development and Management in India and Role of Government and NGOs in Small Scale Watershed Management. Paper presented at the Workshop Organised by the Society for Promotion of Waste-Land Development, New Delhi.

Chopra, Kanchan, (1996). The Management of Degraded Land : Issues and an Analysis of Technological and Institution Solutions. *Indian Journal of Agricultural Economics,* Vol. 51, Nos. 1 and 2, Jan.-June 1996, pp. 238-48.

Crosson, P. and Brubaker, S. (1992), *Resource and Environmental Effects of U.S. Agriculture*, John Hopkins University Press, Baltimore.

FAO, (1977). *Assessing Soil Degradation.* (FAO Soils Bulletin No. 34). Rome.

FAO, (1979). *A Provisional Methodology for Soil Degradation Assessment.* Rome.

Government of India (1995). *Waste Land Atlas of India : Part II*, Department of Space, National Remote Sensing Agency, Balanagar, Hyderabad.

Government of India (1998) Water and Related Statistics, Central Water Commission, p. 268.

Government of India (2000). *Water and Related Statistics.* Central Water Commission, New Delhi.

Hudson, N.U. (1992). *Soil Conservation,* London : Bastford.

Kampen, J. (1974). *F.A.O, Conference on Food Production of Rainfed Areas of Tropical Asia*, Hyderabad.

Sehgal, J, (1995). Soil Resource and Agroecological Zoning for Sustainable Agriculture. *Proc. Conference on Sustainable Management of Resources,* Indian National Science Academy (INSA), New Delhi, Jan. 9-10, 1995.

Sekhon, G.S. and D.N. Puri (1986). The Input-Output Balance of Plant Nutrients in Some Intensive Multiple Cropping System in India, Quoted in M.K., Shekhon and Manjeet Kaur (1996). The Green Revolution in Punjab : Other-side of the Coin. *Agricultural Situation in India,* L. 11(11): 713. 1966.

Singh, D.R. (1996). Mapping of Degraded Lands. *Proc. All India Soil and Land Use Survey,* Nagpur Regional Centre. Reorientation Training Programme, September 23-27, 1996.

Singh, Tapeshwar (1995). *Drought Disaster and Agricultural Development in India.* PPH, New Delhi.

Sharma, R.A. (1992). Management Options for Vertisols and Associated Soil of Central and Peninsular India. In Somani, L.L. *et al.* (Ed.) *Dryland Resource Technology,* Vol. 8, Agroteck Publishing Academy, India.

9

INTEGRATED EVALUATION OF NATURAL RESOURCES FOR SUSTAINABLE DEVELOPMENT

A Study of Ghanpur Area, Nizamabad District, A.P.

H.N. MISRA and ARCHANA GUPTA

INTRODUCTION

The natural resources such as land, water, minerals and forests etc. play significant role in the economic well-being of a country or a region. These resources provide the basic fabric for growth and development and work as catalyst in human ecological settings. However, our natural resources are under great pressure due to increasing human and livestock population. The present scenario is not conducive to sustained development. The resources need to be judiciously used and conserved for sustainable development. The integrated evaluation of natural resources, thus, is the most vital aspect in chalking out the sustainable development action plan. The purpose of this chapter is to suggest the action plan for Ghanpur area which forms part of Machareddy and Kamareddy Mandal and lies in the eastern part of Nizamabad district of Andhra Pradesh (Fig. 9.1). This area is located between 18° 19.85' N to 18° 26. 7' N latitudes and 78° 23' E to 78° 30' E longitudes. This is basically part of the plateau where the major area is covered by less than 1° slope. However, there are some patches where slope is between 15°-35°.

OBJECTIVES OF STUDY

The main objectives of the study are:

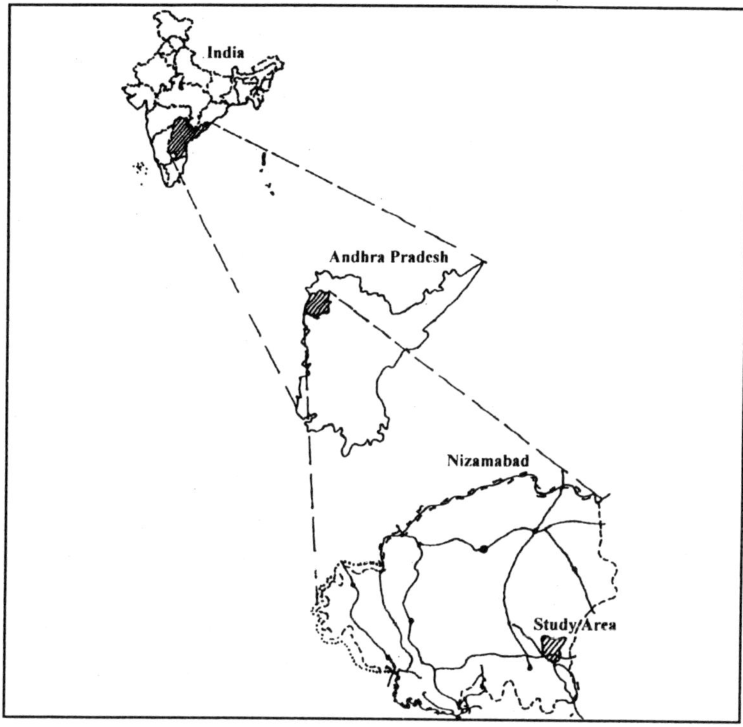

Fig. 9.1 : Location Map

(i) To prepare various thematic maps on 1: 50,000 scale using three seasons data of IRS 1B LISS II for the year 1992-93 and toposheets.

(ii) To make the diagnostic analysis of the project area with the help of various thematic maps.

(iii) To prepare the appropriate and sustainable action plan for developing the area to prevent/mitigate the adverse environmental impacts, such as land degradation, deforestation etc.

METHODOLOGY

The Standard False Colour Composite (FCC) imagery of IRS 1B pertaining to the study area have been interpreted. Three season Standard FCC *i.e.* October, February and May have been taken. Based

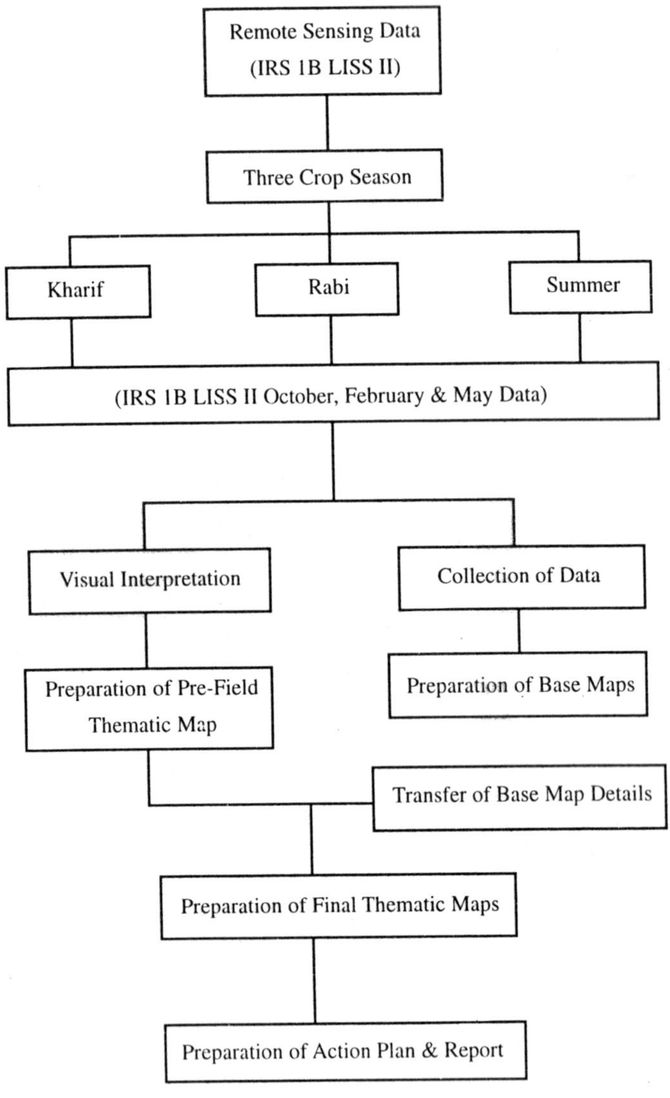

Fig. 9.2: Flow Chart of Methodology

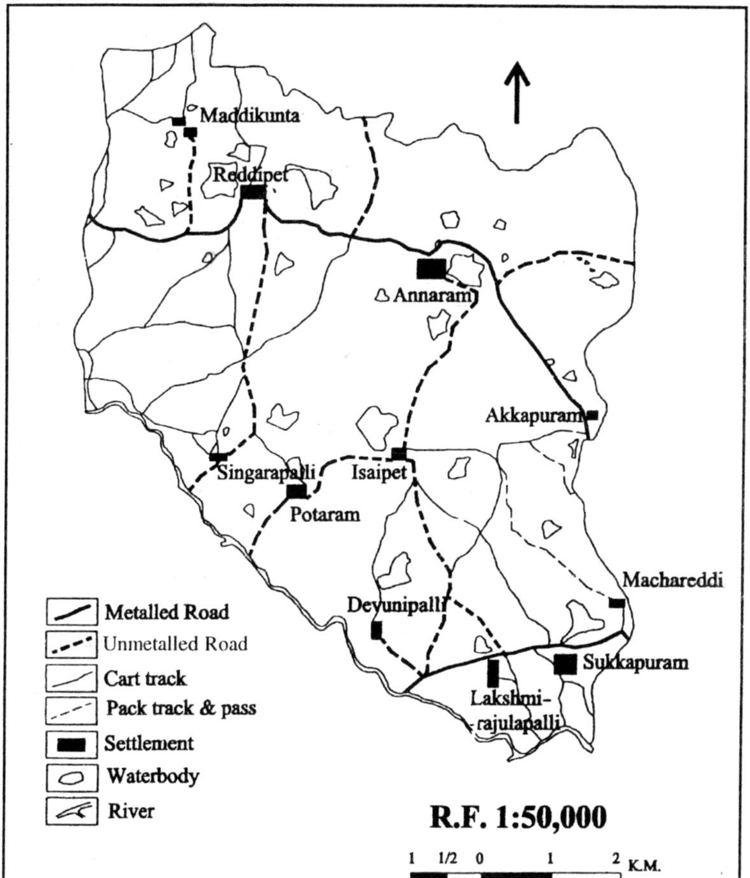

Fig. 9.3 : Transport Network and Settlements—Part of Ghanpur Watershed, Nizamabad A.P.

on the visual interpretation of the satellite data, SOI toposheet and other available data, six thematic maps have been prepared. These are Transport network and settlement location map, Surface waterbodies and Drainage network map, Slope map, Soil map, Hydrogeomorphological map, Land-use and land cover map.

In addition to visual interpretation, GIS for digitization and data base creation has been used. GIS is a powerful tool for storing, manipulation and integration of the spatial and non-spatial data. Different themes prepared by the visual interpretation from satellite imageries and SOI toposheets etc. have been entered into computer

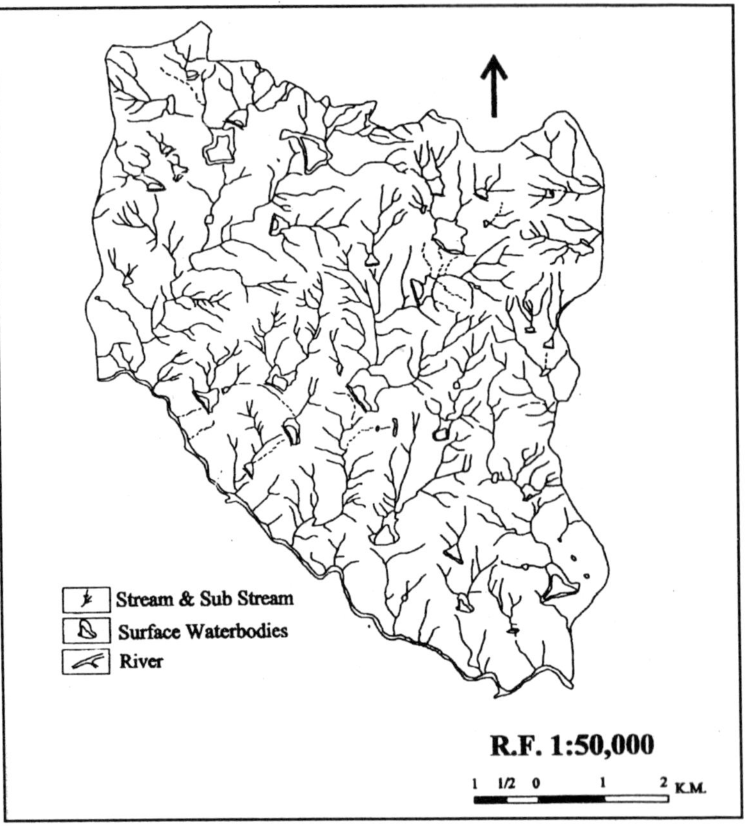

Fig. 9.4 : Drainage Network and Surface Waterbodies—Part of Ghanpur Watershed, Nizamabad (A.P.)

system through digitization. The WS ARC/INFO software has been used for GIS. After digitization, editing and annotation, map composition has been performed. The flow chart (Fig. 9.2) shows the methodology followed in this study.

NATURAL RESOURCES AND THEMATIC MAPS

As stated earlier there are six thematic maps. The transport network and settlement map (Fig. 9.3) shows the different types of road and settlement locations. The drainage network and surface waterbodies have been displayed on the map (Fig. 9.4). The slope map has been

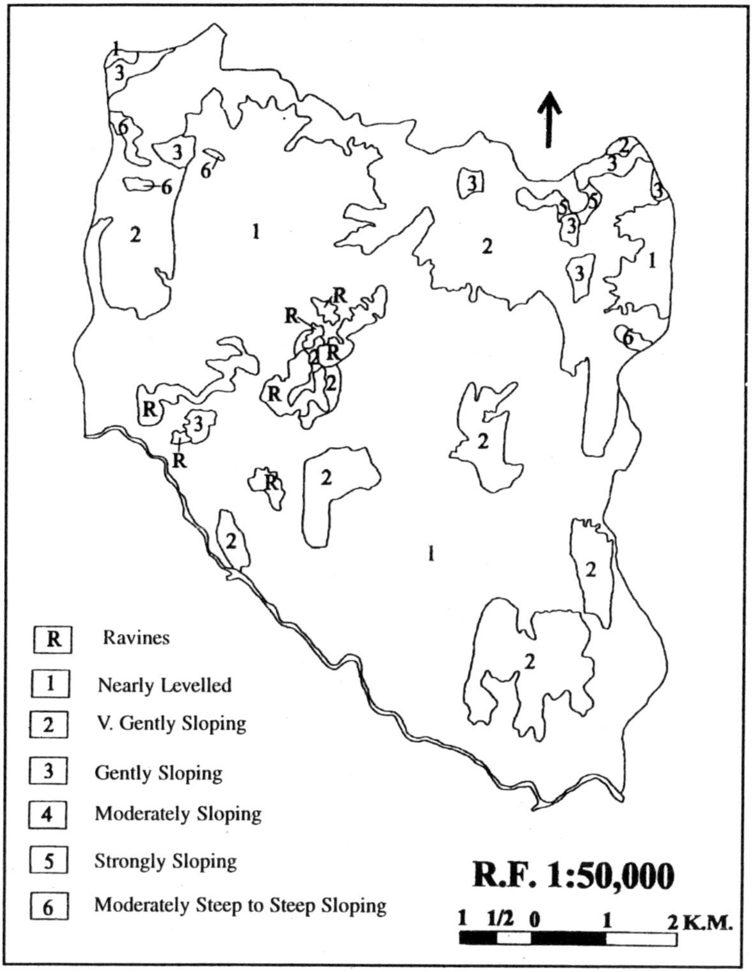

Fig. 9.5 : Slope Map—Part of Ghanpur Watershed, Nizamabad, A.P.

prepared with the help of contour lines given in the SOI toposheet at 20 metres vertical interval (Fig. 9.5). The slope of any area is vital element for land irrigability and capability assessment. On the basis of All India Soil and Land-Use Survey guidelines, slope categories of the study area are shown in Table 9.1.

Table 9.1: Slope Categories of Ghanpur Area

Slope Categories	*Slope Details*	*Slope (%)*
1.	Nearly levelled	0-1
2.	Very gently sloping	1-3
3.	Gently sloping	3-5
4.	Moderately sloping	5-10
5.	Strongly sloping	10-I5
6.	Moderately steep to steep sloping	I5-35

Source: IMSD Technical Guidelines NRSA, Hyderabad.

The semi detailed soil map has been prepared with the help of SOI toposheet and summer season data of IRS 1B LISS II (Standard FCC, 2, 3, 4 bands) of 20th May 1992 and limited field checks. The soils as classified and represented on the map are presented in Table 9.2 (Fig. 9.6).

Table 9.2 : Soil Classification of Ghanpur Area

Map Symbol	*Soil Series/ Association*	*Classification and Association*	*Soil/Land Capability*	*Soil / Land Irrigability*
02	Devaipalli	Loamy skeletal mixed, Hypothermic Typic Ustochrepts	IV es	D, 4st
	Burguda	Loamy skeletal mixed, Hypothermic Typic Ustorthents		
03	Yerrapadu	Fine, Montmorrillomitic, Hyperthermic Typic Ustochrepts	III es	C, 3st
	Baswannapalli	Fine, Montmorrillomitic, Hyperthermic Vertic Ustochrepts		
04	Nandiwada	Fine loamy mixed, Hyperthermic Typic Ustochrepts	III es	C, 3st
	Somavaram	Lomay skeletal mixed, Hyperthermic Typic Ustochrepts		
05	Burguda	Loamy skeletal mixed, Hyperthermic Typic Ustorthents	IV es	D, 4st
	Argonda	Fine loamy mixed, Hyperthermic Typic Ustochrepts		

Contd.

Table 9.2: (Contd.)

Map Symbol	*Soil Series/ Association*	*Classification and Association*	*Soil/Land Capability*	*Soil/Land Irrigability*
07	Borlam	Fine, Montmorrillomitic, Hyperthermic Leptic Ustochrepts	II w	B, 2d
	Padmajiwadi	Fine, Montmorrillomitic Hyperthermic Vertic Ustochrepts		
09	Kalledi	Loamy skeletal mixed, Hyperthermic Lithic Ustorthents	VII es	E, 6st
	Kummarpalli	Loamy skeletal mixed, Hyperthermic Typic Ustorthents		
10	Kondapur	Loamy skeletal mixed, Hyperthermic Typic Ustochrepts	IV es	D, 4st
	Kottur	Coarse loamy mixed, Hyperthermic Typic Ustorthents		
11	Kottur	Coarse loamy mixed, Hyperthermic Typic Ustorthents	IV es	D, 4st
	Dichpalli	Loamy skeletal mixed, Hyperthermic Typic Ustochrepts		
12	Banswada	Loamy skeletal mixed, Hyperthermic Typic Rhodustalfs	III es	C, 3st
	Nandipeta	Loamy skeletal mixed, Hyperthermic Typic Ustochrepts		
13	Uppalvai	Loamy skeletal mixed, Hyperthermic Typic Haplustalfs	III es	C, 3st
	Sompalli	Fine loamy mixed, Hyperthermic Typic Haplustalfs		
14	Uplur	Fine loamy mixed, Hyperthermic Typic Haplustalfs	II e	B, 2st
	Kotapalli	Fine loamy mixed, Hyperthermic Typic Ustochrepts		
18	Bhavanipeta	Fine loamy mixed, Hyperthermic Typic Ustochrepts	II w	B, 2d
	Gogalpet	Fine loamy mixed, Hyperthermic Typic Ustorthents		

Source : IMSD Technical Guidelines, NRSA, Hyderabad.

Description of sub-classes

Land capability : e-erosion, s-soil limitations, w-wetness

Land Irrigability : s-limitations of soil characteristics,
t-topography,
d-drainage

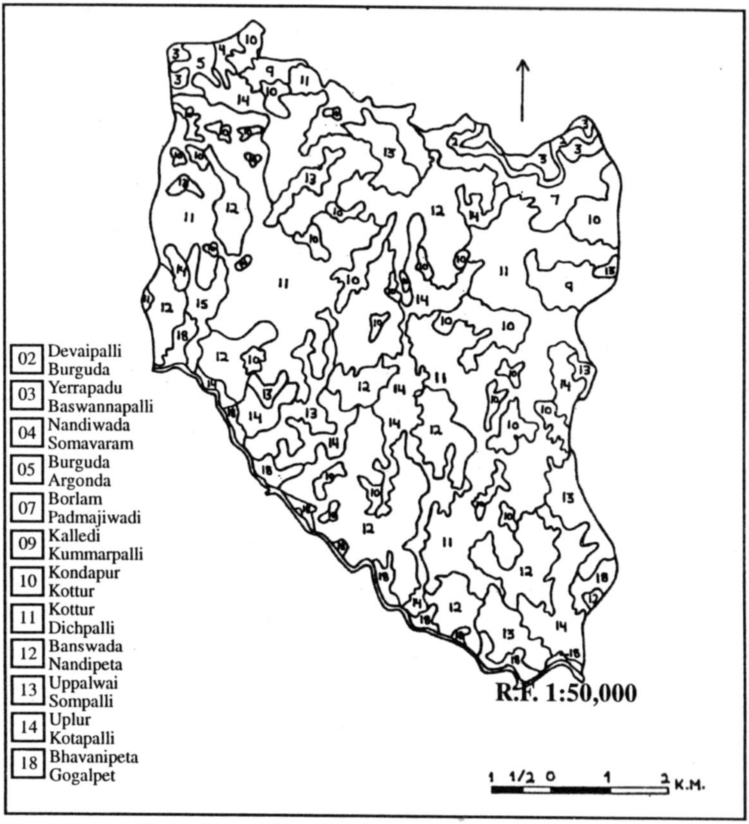

Fig. 9.6 : Soil Map— Part of Ghanpur Watershed, Nizamabad (A.P.)

The hydrogeomorphological map has been prepared using SOI toposheet and satellite data. The details represented on the map are shown in Table 9.3 (Fig. 9.7).

The land-useland cover map is important for preparing the action plan. This map has been prepared using SOI toposheet and three season IRS 1 B LISS II data for the year 1992-93. The description of the land-use/land cover type is given in Table 9.4 (Fig. 9.8).

PROBLEMS IDENTIFICATION AND RECOMMENDATIONS

On the basis of different thematic maps prepared and detailed analysis of the available data and maps individually and in combination on

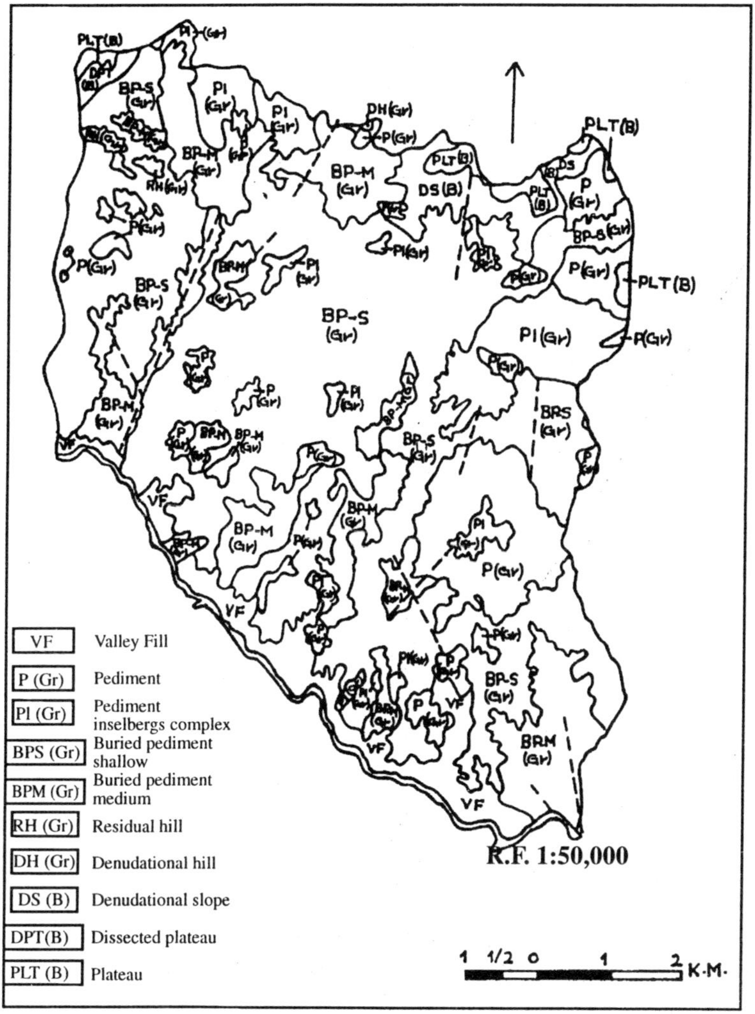

Fig. 9.7 : Hydrogeomorphology Map—Part of Ghanpur Watershed Nizamabad (A.P.)

various aspects, the following problems have been identified in the study area :

(i) Land degradation due to soil erosion.

(ii) Degradation of natural vegetation cover because of over exploitation/over grazing and other biotic interference.

Table 9.3 : Hydrogeomorphological Classification of Ghanpur Area

Sl. No.	*Map Symbol*	*Geomorphic Unit*	*Structure*	*Materials/ Lithology*	*Description*	*Groundwater Potential*
A. Fluvial Origin						
1.	VF	Valley Fill	Valleys are mostly fracture controlled	Constitutes boulders, cobbles, pebbles	The unconsolidated sediments deposited by streams/rivers normally in a narrow	Very good, depending on the thickness of fill the fracture / lineaments
B. Denudational Origin						
1.	P (Gr)	Pediment	Some-times fracture controlled	Varying lithology	Gently sloping, smooth surface of erosion bedrock between hill and plain with thin veneer of detritus	Moderate to poor, varies with underlying lithology, presence of
2.	IP (Gr)	Pediment inselbergs complex	Sometimes controlled by joints, fracture lineaments etc.	Varying lithology	Isolated low relief/ hill surrounded by gently sloping, smooth, erosional bedrock with veneer of detritus	Moderate to Poor
3.	BP-S (Gr)	Buried pediment Shallow	May be crisscrossed by lineaments/	With shallow overburden of	Flat and smooth surface of buried pediment with (0-5 thickness) shallow	Moderate to Poor

Contd.

Table 9.3 *(Contd.)*

Sl. No	*Map Symbol*	*Geomorphic Unit*	*Structure*	*Materials/ Lithology*	*Description*	*Groundwater Potential*
			fractures	weathered material of varying lithology	overburden	
4.	BP-M (Gr)	Buried pediment medium	May be crisscrossed by lineaments / fractures	With moderately thick overburden of weathered material of varying lithology	Flat and smooth surface of buried pediment with (0-20 m thickness) medium overburden	Good to Moderate
5.	RH (Gr)	Residual hill	Joints, fractures lineaments etc.	Varying lithology	Isolated low relief formed due to differential erosion so that a more resistant formation stand as residue like small hills	Poor to Nil
6.	DH (Gr)	Denudational joints, fractures lineaments etc.	Joint, fractures lineaments etc.	Varying lithology	Formed due to differential erosion and weathering so that a more resistant	Poor to Nil

Contd.

Table 9.3 *(Contd.)*

Sl. No	*Map Symbol*	*Geomorphic Unit*	*Structure*	*Materials/ Lithology*	*Description*	*Groundwater Potential*
					formation or intrusion stand as mountains/hills	
C.	*Structural Origin*					
I .	DS (B)	Denudational Slope	Joints, fractures lineaments etc.	Basalt	Side slope of plateau & mesa of Deccan basalt	Poor to Moderate
2.	DPT (B)	Dissected Plateau	May be crisscrossed by fractures / joints lineaments	Varying lithology	Deep valleys / gullies with gently sloping land developed due to stream / river erosion on plateau	Moderate to Poor
3.	PLT (B)	Plateau	May be crisscrossed by fractures/ joints lineaments	Forms over various litho units	Essentially formed over horizontally layered rocks marked by extensive flat top and steep slopes	Good to Moderate along fractures /joints / lineaments especially their intersections

Source : IMSD Technical Guidelines, NRSA, Hyderabad.

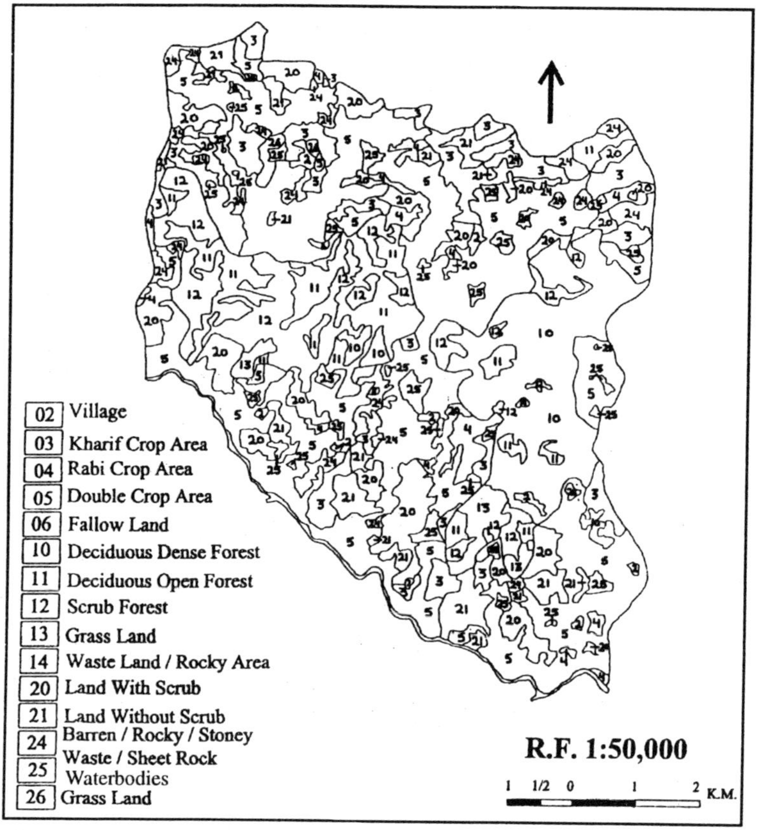

Fig. 9.8 : Land-use/Land Cover Map—Part of Ghanpur Watershed, Nizamabad (A.P.)

(iii) Less crop productivity due to continuous degradation of soil fertility, irregular rainfall and traditional agricultural practices.

(iv) Decrease in surface water resources as well as groundwater due to lack of soil conservation practices and continuous exploitation of groundwater.

(v) Reluctance of the farmers towards adoption of improved agricultural package and practices.

(vi) The local livestock are poor in quality as well as in production. The pastures and grazing fields have been degraded and, therefore, sufficient fodder is not available.

(vii) For want of proper communication facilities like roads,

Table 9.4 : Land use/Land Cover Type in Ghanpur Area

Sl. No.	*Map Symbol*	*Type of Land*	*Description*
1.	02	Built up land	Village
2.	03	Agricultural land	Kharif crop area
3.	04	Agricultural land	Rabi crop area
4.	05	Agricultural land	Double crop area
5.	06	Agricultural land	Fallow land during both the seasons
6.	10	Forest area	Deciduous dense forest
7.	11	Forest area	Deciduous open forest
8.	12	Forest area	Scrub Forest
9.	13	Forest area	Grass land
10.	14	Forest area	Waste land/Rocky area
11.	20	Waste land	Land with scrub
12.	21	Waste land	Land without scrub
13.	24	Waste land	Barren/Rocky/Stony waste/Sheet rock
14.	25	Waste land	Waterbodies
15.	26	Others	Grass land

Source: IMSD Technical Guidelines, NRSA, Hyderabad.

most of the village community could not adopt latest improved technology.

Keeping in view these problems, an action plan based on broad decision rules (Table 9.5) has been developed (Fig. 9.9). The recommendation for the implementation of the action plan are as follows :

A. Soil Conservation and Water Resources Development

I. Conservation measures
II. Production measures.

I. Conservation measures

Since land and the water resources are most important factors for the development of any area, the conservation of these resources is

Table 9.5 : Broad Decision Rules for Action Plan

Map symbol	*Land form*	*Soil Association unit*	*Slope categories*	*Land-use / Land cover*	*Proposed Treatment*
01	VF, BP-M (Gr)	7, 9, 8, 12, 14, 11	01	DCA & SCA	Intensive agriculture
02	BP (Gr) BP-S (Gr)	7, 14, 13,	01 & 02	DCA & SCA	Agrohorticulture (Irrigated)
03	BP-S (Gr), P Gr	4, 11 ,12	01, 02, & 03	SCA	Agrohorticulture, Agroforestry
04	P (G), BP-S (Gr) PI (Gr)	4, 12, 10, 11, 5, 13	02 & 03	Land with or without Scrub Forest	Silvi pasture development
05	RH (Gr), PI(Gr), DS (Gr)	2, 3, 4, 5, 10, 11	05, 06 & 03	Open or Scrub Forest	Afforestation or rehabilitation of degraded forest
06	P (Gr), PI(Gr), BP-M (Gr), BP-S (Gr)	2, 3, 4, 5, 10, 11	05 & 03	Dense Forest	Forest Conservation
07	P (Gr), PI (Gr), DS	3, 14, 4, 12, 5, 10 , 11, 13	01,02 & 03	Scrub Area / Stony waste area	Fuel and Fodder plantation

Source : IMSD Technical Guidelines, NRSA, Hyderabad.

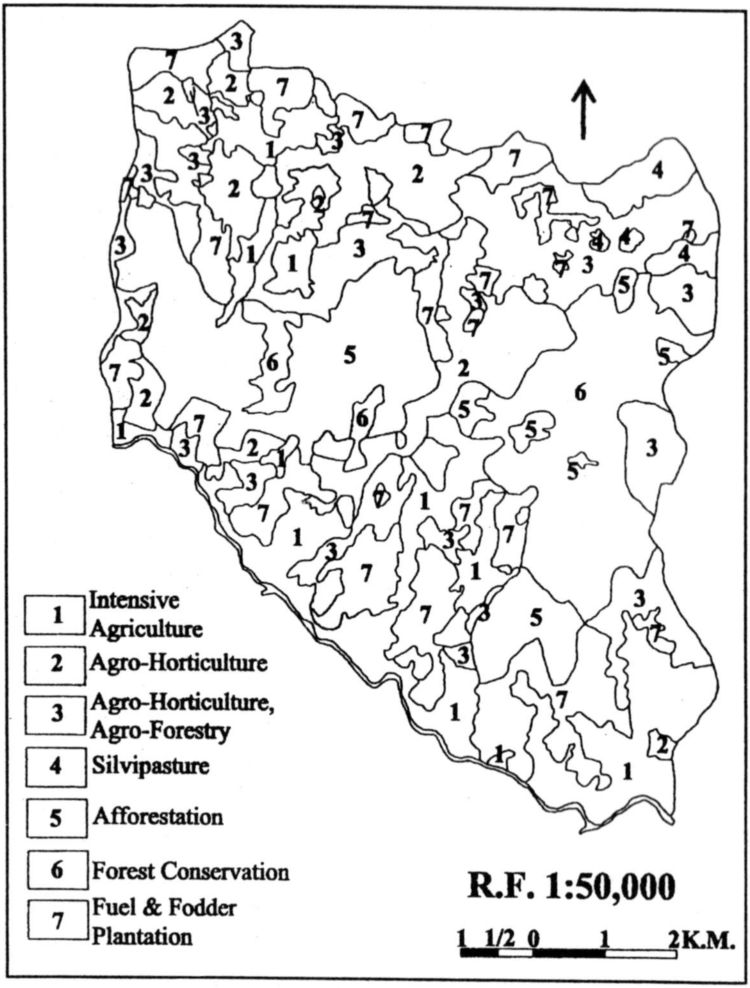

Fig. 9.9 : Action Plan— Part of Ghanpur Watershed, Nizamabad (A.P.)

essential. Keeping in view rainfall of the area, topography, soils and groundwater potential of the area, the following measures have been recommended for soil and water conservation:

(a) On Farm Development

(*i*) Establishment of contour vegetative barrier.

(ii) Construction of contour 'V' ditches.
(iii) Dry farming agronomical practices.

(b) Off Farm Development

(i) Contour vegetative barrier.
(ii) Construction of contour 'V' ditches.
(iii) Establishment of shrub line.
(iv) Chiselling along the contour.

(c) Drainage Line Treatment

(i) Loose stone check dams.
(ii) Live vegetative check dams.
(iii) Dry stone check dams.
(iv) Medium masonry structures.
(v) Minor masonry structures.
(vi) Water harvesting structures.

II. Production Measures

For optimum use of conserved moisture certain production measures have to be proposed so that productivity of biomass can be maximized. The production measures may be as under:

(a) On Farm Development

(i) Intensive agriculture/crop production.
(ii) Agroforestry.
(iii) Agrohorticultre/mixed horticulture.

(b) Off Farm Development

(i) Silvi pasture development
(ii) Multi-tier canopy /afforestation
(iii) Rehabilitation of degraded forest

(c) Drainage Line Treatment

(i) Nalah plantation.

B. Livestock Development

(i) Health coverage
(ii) Breed Improvement
 (a) Artificial insemination.
 (b) Improved breed bull and ram distribution.

C. Land Management

Land management decision rules are based on study of different thematic maps and arriving of decision rules as given in Table 9.5 (Fig. 9.9). These are based on the actual combinations of various items of the theme maps observed in the field.

I. Intensive Cultivation with Groundwater Exploitation

Intensive cultivation has been proposed in such areas where either double cropping or single cropping is practised with groundwater exploitation. The selection of crop will be based on as per feasibility of the area and local needs. In these areas high yielding varieties along with improved agronomical practices have to be grown.

II. Agrohorticulture

The sites recommended for agrohorticulture have been found to possess good groundwater potential. In these areas cultivators will grow fruit plant with crop cultivation. The following fruit plants are proposed to be grown along with the crops:

1. Mango
2. Guava
3. Sapota
4. Orange

The horticulture species provide additional income to the farmers.

III. Agrohorticulture-cum-Agroforestry

In such areas where soil depth as well as groundwater potential is poor agrohorticulture (rained) or agroforestry has to be proposed so that cultivator could get extra benefit from these areas. For agrohorticulture and agroforestry, the following plants are proposed along with the crop cultivation:

Fruit Plants	*Forest Plants*
Mango	*Leucaena leucocephala*
Guava	*Acacia nilotica*
Pomegranate	*Albizia lebeck*
Sapota	*Ailanthus excelsa*

IV. Fuel and Fodder Plantation

This activity is mainly recommended for those areas having scrub vegetation. The tree species which meet out fuel and fodder requirements are proposed as under:

1. *Leucaena leucocephala*
2. *Acacia nilotica*
3. *Albizia lebeck*
4. *Ailanthus excelsa*
5. *Azadirachta indica*
6. *Cassia fistula*

V. Silvi Pasture Development

Silvi pasture is grass farming in trees enclosures. It is recommended in such areas having some scrub vegetation, shallow soils and land degradation problems. The following trees and grass species are proposed for fuel and fodder production.

Trees	*Grasses*
Leucaena leucocephala	*Dicanthium annulatum*
Sesbania sesban	*Cenchrus celiaris*
Albizia amara	*Stylosenthes gracilis*
Cassia siamea	*Penicum maximum*

VI. Forest Enrichment/Rehabilitation of Degraded Forest

Some of the notified forest have become degraded due to increased biotic interference and corrective measures have to be initiated in such areas. These areas have little density of forestry plants. Following plants are proposed for enrichment of these forest areas:

1. *Tectona grandis*
2. *Prosopis juliflora*
3. *Acacia catechu*
4. *Butea monosperma*
5. *Acacia nilotica etc.*

VII. Forest Conservation

This item is recommended in those patches of notified forest areas having fairly dense vegetation cover. These areas need to be protected from illegal encroachments, deforestation and intrusion of cattle for grazing.

REFERENCES

Indian Resources Information & Management Technology Pvt. Ltd. (1996), IMSD, *Role of Socio-economic Data Analysis in IMSD Programmes with particular reference to Machareddy Mandal of Nizamabad District (A.P.).*

NRSA (1986), *Manual of Procedure of Watershed Mapping using Remote Sensing Techniques*, National Remote Sensing Agency, Department of Space, Hyderabad.

NRSA (1989), *Manual of Nationwide Landuse / Landcover Mapping Using Satellite Imagery,* National Remote Sensing Agency, Department of Space, Hyderabad.

NRSA (1990), *Description and Classification of Landuse/Landcover,* National Remote Sensing Agency, Department of Space, Hyderabad.

NRSA (1991), *Technical Guideline: Integrated Study to Combat Drought for Sustainable Development*, National Remote Sensing Agency, Department of Space, Hyderabad.

NRSA (1995), *Technical Guideline: Integrated Mission for Sustainable Development*, National Remote Sensing Agency, Department of Space, Hyderabad.

10

SOCIO-CULTURAL AND GEO-ECOLOGICAL DIMENSIONS OF MINING AREAS

M. AHMAD and KUMAR NIKHIL

Introduction

Society's interaction with nature is so extensive that it has given rise to various environmental questions affecting all humanity. Environmental deterioration can be attributed to industrialisation and organisation, the depletion of traditional sources of energy and raw materials, constant population growth, disruption of natural ecological balances, destruction for economic ends of various animals and plant species and negative genetic consequences of industrial and other pollutants which include the danger of man's genetic degeneration.

Whereas non-intervention in nature is no solution to human civilisation which is a socio-humanitarian in realisation process, the way out of this contradiction is the new ecologised understanding of the world under the impact of practice. The new ecologisation of the world outlook of the mankind will become a global force preserving and transforming nature. Harmonisation of humanity-environment relationship will require restructuring of the social system and ownership of modern technology with a need to learn to link a new.

Principles of Ecosystem

The following principles govern the ecosystem of which man and social systems are integral parts (Srivastava, 1990):

1. Everything affects everything, all living things exist as part of the system; and in each system there are factors that limit the growth; each system has a definite capacity to carry the organism of that system, in these systems all material cycles are used, but the energy to the system flows in one direction—downwards to uselessness, stability of system increases with specialisation.
2. Environmental crisis sets in if these governing principles are disturbed. By definition, an environmental crisis is : an imbalance among the physical, chemical and biological elements and/or process of an ecosystem that threatens the vitality and productivity of some or all species of the system; disruption of a biological system to the point where it no longer can support some of or all the organism in it, several of ecological processes of an ecosystem to a point where several to many generations are required to restore the productivity, complexity and stability of the system.

Geo-Environment and Mining

Various changes occur in the geological environment as a result of mining, underground coal gas production, sulphur melting and other activities. All types of surface mining of coal has brought widespread environmental damage. Its impact on air and water, land-use, health and safety, wild life habitat and aesthetics. It has made farm and forested lands unsuitable to vegetations unstable mountainous areas have gone prone to landslide hazards. The water quality of streams has been degraded by acid mine drainage from abandoned strip mines. Thousands of acres of land have been degraded by strip mining methods (area strip, contour strips or augering). All types of surface mining—open pit mining, quarrying and dredge mining involves the excavation of extensive areas of land and thereby rendering them useless for agricultural purposes or foresting.

Surface mining laws and regulatory programmes can protect environmental quality by adequate planning and reclamation plans, adequate performance standards and adequate enforcement of regulatory programmes. The earth's mineral resources are not inexhaustible. They are not renewable in historic time.

Socio-Economic Impact of Mining Activities

Socio-economic impacts relate to changes to the social structure and characteristics of a community. They may initiate secondary environmental impacts which are very difficult to predict because they are somewhat removed from the original cause. Socio-economic impacts are linked to the economic, environmental and health impacts. Socio-economic study is essential to estimate the socio-economic effects such as changes in employment, transportation, recreation, changes in aesthetic, in landscape etc. The impacts on society because of various bio-physical effects caused due to mining activities can also be estimated.

Coal mining regions in India remain economically depressed areas. They are the pockets of ignorance, unemployment and poverty. The socio-economic impacts of coal mining are intrinsically linked with its economic, environmental and health impacts. The distinction between socio-economic and other impacts is not always clear cut. Fig. 10.1 (Chadwick *et al.,* 1987) illustrates by a stylised input-output model for coal mines, the relationships between coal mine and its socio-economic environment.

Munn (1979) has given spatial dimensions of socio-economic impacts and socio-economic indicators categorised for desertification and socio-economic data collection and areas of human concern. Some of the broad categories of socio-economic data collection method include :

— using existing data, and
— asking questions, observing individual and group behaviour (direct and indirect observations).

The process of gathering the required socio-economic informations called as social profiling is done only for communities in the immediate areas in the adjacent region. Those most affected by the mining activity include people displaced or communities which undergo vast changes and groups or affected people in adjacent region are identified when trained, transportation route, recreational, cultural and ethnic linkage between the regions is studied. The socio-economic data could be objective or subjective in types.

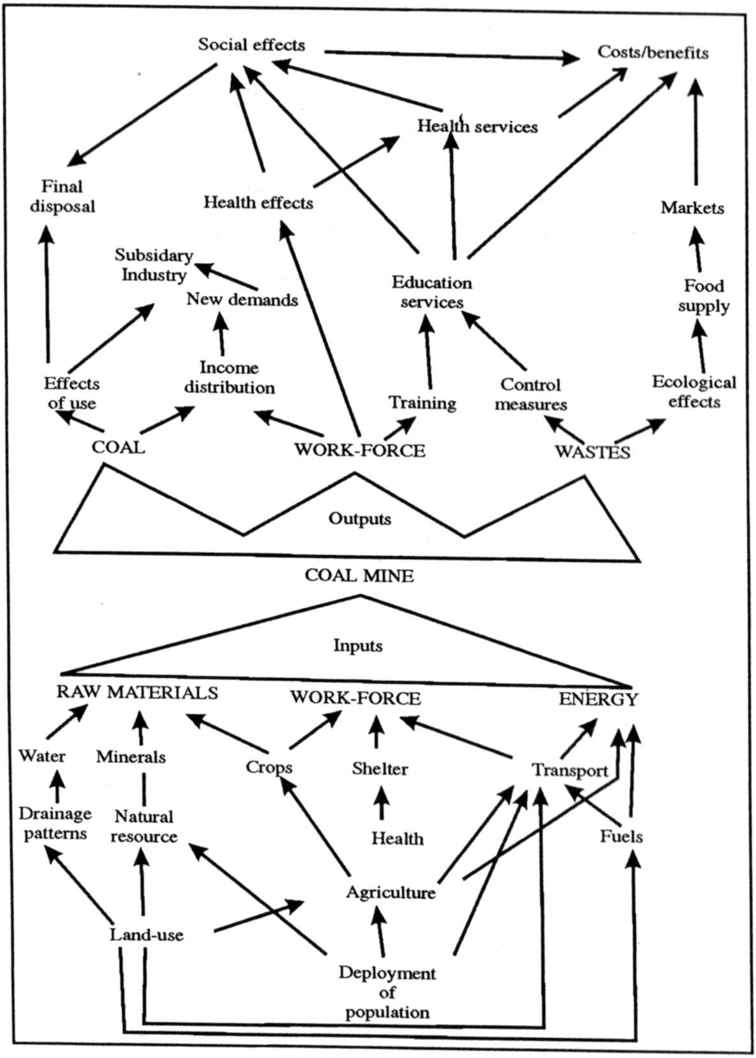

Fig. 10.1 : Input-output model showing relationships between a coal mine and its socio-economic environment

Social Impacts

The mining activity has definite impacts on socio-economic fronts in the core as well as buffer zones. The social impacts are :

(i) *Population growth :* Direct and secondary employment opportunity both as short and long-term basis in any mining project which results in substantial population growth within the area and adjoining areas.

(ii) *Immigration :* There have been migration of population both within and outside the State in the form of skilled personnel who are not available from the local inhabitants.

(iii) *Educational facilities :* Steps to promote educational facilities in the area result in developing various types of educational institutions in the area.

(iv) *Health care facilities:* Hospital and health care are needed in the projects and are set up for the general population and for the various emergency cases arising from various mine accidents.

Economic Impacts

Mining projects create additional direct jobs in the mines. During construction period, a number of semi-skilled and unskilled workers are provided jobs for construction of residential houses, office buildings, infrastructures and service sectors are generated. Persons who lose their agricultural lands are given training for alternative jobs in addition to payment of compensation.

The land oustees from the inhabitants are adopted for community development programme of villages and neighbouring residential and under the rehabilitation scheme they are provided with developed land and cash grant for shifting.

Impact during Initiation of Mining Projects

Mining operations have major consequences for communities and populations in their vicinity in several categories. The major social costs of mining and some of the methods for assessing their magnitude are (Sivudu, 1988) :

(a) *Social cohesion :* Mining operations may generate new residents into settlements with history of close interaction within kin and neighbourhood. Newcomers may not be aware of the local customs and styles of life and communications and may feel contemptuous of them.

(b) *Socio-economic structure :* Newcomers can modify the hierarchy of income, wealth and authority in the community, which may result in changes in socio-economic structure of the community but the disruptive effects of changing the economic patterns in a stable society need to be dealt with immediately,

(c) *Disturbance of settlement pattern and special arrangements :* Relocation of residence may come out and new sub-divisions for newcomers may be created. Also the existing transportation system may be inadequate,

(d) *Quality of life and economic well-being :* The effect of increased population numbers and density, noise and dirt, decline in social services may result in degradation of quality of life.

(e) *Aesthetic sensibilities :* Disturbance of existing landscape accompanied by dust, noise or traffic are source of irritation. Often protests are more vocal over aesthetic issues than over material issues.

(f) *Personal disorganisation :* Disruption of accepted cultural norms and breakdown of conventional systems of social relations and communications affect the children and the aged strongly. This often results in out-migration to escape the disorienting conditions.

(g) *National integration :* Mining projects do help in national integration. Newcomers with various cultural affinities come together and find opportunity to understand cultural, socio-economic systems of various peoples and parts of the country.

Impact after the Initiation of the Mining Projects

1. Communities may well be split into those who favour change and those who do not. Community leaders are often unable to make the adjustments required owing to lack of funds or experience.
2. Increased rate of alcoholism, broken homes are among the personal manifestations of breakdown of social order.
3. Relationship between the local population and mining companies can be considered as instances of adaptive coping method that local people devise in order to defend themselves against the cost of mining and the associated developments.

4. Often sense of helplessness prevails over the local residents and they take to tolerate extreme conditions. In such cases outsiders have to speak for them. A case must be made by others on universalistic or ideological grounds. Yet such interference can result in well meaning but ecologically or culturally unsound attempts at restoration of the premining, physical or social resources.

Secondary Environmental Impacts

Secondary environmental impacts fall into two categories:

(i) those which result from a change in people's activities or an increase in population pressure because of new opportunities offered by the existence of the project, and

(ii) those which are passed by programmes intended to develop specific areas and which are catalyzed by the industrial project.

Many of the secondary environmental impacts that can occur and their linkages are illustrated by the example of unplanned migration to an area as a result of the development of an industrial project and is shown in Fig. 10.2 (Chadwick *et al.*, 1987). These impacts can be very severe with potential long-term effects. Because the linkage and the cause of impact are complex and sometimes obscure, their control necessitates some regulation of people's activities. Effective control may be exceedingly difficult. Control measures may fall outside the scope of activities for which the developer believes he is responsible.

Impact on Trial Communities—A Case Study

In the assessment of the environmental impact by mining projects, sociological and ethnic aspects attract greater attention in the tribal belt due to sensitivity of the nature and direct influence of urbanisation on society. The adverse impact of mining activities can be enumerated as :

— imbalance in the economic developments in the entire section of the society,

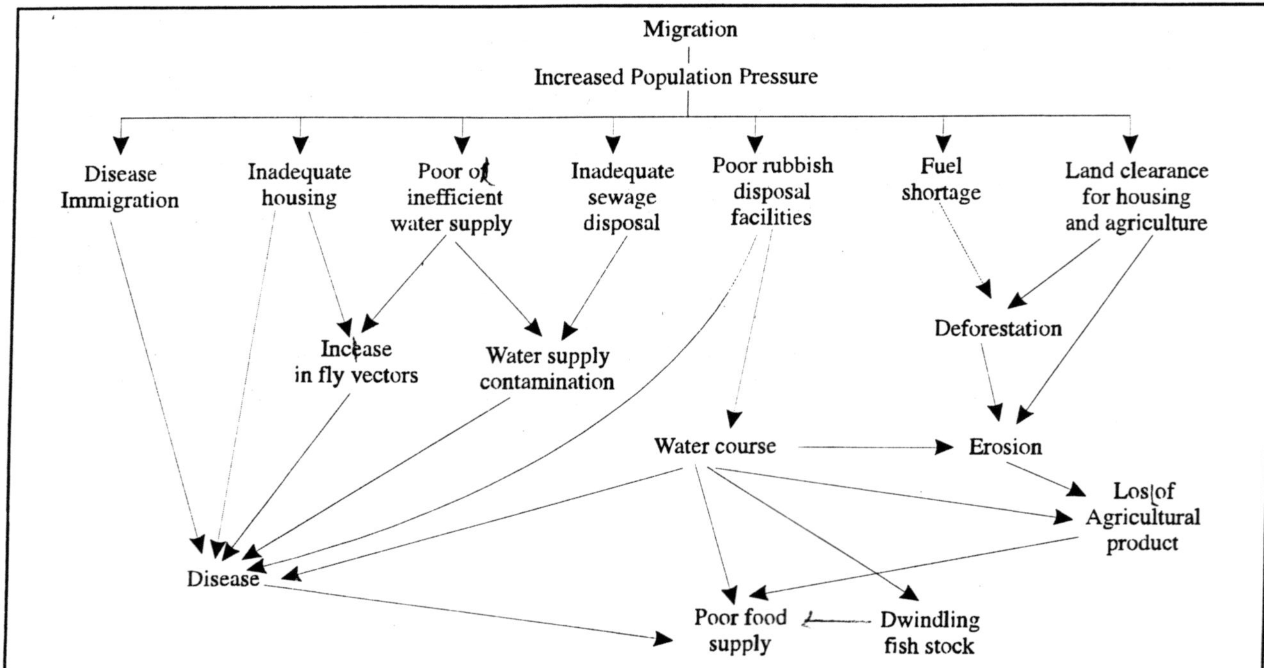

Fig. 10.2: Some possible effects of unplanned migration to a new industrial area

- — inadequate mental make up commensurate with economic rise,
- — privilege to cut the trees and do hunting/exploitation by the unorganised migrants or traders to encash on the above shortcomings. This leads to deforestation for trading and increase in urban-oriented crimes etc. and
- — exploitation of weaker sex who are more lured to modern life by unorganised sector, contractors etc.

Tribals are traditional race with their own cultural and community discipline, they are truthful, honest, eager to learn and do more and adherent to principles with proper understanding of their cultural and anthropological background, they can be handled and trained to tap their hidden talents for developmental work and mental make-up for their own protection from evil forces.

All welfare measures created at the projects are extended not only to the employees but to the tribal inhabitants in the adjoining areas. It is of utmost importance to know whether the transformation of tribals by urbanization had been beneficial to them or led to erosion of their identity or cultural heritage. An interesting case study concerning tribals of Bastar Region of Madhya Pradesh is quoted herewith.

An indepth study was carried out on the tribals of Bastar (*i.e.* Murias, Halbas, Koyas etc.). Salient factors of this study are (Rangachari, 1988):

- *(i)* land acquired from tribal owners is 119.7 hectares for a project complex having lease of over 4280 hectares, is only 2.8 per cent. Due compensation has been paid,
- *(ii)* out of total 3602 workers, 1061 are from Bastar (593 local tribals, 240 scheduled caste). Apart from this around 1000 tribals are employed in other satellite service industries, shops etc.,
- *(iii)* better communication by road, rail and postal facilities have provided assured transport, mobility, marketing by local tribals and also smooth development of State (school, hospital, farm, research centre). It has provided electricity to nearby villages and banking facilities at the market centre,
- *(iv)* educational institutions for 4000 pupils have been created with free education to tribal students,
- *(v)* all special medical facilities for chronic diseases such as TB,

leprosy, abnormal tumours, burn-injuries, skin grafting, etc. are given free of charge and diet. This facility is unparallel in entire tribal belt. Though tribals have their own herbal therapy, they use it for minor ailments and go to project hospital for severe or chronic diseases,

(vi) entertainment by radio, television, cultural programme, subsidy in purchase of items for promotion of cultural activities by educational institutions,

(vii) tribals could switch over to a variety of jobs, diversification of their livelihood, agricultural practices for better yield, and

(viii) growing tendencies to resort to achieve productive work. Urge to earn and save more and invest in their jobs, houses and other facilities. In spite of urban style of dress, food and other habits, the tribals retain their traditional way of life and community bond at their villages.

The Dilemma

There is no solution of the dilemma : if unregulated mining and quarrying is allowed, they will enlarge problems of environmental degradation, their closure will create unemployment and organised oppression of labour. This dilemma has been emerged powerfully, recently in Doon Valley environmental movement after partial closure of limestone quarrying. Doon limestones are in great demand for their 99 per cent purity. Ruthless exploitation, dynamite blasting and spoilage of debris put an end to agricultural activities in the vicinity of the quarries.

The destruction of limestone hills lead to floods during heavy rains and severe drought during the summer. The closure of quarries resulted in about 15,000 workers being thrown out of job and labour unions starting opposing the elitist environmental movement. Two mutually opposed trends are observed :

1. the environmental activists movement is seen as a threat to their jobs; and
2. the vulgar environmentalists want to create an idyllic life in the undisturbed natural terrain.

the negative impacts, so that positive impact can be seen by having good quality of life of local habitant.

REFERENCES

Chadwick, M.J. *et al.* (1987). *Mining Projects in Developing Countries*, A Manual prepared by the BEIJER Institute Center for Resource Assessment and Management and the BEIJER Institute, New York, United Kingdom and Stockholm, Sweden, p. 303.

Munn, R.E. (ed.) (1979). *Environmental Impact Assessment. How can Socio-economic Methods be used.* Scope - 5, John Wiley & Sons, New York, pp. 81-90.

Rangachari, G. (1988). Strategy of Environmental Impact Assessment and Management at Iron Ore Mining Complex of Bailadila in Tribal Belt of South Bastar, *Proc. Environmental Problems in Mining and Their Management,* eds. Kurnar, *et al.* Published by Environmental Planning and Coordination Organisation, Bhopal, 1989, pp. 45-63.

Sivudu, K.V. (1988), Socio-economic Problems in Mining and Their Management, *Proc. Environmental Problems in Mining and Their Management,* eds. Kumar, *et al.* Published by Environmental Planning and Coordination Organisation, Bhopal, 1989, pp. 234-243.

Srivastava, H.C. (1990). *Society and Environment in Context to Mining and Pollution.* Environmental Management of Mining Operations, Dhar, B.B. (ed), Ashish Publishing House, New Delhi, pp. 368-376.

11

GIS AND WATER QUALITY MANAGEMENT

A Study of Eastern Uttar Pradesh, India

RAJANISH KUMAR

GIS AN INTRODUCTION

GIS is not a new concept or technique, it was introduced in Europe and America in the early seventies, much before the availability of GIS softwares. In the seventies some of the GIS projects were carried out successfully in Europe ard North America. In the beginning GIS suffered from various technical (absence of intelligent software), economic (costly hardware) and cultural problems. GIS attracted the attention of the software companies in the early nineties when various softwares like archview, elbis, Span, MapInfo, ArcInfo, Microstation, etc. The success of GIS is due to the North American and European countries like U.S.A., Germany, Holland, Brazil and Canada etc. GIS projects in most of the developing countries are suffering from various shortcomings, including lack of knowledge, high cost of software, lack of qualified staff and digital database.

GIS has been defined in many ways by many people. It has been defined as a systematic integration of computer hardware, software, and spatial data for capturing, updating, manipulating and analysing in order to solve complex management problems. It is not simply a computer system for preparing map in different scales, projections, colours, and patterns etc. GIS can be regarded as high technology equivalent of maps along with the analytical ability and many more. It is an organised combination of technology and human efforts to retrieve, store, transfer, identify, analyse, assess and even to monitor the spatio-temporal relationship between the physical entities and statistical data related to socio-economic aspects, natural resources, demographic parameters etc. This linking facilities has provided great help to the geographers, planners and policy makers to enhance the process of decision-making and allocation of resources adequately. This will expedite the process of decision-making so that the developmental process can be channelised in proper and effective manner.

Uses of GIS

A GIS user is facilitated to interface, making queries for input output at a very high speed. It is versatile user friendly technique which is being utilised not only by the geographers but also by the hydrologists, geologists, market managers, planners and policy makers as well. GIS reduces the redundancy by relevant and updated data, improves standardisation and also identifies data gaps. In GIS the data available in statistical form from various agencies can be interfaced easily with the geometric features, to show different aspects and attributes of the study area both spatially and temporally. The graphical spatio-temporal representation of attributes provides a better insight and explanation to a particular phenomenon. By this GIS enhances the role of physical planners in decision-making and policy intervention process.

- — Geographers and planners are using GIS for making Spatial Plans etc.
- — Environmentalists are using GIS for Environmental management plans.
- — Exploration and mining companies use GIS information to find out prospective areas for exploration and mining and also to analyse the orebody.
- — Power companies use GIS to monitor and analyse the electricity load on the grid network for a particular area.
- — Transportation companies use GIS to help determine the best route for delivering there good at less time.
- — Retail trade companies using GIS to determine the best location for a new outlet.
- — GIS data is now-a-days also used on the Internet for giving a graphical representation of the users querying.

GIS and Environmental Management

GIS plays a very important role in the task of monitoring and controlling the activities of potential polluters and creating an inventory of Water Quality. It helps in the formation of Environmental Information System (EIS). Development of generic decision support system helps using data bases and models for environmental planning and management at micro, meso and at macro levels. GIS helps in identifying the protected areas from environmental point of view, in perusing environmental risk assessment analysis, identifying high, medium, low risk areas. It also provides the grounds to find out the potential areas and location points for monitoring, day to day situation of environmental degradation, and remedial action to be taken to minimise the environmental hazards.

With the help of GIS one can map the existing natural resources, soil texture, soil reaction, soil degradation, soil taxonomy, organic carbon status, fertility status, land forms analysis, slope analysis, availability of water capacity, level of pollution, extension of water pollution, aquifer recharge zones, catchment areas, watersheds, agro ecological zoning, forest extension. Forest types, environmental impact assessment and existing economic activities etc.

Developmental activities as a whole, irrespective of their scale and magnitude affect the environment in short-term as well as long term. Environmental problems have attracted the attention of planners, policy makers, and politicians. It is almost a matter of concern to each and everyone to think about the increasing water pollution. The many hue and cries regarding environment pollution has started during the last three decades. India is in the grip of serious environmental problems resulting from the urban sprawl. Industrial development, increasing use of insecticides and pesticides. Water pollution from untreated domestic as well as industrial waste has drawn the attention of the planners and policy makers. The untreated wastewater is discharged into various rivers and streams resulting into the water pollution both the sources of water (Surface as well as Ground) are polluted by this. Rivers are the main source of surface water in a country iike India. In older days the human society has settled on the banks of various rivers. Due to this their religious feelings were also very much attached to these rivers. These rivers are treated as goddesses in our religious books. There are old age customs of community bathing in the rivers at various occasions. Human population has relied on these streams to fulfil their drinking, agricultural and industrial needs. But man has also used the rivers to flush out the domestic wastes and industrial effluents.

Study Area Profile

The present study is confined to the Eastern Region of Uttar Pradesh. Eastern Uttar Pradesh Region is located in the eastern part of the mother State *i.e.* Uttar Pradesh. It lies between the latitudes of 23° 52" to 28° 24" north, and between the longitudes of 81° 5" to 83°58"east. Administratively the region is divided into five divisions, namely Allahabad, Varanasi, Azamgarh, Faizabad and Gorakhpur. There were only 19 districts in the region in 1981. According to 1991 census the population of Eastern U.P. Region is 52.72 million (37.9% of total population). The population density is reported at 611 persons/sq. km which is higher than the State average of 473 persons/sq. km and much higher than the national average of 267 persons/sq. km. In this chapter an attempt is made to answer the basic questions related to GIS and how it is applicable to water quality assessment, monitoring and environmental management. The chapter also attempts to study the quality of surface and groundwater with respect to environmental pollution in Eastern Uttar Pradesh and examines the quality of river water on the basis of designated best use criteria adopted by the Central Pollution Control Board.

WATER RESOURCES

Surface Water

The study area is marked by six major rivers systems, namely, Ghaghara draining about 43.5 per cent area, Gomti covering 25.47 per cent Ganga 21.5 per cent area, Son 5.61 per cent area and Yamuna 2.61 per cent, of area. In addition to these rivers, there are small rivers and narrow perennial tributaries as well. These include Sai, Tons, Little Gandak, Sarju, Rapti, Terhi, Rohini, Manwar, Mangai, Belun, Basuhi, Chandraprabha, Rihand and Kanhar. All other drainage channels flowing through the region are seasonal in nature. Apart from the surface water the Eastern Uttar Pradesh Region is characterised by numerous lakes. Some of these lakes are large enough to form a valuable addition to the water supply of the region. There are two groups of lakes formed in this region.

(i) Lakes formed through fluvial action of rivers, and
(ii) Natural depressions.

The rivers present in this region generally form loops and bends or change their courses during rainy season. After rainy season these loops and bends are detached aside due to lack of water in these rivers and form ox-bow lakes resembling a horseshoe. These lakes are locally known as *naukhans*. Some of these ox-bow lakes are connected with the rivers by small channels and the excessive rainwater during the wet monsoon period is discharged through these channels to the main river. The *naukhans* play a dominant role to protect the surrounding areas from the danger of floods.

The second types of lakes are the natural depressions. These depressions work as the reservoirs in which the rainwater accumulates. These depressions over flow during the period of heavy rainfall form lagoons and again shrink to almost dry beds at the end rainy season. Sometimes these depressions are found in a series to form an ill-defined drainage line, which may turn into a small stream when the rainfall is abnormally high. The Chillua Tal in Gorakhpur is one such example. The major lakes found in the study area are Govind Ballabh Pant Sagar in the southern part of Sonbhadra district, Chillua Tal in the northern part of Gorakhpur district, Bakhira, Ramgarh, Domriagarh and Amical in other central districts.

The Eastern Uttar Pradesh is a part of the Ganga Basin. Only a small part of the Allahabad district falls in the Yamuna Basin. On the basis of drainage characteristics the region can be further divided into three distinct divisions. These three divisions signify three major river catchments. Apart from the three major catchments, a part of the Son Catchment also falls in the study area covering Sonbhadra district at the southern end of the study area. These catchments are further divided into sub-catchments. A list of the major catchments and sub-catchments and area covered by these catchments is given in Table 11.1.

Ground Water

Even after the enactment of water (Prevention and Control of Pollution) Act in 1974, water quality continues to deteriorate in the country. Rivers, streams and ground water are severely polluted. Ground water was supposed to be clean water and was used to provide guarantee

Table 11.1 : Major Rivers and their Catchments in Eastern Uttar Pradesh

Catchment/Sub-catchment	*Parts of the Districts*
A. The Ghaghara Catchment	
(a) River Rapti Catchment	N. Babraich, Gonda, Siddharthnagar and Maharajganj
(b) River Tehri Catchment	Bahraich and Gonda
(c) River Manwar Sub-catchment	Basti and Gonda
(d) River Rohini Sub-catchment	Maharajganj and Gorakhpur
(e) River Little Gandak Sub-catchment	Extreme East of Deoria
(f) River Ghaghra Sub-catchment	
B. The Ganga Catchment	
(a) Sub-catchment of River Belen	Allahabad, Mirzapur and Sonbhadra
(b) River Chandraprabha Sub-catchment	Eastern Part of Varanasi
(c) The Son Catchment	
(i) Rihand Sub-catchment	Sonbhadra Western Part
(ii) Kanhar Sub-catchment	Sonbhadra Eastern Part
(d) The Ganga Sub-catchment	Allahabad and Varanasi
C. The Gomti Catchment	
(a) Sub-catchment of River Sai	Pratapgarh and Jaunpur
(b) Gomti Sub-catchment	Sultanpur and Jaunpur
(c) Tons Sub-catchment	Faizabad, Azamgarh and Mau
(d) Mangai Sub-catchment	Azamgarh and Ghazipur
D. Son Catchment	Sonbhadra

against pollution, has not been spared. Groundwater is a natural resource, which plays a vital role in regional economy. Evaluation, scientific management and proper planning is a prerequisite for its judicious utilisation and conservation. Not only the pollutants are leaching in the aquifers below and polluting the groundwater but the increasing groundwater exploitation, inorganic like fluorides and arsenic present under the ground find their way in the extracted water causing thereby fluorosis and arsenic poisoning. Groundwater pollution is highly dangerous to humanity since 80 per cent of domestic water needs are fulfilled from this source.

There is adequate rainfall and porous alluvial soil cover in the Ganga Plain, which gives a perception that there is adequate storage of ground water. Groundwater reservoir occurs in the aquifers and water-bearing sediments. The watertable is lower in the newer alluvium areas, particularly in the river levee zones where the sandy nature of the sub-soil allows greater sinking of the water than in the clayey inter-fluvial areas of older alluvium. The watertable varies from 3 m to 20 m, being particularly high in terai zone. At some places artesian conditions also exist probably at a depth of 60-65 m below the surface.

Ground Water Condition

Vast resources of groundwater are available in the alluvial formation of the middle Ganga plain. The highest yield (more than 150 m^3/hour) can be observed in some isolated patches. The greater part of the extensive Gangetic plain lying mainly to the north of Ganga in Eastern Uttar Pradesh yields groundwater at a rate of more than 150 m^3 per hour. The lower part of the Ganga Yamuna Doab in Uttar Pradesh has equally high yield of groundwater. Aquifer zones of moderate yield (between 50 m^3 and 150 m^3 per hour) are found in several areas such as the upper Ganga-Yamuna tract and Ganga-Ghaghara Doab. On the basis of the level of groundwater recharge, the aquifer recharge zones can be classified into 5 major groups which are as follows:

(i) Aquifer Recharge Zone 1

This is the zone of thick aquifers with very high recharge ranging between 150 m^3 per hour or more. It can be observed from the aquifer recharge map that most part of the districts of Faizabad, Sultanpur, Pratapgarh, Varanasi, Maharajganj and Azamgarh have thick aquifers. Small patches of aquifer recharge are present in the Northern part of Gonda and Siddharthnagar districts and southern parts of Ghazipur district.

(ii) Aquifer Recharge Zone 2

This zone is characterised by the aquifers with limited extent within the capacity of 100-150 m^3/hour. The districts under this capacity are Bahraich, Gonda, Siddharthnagar, Basti, Deoria, Ballia, Mau and Ghazipur.

(iii) Aquifer Recharge Zone 3

The zone of less than 100 m^3/hour covers the small parts of Allahabad, Mirzapur and Varanasi districts.

(iv) Aquifer Recharge Zone 4

The fourth zone is characterised by the absence of aquifers in this zone. There are a very few aquifers with a very-very low recharging capability or no aquifers at all. This zone extensively covers the districts of Mirzapur and Sonbhadra. A very small portion of Allahabad and Varanasi districts also lies in this zone.

Table 11.2 : District-wise Underground Water Resources in EUP

(as on 31.3.1995)

	District	*Net Recoverable Recharge*	*Net Draft*	*% of Ground Water Utilization*
1.	Azamgarh	1261	539	42.70
2.	Allahabad	1558	515	33.10
3.	Baharaich	1949	636	32.60
4.	Ballia	1075	327	30.40
5.	Basti	1211	522	43.10
6.	Deoria	2493	576	23.10
7.	Faizabad	1568	593	37.80
8.	Gonda	2100	644	30.70
9.	Gorakhpur	1128	430	38.10
10.	Ghazipur	1295	381	29.50
11.	Jaunpur	1259	580	44.50
12.	Maharajganj	1662	228	13.70
13.	Mau	527	213	40.40
14.	Mirzapur	724	94	13.00
15.	Pratapgarh	1253	440	35.10
16.	Siddharthnagar	1139	341	29.90
17.	Sonbhadra	674	68	10.10
18.	Sultanpur	1584	413	26.10
19.	Varanasi	1334	450	33.70
	Total	25794	7990	29.38

Source : Economics and Statistics Division, State Planning Institute, Uttar Pradesh.

(v) Aquifer Recharge Zone 5

This zone covers aquifer recharge areas other than the surface soil recharge such as the riverbeds, reservoirbeds and lakes etc.

In the remaining part of the Central Indian uplands and on the higher slopes of Himalayas within the Ganga basin the available aquifers are found scattered in local pockets where the yield of groundwater is normally less than 20 m^3 per hour. A special zone yielding water under artesian condition at a high rate of more than 150 m^3 an hour is found in the lower Ganga-Yamuna doab.

WATER POLLUTION

Water pollution can be defined as physical or chemical changes in water that can adversely effect organisms. Water pollution is characterised by the deterioration of the quality of water, primarily due to human activities. The activities such as mining, agriculture, stock breeding, fisheries, forestry, urbanisation, industrialisation, construction etc. are all responsible for water pollution to some or the other extent. One of the most significant environmental problems relates to the pollution of ground and surface water resources. The primary sources of water pollution include the following:

— domestic sewage,
— industrial waste water, and
— run-off from agricultural fields.

The critical nature of the problem was realised by the government in 1980s following which the 'Ganga Action Plan' was formulated. Later, a similar action plan was made for River Gomti as well. Yet, the quality of water in these rivers is far below the best designated uses. Thus, the problem of water pollution is by far one of the most critical areas of environmental degradation, which needs to be urgently addressed because it effects areas outside the regional boundaries as well. If not managed comprehensively, most of the areas are likely to suffer damage that investments will not be able to reverse. Lack of resources and ineffective management are more pronounced in smaller urban municipalities, which have smaller financial bases, limited management capacity, low technical expertise and little motivation. The environmental status of urban areas highlights a similar scenario, where in the smaller class towns represent a more critical scenario.

Major Water Pollutants

The following are the major water pollutants:

1. Oxygen demanding wastes (domestic sewage, cattle waste, industrial wastes).
2. Disease causing agents (bacteria, parasites, virus).
3. Inorganic chemicals and minerals (acids, salts and toxic metals).
4. Organic chemicals (pesticides, plastics, detergents, industrial wastes and oil).
5. Plant nutrients (nitrates and phosphates).
6. Sediments (soils, silt and other solids from land erosion).
7. Radioactive substances.
8. Heat (from industrial and power plants).

The water pollutants can be classified into

1. Rapidly degradable (non-persistent),
2. Slowly degradable (persistent), and
3. Non-degradable pollutants.

The industrial plants discharge their wastes directly into streams and nearby rivers creating a major hazard for both natural and man-made environment. Water pollution due to discharge of industrial effluents changes the water quality drastically. Industrial activities generate large quantities of solid wastes, which too are let off into the rivers thereby increasing the suspended particulate matter in the water.

Use of River Water

Rivers are the main sources of surface water in a country and so human beings not only preferred to settie along it but also developed a sentimental bond with it. In India, divinity is attached to a river—be it the Ganga in the north or Godavari or Cauvery in the south. Community and ritual bathing in rivers is an age-old function. At various points many towns grew up to facilitate temporary and permanent stay for religious purposes. Human beings also relied on the river as dependable sources for supplying water for drinking, irrigation and industrial purposes. The river is a natural channel for draining out the excess water on the land including domestic as well as industrial effluents. It also endows fish as a protein-rich food for man. The non-turbulent stretches provide for inland navigation as well. Thus management of water resources can effectively help in solving some of the problems, which often transcend regional boundaries.

Table 11.3 : Major Sources of Water Pollution

Content	*Source (Industries)*
BOD content	Food processing industry, paper industry, ceramic industries, acids or alkalis, mining, paper and pulp industry and other organic chemical industry.
Sulphides	Pulp and gas industries, sewage treatment plants, dyeing and other leather industries.
Colour	Dyestuff, dyeing, paint, leather, chemical industries.
Odour	Chemical, pulp, marine product processing, sugar and meat processing.
Detergent	Maling leatrerint sewerage, textile industries.
Hot water	Steam and nuclear power plants.

Source : Central Pollution Control Board Report, 1991.

Quality of River Water

Surface water is the main source for drawing water and discharging effluents for industries. The water quality in surface waterbodies depends upon the discharges at various points. Based on the water quality criteria, Central Pollution Control Board (CPCB) has designated water uses for all the Indian rivers. The surface water has been classified into five categories for various uses, which are given in Table 11.4.

Analysis of the water quality status in terms of designated best use shows that :

— River Ganga at Allahabad (Rasoolabad) was noted to have designated best use as class B during 1993 and 1994 it was the same as recommended while, at downstream Allahabad (Sangam) it was class B in 1993 and recorded as class C in 1994. Similar case was observed at the third and fourth stations of Ganga which were taken at Varanasi U/S

(Assighat) and Varanasi D/S. It was all due to the confluence of Ganga and

Table 11.4 : Primary Water Quality Criteria for Various Uses of Fresh Waters

Designated Best Use	*Class*	*Criteria*
Drinking water source without conventional treatment but after disinfection	A	1. Total Coliform Organism MPN/100 ml. shall be 50 or less 2. *p*H between 6.5 and 8.5 3. Dissolved Oxygen 6 mg/*l* or more 4. Biochemical Oxygen Demand 5 days 20° C 2 mg/*l* or less.
Outdoor Bathing (Organised)	B	1. Total Coliform Organism MPN/100 ml. shall be between 50-500. 2. *p*H between 6.5 and 8.5 3. Dissolved Oxygen 5 mg/*l* or more 4. Biochemical Oxygen Demand 5 days 20° C 3 mg/*l* or less.
Drinking Water Sources with conventional treatment and disinfection	C	1. Total Coliform Organism MPN/100 ml. shall be between 500-5000. 2. *p*H between 6.0 & 9.0 3. Dissolved Oxygen 4 mg/*l* or more 4. Biochemical oxygen Demand 5 days 20° C 3 mg/l or less.
Propagation of Wild life		1. *p*H between 6.5 and 8.5 2. Dissolved Oxygen 4 mg/*l* or more 3. Free Ammonia (as N) 1.2 mg/*l* or less.
Irrigation, Controlled Cooling, Controlled Waste		1. *p*H between 6.0 and 9.0 2. Electrical Conductivity at 25°C micro mhos/cm maximum 2250 3. Sodium Absorption Ratio, maximum 26 4. Boron, maximum 2 mg/*l*

Source : Water Quality—Status and Statistics (1993-1994), Central Pollution Control Board.

Yamuna , since the Yamuna has C class water which is in larger quantity than the Ganga River Water.

— The River Ghaghra was observed at two places, namely Jaunpur D/S and Varanasi. It was found that the quality of water of the Ghaghra river has declined from Jaunpur to

Table 11.5 : Water Quality Status of the Rivers, monitored in 1993 and 1994

River/Place of Monitoring	*Desired Class*	*Existing Class*		*Parameters 1994*
		1993	*1994*	
River Ganga				
Allahabad (Rassolabad)	B	B	B	T. Coliform
D/S Allahabad	B	B	C	T. Coliform
U/S Varanasi (Assighat)	B	B	C	T. Coliform
D/S Varanasi	B	B	C	T. Coliform
Trighat (Ghazipur)	B	B	B	T. Coliform
River Ghaghra				
D/S Jaunpur	C	C	C	T. Coliform
Varanasi	C	D	D	T. Coliform
River Sai				
Unnao After Drain Outfall	C	D	D	T. Coliform
River Rapti				
After Confluence of River	B	E	E	T. Coliform
River Yamuna				
Allahabad	B	B	C	T. Coliform
D/S Allahabad (Balua Ghat)	B	B	D	T. Coliform

Source : Water Quality — Status and Statistics (1993-1994),Central Pollution Control Board.

Varanasi. C class water was found at Jaunpur and it deteriorated to D class at Varanasi during both the periods *i.e.* 1993-1994. At Jaunpur the water quality was found as designated.

— The water of river Sai has also been polluted and its quality degraded from designated best use *i.e.,* from class C to class D.

— A very high level of pollution was observed in the River Rapti after the confluence of River Rohin. Water quality in the river degraded highly *i.e.* from B class as designated to E class during 1993 and 1994.

— Water of the river Yamuna contributing a larger volume to

the river Ganga has reflected a very high level of pollution. Its class has been deteriorated from designated class B to class C at Allahabad U/S and class D at Allahabad D/S during 1994.

Pollution Generation in Eastern Uttar Pradesh

The weather in the Ganga basin is characterised by a distinct wet season, during which more than 70 per cent of the total annual rainfall is received (June to September). The water flow in the streams witnesses great fluctuation in the flow pattern during the wet and dry seasons. In the pre-monsoon months, the flow of even major tributaries falls to such a low level that concentration of pollutants rises to such a high degree that it crosses all critical limits. For this reason, the smaller tributaries especially those originating in the Peninsular upland and the upper reach of Ganga are effected intensely by pollutant discharges. Some of Ganga's major tributaries like Yamuna, Ghaghra, Kosi, Gandak, and Son are large enough by themselves to effectively bring down the pollution level by dilution alone.

Pollution load of the districts can be attributed to both Urban as well as Rural pollution. District-wise contribution of urban domestic and urban industrial pollution load show that the total urban load is highest from Deoria district (86,559 kg/day) followed by the district of Varanasi (34,705 kg/day). The districts of Allahabad and Gorakhpur have similar pollution load by urban areas (*i.e.* 27,500 kg/day) Ghazipur district has about 20,191 kg/day of load suburban centres district of Gonda, Basti and Faizabad have less urban contribution of pollution *i.e.* 9,289, 9,587 and 9,767 kg/day). The District of Ballia has the lowest urban contribution to the pollution load *i.e.* 4,837 kg/day, Azamgarh, Jaunpur and Mirzapur have urban load geration of 4985, 5776 and 6604 kg/day respectively.

Share of domestic and industrial wastes in the total waste load has been analysed. The analysis shows that 97 per cent of industrial pollution load is generated by the district of Deoria, followed by the

Table 11.6: Contribution of Urban Domestic and Industrial Load

	District	*Waste water Kl /day*	*Urban BOD Load, Kilograms/day*	
			Domestic	*Industrial*
1.	Azamgarh	7357	3920	1065
2.	Allahabad	86639	22991	4849
3.	Baharaich	7930	3046	5246
4.	Ballia	3576	1837	3021
5.	Basti	5503	2165	1877
6.	Deoria	7288	2776	83783
7.	Faizabad	17001	5969	3798
8.	Gonda	5003	2243	7046
9.	Gorakhpur	29828	8826	18511
10.	Ghazipur	5043	1932	18259
11.	Jaunpur	7221	3676	2101
12.	Maharajganj	N.A	N.A.	N.A.
13.	Mau	N.A	N.A.	N.A.
14.	Mirzapur	12680	3532	772
15.	Pratapgarh	2485	981	36
16.	Siddharthnagar	N.A.	N.A.	N.A.
17.	Sonbhadra	N.A.	2501	1892
18.	Sultanpur	2728	1055	—
19.	Varanasi	114136	30563	4142
	Total	116864	34119	4142

Source: CPCB. Ganga Basin Report 1993-94.

districts of Ghazipur which generated 90 per cent of industrial pollution load 76 per cent and 77 per cent of industrial load is generated by the district of Basti and Gonda, Gorakhpur and Ballia generate industrial load ranging from 62 per cent to 68 per cent. The districts of Allahabad, Varanasi and Mirzapur generates only about 12 per cent and 17 per cent of industrial load study of the domestic pollution load generation in various districts shows that 88 per cent of domestic pollution load is generated by the districts of Allahabad and Varanasi followed by 83 per cent contribution by Mirzapur district. Jaunpur and Faizabad districts shows 61 per cent and 64 per cent of domestic load contribution, Gonda and Basti have similar contribution, 23 per cent and 24 per cent respectively. Least contribution by domestic load is observed in Deoria district followed by 10 per cent Ghazipur district.

From the perusal of Table 11.6 it is clear that Varanasi district has highest percentage of waste water generation from urban areas followed by Allahabad district. Low waste water generation is observed for the district of Pratapgarh, Sultanpur followed by Gonda, Ghazipur and Gorakhpur.

Industrial Pollution Load

Eastern Uttar Pradesh has various categories of industries within its jurisdiction. The contribution of small-scale industrial units is more than 90 per cent in the total industrial production of the region. In terms of typology, textile and food based industries overshadow the total industrial scenario of Eastern Uttar Pradesh. As mentioned previously, the Uttar Pradesh Pollution Control Board has listed down the polluting industries in the State. There are various standards that have been laid down by the agency to control massive, haphazard and unplanned growth of the industries in the State. Some of the major pollution generating industries in the region are textile industries, food processing and mechanical engineering units.

(a) Pollution Generation — Small Scale Industrial Units

The SSI units generate maximum pollution among all industries — the load being highest from textile and food processing based industries. The districts of Allahabad (50.17%), Gorakhpur (19.55%), Maharajganj (27.7%), Mirzapur (55.6%), Sonbhadra (33.28%), Varanasi (32.5%) and Sultanpur (26.5%) account for highest share in the small scale units and contribute to pollution load quite significantly through them. The food processing small-scale units are maximum in Bahraich (40% of the total industrial units), Deoria (44.4%), Basti (24.5%), Faizabad (39.4%), Gonda (39.8%) and Maharajganj (39.6%), which forms one of the main source of pollution in these districts.

(b) Pollution Generation from Large and Medium Scale Industrial Units

The pollution generating industries among large and medium scale units include mechanical engineering, chemical, textile and food processing units. The districts of Allahabad, Sonbhadra, Basti, Deoria, Faizabad, Gonda, and Jaunpur account for highest proportion of large-scale industries. The maximum percentage of engineering units is concentrated in Allahabad district while chemical industries are concentrated in the districts of Allahabad, Sonbhadra, Sultanpur and Varanasi. Food processing units are concentrated in Varanasi, Jaunpur, Maharajganj, Ghazipur, Gonda, Bahraich and Basti. Large and medium scale textile industries are maximum in the district of Varanasi.

The pollution generated from LSI/MSI and SSI units has significant impact on the environment. The textile industries are hazardous in the sense that water effluents are being directly discharged into the river channels—specially in the district of Varanasi, Allahabad and Gorakhpur. Large and medium scale, mechanical-engineering industries are maximum in Allahabad and Sonbhadra create major problems of water, air and noise pollution in the districts. The chemical industries are maximum in the district of Sonbhadra creating severe water, land and air pollution problems. Textile and food processing small-scale units are responsible for immense pollution problem in the region. Though LSI/MSI units are the ones which discharge large amount of industrial effluents into the water channels specially in Allahabad, Varanasi, Gorakhpur, Ghazipur and Faizabad, the SSI units have shown a much greater impact on the environment primarily due to their unplanned, illegal and dispersed growth.

(c) Pollution Generation from Mining

The mining activities in the study area are concentrated in Mirzapur and southern part of Sonbhadra district. Kakri, Bina and Marrak mine blocks are in Mirzapur district and 8 mining blocks are west of Sonbhadra district (beyond the State and the study area boundary), which have an impact on the water. The pollution from mining activity is maximum in Mirzapur and Sonbhadra only, as other districts of the study area do not have any extensive mineral reserve that can be commercially exploited.

Large scale mining operation brings about changes in the surface drainage pattern—affecting significantly both quantity and quality of the surface water. Due to deforestation, soil becomes prone to erosion and increased run off takes place. The surface water channels get clogged due to boulders, stones, and other surface material along with toxic mine wastes. In the study area, the major water body severely affected by mining activity is Govind Ballabh Pant Sagar (Rihand river which drains out from the Govind Ballabh Pant Sagar is also affected).

Overburden dumps are prone to gully and rill erosion during the rainy season as a result of which large quantities of sediments are carried away to the lake. Mine roads and coal piles also seriously disrupt the natural drainage and they are often an important source of sediments. Moreover, run off water from the mine fields carry very high loads of suspended particles and minerals. Apart from this, discharge waters from mines become acidic due to their contact with iron sulphide and pyrites. Therefore, discharge of acidic water in large volumes into the natural drainage system contaminates surface water. Run off water containing fine coal particles increase turbidity, which reduces light penetration and thus adversely effects photosynthesis in aquatic plants. Presence of large amount of suspended particulate matter adversely affects the aquatic organisms.

Thus, the natural drainage and waterbodies are affected through–

• clogging of stream / channels • reduces storage capacity of the resources • destruction of aquatic habitats	• increase in water treatment cost • increase in suspended particulate matter load

(d) Pollution from Thermal Power Plants

Sources of water pollution from the thermal power plants mainly include—the ash dykes and discharge of hot water. It is estimated that 18,865 tonnes of fly ash is being produced in Sonbhadra district. For this purpose, fine ash ponds covering the area of more than 3319 acres have been constructed in the area around lake and ash slurry is transported by water slicking into these ponds.

The silt from ash ponds on hill slopes along the lake gets washed off into the lakes with the rain, thereby increasing the problem of water pollution. Increased quantities of fly ash in water increases turbidity and limits sunlight penetration which affects aquatic plants and organisms. Leached water from the ash ponds seeps through the soil carrying toxic heavy metal ions and poses danger to the Pant Sagar lake, which is the only source of drinking water in the area.

Along with fly ash, thermal power plants also produce varying amounts of low density minute hallow sphere ash called Ionosphere, which normally floats on water adding to the water pollution in the reservoir. Besides fly ash disposal, the lake is also used as a cooling pond and waste heat is discharged into it by almost all the thermal power plants of the region. The waste heat discharge into the reservoir changes the thermal characteristics of the waterbody, producing detrimental effect on aquatic flora and fauna and leads to change in the micro-climate. Increase in temperature of the lake reduces dissolved oxygen in water which influences biological activities such as feeding habits and reproductive behaviour in fish and other aquatic organisms. Many harmful chemicals and substances like oil and grease too find their way into the lake adding to water pollution. The water of the lake is also withdrawn for various operations and this large-scale withdrawal of water reduces the total volume and thus, the concentration of pollutants cause a detrimental impact on the aquatic life.

CONCLUSIONS

Considering the pollution load into account, it can be concluded that few districts are more critical than others in terms of pollution from domestic as well as industrial and mining activities. These are the districts which need careful planning and close monitoring so as to cause no or the least possible damage to environment.

- Comparative analysis of district-wise pollution generation from industrial activities shows that Deoria district have very high level of pollution followed by Ghazipur, Gonda and Gorakhpur. The districts of Allahabad, Bahraich and Varanasi show moderate level of pollution.
- Pollution generation from thermal power generation and mining activities is very high in Sonbhadra and Mirzapur districts.
- The districts of Allahabad and Varanasi have very high domestic pollution, followed by the district of Gorakhpur. Moderate level of pollution can be identified in case of Deoria, Mirzapur and Sonbhadra districts.

The following districts are found to be critical on the basis of pollution generation load from various economic activities like industrial, mining and power generation etc.

Very Critical	Allahabad, Varanasi, Sonbhadra and Varanasi
Moderately Critical	Bahraich, Gonda, Deoria, Mirzapur and Gorakhpur
Critical	Ghazipur and Faizabad

Impact on water system is found to be very high in Varanasi and Gorakhpur and high in the districts of Allahabad and Ghazipur. Impact on groundwater is high in case of Maharajganj, Gonda and Bahraich districts.

ENVIRONMENT, AGRICULTURE
AND
POVERTY

About the Editors

Ali Mohammad, M.A., PG. D. is former Chairman, Department of Geography, Aligarh Muslim University, Aligarh. His field of specialization is agricultural geography, food security and nutritional problems in rural India. Earlier he has worked as a Senior Research fellow in Jawaharlal Nehru University; as an Associate Professor at the Centre of West Asian Studies, Aligarh Muslim University, Aligarh; Associate Professor in Egerton University, Kenya; and Professor in Asmara University East Africa, Eritrea, Dr. Ali Mohammad has to his credit numerous research papers and more than one dozen books on agriculture, food security and nutritional deficiency.

Abdul Munir, Senior Reader, Department of Geography, Aligarh Muslim University, Aligarh, U.P.

S.H. Siddiqui, Senior Reader, Department of Geography, Aligarh Muslim University, Aligarh, U.P.

Rocket Ibrahim, Director, Jamia Millia Islamia, (Central University), Jamia Nagar, New Delhi.

ENVIRONMENT, AGRICULTURE
AND
POVERTY

VOLUME TWO

Editors
Ali Mohammad
Abdul Munir
Shamsul Haque Siddiqui
Rocket Ibrahim

CONCEPT PUBLISHING COMPANY, NEW DELHI-110059

ISBN

978-81-8069-270-3 (Set)
978-81-8069-285-7 (Vol. I)
978-81-8069-286-4 (Vol. II)
978-81-8069-287-1 (Vol. III)

Edition 2016

Published and Printed by

Concept
Since 1974

Concept Publishing Company Pvt. Ltd.
A/15-16, Commercial Block, Mohan Garden
New Delhi-110059 (India)
T: 25351460, 25351794, *F*: 091-11-25357109
E: publishing@conceptpub.com, W: www.conceptpub.com
Editorial Office: H-13, Bali Nagar, New Delhi-110 015, India.

Cataloging in Publication Data--Courtesy: D.K. Agencies (P) Ltd. <docinfo@dkagencies.com>

Environment, agriculture and poverty / editors, Ali Mohammad ... [et al.].
3 v. cm.
Papers presented at a conference.
Includes bibliographical references.
Includes index.
ISBN 9788180692703

1. Agriculture—Environmental aspects—India—Congresses. 2. Agriculture—Environmental aspects—Asia—Congresses. 3. Poverty—Government policy—India—Congresses. 4. Poverty—Government policy—Asia—Congresses. I. Ali Mohammad, 1946-

DDC 630.954 22

INTRODUCTION

The future of humanity is at critical juncture as the population continues to grow at a high rate in most of the regions surrounding South Asia, which is characterized as world's most deficit region with high level of poverty. The looming shortage of basic needs has compelled the man to irrational exploitation of his immediate surrounding, *i.e.*, natural resources available which has adversely affected the environment. In regional context increasing greenhouse gas emission, ozone depletion and acid rain, loss of bio-diversity, deforestation and loss of top soil and fertility, shortage of water, food and fuel indicate how the natural systems are being pushed even closer to their limits of exploitation.

In order to fulfil the basic needs and to have a hunger free society, we should harness the available resources and the public distribution system in a judicious manner.

In this country and other surrounding countries agriculture is still a main source of survival but highly affected by environmental phenomenon because of the poor infrastructure and poor socio-economic base. With the increase of population and other demographic characteristics we want to raise more food, raw material and for other purpose such as food security and changing demand. Therefore, we have to produce very much more per hectare because we have very little more land, which we can bring under cultivation. Therefore, multiple cropping and intensive input based agriculture is going to be surely a characteristic of the future. This agricultural system require intensive use of chemical fertilizers and irrigation that leads to ecological imbalances. What are going to be the long-term effects of some of these changes which we try to introduce? The high yielding varieties have made a great impact which has aroused the enthusiasm of the farmers but it needs high amount of modern inputs. Probably under these conditions there will be more outbreaks of pests and diseases in the severe form than in the past. What will be the long range effects of some of these practices?

In spite of substantial increase in global food production, FAO has estimated that about 790 million people in the developing countries and about 34 million people in developed and transition economies were suffering from under-nourishment during 1995-1997.

Approximately, 75 per cent of these food deficit population is located in Asia, the Pacific and sub-Sahara region. The largest number of food insecure people is in India (about 200 million).

Agriculture alone cannot solve the problem of hunger and poverty unless and until other supporting factors are not applied. The agricultural production should reach to the weaker sections of the society. Moreover, efforts should be made to generate employment in rural areas. This will improve the purchasing power of the rural masses, particularly the landless labours and artisans.

The regional dimension of environmental degradation is also very dangerous for its inhabitants throughout the world. Change of the environment in one region of country affected others and the problems had to be tackled on a global or regional scale. The issue of environmental diversity with diversified pattern of resource endowments should be understood according to the need of the people living in the region.

An integrated approach is needed to have a sustainable development of all the resources available in the diversified environment, so that it should be viable for man in anchoring the poverty.

About 65 per cent of the livelihood of the people is dependent on agriculture in India by providing employment but poverty has become a way of life for those totally dependent on agriculture because of the small and marginal size of landholding. It is also observed that per hectare productivity has increased two and half times after independence and degradation of environment, especially the soil nutrients depletion has become a major issue for the sustainable agriculture. Input-output ratio is being disturbed in intensive agricultural pockets and it is observed that the model of Punjab agricultural development is not favourable to continue. It shows that environment should be protected in a situation where India is characterised as the home of the largest number of poorest population which is not at all desirable. Therefore, poverty alleviation programme, schemes and planning will affect the environment. But at the cost of development, environment should be protected or not, poverty will prevail at mass level in India.

ALI MOHAMMAD

CONTENTS

Volume One

ENVIRONMENT

VOLUME TWO

AGRICULTURE

Volume Three

POVERTY

CONTRIBUTORS

A. Taheri Tizro
College of Agriculture,
Razi University,
Kermanshah, Iran.

A.K. Srivastava
Haryana Institute of Public Administration,
Sector 18, Gurgaon,
Haryana.

Abdul Munir
Senior Reader,
Department of Geography,
Aligarh Muslim University,
Aligarh.

Abdul Rauf Shah
Research Associate,
Geography Unit,
Center of Central Asian Studies,
University of Kashmir,
Srinagar (J&K).

Agha M. A. Siddiqui
Department of Architecture and Planning,
University of Roorkee,
Roorkee-247667.

Ahmad Ali Bozorgzad
Islamic Azad University,
Najafabad, Esfahan,
Iran.

Ahmad Mujtaba Siddiqui
Research Scholar,
Department of Geography,
Aligarh Muslim University,
Aligarh.

Ali Mohammad,
Professor, Department of Geography,
Aligarh Muslim University,
Aligarh (U.P.)

Alpana Chandra Scott Lecturer,
I.T. Degree College,
Lucknow.

Archana Gupta
Department of Grography,
Allahabad University,
Allahabad.

B.R. Kareriya
Research Scholar,
Department of Geography,
Banaras Hindu University,
Varanasi.

B.R.K. Sinha
Reader, Department of Geography,
Visva Bharti University,
Shantiniketan-731235.

Balak Ram
Central Arid Zone Research Institute,
Jodhpur-342003.

Beik Mohammadi
Department of Geography,
University of Isfahan,
Iran

D.D. Vishwakarma
Assistant Professor,
Department of Geography,
Government P.G. College,
Chhindwara, (M.P.).

D.K. Mishra
Department of Grography,
Delhi School of Economics,
University of Delhi,
Delhi.

G.M. Mir
Reader, Geography Unit,
Centre of Central Asian Studies,
University of Kashmir,
Srinagar (J&K).

G.S. Chauhan
B.R. Ambedkar College,
(University of Delhi),
Geeta Colony,
Delhi-110 003.

Gheesa Lal
Central Arid Zone Research Institute,
Jodhpur-342003.

H.N. Mishra
Department of Geography,
Allahabad University,
Allahabad.

J.S. Chauhan
Central Arid Zone Research Institute,
Jodhpur-342003.

K.V. Sundaram
Founder Chairman,
Bhoovigyan Vikas Foundation,
641, Sector-A,
Pocket-C, Vasant Kunj,
New Delhi.

Kumar Nikhil
Scientist,
Environmental Management Group,
Central Mining Research Institute,
Dhanbad, Bihar.

M. Ahmad
Scientist,
Environmental Management Group,
Central Mining Research Institute,
Dhanbad (Bihar).

M.P. Tripathi
Department of Geography,
Tribhuwan University,
P.N. Campus, Pokhra,
Nepal

Mehdi Momeni
Najafabad Islamic Azad University,
Najafabad, Iran.

Mohammad Jafar Zomorrodian
Department of Geography,
Ferdowsi University Mashhad,
Mashhad, Iran.

Mohammad Shafi
(Prof. Emeritus) (Padam Shri)
Department of Geography,
Aligarh Muslim University,
Aligarh.

Mohd. Taufique
Research Scholar,
Department of Geography,
Aligarh Muslim University,
Aligarh.

Mumtaj Ahmed
Research Scholar,
Department of Geography,
Aligarh Muslim University,
Aligarh.

Nizamuddin Khan
Senior Lecturer,
Department of Geography,
Aligarh Muslim University,
Aligarh.

R.B. Singh
Department of Geography,
Delhi School of Economics,
University of Delhi,
Delhi-110007.

R.S. Dixit
Reader-Head,
Post-Graduate Department of Geography,
Isabella P.G. College,
Lucknow University,
Lucknow, U.P.

R.S. Sangwan
Department of Geography,
University College (M.D.U.),
Rohtak, Haryana.

Rais Ahmad
Reader,
Department of Agricultural Economics and Business Management,
Aligarh Muslim University,
Aligarh (U.P.).

Rajanish Kumar
G.B. Pant Social Science Institute,
Jhusi-Allahabad.

Rocket Ibrahim
Professor,
Department of Geography,
Aligarh Muslim University,
Aligarh (U.P.)

S. Hasan Ansari
Reader,
Department of Geography,
M.D. University,
Rohtak (Haryana).

S.B. Singh
Head,
Department of Geography,
Banaras Hindu University,
Varanasi.

S.W.A. Ashraf
Professor,
Department of Geography,
Aligarh Muslim University,
Aligarh (U.P.)

Saifuddin
Remote Sensing Application Centre,
7-B, Old Shibli Road,
Aligarh Muslim University,
Aligarh.

Sayed Hassan Hossaini Abari
Department of Geography,
Esfahan University,
Esfahan, Iran.

Shamsul Haque Siddiqui
Senior Reader,
Department of Geography,
Aligarh Muslim University,
Aligarh.

Soorya Lal Amatya
Department of Geography,
University of Kathmandu,
Kathmandu, Nepal.

Sumanta Chatterjee
Research Scholar,
Department of Geography,
Visva Bharti University,
Shantiniketan-731235.

Tapeshwar Singh
Reader,
Department of Geography,
M.M.H. College (C.C.S. University),
Ghaziabad (U.P.).

Yasir Saeed Hanafi
Research Scholar,
Department of Geography,
Aligarh Muslim University, Aligarh.

AGRICULTURE

12

APPLICATION OF GIS TECHNIQUE FOR SUSTAINABLE AGRICULTURAL DEVELOPMENT IN SEMI-ARID REGION

A Study of Bharatpur, (Rajasthan), India

SAIFUDDIN

INTRODUCTION

Out of the total geographical area of India of 329 million hectares, the net cultivated area is 143 million hectares which may shrink to 120 million hectares by the turn of the century due to the process of land degradation (Sehgal and Sharma, 1994). The main themes addressed in the project are : *(a)* Groundwater, *(b)* Soil surface texture, and *(c)* Geomorphology.

LOCATION

The area forms the western most district of Rajasthan. It lies in the semi-arid interface and lies between the 26°40' to 27°50' N latitude and 76°55' and 77°45' E longitude. There are ten tehsils in the district *viz.* Pahari, Kaman, Nagar, Deeg, Bharatpur, Bayana, Nadbai, Weir, Rupbas and Kumher consisting of 1,454 villages.

PHYSIOGRAPHY, RELIEF AND CLIMATE

The general slope of the area is from northwest to southeast. The climate of the region is semi-arid. The average annual rainfall is 680 mm. The rainfall is mainly varies from June to August. The highest temperature varies from 49°C during the month of June to lowest of 2°C during the month of January.

GEOLOGY AND GEOMORPHOLOGY

The geology of the area had been described after systematic geological mapping in the study area, the succession given by Singh has been modified by Saifuddin, 1998. The geological succession of the area is given as:

Banganga Recent Alluvium Formation
Bharatpur Loam Deposit
Aligarh Older Alluvial Formation
Vindhyans
—Upper Bhandar Sandstone Formation
—Upper Rewa Sandstone Formation
Arauli Formation
Weir Formation (Quartzite)
Damdama Formation
—Lakhanpur Sandstone
—Kanawar Quartzite
—Umraind Conglomerate
Bayana Formation
—Mahloni Conglomerate
—Mortalab Quartzite
Badalgarh Formation
—Alapuri Quartzite
—Bagrain Sandstone
Jogipura Formation
—Sita Conglomerate
—Quartzite
Jahaz Govindpura Volcanics
Nithar Formation
Delhis
—Bharkol Formation
—Pratapgarh Formation
Pre-Delhi

The Pre-Delhis underlie the Nithar formation exposed in Nithar area in Bayan tehsil of the district which are overlain by Jahaz Govindpura Volcanics. These formations indicate a massive volcanism the tuffs 50 to 100 m thick are observed at Jahaz village the volcanic bombs and lapalli of 3 to 5 cm size are reported (Iqbaluddin, 1996).

The Delhi outcrop are seen in Pahari and Kaman tehsils and in Weir. The southern part of the Bayana tehsil is covered by Upper Rewa Sandstone. The hard rock geology is restricted to Pahari, Kaman and Bayana. The major portion of the district is under thick alluvial cover of Banganga and Ghambhir rivers referred to as Banganga Recent Alluvial Formation. The major part of the district is under Aligarh Older Alluvial Formation (occupying the northeastern part of the district) and Bharatpur Loam Deposit (occupying the southwestern part of the district).

Geomorphologically the area consists of Vindhyan Plateau, Cuesta and Hog Back ridges belonging to Delhis and Bayan and Erosional hills. Erosional Valleys, Alluvial Plain of Ghambhir and Banganga river, Aligarh Older Alluvial Plain consisting of Paleo-channels, Buried Pediment (Fig. 12.1).

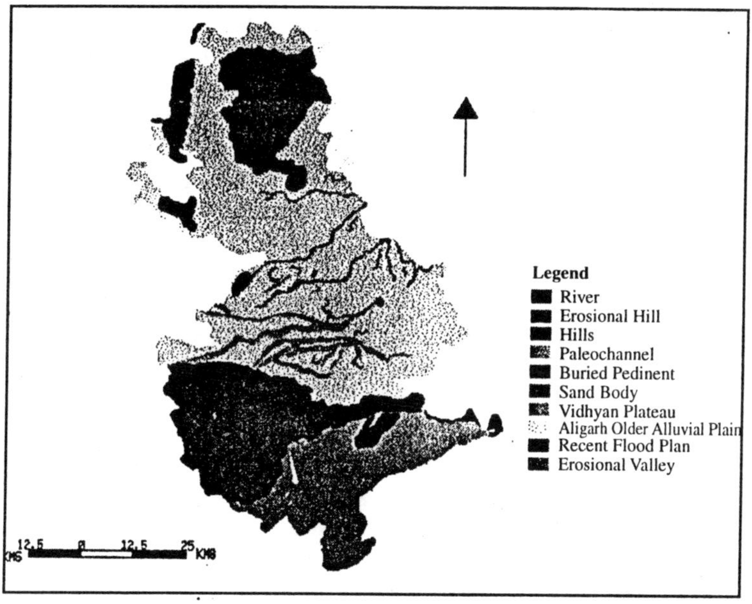

Fig. 12.1 : Geomorphological Map of Bharatpur District, Rajasthan

The surface soil texture was studied and it was classified as under as Loamy Sand, Sandy Loam, Loam, Silt Loam, Clay Loam, Silty Clay Loam and Hills and Ridges.

SURVEY AND MAPPING PROCEDURES

(i) LANDSAT TM FCC path-row 146-041 band 2, 3 and 4 on 1:250,000 scale of 9th March, 1990.

(ii) Survey of India toposheets pertaining to the area on 1:250,000.

(iii) Geological and Geomorphological maps of the study area of Geological Survey of India.

The geological and geomorphological maps of the area were prepared using the LANDSAT TM FCC. The pre-field maps were corrected by selective ground check.

The soils were interpreted by visual interpretation and the indirect relationship between the individual photo elements led to recognization of different soil types. The prepared map was used as a pre-field map for ground truthing thus selective reconnaissance soil survey of the area was undertaken following methodology outlined in Soil Survey Manual (All India Soil and Landuse Survey, 1972). Soil profiles were studied by exposing upto 1.5 metres depth. The surface soil texture map was prepared which was used during the present study.

Groundwater map was generated by taking 168 samples randomly from dug wells and the depth to watertable map was generated by extrapolation of point data by inverse distance method. The computer generated output maps were produced.

The soil and water samples were analysed in the Remote Sensing Applications Centre for Resource Evaluation and Geo-engineering, Aligarh Muslim University, Aligarh.

GEOGRAPHICAL INFORMATION SYSTEM (GIS)

GIS is considered as a tool for effective and efficient storage and manipulation of spatial and non-spatial data. It is defined as a potential set of tools for collecting, storing, retrieving at will, transforming and display spatial data from real world (Burrough, 1987). The word 'geographic' implies that the location of data items are known, or can be calculated in terms of geographic coordinates, 'information' implies that the data are organised to yield useful knowledge and word 'system' implies that it is made up from several linked components with different functions.

Modular GIS Environment (MGE) was used in the present study.

The systematic procedure for creating a project in MGE is given below:

(*a*) Creating directory
(*b*) Create Schema
(*c*) Locate Schema
(*d*) Creating Project
(*e*) Defining MSCO Working units and Registration of maps
(*f*) Digitization of maps
(*g*) Feature building
(*h*) Map cleaning
(*i*) Graphic processing
(*j*) Attribute processing
(*k*) Non-Spatial data entry
(*l*) Building a topological file
(*m*) Creating query set
(*n*) Query display
(*o*) Overlay analysis

The study was carried out under Natural Resource Informatics System programme of National Informatics Centre, Planning Commission, Government of India.

The Thematic maps were prepared on 1:250,000 scale using LANDSAT TM FCC 2, 3, 4 path-row 146-041 of 9th March, 1990.

The base map was prepared using Survey of India toposheets No. 54A, 54E and 54F on 1:250,000 scale.

METHODOLOGY

Thematics maps were generated using LANDSAT TM FCC 2, 3 and 4 path-row 146-041 on 1:250,000 scale. The maps were generated for Geology, Geomorphology, Soils, Landuse, Wastelands and Groundwater.

The Natural Resource Data Base created for various files on geology, geomorphology, wastelands, groundwater, soil, etc. have facilitated the synoptic viewing of the district into Terrain Mapping Units which constitute geographically homogeneous natural entities (Saifuddin, 1998; Saifuddin, 1999; Iqbaluddin *et al.*, 1999). The conventional approach for GIS in India try to map out area for planning and development on the basis of co-ordinates or village boundaries etc. which are artificial and man-made. Hence, the area development is burden with artificial fences. The methodology of Terrain Mapping Units based on natural geographic boundaries will help in planning agricultural inputs for developmental needs of a basically agro-economic eco-system like Bharatpur.

Conventional methodology of inputs on the basis of tehsil and district will require to be changed, at the micro-level individual TMU require separate treatment by way of agricultural inputs in terms of seed, fertilizer and irrigation inputs for optimum outputs.

The concept of spatial data was introduced in India through NRDMS, NNRMS and NICNET (Anon, 1983; Khan and Qureshy, 1983; DISNIC Highlights: DISNIC Handbook and Anon, 1989). The data bases were created using GIS technology, where the Natural Resources Informatics were graphically transferred in relation to geographic coordinates in terms of Grid Cells (Khan and Qureshy, 1983). In order to meet the spatial requirements of micro-level planning later attempts developed spatial information system in polygonal mode with village as a unit for Natural Resources Informatics. Most of the attempts made for GIS taking grid cells co-ordinates or village boundaries as a basis for geographic display of spatial data are burdened with artificial boundaries, which are cultural and arbitrary. To generate scenarios to assist decision makers, planners and common farmers, in an eco-system, it is desirable the area be looked in terms of natural geographic entities characterised by homogeneity of lithology soil, moisture, drainage characteristics of micro-watersheds, etc. So that inputs in terms of seed, fertilizers, irrigation, etc., are planned for optimum outputs.

For Natural Resources Informatics in Bharatpur district the concept of TMU has been adopted. Unlike the grid cells of conventional GIS, the TMU are natural entities characterised by homogeneity of lithology, soil types, groundwater conditions, slope characteristics and uniformity of landforms. Thus the land system management for agriculture, forestry, etc. in the district will be environmentally compatible, if the TMU are developed as management unit for deciding agricultural inputs and practices instead of district, tehsil or village which have artificial boundaries.

In Bharatpur district, the areas with similar geo-environmental conditions were mapped into separate zones or units through registration of overlays of different themes, natural geographic boundaries, characterised by homogeneity of lithology, soil, groundwater and geomorphology were mapped as Terrain Mapping Units. The above themes are important as they have crucial role in the pedogenic process of the land-system of the area.

The Natural Resources data in the country are present either in the Vector or Raster mode in the forms of reports, maps, statistics and space data products, were integrated and registered, and various overlays were prapared depicting areas of homogeneity of landforms (Fig. 12.1), groundwater conditions (Fig. 12.2) and soil types (Fig. 12.3). The lithology has been eliminated as constrained for TMU categorisation, due to the fact that the soil in the area is generally transported, *in-situ* soils are developed in the hilly terrain. The landform overlay has provided coverage for hard rocks contribution to soil characteristics and the kindred relationship of lithology, geomorphology and soils in TMU classification. However for finer tuning of the TMU's is with resolution of individual land-holdings, the overlaying of geological formations may be essential. The data base with village level resolution has been found to be well constrained by landforms overlay, which reflect the lithology of the area in the TMU classification in the present study.

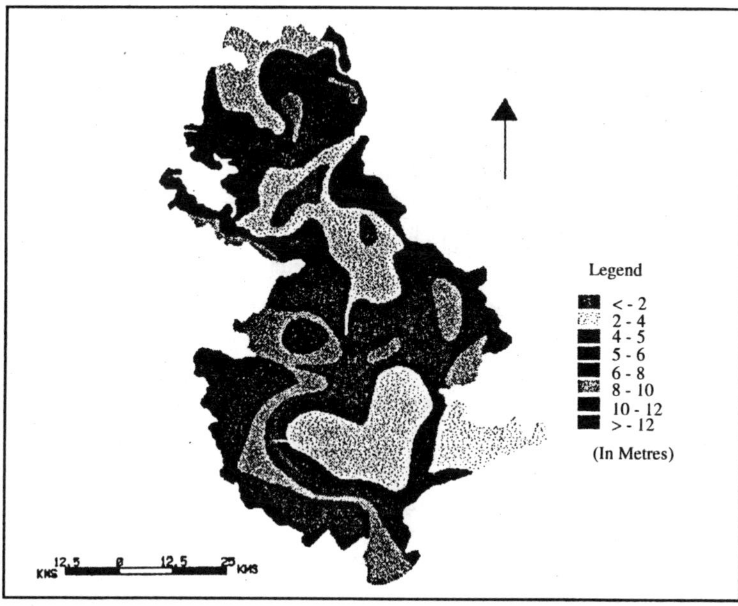

Fig. 12.2 : Groundwater Level Map of Bharatpur District, Rajasthan

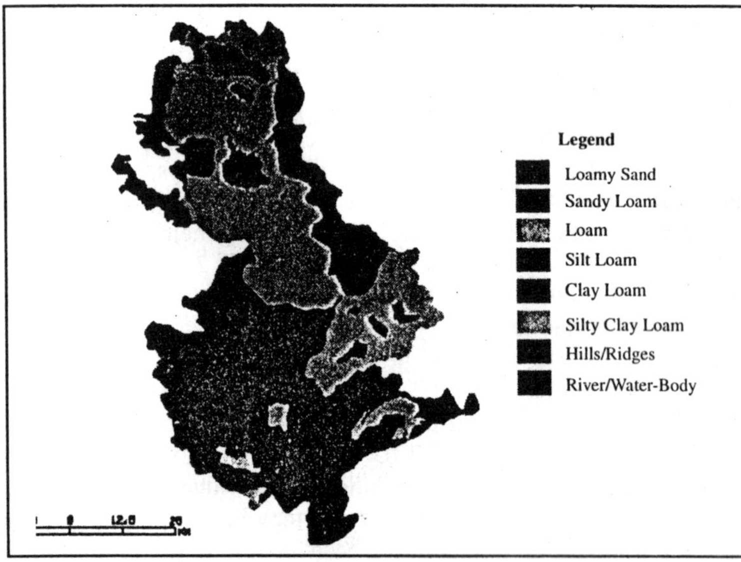

Fig. 12.3 : Surface Soil Texture Map of Bharatpur District, Rajasthan

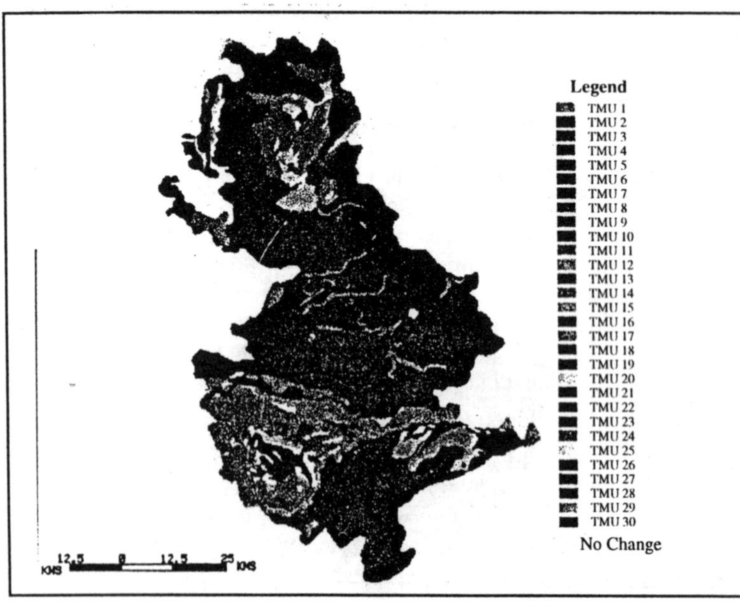

Fig. 12.4 : Terrain Mapping Units of Bharatpur District, Rajasthan

Table 12.1 : Classification of Terrain Mapping Units of Bharatput District

Category of TMU	*Geomorphic Unit*	*Groundwater Condition*	*Soil Type*
TMU-1	H	DWT	SL
TMU-2	VP	DWT	SL
TMU-3	VP	DWT	LS
TMU-4	VP	SWT	SL
TMU-5	VP	SWT	L
TMU-6	VP	DWT	L
TMU-7	VP	SWT	CL
TMU-8	OAP	DWT	SL
TMU-9	OAP	DWT	L
TMU-10	OAP	SWT	L
TMU-11	OAP	DWT	SIL
TMU-12	OAP	SWT	SIL
TMU-13	OAP	DWT	CL
TMU-14	RFP	DWT	LS
TMU-15	RFP	SWT	LS
TMU-16	RFP	DWT	SL
TMU-17	RFP	SWT	SL
TMU-18	BP	DWT	SL
TMU-19	BP	SWT	SL
TMU-20	BP	DWT	L
TMU-21	BP	SWT	L
TMU-22	EV	DWT	LS
TMU-23	PC	DWT	SL
TMU-24	PC	DWT	L
TMU-25	PC	SWT	L
TMU-26	OAP	SWT	SL
TMU-27	H	DWT	LS
TMU-28	H	SWT	LS
TMU-29	EV	DWT	H
TMU-30	BP	DWT	LS

H = Hills/Ridges, VP = Vindhyan Plateau, OAP = Older Alluvial Plain, RFP = Recent Alluvial Plain, BP = Buried Pediment, EV = Erosional Valley, PC = Paleochannel, DWT = Deep watertable (> 4 m b.g.l.), SWT = Shallow watertable (<4 m b.g.l.) LS = Loamy Sand, SL = Sandy Loam, L = Loam, SIL = Silt Loam, CL = Clay Loam.

The areas of homogeneity of geomorphology, soil texture and depth to watertable for Bharatput district, were delineated as TMUs. The final map of TMU was prepared by giving various symbols to each unit (Fig. 12.4).

CLASSIFICATION AND DESCRIPTION OF TMUS

The Bharatpur district has 10 tehsils, comprising 1,454 villages which are represented by thirty TMUs (Fig. 12.4).

To facilitate description, these have been designated as TMU-1, TMU-2, etc. (Table 12.1), using alpha-numeric code of nomenclature.

The description of these TMUs is as follows :

TMU-1 (H + DWT + SL)

The TMU-1 comprise Hills and Ridges of Bharatpur district is associated with Hogback and Cuesta of Delhi Supergroup. This is characterised by deep watertable and moderate to very steep slope. The unit has sandy loam soil texture and is a member of coarse loamy, mixed hyperthermic family of Lithic Ustorthents of Entisols Order. The soils of this unit are non-calcareous, excessively drained with high permeability and suffer from severe erosion, and are mostly under bushes and shrubs forest.

The soils of this unit are very shallow, yellowish to dark reddish brown in colour and gravelly sandy loam in texture, comprising sand 70.0 per cent, silt 20.0 per cent and clay 10.0 per cent. It has organic carbon (OC) 0.045 per cent hydrogen ion concentration (*p*H) 7.4 and electrical conductivity (EC) 0.24 mmhos/cm. The fertility in terms of macronutrients is nitrogen (N) 0.004 per cent, phosphorus (P) 0.055 per cent and potassium (K) 1.171 per cent. The soils of this unit have exchangeable bases like calcium (Ca^{++}) 3.47 me/100g, magnesium (Mg^{++}) 2.89 me/100g, sodium (Na^{+}) 0.51 me/100g and potassium (K^{+}) 0.18 me/100g. It has cation exchange capacity (CEC) 7.05 me/100g.

The soils of TMU-1 can be used for limited grazing, wildlife and forestry. The lands of this unit are not suitable for irrigation due to soil (shallow) and topography (moderate to steep slope) limitations.

TMU-2 (VP+DWT+SL)

The TMU-2 comprise Vindhyan Plateau, deep watertable and sandy loam in Bharatpur district. The unit has gentle to moderate slopes and suffer from moderate to severe erosion. The soil of this unit are member of coarse loamy mixed hyperthermic family of Lithic Ustorthents of Entisols Order. It is non-calcareous, excessively drained with high permeability and deep watertable and the landcover is mostly under bushes and grazing lands.

The soils of the TMU-2 are shallow, yellowish brown in colour and sandy loam in texture. Granulometrically, the surface soil of this unit comprise sand 55.36 per cent, silt 28.92 per cent, and clay 15.72 per cent. It has organic carbon (OC) 0.29 per cent, calcium carbonate ($CaCO_3$) 0.28 per cent, hydrogen ion concentration (*p*H) 7.5 and electrical conductivity (EC) 0.28 mmhos/cm. The fertility in terms of macronutrients in nitrogen (N) 0.024 per cent, phosphorus (P) 0.049 per cent and potassium (K) 0.78 per cent. It has exchangeable bases like calcium (Ca^{++}) 5.94 me/100g, magnesium (Mg^{++}) 4.96me/100g, sodium (Na^{+}) 0.69 me/100g, and potasium (K^{+}) 0.32me/100g and cation exchange capacity (CEC) 11.01 me/100g.

The soils of TMU-2 can be used for moderate grazing, wildlife and forestry.

TMU-3 (VP + DWT + LS)

The TMU-3 comprise Vindhyan Plateau, deep watertable and loamy sand in Bharatpur district. The unit is nearly level to very gently sloping, and suffer from severe erosion. It is a member of coarse loamy mixed hyperthermic family of Typic.

Ustipsamments of Entisols Order

The soils of this unit are excessively drained with rapid permeability and deep watertable and mostly support grazing lands.

The soils of the TMU-3 are very deep dark to yellowish brown in colour and loamy sand in texture. Granulometrically, the surface soil of this unit comprise sand 79.0 per cent, silt 10.56 per cent and clay 10.44 per cent. It has organic carbon (OC) 0.21 per cent, calcium carbonate ($CaCO_3$) 1.02 per cent, hydrogen ion concentration (*p*H) 8.5 and electrical conductivity (EC) 0.18 mmhos/cm. The fertility in terms of macronutrients is nitrogen (N) 0.017 per cent, phosphorus (P) 0.062 per cent and potassium (K) 0.781 per cent. It has exchangeable bases like calcium (Ca^{++}) 7.68 me/100g, magnesium

(Mg^{++}) 6.41 me/100g, sodium (Na^{+}) 0.17 me/100g, potassium (K^{+}) 0.23 me/100g and cation exchange capacity (CEC) 14.49 me/100g. The soils of this unit can be used for moderate grazing, wildlife and forestry.

TMU-4 (VP + SWT + SL)

The TMU-4 comprise Vindhyan Plateau, shallow water-table and sandy loam is in Rupbas tehsil of Bharatpur district. It occur on nearly level to moderately slopping, lands and suffer from slight to severe erosion. It is a member of coarse loamy mixed hyperthermic family of Tpic Ustochrepts of Inceptisols Order. The soils of this unit are calcareous, poorly drained with low permeability and mostly under Rabi crops.

The soils of the TMU-4 are very deep, greyish brown in colour and sandy loam in texture. Granulometrically, the surface soil of this unit comprise sand 71.00 per cent, silt 15.00 per cent and clay 14.00 per cent. It has organic carbon (OC) 0.17 per cent, calcium carbonate ($CaCO_3$) 0.55 per cent, hydrogen ion concentration (*p*H) 8.0 and electrical conductivity (EC) 1.30 mmhos/cm. The fertility in terms of macronutrients is nitrogen (N) 0.014 per cent, phosphorus (P) 0.040 per cent and potassium (K) 2.02 per cent. It has exchangeable bases like calcium (Ca^{++}) 8.5 me/100g, magnesium (Mg^{++}) 6.5 me/100g, sodium (Na^{+}) 1.20 me/100g, potassium (K^{+}) 0.75 me/100g and cation exchange capacity (CEC) 16.95 me/100g.

The TMU-4 can be used for moderate cultivation, intense grazing, wildlife and forestry. It is suitable for wheat and barely of Rabi crop and sugarcane and millet of Kharif crop.

TMU-5 (VP + SWT + L)

The TMU-5 comprise Vindhyan Plateau, shallow watertable and loam also occur in Rupbas tehsil of Bharatpur district, and form nearly level to moderately sloping, depressional lands. It suffers from slight to moderate erosion. It is a member of fine loamy mixed hyperthermic family of Typic Ustochrepts of Inceptisols Order. The soils of this unit are strongly calcareous, poorly drained with low permeability and are mostly under Rabi crops.

The soils of this unit are very deep, very dark greyish brown in colour and loam in texture. Granulometrically, the surface soil of this unit comprise sand 48.28 per cent, silt 30.28 per cent and clay 21.44 per cent. The soil has organic carbon (OC) 0.17 per cent, calcium carbonate ($CaCO_3$) 1.32 per cent, hydrogen ion concentration (*p*H) 8.5 and electrical conductivity (EC) 1.35 mmhos/cm. The fertility of the unit in terms of macronutrients is nitrogen (N) 0.014 per cent, phosphorus (P) 0.021 per cent and potassium (K) 2.669 per cent. It has exchangeable bases like calcium (Ca^{++}) 9.42 me/100g, magnesium (Mg^{++}) 6.28 me/100g, sodium (Na^{+}) 1.22 me/100g, potassium (K^{+}) 0.20 me/100g and cation exchange capacity (CEC) 17.12 me/100g.

The TMU-5 can be used for moderate cultivation, intense grazing, wildlife and forestry. It is suitable for wheat and barely of Rabi crop and sugarcane and paddy of Kharif crops.

TMU-6 (VP + DWT + L)

The TMU-6 comprise Vindhyan Plateau, deep water-table and loam in Rupbas tehsil in Bharatpur district. It occurs on nearly level to moderately sloping lands and suffer from slight to moderate erosion. It is a member of fine loamy mixed hyperthermic family of Typic Ustochrepts of Inceptisols Order. The soils of this unit are calcareous, medium permeability, and are mostly under Rabi crops.

The soils of this unit are very deep, greyish brown in colour and loam in texture. Granulometrically, the surface soil of this unit comprise sand 48.20 per cent, silt 30.00 per cent and clay 21.80 per cent. The soil has organic carbon (OC) 0.18 per cent calcium carbonate ($CaCO_3$) 1.30 per cent, hydrogen ion concentration (*p*H) 8.3 and electrical conductivity (EC) 1.05 mmhos/cm. The fertility of the unit in terms of macronutrients is nitrogen (N) 0.015 per cent, phosphorus (P) 0.020 per cent and potassium (K) 0.660 per cent. The soils of this unit have exchangeable bases like calcium (Ca^{++}) 9.00 me/100g, magnesium (Mg^{++}) 6.20 me/100g, sodium (Na^{+}) 1.20 me/100g, potassium (K^{+}) 0.25 me/100g and cation exchange capacity (CEC) 16.65 me/100g.

The TMU-6 can be used for moderate cultivation, intense grazing, wildlife and forestry. It is suitable for wheat and barley of Rabi crop and sugarcane and millet of Kharif crop.

TMU-7 (VP + SWT + CL)

The TMU-7 is developed over Vindhyan Plateau and has shallow watertable and clay loam is restricted to Rupbas tehsil of Bharatpur district. The unit has nearly level to gentle slope and suffer from slight to moderate erosion. Its soil is a member of clayey mixed hyperthermic family of Vertic Haplaquapts of Inceptisols Order. This unit has low permeability. The soils are calcareous and are under Rabi crops only.

The soils of this unit are very deep, dark brown to very dark brown in colour and clayey in texture. Granulometrically, the surface soil of this unit comprise sand 31.28 per cent, silt 29.28 per cent and clay 29.44 per cent. It has organic carbon (OC) 0.05 per cent, carbonate ($CaCO_3$) 1.88 per cent, hydrogen ion concentration (*p*H) 8.5 and electrical conductivity (EC) 2.8 mmhos/cm. The fertility of the unit in terms of macro-nutrients is nitrogen (N) 0.004 per cent, phosphorus (P) 0.05 per cent and potassium (K) 0.945 per cent. The soils of this unit have exchangeable bases like calcium (Ca^{++}) 10.23 me/100g, magnesium (Mg^{++}) 5.97 me/100g, sodium (Na^{+}) 27.00 me/100g, potassium (K^{+}) 0.295 per cent me/100g and cation exchange capacity (CEC) 43.49 me/100g. The TMU-7 can be used for moderate cultivation, intense grazing, wildlife and forestry. It is suitable for wheat, barley and peas of Rabi crop and sugarcane and paddy of Kharif crops.

TMU-8 (OAP + DWT + SL)

The TMU-8 is developed over Older Alluvial Plain and comprise sandy loam with deep watertable conditions. The unit occurs on nearly level to gently sloping plain lands. Its soil is a member of coarse loamy, mixed hyperthermic family of Fluventic Ustochrepts of Inceptisols Order. The soils of this unit are well drained with moderate permeability and suffer from slight to severe erosion. These soils are non-calcareous and are mostly under Kharif and Rabi Crops.

The soils of this unit are very deep, yellowish brown dark brown in colour and sandy loam in texture. Granulometrically, the surface soil of this unit comprise sand 55.32 per cent, silt 32.92 per cent and clay 11.76 per cent. It has organic carbon (OC) 0.05 per cent, calcium carbonate ($CaCO_3$) 1.93 per cent, hydrogen ion concentration (*p*H) 7.8 and electrical conductivity (EC) 0.85 mmhos/cm. The fertility of this unit in terms of macronutrients is nitrogen (N) 0.004 per cent, phosphorus (P) 0.045 per cent and potassium (K) 2.892 per cent. It has exchangeable bases like calcium (Ca^{++}) 4.03 me/100g, magnesium (Mg^{++}) 3.37 me/100g, sodium (Na^{+}) 0.32 me/100g, potassium (K^{+}) 0.32 me/100g and cation exchange capacity (CEC) 7.89 me/100g.

The TMU-8 can be used for moderate cultivation, intense grazing, wildlife and forestry. It is suitable for sorghum, millet, pulses and groundnut of Kharif crop and gram, mustard and barley of Rabi crop.

TMU-9 (OAP + DWT + L)

The TMU-9 of the Bharatpur district is developed over Older Alluvial Plain of Aligarh Older Alluvial Formation. The soils of the unit occur on nearly level to gentle sloping plain lands and suffer from non to moderate erosion. The unit is a member of loamy, mixed hyperthermic family of Haplustafs of Alfisols Order. The unit has deep watertable with medium permeability. The soils of this unit are slight to moderate calcareous and are under Rabi and Kharif crops.

The soil of TMU-9 are very deep, brown to dark brown in colour and loam in texture. Granulometrically, the surface soil of this unit comprise sand 43.60 per cent, silt 24.92 per cent and clay 31.48 per cent. It has organic carbon (OC) 0.30 per cent, calcium carbonate ($CaCO_3$) 0.18 per cent, hydrogen ion concentration (pH) 7.6 and electrical conductivity (EC) 1.60 mmhos/cm. The fertility of this unit in terms of macronutrients is nitrogen (N) 0.031 per cent, phosphorus (P) 0.046 per cent and potassium (K) 1.970 per cent. It has exchangeable bases like calcium (Ca^{++}) 12.40 me/100g, magnesium (Mg^{++}) 7.30 me/100g, sodium (Na^{+}) 2.1 me/100g, potassium (K^{+}) 0.42 me/100g and has cation exchange capacity (CEC) 22.22 me/100g. The TMU-9 can be used for intensive cultivation, intense grazing, wildlife and forestry. It is suitable for sugarcane and millet of Kharif crop and wheat, barley and peas of Rabi crop.

TMU-10 (OAP + SWT + L)

The TMU-10 of the Bharatpur district is developed over Older Alluvial Plain characterised by loam and shallow watertable. It forms nearly level plain lands and suffer from non to slight erosion. Its soil is a member of fine loamy, mixed hyperthermic family of Typic Haplustafs of Alfisols Order. The soils of this unit are poorly drained, strongly calcareous with low permeability. The soils are under Rabi and Kharif crops.

The soils of this unit are very deep, dark brown in colour and loam in texture. Granulometrically, the surface soil of this unit comprise sand 41.60 per cent, silt 25.90 per cent and clay 32.50 per cent. It has organic carbon (OC) 0.39 per cent, calcium carbonate ($CaCO_3$) 0.19 per cent, hydrogen ion concentration (*p*H) 7.8 and electrical conductivity (EC) 1.66 mmhos/cm. The fertility of the unit in terms of macronutrients is nitrogen (N) 0.032 per cent, phosphorus (P) 0.048 per cent and potassium (K) 1.979 per cent. It has exchangeable bases like calcium (Ca^{++}) 12.49 me/100g, magnesium (Mg^{++}) 7.35 me/100g, sodium (Na^{+}) 2.0 me/100g, potassium (K^{+}) 0.43 me/100g and cation exchange capacity (CEC) 22.27 me/100g.

The TMU-10 can be used for intensive cultivation, intense grazing, wildlife and forestry. It is suitable for wheat, barley and peas of Rabi crop and sugarcane and paddy of Kharif crop.

TMU-11 (OAP + DWT + SIL)

The TMU-11 of the Bharatpur district is developed over Older Alluvial Plain, characterised by silt loam and deep watertable. It forms nearly level to very gently sloping plain lands and suffer from non to slight erosion. The soils of this unit are moderately well drained with moderate permeability and are mostly under Rabi and Kharif crops. It is a member of fine loamy, mixed hyperthermic family of Udic Ustochrepts of Inceptosols Order.

The soils of this unit are very deep, dark yellowish brown to brown in colour and fine loamy (silt loam) in texture. Granulometrically, the surface soil of this unit comprise sand 34.32 per cent, silt 49.92 per cent and clay 15.76 per cent. The soil has organic carbon (OC) 0.16 per cent, calcium carbonate ($CaCO_3$) 0.82 per cent, hydrogen ion concentration (*p*H) 8.1 and electrical conductivity (EC) 1.44 mmhos/cm. The fertility of this unit in terms of macronutrients in nitrogen (N) 0.013 per cent, phosphorus (P) 0.037 per cent and potassium (K) 1.004 per cent. It has exchangeable bases like calcium (Ca^{++}) 6.11 me/100g, magnesium (Mg^{++}) 5.09 me/100g, sodium (Na^{+}) 0.80 me/100g and potassium (K^{+}) 0.26 me/100g and cation exchange capacity (CEC) 12.26 me/100g. The TMU-11 can be used for moderate cultivation, intense grazing, wildlife and forestry. It is suitable for wheat, gram, mustard, barley and peas of Rabi crop and maize, sorghum, millet, cotton, pulses, sugarcane and groundnut of Kharif crops.

TMU-12 (OAP + SWT +SIL)

The TMU-12 of the Bharatpur district is developed over Older Alluvial

Plain characterised by silty loam and shallow watertable. It forms nearly level to plain lands and suffer from non to slight erosion. The soils of this unit are strongly calcareous, poorly drained with low permeability and are mostly under Rabi crops. It is a member of fine loamy mixed hyperthermic family of Typic Halaquepsts of Inceptosols Order.

The soils of this unit are very deep, very dark greyish brown to greyish brown in colour and silty loam in texture. Granulometrically, the surface soil of this unit comprise sand 19.84 per cent, silt 52.20 per cent and clay 27.96 per cent. The soil has organic carbon (OC) 1.15 per cent, calcium carbonate ($CaCO_3$) 0.16 per cent, hydrogen ion concentration (*p*H) 8.0 and electrical conductivity (EC) 0.75 mmhos/cm. The fertility of this unit in terms of macronutrients is nitrogen (N) 0.096 per cent, phosphorus (P) 0.084 per cent and potassium (K) 3.568 per cent. It has exchangeable bases like calcium (Ca^{++}) 9.37 me/100g, magnesium (Mg^{++}) 7.81 me/100g, sodium (Na^{+}) 3.80 me/100g and potassium (K^{+}) 0.70 me/100g and cation exchange capacity (CEC) 21.68 me/100g.

The soils of this unit are suitable for moderate cultivation, intense grazing, wildlife and forestry. It is suitable for wheat, barley and peas of Rabi crop and sugarcane and paddy of Kharif crops.

TMU-13 (OAP + DWT + CL)

The TMU-13 is distributed in Pahari and Nagar tehsils of the Bharatpur district, and is developed over Older Alluvial Plain characterized by clay loam and deep watertable. It forms level to very gently sloping lands and suffer from slight erosion. The soils of this unit are strongly calcareous, with low permeability and are mostly under Rabi and Zayed crops.

The soils of this unit are very deep, very dark greyish brown to dark yellowish brown in colour and fine loamy (clay loam) in texture. Granulometrically, the surface soil of this unit comprise sand 22.24 per cent, silt 45.56 per cent and clay 32.24 per cent. The soil has organic carbon (OC) 0.22 per cent, calcium carbonate ($CaCO_3$) 0.61 per cent, hydrogen ion concentration (*p*H) 8.05 and electrical conductivity (EC) 0.4 mmhos/cm. The fertility of this unit in terms of macronutrients in nitrogen (N) 0.018 per cent, phosphorus (P) 0.034 per cent and potassium (K) 2.732 per cent. It has exchangeable bases like calcium (Ca^{++}) 25.33 me/100g, magnesium (Mg^{++}) 21.13 me/100g, sodium (Na^+) 1.08 me/100g and potassium (K^+) 0.83 me/100g and cation exchange capacity (CEC) 48.37 me/100g.

The TMU-13 can be used for moderate cultivation, intense grazing, wildlife and forestry. It is suitable for wheat, gram, mustard of Rabi crop and paddy and sugarcane of Kharif crop.

TMU-14 (REF + DWT + LS)

The TMU-14 is developed over Recent Flood Plain of Banganga, characterised by loamy sand and deep watertable in Bharatpur district. It forms level to moderately sloping lands and suffer from slight to severe erosion. The soils of this unit are strongly calcareous, excessively drained with high permeability and are mostly under Kharif and Rabi crops. It is a member of coarse loamy mixed calcareous hyperthermic family of Typic Ustipsamments of Entisols Order.

The soils of this unit are very deep, dark brown to dark yellowish brown in colour and sandy in texture. Granulometrically, the surface soil of this unit comprise sand 81.0 per cent, silt 9.56 per cent and clay 10.44 per cent. The soil has organic carbon (OC) 0.17 per cent, calcium carbonate ($CaCO_3$) 0.30 per cent, hydrogen ion concentration (pH) 7.55 and electrical conductivity (EC) 0.16 mmhos/cm. The fertility of this unit in terms of macronutrients is nitrogen (N) 0.016 per cent, phosphorus (P) 0.149 per cent and potassium (K) 0.669 per cent. It has exchangeable bases like calcium (Ca^{++}) 7.36 me/100g, magnesium (Mg^{++}) 6.14 me/100g, sodium (Na^+) 0.15 me/100g and potassium (K^+) 0.37 me/100g and cation exchange capacity (CEC) 14.02 me/100g.

The TMU-14 can be used for moderate cultivation, intense grazing, wildlife and forestry. It is suitable for sorghum, millet, cotton and groundnut of Kharif crop and gram, mustard and barley of Rabi crop.

TMU-15 (REF + SWT + LS)

The TMU-15 is developed over Recent Flood Plain of Banganga in Bharatpur district, it is characterised by loamy sand and shallow water-table. It forms nearly level to moderately sloping lands and suffer from slight to severe erosion. The soils of this unit are strongly calcareous, poorly drained with medium permeability and are mostly under Rabi and Kharif crops. It is a member of coarse loamy mixed calcareous hyperthermic family of Typic Ustipsamments of Entisols Order.

The soils of this unit are very deep, dark brown to dark yellowish brown in colour and sandy in texture. Granulometrically, the surface soil of this unit comprise sand 80.0 per cent, silt 9.50 per cent and clay 10.50 per cent. The soil has organic carbon (OC) 0.18 per cent, calcium carbonate ($CaCO_3$) 0.32 per cent, hydrogen ion concentration (*p*H) 7.6 and electrical conductivity (EC) 0.19 mmhos/cm. The fertility of this unit in terms of macronutrients is nitrogen (N) 0.017 per cent, phosphorus (P) 0.15 per cent and potassium (K) 0.66 per cent. It has exchangeable bases like calcium (Ca^{++}) 7.40 me/100g, magnesium (Mg^{++}) 6.10 me/100g, sodium (Na^{+}) 0.18 me/100g and potassium (K^{+}) 0.36 me/100g and cation exchange capacity (CEC) 14.04 me/100g.

The TMU-15 can be used for moderate cultivation, intense grazing, wildlife and forestry. It is suitable for wheat, mustard and gram of Rabi crop and millet, sorghum and cotton of Kharif crop.

TMU-16 (RFP + DWT + SL)

The TMU-16 is developed over Recent Flood Plain of Banganga and is characterised by sandy loam and deep watertable in Bharatpur district. It forms level to very gentle sloping lands and suffer from slight to moderate erosion. The soils of this unit are strongly calcareous, excessively drained with high permeability. It is a member of coarse loamy mixed hyperthermic family of Typic Ustifluvents of Entisols Order.

The soils of this unit are very deep, dark brown to dark yellowish brown in colour and coarse loamy in texture. Granulometrically, the soil of this unit comprise sand 76.36 per cent, silt 9.92 per cent and clay 11.72 per cent. The soil has organic carbon (OC) 0.17 per cent, calcium carbonate ($CaCO_3$) 1.78 per cent, hydrogen ion concentration (*p*H) 8.4 and electrical conductivity (EC) 0.19 mmhos/cm. The fertility of this unit in terms of macronutrients in nitrogen (N) 0.014 per cent, phosphorus (P) 0.037 per cent and potassium (K) 0.613 per cent. It has exchangeable bases like calcium (Ca^{++}) 8.34 me/100g, magnesium (Mg^{++}) 6.96 me/100g, sodium (Na^{+}) 0.43 me/100g and potassium (K^{+}) 0.42 me/100g and cation exchange capacity (CEC) 16.15 me/100g.

The TMU-16 can be used for moderate cultivation, intense grazing, wildlife and forestry. It is suitable for sorghum, millet, pulses and groundnut of Kharif crop and gram, mustard and barley of Rabi crops.

TMU-17 (RFP + SWT + SL)

The TMU-17 is developed over Recent Flood Plain of Banganga and is characterised by sandy loam and shallow watertable and forms nearly level to very gentle sloping lands and suffer from slight to moderate erosion in Bharatpur district. The soils of this unit are strongly calcareous, poorly drained with low to medium permeability. It is a member of coarse loamy mixed hyperthermic family of Typic Ustifluvents of Entisols Order.

The soils of this unit are very deep, dark brown to dark yellowish brown in colour and sandy loam in texture. Granulometrically, the soil of this unit comprise sand 76.00 per cent, silt 10.22 per cent and clay 11.78 per cent. The soil has organic carbon (OC) 0.18 per cent, calcium carbonate ($CaCO_3$) 1.75 per cent, hydrogen ion concentration (*p*H) 8.5 and electrical conductivity (EC) 0.20 mmhos/cm. The fertility of this unit in terms of macronutrients is nitrogen (N) 0.015 per cent, phosphorus (P) 0.035 per cent and potassium (K) 0.610 per cent. It has exchangeable bases like calcium (Ca^{++}) 8.35 me/100g, magnesium (Mg^{++}) 6.98 me/100g, sodium (Na^{+}) 0.45 me/100g and potassium (K^{+}) 0.44 me/100g and cation exchange capacity (CEC) 16.22 me/100g.

The TMU-17 can be used for moderate cultivation, intense grazing, wildlife and forestry. It is suitable for sugarcane, millet and pulses of Kharif crop and barley, mustard and gram of Rabi crop.

TMU-18 (BP + DWT + SL)

The TMU-18 is developed over Buried Pediments characterised by sandy loam and deep watertable and occur on nearly level to gently sloping plain lands and suffer from slight to moderate erosion in Bharatpur district. The soils of this unit are non-calcareous, moderately well drained with moderate permeability, and are mostly under grass and grazing lands. It is a member of coarse loamy mixed hyperthermic family of Fluventic Ustochrepts of Inceptisols Order.

The soils of this unit are very deep, dark brown in colour and sandy loam in texture. Granulometrically, the surface soil of this unit comprise sand 51.76 per cent, silt 29.92 per cent and clay 18.32 per cent. The soil has organic carbon (OC) 0.33 per cent, calcium carbonate

($CaCO_3$) 0.60 per cent, hydrogen ion concentration (*p*H) 7.5 and electrical conductivity (EC) 0.31 mmhos/cm. The fertility of this unit in terms of macronutrients is nitrogen (N) 0.028 per cent, phosphorus (P) 0.051 per cent and potassium (K) 2.09 per cent. It has exchangeable bases like calcium (Ca^{++}) 5.14 me/100g, magnesium (Mg^{++}) 4.29 me/100g, sodium (Na^{+}) 0.49 me/100g and potassium (K^{+}) 0.51 me/100g and cation exchange capacity (CEC) 10.43 me/100g.

The TMU-18 can be used for intensive cultivation, intense grazing, wildlife and forestry. It is suitable for sorghum, millet, gwar, cotton, pulses and groundnut of Kharif crops and wheat, barley, gram, mustard and peas of Rabi crop.

TMU-19 (BP + SWT + SL)

The TMU-19 is developed over Buried Pediments characterized by sandy loam and shallow watertable in Bharatpur district forming level to gently sloping plain lands and suffer from slight to moderate erosion. The soils of this unit are calcareous, poorly drained with moderate permeability and are under mostly Rabi and Kharif crops. It is a member of coarse loamy calcareous mixed hyperthermic family of Typic Haplustalfs of Alfisols Order.

The soils of this unit are very deep, dark yellowish brown to dark brown in colour and sandy loam in texture. Granulometrically, the surface soil of this unit comprise sand 55.60 per cent, silt 25.92 per cent and clay 18.48 per cent. The soil has organic carbon (OC) 0.22 per cent, calcium carbonate ($CaCO_3$) 0.71 per cent, hydrogen ion concentration (*p*H) 8.2 and electrical conductivity (EC) 0.56 mmhos/cm. The fertility of this unit in terms of macronutrients is nitrogen (N) 0.018 per cent, phosphorus (P) 0.057 per cent and potassium (K) 1.45 per cent. It has exchangeable bases like calcium (Ca^{++}) 13.37 me/100g, magnesium (Mg^{++}) 11.15 me/100g, sodium (Na^{+}) 1.17 me/100g and potassium (K^{+}) 0.32 me/100g and cation exchange capacity (CEC) 26.01 me/100g.

The TMU-19 can be utilised for intensive cultivation, intense grazing, wildlife and forestry. It is suitable for wheat, gram, mustard, barley and peas of Rabi crop and maize, sorghum, millet, cotton, pulses, sugarcane and groundnut of Kharif crop.

TMU-20 (BP + DWT + SL)

The TMU-20 is developed over Buried Pediments, characterised by loam and deep watertable in Bharatpur district. It forms nearly level to sloping lands and suffer from none to slight erosion. The soils of this unit are calcareous, moderately drained with medium permeability. In this unit the cultivation is in patches. It is a member of loamy, mixed hyperthermic family of Aeric Haplustalfs of Inceptisols Order.

The soils of the TMU-20 are very deep, yellowish brown in colour and loam in texture. Granulometrically, the surface soil of this unit comprise sand 45.80 per cent, silt 37.00 per cent and clay 17.20 per cent. The soil has organic carbon (OC) 0.27 per cent, calcium carbonate ($CaCO_3$) 0.75 per cent, hydrogen ion concentration (*p*H) 9.5 and electrical conductivity (EC) 3.0 mmhos/cm. The fertility of this unit in terms of macronutrients is nitrogen (N) 0.022 per cent, phosphorus (P) 0.060 per cent and potassium (K) 1.530 per cent. It has exchangeable bases like calcium (Ca^{++}) 10.45 me/100g, magnesium (Mg^{++}) 2.90 me/100g, sodium (Na^{+}) 5.85 me/100g and potassium (K^{+}) 0.415 me/100g and cation exchange capacity (CEC) 19.615 me/100g.

The TMU-20 can be utilised for limited cultivation, intense grazing, wildlife and forestry. Sorghum and gwar of Kharif crop and gram and pulses of Rabi crop can be grown in patches of this unit.

TMU-21 (BP + SWT + L)

The TMU-21 is developed over Buried Pediments characterised by loam and shallow watertable in Bharatpur district. It forms nearly level sloping basinal lands and suffer from none to slight erosion. The soils of this unit are calcareous, alkaline and poorly drained with low permeability. The soils of this unit are mostly under waste (*usar*) which are not suitable for cultivation. It is a member of loamy, mixed hyperthermic family of Aeric Halaquepts of Enceptisols Order.

The soils of TMU-21 are very deep, dark yellowish brown to yellowish brown in colour and loam in texture. Granulometrically, the surface soil of this unit comprise sand 43.84 per cent, silt 36.20 per cent and clay 19.96 per cent. The soil has organic carbon (OC) 0.28 per cent, calcium carbonate ($CaCO_3$) 0.76 per cent, hydrogen ion concentration (*p*H) 9.9 and electrical conductivity (EC) 3.6 mmhos/cm. The fertility of this unit in terms of macronutrients in nitrogen (N) 0.023 per cent, phosphorus (P) 0.061 per cent and potassium (K) 1.533 per cent. It has exchangeable bases like calcium (Ca^{++}) 10.48 me/100g, magnesium (Mg^{++}) 2.96 me/100g, sodium (Na^{+}) 5.90 me/100g and potassium (K^{+}) 0.416 me/100g and cation exchange capacity (CEC) 19.756 me/100g.

The TMU-21 can be utilised for intense grazing, wildlife and forestry. It is unsuitable for cultivation due to salinity/alkalinity.

TMU-22 (EV + DWT + LS)

The TMU-22 is developed in Erosional Valley characterised by loamy sand and deep watertable in Bharatpur district. It forms nearly level to moderately sloping undulating lands and suffer from slight to severe erosion. The soils of this unit are excessively drained with high permeability and are under Kharif crops. It is a member of coarse loamy mixed hyperthermic family of Typic Ustipsamments of Entisols Order.

The soils of the TMU-22 are very deep, dark yellowish brown to dark brown in colour and sandy in texture. Granulometrically, the surface soil of this unit comprise sand 83.76 per cent, silt 6.92 per cent and clay 9.32 per cent. The soil has organic carbon (OC) 0.14 per cent, calcium carbonate ($CaCO_3$) 0.33 per cent, hydrogen ion concentration (*p*H) 7.9 and electrical conductivity (EC) 0.66 mmhos/cm. The fertility of this unit in terms of macronutrients is nitrogen (N) 0.012 per cent, phosphorus (P) 0.142 per cent and potassium (K) 1.087 per cent. It has exchangeable bases like calcium (Ca^{++}) 4.88 me/100g, magnesium (Mg^{++}) 4.07 me/100g, sodium (Na^{+}) 1.57 me/100g and potassium (K^{+}) 0.25 me/100g and cation exchange capacity (CEC) 10.77 me/100g.

The TMU-22 is not suitable for cultivation, it can be used for intense grazing, wildlife and forestry. Only sorghum and millet of Kharif crop can be grown on this unit.

TMU-23 (PC + DWT +SL)

The TMU-23 is developed in Palaeochannels characterised by sandy loam and deep watertable in Bharatpur district. It forms nearly level to gently sloping plain lands and suffer from slight to moderate erosion. The soils of this unit are well drained with moderate permeability. The soils are mostly non-calcareous and best for Rabi and Kharif crops. It is a member of coarse loamy mixed hyperthermic family of Typic Ustochrepts of Inceptisols Order.

The soils of the TMU-23 are very deep, brown to yellowish brown in colour and sandy loam in texture. Granulometrically, the surface soil of this unit comprise sand 52.0 per cent, silt 34.28 per cent and clay 13.72 per cent. The soil has organic carbon (OC) 0.24 per cent, calcium carbonate ($CaCO_3$) 2.81 per cent, hydrogen ion concentration (pH) 8.45 and electrical conductivity (EC) 0.16 mmhos/cm. The fertility of this unit in terms of macronutrients is nitrogen (N) 0.002 per cent, phosphorus (P) 0.24 per cent and potassium (K) 0.613 per cent. The exchangeable bases like calcium (Ca^{++}) 6.05 me/100g, magnesium (Mg^{++}) 5.05 me/100g, sodium (Na^{+}) 0.30 me/100g and potassium (K^{+}) 0.20 me/100g and cation exchange capacity (CEC) 11.05 me/100g.

The TMU-23 can be used for moderate cultivation, intense grazing, wildlife and forestry. It is suitable for sorghum, millet, pulses and groundnut of Kharif crop and gram, mustard and barley of Rabi crop.

TMU-24 (PC + DWT + L)

The TMU-24 is developed in Paleochannels characterised by loam and deep watertable in Bharatpur district. It forms nearly level to very gently sloping basinal lands and suffer from slight erosion to no erosion. The soils of this unit are not calcareous, moderately well drained with moderate permeability and are mostly under Rabi cultivation. It is a member of loamy mixed hyperthermic family of Typic Ustochrepts of Inceptisols order.

The soils of the TMU-24 are very deep, brown to very dark brown in colour and loam in texture. Granulometrically, the surface soil of this unit comprise sand 42.08 per cent, silt 34.28 per cent and clay 23.64 per cent. The soil has organic carbon (OC) 0.21 per cent, calcium carbonate ($CaCO_3$) 0.31 per cent, hydrogen ion concentration (pH) 7.30 and electrical conductivity (EC) 0.29 mmhos/cm. The fertility of this unit in terms of macronutrients is nitrogen (N) 0.018 per cent, phosphorus (P) 0.303 per cent and potassium (K) 0.728 per cent. It has exchangeable bases like calcium (Ca^{++}) 6.6 me/100g, magnesium (Mg^{++}) 5.5 me/100g, sodium (Na^{+}) 0.45 me/100g and potassium (K^{+}) 0.43 me/100g and cation exchange capacity (CEC) 12.98 me/100g.

The TMU-24 can be used for intensive cultivation, intense grazing, wildlife and forestry. It is suitable for wheat, gram, mustard, barley and peas of Rabi crop and sugarcane, paddy and groundnut of Kharif crop.

TMU-25 (PC + SWT + L)

The TMU-25 is developed in Palaeochannels characterised by loam and shallow watertable in Bharatpur district. It forms nearly level to very gently sloping basin and suffer from non to slight erosion. The soils of this unit are saline in nature and poorly drained with low permeability. Due to salinity/alkalinity these soils are not suitable for cultivation

The soil of the TMU-25 are very deep, dark yellowish brown in colour and loam in texture. Granulometrically, the surface soil of this unit comprise sand 44.84 per cent, silt 38.20 per cent and clay 16.97 per cent. The soil has organic carbon (OC) 0.23 per cent, calcium carbonate ($CaCO_3$) 1.28 per cent, hydrogen ion concentration (*p*H) 10.20 and electrical conductivity (EC) 2.8 mmhos/cm. The fertility of this unit in terms of macronutrients is nitrogen (N) 0.019 per cent, phosphorus (P) 0.362 per cent and potassium (K) 1.84 per cent. It has exchangeable bases like calcium (Ca^{++}) 5.07 me/100g, magnesium (Mg^{++}) 3.64 me/100g, sodium (Na^+) 10.25 me/100g and potassium (K^+) 0.48 me/100g and cation exchange capacity (CEC) 19.74 me/100g. It can be used for intensive cultivation.

TMU-26 (OAP + SWT + SL)

The TMU-26 is developed over Older Alluvial Plain and comprise sandy loam with shallow watertable conditions. The unit occurs on nearly level to gently sloping plain lands. Its soil is a member of coarse loamy, mixed hyperthermic family of Fluventic Ustochrepts of Inceptisols Order. The soils of this unit are well drained with moderate permeability and suffer from slight to severe erosion. These soils are non-calcareous and are mostly under Kharif and Rabi crops.

The soils of this unit are very deep, yellowish brown dark in colour and sandy loam in texture. Granulometrically, the surface soil of this unit comprise sand 55.32 per cent, silt 32.92 per cent and clay 11.76 per cent. The soil has organic carbon (OC) 0.05 per cent, calcium carbonate ($CaCO_3$) 1.93 per cent, hydrogen ion concentration (*p*H) 7.8 and electrical conductivity (EC) 0.85 mmhos/cm. The fertility of this unit in terms of macronutrients is nitrogen (N) 0.004 per cent, phosphorus (P) 0.045 per cent and potassium (K) 2.892 per cent. It has exchangeable bases like calcium (Ca^{++}) 4.03 me/100g, magnesium (Mg^{++}) 3.37 me/100g, sodium (Na^+) 0.32 me/100g and potassium (K^+) 0.32 me/100g and cation exchange capacity (CEC) 7.89 me/100g.

The TMU-26 can be used for moderate cultivation, intense grazing, wildlife and forestry. It is suitable for sorghum, millet, pulses and groundnut of Kharif crop and gram, mustard and barley of Rabi crop.

TMU-27 (H + DWT + LS)

The TMU-27 comprise Hills and Ridges of Bharatpur district and is associated with Hogback and Cuesta of Delhi Supergroup. This is characterised by deep watertable and moderate to very steep slope. The unit has loamy sand soil texture and is a member of coarse loamy, mixed hyperthermic family of Lithic Ustorthents of Entisols Order. The soils of this unit are non-calcareous, excessively drained with high permeability and suffer from severe erosion, and are mostly under bushes and shrubs forest.

The soils of this unit are very shallow, yellowish to dark reddish brown in colour and gravelly sandy loam in texture, comprising sand 70.0 per cent, silt 20.0 per cent and clay 10.0 per cent. The soil has organic carbon (OC) 0.045 per cent, hydrogen ion concentration (*p*H) 7.4 and electrical conductivity (EC) 0.24 mmhos/cm. The fertility of this unit in terms of macronutrients is nitrogen (N) 0.004 per cent, phosphorus (P) 0.055 per cent and potassium (K) 1.171 per cent. The soils of this unit have exchangeable bases like calcium (Ca^{++}) 3.47 me/100g, magnesium (Mg^{++}) 2.89 me/100g, sodium (Na^{+}) 0.51 me/100g and potassium (K^{+}) 0.18 me/100g and cation exchange capacity (CEC) 7.05 me/100g.

The TMU-27 can be used for limited grazing, wildlife and forestry. The lands of this unit are not suitable for irrigation due to soil (shallow) and topography (moderate to steep slope) limitations.

TMU-28 (H + SWT + LS)

The TMU-28 comprise hills, shallow watertable and loamy sand is in Rupbas tehsil of Bharatpur district. It occurs on nearly level to moderately sloping lands and suffer from slight to severe erosion. It is a member of coarse loamy mixed hyperthermic family of Typic Ustochrepts of Inceptisols Order. The soils of this unit are calcareous, poorly drained with low permeability and mostly under Rabi crops.

The soils of the TMU-28 are very deep, greyish brown in colour and sandy loam in texture. The fertility in terms of macronutrients in nitrogen (N) 0.014 per cent, phosphorus (P) 0.040 per cent and potassium (K) 2.02 per cent. Cation exchange capacity (CEC) 16.95 me/100g.

The TMU-28 can be used for moderate cultivation, intense grazing, wildlife and forestry. It is suitable for wheat and barley of Rabi crop and sugarcane and millet of Kharif crop.

TMU-29 (EV + DWT + H)

The TMU-29 is developed in Erosional Valley characterised by hills and deep watertable in Bharatpur district. It forms moderately to highly sloping undulating lands and suffer from slight to severe erosion. The soils of this unit are excessively drained with high permeability and are mostly under kharif crops. It is a member of coarse loamy mixed hyperthermic family of Typic Ustipsamments of Entisols Order.

The soils of the TMU-29 are very deep, dark yellowish brown to dark brown in colour and sandy in texture. Granulometrically, the surface soil of this unit comprise sand 83.76 per cent, silt 6.92 per cent and clay 9.32 per cent. It has organic carbon (OC) 0.14 per cent, calcium carbonate ($CaCO_3$) 0.33 per cent, hydrogen ion concentration (*p*H) 7.9 and electrical conductivity (EC) 0.66 mmhos/cm. The fertility of this unit in terms of macronutrients is nitrogen (N) 0.012 per cent, phosphorus (P) 0.142 per cent and potassium (K) 1.087 per cent. It has exchangeable bases like calcium (Ca^{++}) 4.88 me/100g, magnesium (Mg^{++}) 4.07 me/100g, sodium (Na^{+}) 1.57 me/100g and potassium (K^{+}) 0.25 me/100g and cation exchange capacity (CEC) 10.77 me/100g.

The TMU-29 is not suitable for cultivation, it can be used for intense grazing, wildlife and forestry. Only sorghum and millet of Kharif crop can be grown on this unit.

TMU-30 (BP. + DWT + LS)

The TMU-30 is developed over Buried Pediments by loam sand, deep watertable and occur on nearly level to gently sloping plain lands and suffer from slight to moderate erosion in Bharatpur district. The soils of this unit are non-calcareous, moderately well drained with modérate permeability, and are mostly under grass and grazing lands. It is a member of coarse loamy mixed hyperthermic family of Fluventic Ustochrepts of Inceptisols Order.

The soils of this unit are very deep, dark brown in colour and sandy loam in texture. The fertility of this unit in terms of macronutrients in nitrogen (N) 0.028 per cent, phosphorus (P) 0.051 per cent and potassium (K) 2.09 per cent and has cation exchange capacity (CEC) 10.43 me/100g.

The TMU-30 can be used for intensive cultivation, intense grazing, wildlife and forestry. It is suitable for sorghum, millet, gwar, cotton, pulses and groundnut for Kharif crops and wheat, barely, gram, mustard and peas of Rabi crop.

Acknowledgements

The work was carried out as part of NRIS project, National Informatics Centre, Planning Commission, Government of India, New Delhi. The authors acknowledge with thanks the financial assistance from National Informatics Centre under guidance and overall supervision of Professor Iqbaluddin, Remote Sensing Applications Centre, Aligarh Muslim University, Aligarh; Dr. N. Seshagiri, Special Secretary, Planning Commission & Director General, National Informatics Centre are thankfully acknowledged. Thanks are due to Dr. (Mrs.) Vandana Sharma, Head GIS Division National Informatics Centre for useful discussions. Thanks are due to Professor Ali Mohammad, Department of Geography for useful discussions.

REFERENCES

All India Soil and Landuse Survey (1972), Soil Survey Manual (Revised), IARI, New Delhi.

Anon (1989), Report of the Committee of the Study Group on Information Gap, Planning Commission, Government of India, New Delhi.

DISNIC Handbook, District Information System of National Informatics Centre, Planning Commission, Government of India, New Delhi.

DISNIC Highlights, District Information System of National Informatics Centre, Planning Commission, Government of India, New Delhi.

Iqbaluddin (1983) Multiple Database Approach to Micro-level Planning, An Integrated Study in Ghaghara-Gandak sub-basin, Detailed project report, Department of Science and Technology, Government of India, New Delhi.

Iqbaluddin (1996) Report Geology of Bharatpur District, Rajasthan, National Informatics Centre, Planning Commission, Government of India, New Delhi.

Iqbaluddin (1996) Report of Geomorphology of Bharatpur District, Rajasthan, National Informatics Centre, Planning Commission, Government of India, New Delhi.

Iqbaluddin (1996) Report Groundwater of Bharatpur District, Rajasthan, National Informatics Centre, Planning Commission, Government of India, New Delhi.

Iqbaluddin (1996) Report of Soils of Bharatpur District, Rajasthan, National Informatics Centre, Planning Commission, Government of India, New Delhi.

Iqbaluddin, Saifuddin, Javed, A. and Moni, M. (1999), Terrain Mapping Unit based approach for sustainable agricultural development in India—An example from plains ecosystem at semi-arid interface, *Jour. Arid Environments*, Academic Press, Vol. 42, pp. 319-330.

Khan, S. and Qureshy, M.A. (1983) *Natural Resource Data Management System, District Database Gurgaon, Concept, Problems and Perspective,* Department of Science and Technology, Government of India, New Delhi.

Saifuddin (1998), Application of GIS Technique in Preparation of Terrain Mapping Units in Parts of Bharatpur District, Rajasthan, Project Report, Geoinformatics Division, Indian Institute of Remote Sensing, National Remote Sensing Agency, Department of Space, Government of India.

Saifuddin, Iqbaluddin and Ahmad, M. (1998), *Terrain Mapping Units of Bharatpur District,* GIS Development, New Delhi, India, July-August, 1998, pp. 35-38.

Saifuddin (1999), Land System Studies using Statistical Image Analysis Techniques for Remotely Sensed Data and GIS Applications in Parts of Yamuna Basin in Aligarh, Mathura and Bharatpur Districts, Ph.D. Thesis, Aligarh Muslim University, Aligarh, 334 p. (Unpublished).

Sehgal, J. and Sharma, J.P. (1994) *Soil Climatic and Resources Characterization of Dryland Ecosystems in India for Sustainable Agriculture, Soil Management for Sustainable Agriculture in Dryland Areas*, India Society of Soil Science, New Delhi.

13

GEOMORPHIC LANDFORMS OF IRAN FROM AGRICULTURAL DEVELOPMENT PERSPECTIVE

An Assessment

MOHAMMAD JAFAR ZOMORRODIAN

INTRODUCTION

The geomorphic landforms and processes, based on their types, nature and potentials, have led to the genesis of various agricultural landscapes, in historical periods; and as one of the fundamental factors have played an important role in agricultural development. Therefore, considering the topographic and geomorphic elements in land-use and rural planning of the farming spaces is indispensable. Furthermore, to conservate and develop agricultural landscapes on national, regional and zonal scale, the geomorphologic studies are very important. The importance of the subject becomes obvious when agricultural spaces occupy first vast lands and second superficial crisis, strikes environmental equilibrium, and the life is threatened (Tricart, 1991).

The importance of the matter in Iran is twofold: because first, over 60,000 rural settlements as well as one-third of the whole population are involved in farming, second, the geomorphology in Iran is structurally and climatically in such a way that most of the roughness lack the required potential for farm and only one-fourth of the geomorphic units over the country is suitable for farming. The spatial dispersion of agricultural lands in Iran is subjected to natural conditions and specially geomorphic potential; and their settling is of course the result of action and reaction as well as long-term adaptability to environment. Taking this matter into account will be the guide to agricultural planning.

BASIC FRAMES OF GEOMORPHOLOGY OF IRAN

Plateau of Iran with the average height of 1000 metres, in view of structural geomorphology and endogen dynamic, has shaped by orogenic and epeirogenic movement of different Alpine phases, and specially its last phase (pasadenien), and forms the connective link between Anatolia plateau (in northwest), and Pamir and Tibet plateau (in northeast and east).

The plateau is a bowl-like one that is surrounded by mountainous rims; and except the connective link, it overlooks to seas and lowlands (plains and coastal plains) from other sides. This characteristic has provided geographical independence of plateau, and access to it in the political realm of Iran is impossible. In the plateau interior also there are independent varied roughness: The folded and broken ranges, single or continuous volcanic masses, and lowland plains with desertic feature, etc. are the most typical.

From the point of view of climatic geomorphology and dynamic exogen, Iran is located at the heart of "desertic belt" of north hemisphere. Air subsidence and dominance of "Azoure" and subtropical high pressure on the one hand and the mountainous rims of plateau, specially in north and west, on the other hand prevent the west low pressure systems and rain production, and has caused desertic arid conditions in the plateau interior.

Of course, roughness trend, extension of Iran in latitude direction, and height factor, have reduced and limited the arid intensity and in some areas have led to its disappearance. Therefore, the mentioned factors have become the origin of environmental differences in Iran so that sometimes two basins with quite distinct human and physical features are located side-by-side (such as green and fertile coastal plain in the north and dry regions on the south of Elborz), whereas in some places, such as "Lut", and Dasht-e-Kavir, an eccentric similarity is observed for hundreds of kilometres.

GENERAL VIEW OF MORPHOLOGY OF IRAN

From roughness point of view, Iran is divided into two main parts, that mark the agricultural realm of this country.

Highlands and Mountainous Areas

Generally, about one-half of the country is mountainous; including north (Elburz), west (Zagross), east and interior scattered ranges. The highest peak is "Damavand" in Elburz, that with 5610.27 ± 0.4 metres high, is one of the highest peak in Alpe-Himalaya mountains; and is the highest peak in West Asia. The height of the rim mountains is often more than 2500 metres (Table 13.1 and Fig. 13.1).

Table 13.1 : Altitude Levels of Iran

Height (m)	*Under zero*	*0-500*	*500-1000*	*1000-2000*	*More than 2000*	*Level of interior lakes*	*Total*
Area (km^2)	11000	332000	154000	879000	260000	14000	1650000
Ratio (%)	0.9	20	9.3	53.3	15.7	0.7	100

Source : A Ivazi, 1995, p. 11.

The figures above introduce Iran as a highland with much complications. It is clear that these altimetric characteristics, affect the climatic geomorphology and exogen dynamic of the country. For example the considerable height of plateau, firstly reduces temperature (maximally 30°C in Damavand peak, and minimally 10°C in most of plateau) and as a result influences the increase of precipitation and snowfall coefficient. These conditions in turn, affect hydrography network, outgoing of streams and erosive mechanism, and finally lead to more permanent streams and normal erosion. Of course, height differences in mountainous areas of Iran suggest that highlands of the country are very young and steep. Therefore, from the point of view of erosion and morphogenesis they are very active. Morphogenesis-pedogenesis balance of these areas is in favour of morphogenesis; hence are not suitable for agriculture.

Flat and Lowlands

The intermount plains at inner plateau and coastal plains in outer plateau portion, face highland of Iran. This portion is more extensive than rough highlands, and the view of 109 sheets of "Slopes Atlas of

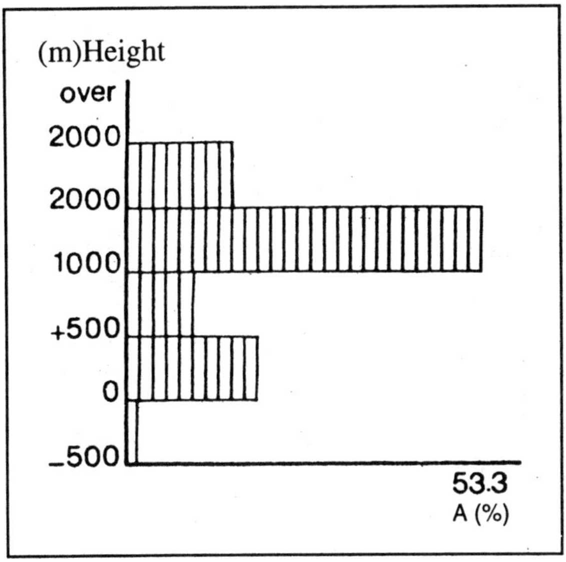

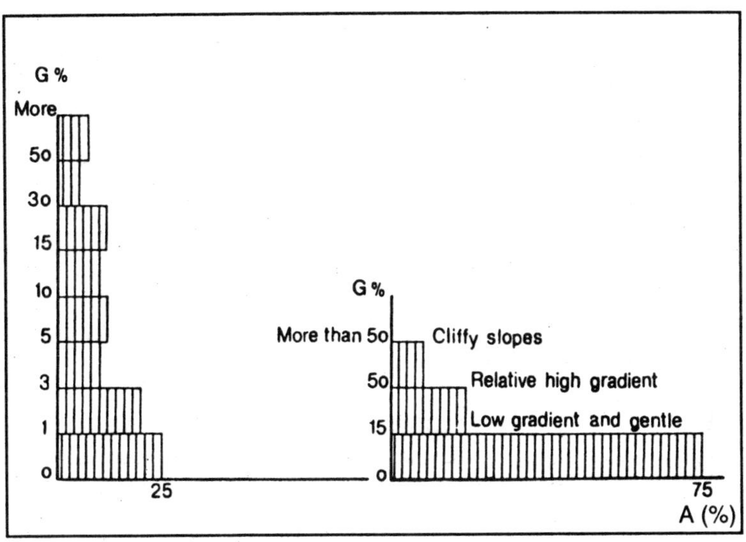

Fig 13.1 : Altimetric and Gradient Frequency of Iran

Iran* suggests that the slope of 54 per cent of the country is between 1 per cent to 5 per cent (Table 13.2 and Fig. 13.2).

It is remarkable that, smooth and gentle areas are very extensive, therefore, about 75 per cent of the country area is cultivable (of course only by considering this parameter). But this is not possible in practice; because remarkable parts of lowlands of Iran are covered by desertic areas and playa such as Lut, Dasht-e-Kavir, Jazmoorian etc. Hence they have no desirable potential for agricultural activities.

Table 13.2 : Slope Position in Iran

Rate of slope %	*Area (km²)*	*Area ratio to total %*	*Form of slope*	*Roughness conforming to slope*
1	352044.9	25	Levelled	Smooth lands: inner plains, flat and coastal plains, flood plains, terraces, alluvial cones, and sedimentation areas
1-3	281457.1	20	and	
3-5	134794.7	9	steep-weak	
5-10	171847.1	12	Gentle steep	Glacis, gentle slope of the plains margin
10-15	126197.2	9	Moderate steep	Erosional areas
15-30	152619	11	Relatively excessive	Mountainous and piedmont areas
30-50	89239.5	6	Excessive	Mountainous areas
More than 50	118268.3	8	Very excessive and cliffy	Prominent and rocky areas of mountains
Total	1426467.2	100		

Source: Slope Atlas of Iran with little change.

CLASSIFICATION OF THE LANDFORMS IN TERMS OF AGRICULTURAL POTENTIAL

With regard to indicators such as; general morphology, relative flatness, slope rate, scale, area, type of the formation and its structure—texture, stability of the geomorphic environment, etc., and in terms of their potential and fertility to agriculture; and land froms of Iran, have been divided into 4 or 5 groups. These groups are marked on the attached map as well. They are as the following:

* This Atlas contains 134 sheets with a scale of 1:250000.

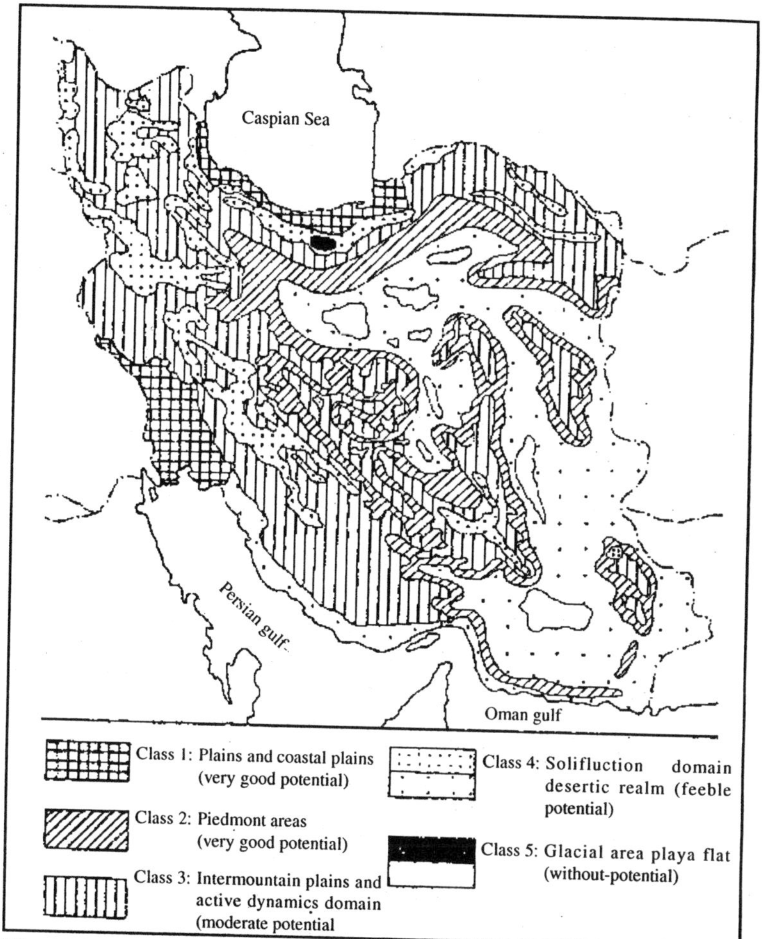

Fig. 13.2 : The Geomorphic Units of Iran in terms of Agricultural Potential

Class 1

The stable landforms with very high potential (extensive plains and coastal plains)—this group contains the north coastal plains (along the shore of Caspian Sea), south coastal plains (specially Khouzestan coastal plain), and some of the inner plains (for example Mahidasht). In terms of the genesis and the origin of the formation, these landforms have been divided into 3 groups :

(i) *Structural plains:* that have been formed on the block, shields, or resisting crystalline cores. There is only structural plain in Iran, and that is Lut.

(ii) *Erosional plains*: that their formation being the result of initial destructive processes (weathering and alteration), and erosional processes (by runoff and sometimes wind). These plains, in general, have been formed in the tectonic depressions or hollows in the central part of Iran plateau, resulting from erosion of Neogen formations or mountains frontier parts. Thus the glacises and some of the piedmont plains are included in this group.

(iii) *Depositional plains*: including most of the plains in Iran, resulting from sedimentation in troughs, vast valleys and depressions adjacent to the mountains. The sedimentation mainly has occurred during the pleistocene erosional phase (Mahmoudi, 1976; 274).

In general, depositional plains, especially, those located in northern and south-western coasts and those situated between the west mountain, are the most suitable places for agricultural activities and development. Coastal plains of Iran have occupied 11,000 km^2 (7%) of the country area. Khouzestan plain with 50 per cent, is the largest of all, and northern coastal plains (Guillan, Mazanderan and Golestan provinces) have occupied 25 per cent of the total area.

These regions, are the most important and largest cultivable lands for agriculture land-use planning and development; because they have a gentle morphology with low uniform gradient (less than 3% or 2°), without any problem or topography obstacles. Gradient of these landforms is desirable for the most activities, and easy for the usage of mechanized cultivation, agriculture, heavy machinery such as weeding, seeding, plough, etc. Furthermore it will be good for making canal, Qanat, etc. to carry water (Clows & Comfort, 1989, reported from Young, 1975 and Croft, 1973).

Geomorphic environment of this class of landforms is essentially stable. Hence morphogenesis-pedogenesis balance is in favour of pedogenesis. So, the soil has thickness, depth perfect horizons, and on the whole relatively light for ploughing. In some places such as Guilan that soil of plain comes from volcanic materials (the volcanic areas of Sahand, Sabalan, Zanjan, Hamedan, Kordestan, etc.) and contains abundant silica, the soil is very fertile and produces the best rice of Iran (for example rice of Astaneh) (Amin Sobhani, Ebrahim, 1986).

Since most of these geomorphic features are the product of hydromorphodynamic processes and alluvium materials, they have been fed from past to now, and have formed important aquifers (with fossil waters and fresh waters). Also the gentle morphology and formation texture of these landforms have led to more easy water penetration.

The whole of above properties, caused by suitable topographic and geomorphic conditions, have provided very good potentials for agricultural activities and cultivation. Consequently, the landforms of this class were occupied by mankind, and many rural and urban settlements were established within them. Even ancient civilizations such as "Shoush" are formed by them. At the present time the large cultivation units that have appeared these landforms, are good evidence to the claim.

Class 2

The semi-stable landforms with good potential (piedmont plains)—piedmont aeas within the plateau such as glacis, Bajada (Bahada), vast alluvial cones (fans), flood plain, extensive deltaic plain (as Sefidrud delta or Arvandrud delta) are included in this class.

Class 2 landforms are more limited than Class 1, but with a more gradient (3-5%). This group also has a flat and uniform surface, however in some cases alluvial cones were dissected by rivers or there are a few marshes and pools on deltaic plains. The presence of piedmont across faults, outgoing of mountainous floods into piedmont portion, relative gradient, and river overflowing on flood plains or deltaic plains, etc., set the Class 2 landforms in semi-stable geomorphic environments. So that the balance of morphogenesis-pedogenesis is almost in equilibrium state, and any wrong policy or activity may lead to instability of the environment.

The mentioned characteristics have brought some limitations and problems for cultivative activities. However, relative rich soil, considerable water resources, relative flatness, especially in alluvial cones have provided good potentials for agriculture of Iran. These factors along with cooler and more humid climatic conditions (than inner plateau) have caused the appearance of several rural and urban settlements (such as Tehran, Karaj, Shahrood, Semnan, Isphahan, etc.) on the inner piedmont strip.

Alluvial cones on the margin of central and southeastern deserts of Iran, have been formed by the fluctuations of cold and humid periods, therefore, these landforms are among heritages of pleistocene climatic and geomorphic. Some of important alluvial cones in Iran are : The alluvial cones of western Lut, the large alluvial cone on the northeast of Dasht-e-Kavir, Tehran alluvial cone, the vast alluvial in Qomrud, Qarachai and Rudshour on the east side at Huz-e-Soltan and the alluvial cones on the north and south at Jazmoorian.

Another landform in this group is deltaic plain. The most important of them has been formed on the mouth at Arvandrud (on the northwest shore at Persian Gulf). The whole of swampy lands in this region should be considered as Holocen deltaic deposits at the present time. In this district there is another delta called Abadan Delta that has been formed by alluvial at rivers such as Shadgan Bahmanshir etc., and Abadan city is founded on it. The Minno island is one of the small triangles of this delta (Darvish Zahed, 1991, pp. 753 and 754).

Class 3

Instable landforms with moderate potential (domain of active dynamics)—the foothills, specially at the northern and western mountains of Iran are set in this group. The most important topographic and geomorphic characteristics of this region, is their slope or gradient, that ranges from 10 per cent or 15 per cent (moderate gradient) to more than 50 per cent (excessive and cliffy).

These landforms include about 35 per cent of the country area. It is noticeable that on the slopes of 19 per cent, local development and the use of wheeled tractors are possible (Clows and Comfort, 1989).

These regions, because of high gradient, rocky areas, shallow soils and limited lands, and instability of their geomorphic environment have not, essentially, provided favourable conditions for cultivation. Farming in these landforms faces active factors of morphogenesis more (Rajaii, 1993). This characteristic is revealed, specially, when farmers, in the lack of terra, have to cultivate on slopes (foothills), and consequently farming activities (as contour plough), irrigation and water transfer, etc., increases the instability of lands. Therefore, occur the mass movement such as landslide, slump, glide, etc. furthermore on the higher areas of foothills and under the snow-line (4000 m in Elborz and Zagros) occurs solifluction and gelifraction that produce problems for cultivation.

In these areas, the use of heavy machinery, mechanized farming and tasks such as remembrement is not possible. Consequently the rate of output per unit decreases (Zomorrodian, 1996, p. 83). Furthermore, with regard to climatic conditions, variety of cultivation is almost limited. Hence, three main operations have been accomplished in the historic periods on the mountains and foothills.

(*i*) The use of relative rich pastures and grasslands for grazing of domestic animal, specially in Elborz and Zagros. This type of operation has led to promotion of animal husbandry economy (in the form of herding, transhumance...) and has formed the tribal (tribesman) structure. Therefore the fixed living residented has flourished less in these regions.

(*ii*) Dry farming on lower altitudes based on relatively favourable rainfall and soft-soiled areas.

(*iii*) Establishment of fruit gardens on the lowest stages of slopes near the valley floors, especially on the alluvial terraces. It is considerable that some of the vallies specially in Zagros are of gap or gorges type, and because of geomorphic circumstance (bottleneck) lack the agricultural potentials.

In this group, the alluvial terraces are the most favourable places for agricultural activities, specially fruit gardens. The existence of the vast stratified and regular alluvial terraces at the margin of present arid vallies, indicates the domination of ancient humid periods in these zones, for example we can refer to the alluvial terraces that are present in the driest regions at Baluchistan (southeast of Iran) and have sometimes kilometric dimensions: Nikoo Jahan in the south of Nikshahr (Baluchistan), Drakhtangan in the north of Shahdad (west of Lut), etc. are example of these terraces.

These features include flat areas, relative fertile soil and water resources, that can be suitable for agricultural activities. But the existence of coarse rubble stones, occurrence of landslide and mass movement (specially on the rivers banks and meanders) are the most important problems in these regions.

Class 4

The land forms with feeble potential (desertic domains, and solifluction zones)—the central plains of Iran that surrounded by rim mountains, are different in height. Their average height is about 1000 m and generally, reduces from west and northwest to east and southeast. The highest of them is Havatoo (north of Sanandaj) with 2200 m high and their lowest is Lut (with a minimum of +56 m) (Mahmoudi, 1975, 1977).

Factors such as height, latitude, physical conditions of the mountain rims etc., has led to climatic changes from west to southeast plains, and domination of desertic and dry climate in the east and southeast. Even in some cases geomorphic features such as salty and saline lands, reg, sand winds and dunes, mud crack, wadi (*oued*), poeriodical and occasional floods etc., have limited and threatened agricultural activities, in these plains. In these regions even a 10° gradient leads to soil erosion (Clows & Comfort, 1989). Therefore, aforementioned regions have not required abilities for cultivation. Hence farming activities are accomplished in a local, scattered, and limited with very low output and confined to oases.

Class 5

The landforms lack of potential—the lowest and most central parts of plains (heart of depression and plains) that covered by playa areas, sand dunes, etc.; including of the highest peaks (the volcans of Damavand, Sahand, Sabalan, Bazman and glaciers summits such as Alam-Kuh in Elborz) that hidden by permanent snows or neve; in substance and naturally cannot be used for agricultural activities.

REFERENCES

Amin Sobhani, (1986) *Regional Geomorphology of Iran.*

Clows, Alan and Comfort Peter (1989), *Process and Landform, Conceptual Framework in Geography,* Third impression, Oliver and Boy.

Darvishzadeh, Ali, (1991), *Geology of Iran*, Published by Amirkabir.

Farhoudi, R. *et al.*, (1982), *Slope Atlas of Iran*, Institute of Geography, Tehran University.

Hooke, J.M., (1994), *Geomorphology in Environment Planning*, Translated by Mohammad Jafar Zomorrodian, Published by Samt.

Jedari, Aivazi, Jamshid (1996), *Geomorphology of Iran*, Published by Payam-e-Nour University.

Mahmoudi Farajollah, (1975), *Physical Geography of Iran*, University of Tehran.

Mahmoudi Farajollah, (1977), *Geography of Iran*, Published by Ministry of Education.

Mahmoudi Farajollah, (1988), *Roughnesses Evolution of Iran in Quaternary*, Publication of Geographical Researches, No. 23, Tehran University.

Rajaii Abdolhamid, (1995), *Application of Geomorphology in Landuse Planning and Environment Management*, Published by Ghoumes, First Print.

Tricart Jean, *et al.*, (1993), *Affairs of Rural Development* (Landscapes, Rural Spaces, and Farming Systems) Translated by Sarvghad Moghaddam A., Published by Astanghods, First edition.

Zomorrodian Mohammad Jafar, (1996) *Application of Physical Geography in Urban and Rural Planning*, Published by Payam-e-Nour University.

14

SMALL FARMER DEVELOPMENT STRATEGIES FOR THE NEXT MILLENNIUM

K.V. SUNDARAM

The Problem in its Setting

A Central issue in Agricultural Development is the necessity to increase productivity, employment and income of poor segments of the agricultural population. Among the rural poors, the small farmers constitute a sizeable portion in the developing countries. Studies by FAO have shown that small farms constitute between 60-70 per cent of total farms in these countries and contribute around 30-35 per cent to total agricultural output (Randhawa and Sundaram FAO, 1990). The inequality in landholding distribution is high in most Asian countries, except China and the Republic of Korea. Small farms generally predominate in several countries in the region. Bangladesh, India, Indonesia and the Republic of Korea have a particularly large number of small holdings below 1 ha, while Pakistan, Thailand and the Philippines have relatively few extremely small holdings. Looking at the countries by farm size, large farms seem to control a disproportionate amount of land. The largest 10 per cent of farms control about 40 per cent of land in Pakistan, 50 per cent in India and 35 per cent of rice land in Indonesia. On the other hand, in the Republic of Korea and Japan, the distribution of farms is less inequitable (FAO, 1990). These percentages underline the inequitable concentration of ownership and operational land-use in these countries. The size of a small farm varies from country to country. In India, a small farm is one which is below 2 ha in size. Those cultivating farms below one hectare in size are classified as marginal farmers. The cultivable land per capita declined in India from 0.48 hectares in 1951 to 0.26 hectares in 1981. It is expected to be 0.11 hectares in 2000. (Swaminathan, 1990) FAO's Study of agriculture at 2000 shows that in most developing countries, the average size of an operation holding will continue to decline (FAO, 1988). This policy analysis Chapter addresses the several problems involved in designing an efficient agricultural strategy, that has an appropriate focus on small farmers, and identifies the configuration of policies needed in the next millennium, to restructure the agricultural sector in order to achieve growth with equity. The policy analysis attempted here is based mainly on two sources.

1. A study of small farmer development in the Asia and the Pacific Region undertaken by this author (with N.S. Randhawa) in FAO, of seven countries *viz.* India, Indonesia, Malaysia, Bangladesh, Sri Lanka, The Philippines and Thailand (Randhawa and Sundaram, FAO 1990).
2. A review of Rural Development activities in 12 countries in the South and South East Asia region, undertaken by this author for the Centre on Integrated Rural Development for Asia and the Pacific (Sundaram, CIRDAP, 1996).

Characteristics of the Small Farmers

Basic to the formulation of strategies for the small farmer development, is an understanding of their behavioural characteristics and the various constraints they face in the real world context. Several farm management studies in India and other countries have brought out the salient characteristics of small farms, some of which may be highlighted. In the small farms, agricultural work is carried out mostly by family labour and a large part of family subsistence requirements are met directly from farm production. This would mean that they have a small hired labour component. At the same time, a number of small farmers also participate in labour markets as wage earners. Thus, they are both producers and workers at the same time. Small farms have a high proportion of land devoted to food crops, which is necessary in order to meet the family subsistence needs, *i.e.*, for their "use value". Small farmers also cultivate other crops for their "exchange value". Crop-mix and intercropping are the characteristics of small-farm agriculture, particularly in rainfed areas. A high proportion of their output is retained for home consumption, that is, they have a lower ratio of marketed surplus to output. Small farms grow a diverse mixture of crops, that allow them to hedge rationally against uncertainties (*e.g.*, agroclimatic conditions and weather fluctuations) and risks (*e.g.*, variation in output prices). Crops are often interplanted. There is ample evidence to show that small farmers generally have higher cropping intensities, more area proportionately under irrigation, and use their resources more efficiently.

In the small-farm household, special emphasis is given to the generation of scarce cash resources, risk avoidance and food security, and in particular to all those activities that will yield fair returns to their most limiting factor of production (land). What is practiced in this context is a complex system of activities — crops, livestock and off-farm activities — in which members allocate the factors of production to each activity in a manner which, given their knowledge, maximizes the goals they strive to reach. Small farms are generally characterized by very low operating capital turnover in production, coupled often with a higher debt burden in relation to resources. This contradiction is due to their need to borrow heavily for consumption purposes, in order to sustain the family from one production cycle to another.

Small farmers are usually unable to make heavy on-farm investments, and in this sense, they are technologically constrained. The small farms are also usually characterized by weak forward and backward linkages except some, which are engaged in commercial enterprises. In regard to the adoption of new technologies by small farmers, three critical lessons emerge from a wide array of studies and approaches. These are :

(i) Small farm families are receptive to change and small farm systems are dynamic. The concept that "traditional" agriculture is static is misleading.

(ii) Small-farm families are selective and adaptive in their adoption and use of recommended practices and technologies.

(iii) No single attitude, trait, factor, or farming condition explains the pattern of small-farm adoption of all new innovations (FAO, 1986).

Constraints of the Small Farmers

The constraints of small farmers may be broadly classified as structural, technological, institutional and in reality, many constraints are interrelated and interdependent. Because of small size of land, small farmers are compelled to augment their operational area through leasing land. The terms and conditions they face in tenancy market are unfavourable. Although seed-fertilizer technology is supposed to be scale-neutral, in practice, various factors like the community's power structure, uneven access to agricultural extension and information services, inability to obtain adequate irrigation water and unavailability of credit act as deterrents, giving large farmers a clear edge over small farmers. Among institutional constraints, special mention must be made of the unequal access to credit that exists between small and large farmers, and the fact that large farmers manage to corner a higher proportion of institutional credit provided by cooperative credit socieities, etc. Small farmers' economic constraints stem mainly from their limited land area, low levels of farm output and income, very little net surplus, and consequently, extremely limited possibility for farm investment out of their own resources. The rural markets in these countries are also functioning in a way unfavourable to small farmers. They do not receive timely market information and experience considerable difficulties in obtaining access to government services, such as extension. The locational inaccessibility of small farms (with respect to irrigation, roads, services etc.) has been cited especially as a constraint in some studies. Policies and programmes pursued for the development of small farmers: A cross-country analysis. The critical policies and programmes affecting small farmers, by and large, fall in seven major areas. These are:

(i) Agrarian Reforms;
(ii) Institutional Credit;
(iii) Infrastructure and Input Policies;
(iv) Special programmes;
(v) Extension and Research;
(vi) Organization and Administrative Strategies; and
(vii) Training.

We may briefly analyse the above policies hitting at some high spots of the evaluation across countries.

Agrarian Reforms

Studies show that in India, agrarian reforms, on the whole, have not made a desired impact on the small and marginal farmers. Their position in the land-lease market remains vulnerable. On the whole, the ownership status of marginal/small farmers has not improved beyond a point, whether due to the failure of tenancy laws or ceiling laws or both (Chadha, 1986). In Indonesia, it is reported that the Land Reform Law (1960) has not been effectively implemented owing to the unholy alliance between land reform officials and the large land owners. In the Philippines, the Agricultural Land Reform Code (1963 as amended in 1971) is reported to be a moderate success (Sirirugsa, 1986). In Thailand, the Land Reform Act (1975) only partially succeeded in achieving the targets. In the South Asian Countries, where any substantial expansion of holding size is ruled out by land expansion, the redistribution of land through agrarian reforms seems to be the only potential solution.

Credit Policies

Since the late 70s, profound changes have taken place in many countries in regard to agricultural credit policies and programmes. These changes include several policy steps on simplification of lending procedures, interest rates, refinance measures, expansion of the banking network, introduction of supervised credit and in some cases, the doing away with collateral security-based credit system for small farmers and landless labourers. Many country studies reveal, that while informal credit sources still continue to hold sway in different degrees, the availability of institutional credit to farmers has grown in volume tremendously over the years. The provision of concessional loans to farmers has been adopted as a major policy instrument for inducing the adoption of improved farm technology. In some countries, it exists as a general policy applicable to all farmers alike, while in certain others, special programmes of credit for small farmers have been devised. More and more countries seem to be veering round to target-group orientation in the matter of disbursement of agricultural credit.

Price Policies

The weight of evidence thrown by the country experiences points to the conclusion, that the price support policies have, by and large, achieved some success. Fluctuation in prices between the peak and lean seasons have considerably narrowed in most of the countries.

There is no evidence to suggest that the small farmers are discriminated against in the matter of prices received by them. "The concept of 'weak bargaining power' of such farmers is disappearing fast. Some policy measures, which might greatly benefit the small farmer from the government support price may be noted. These are:

(i) Purchases being made directly from the farmer than through intermediaries.

(ii) Ensuring that "procurement centres" and "buying points" are located within easy access of the farms.

(iii) Price support being geared to the nature of crops grown by the farmer, bearing a favourable relation to input prices.

(iv) *Infrastructure and input policies :* There is, no doubt, that the provision of subsidy *e.g.* for fertilizers has encouraged the small farmers to increasingly adopt the new technology and enhance their productivity. Evidence from Bangladesh shows that any withdrawal of the subsidy will have an adverse effect on the small farmers. Since subsidies are always related to the fulfilment of certain objectives, they need to be gradually withdrawn once those objectives are fulfilled. However, in many countries, this has proved to be difficult. An equally important aspect is the working of the public distribution system. Country experiences have underlined the importance of targetting the supplies and the use of stronger cooperative management to ensure a coordinated and timely delivery of inputs.

(v) *Special Programmes:* Special programmes for small farmers have been functioning in many countries *e.g.* India, Bangladesh, Indonesia, Nepal, the Philippines and Thailand. The major problem about such programmes is their inadequate coverage. Further, they have been confined to the delivery of one or two inputs only *e.g.* delivery of credit inputs ane subsidies. They have been essentially piecemeal interventions and have tended to function as "isolated" or disembodied" programmes.

(vi) *Extension and Research :* Many evaluation studies have emphasised that the extension functions cannot be effectively performed by paid professional workers alone. For better spread effect and effective impact, they have to be organised

through the trained leaders elected by the farming community itself. In this context, it is necessary to encourage the process of training the villagers themselves, so that they become the effective torch-bearers of essential knowledge to their farming community. On the research side, it should be emphasised that the single crop focused research system must be matched to a far higher degree, by farm management research, dealing with the small farm enterprise as a whole. This will be replacing research "by product" by "research by agricultural system", which will have more pay-off for small farmers. Research has also to focus more and more on the problems of rainfed agriculture and of poor farmers in their environmental setting. This is because more poor farmers live in rainfed areas and research has, so far, tended to pay less attention to their specific needs.

(vii) *Organisational and Administrative Strategies :* Country studies indicate that numerous organisations are involved in providing services to the agricultural system. They range from local government institutions, extension network, cooperative organizations, commercial banks, to autonomous bodies set up for specific purposes and a host of private and voluntary agencies. In making appropriate choices of organisational arrangements to help the small farmer, it is necessary that the problem has to be addressed at both ends *i.e.* at the delivering end, through several governmental/parastatal agencies, and at the receiving or utilizing end, through beneficiary groups and/or local organisations. The latter arrangement is necessary to ensure that the benefits do reach the intended target groups. Apart from constituting specific "receiving mechanisms", it is also necessary to provide for the active participation of small farmers in all elected local bodies by powering them, so that their voices could be heard in the appropriate forums and that they derive due share of benefit from the general and specific programmes of development. This will be best served by encouraging the formation of "small farmer groups", to increase the bargaining strength of small farmers.

(viii) *Training :* In a small farmer development strategy, both the

government functionaries as well as the small farmers themselves have to be trained. Government functionaries need training so as to become more responsive to the needs of the small farmers. For the small farmers, the training is necessary to make them aware of the government programmes/projects that have been intended for their benefit and to enable them to shed their shyness and reluctance for approaching development functionaries. In the case of small farmers, since the target group is very large, training is to be imparted in a "cascading" mode, by first training those who have leadership qualities and through them the larger community of small farmers, on the principle of "one candle lights the others".

Past Strategies and their Impact

In the past, agricultural development philosophy, strategy and practice have all contributed to the relative neglect of agricultural development in general and of small farmer development in particular. With the current emphasis in development theory being "Growth-with-equity", "adjustment with a human face" and "sustainable development", it is being increasingly recognized that small farmer development is not only ethical, but also economically efficient for creating a broad-based and long-term agricultural development. The small farmer development has suffered on three accounts: first, from the low priority accorded to the agricultural sector in development; second from several insidious myths, such as the small farmers being tradition-bound, irrational and fatalistic; and three, from a widespread belief that economies of scale are extremely important. In the past, all these notions led to relatively greater emphasis on large and medium farmer development. Modernization and technological development have, however, brought some changes in productivity relations of small farms *vis-a-vis* large farms. According to farm management studies conducted in India, an inverse relationship between farm size and output per acre was discernible in the 1950s. That is, small farms had higher productivity per unit of land due to more intensive cultivation and greater use of traditional inputs, particularly human and animal labour power. In the early 1970s, the inverse relationship weakened. The analysis of extensive 1974-75 farm data indicated that the marginal and small farmers of the Punjab, despite their limited land base, had been able to record as much total crop output and farm business income per acre as their larger counterparts (Bhalla and Chadha; 1983). Thus the small farmers have demonstrated that they are equally as efficient or are indeed more so than even large farmers, given an appropriate environment and suitable technologies. The economies of scale argument, to deprive small farmers of appropriate attention in development, is thus challengeable. In the past, the general Agricultural Development strategy was mostly growth-oriented and has had a built-in bias in favour of large farmers over small farmers. This bias has operated at almost every stage, in each component of planning, investment allocation, policies, programmes and projects. It has ultimately determined the pattern of growth as well as the distribution of its benefits. Eventhough special programmes have been undertaken in several countries to correct these deficiencies, their impact has been limited. Correcting the bias: The bias in the current policies has to be corrected by a conscious restructuring of the agricultural development strategy and planning, which would accord appropriate weight to the development of small farmers, leading to well-balanced agricultural growth, that embodies growth-with-equity and has a decisive poverty-reducing impact. The implications of this restructuring are manifold and extensive. It has to keep in view several dimensions, such as the economic reforms and liberalisation, the globalisation of the economy, the technological developments, economic viability and the role and attitudes of government and NGOs.

Facing the New Millennium: The Agenda for Small Farm Reforms

As we are entering the new millennium, three factors must be recognised for understanding the directions of change and to provide a framework to assess the development of small farms. These are :

1. Economic reforms and liberlisation;
2. Globalisation of the economy; and
3. Technological progress.

The reforms initiated in June 1991, sought to make the Indian economy a fast growing one. Despite the good performance of the economy (7.5% growth rate recorded during three years in succession, 1994-96), our hopes of a sustained higher growth path for the economy were belied and the economy slowed down. However, the present growth is still at over 5 per cent. Inflation is under reasonable control. Exports seem to be turning around. Many new initiatives are under way for the infrastructure sector. The incidence of hunger has been considerably reduced. The ushering in of the information technology has led to the growth of a new breed of skills in the areas of technology and management, which holds out great possibilities for increasing the standards of efficiency and competitiveness among India's firms (Kirit Parikh, 1999). The process of economic reforms and the gradual opening up of Indian agriculture to world markets would create a better incentive environment for agriculture. To fully exploit this opportunity, a major reform in supply-side factors *i.e.* technology, fertilizers, irrigation, infrastructure and credit is needed as well as a total dismantling of all export controls (Hanumantha Rao and Ashok Gulati, 1994). The other factor to be reckoned with is globalisation, which is now a fact, not an option. The highly inequitous nature of the on-going processes of globalization — both among nations and among social classes in most nations — has been focused in the recent Human Development Report (HDR) 1999, brought out by the UNDP. Both the technologies (Information and communications technology and biotechnology), which are the "drivers" of globalisation with their complementarities of liberalisation, privatisation and tighter intellectual property rights, are bound to create new risks of marginalisation and vulnerability. The HDR admits that "elite groups and other interest groups benefit from the present nature and structure of globalisation" (HDR 1999, p. 95). Although the report calls for "globalisation with a human face", it does not spell out the new "humane" rules that will make capitalist globalisation serve human development. The reforms of the 1990s have undoubtedly generated better terms of trade for agriculture by the increase in food prices. But the rural poor, including the small farmers, are not going to be benefited by the higher food prices. Both in the short run as well as in the long run, they will be greatly hurt by higher food prices, if real wages do not rise as fast as relative food prices. Hence the problems of the small farmers are bound to persist in the next millennium. How can they be helped to improve their yield and income in an ecologically sustainable manner and face the new challanges of the millennium is an intriguing question. In this context, two scenarios of change may be visualised. It may be noted that the scenarios depicted here not mutually exclusive. As may be seen in our later discussion, the one may gradually merge into the other. But the changes themselves will be traumatic, as far as the small farmers are concerned.

SCENARIO I

The Decline and Disappearance of Small Farms

Economic reforms and liberalisation of agriculture with their implications of open market operations, will enhance competition in marketing of agricultural production. The small and marginal farmers who have limited staying power, will no doubt find the competition stiff and are bound to suffer to some extent. If economic reforms also imply a reduction in or elimination of subsidies and the government support price programme, they will be affected all the more. If the modern on-going slogans on Population Control like :

1. "Hum Do, Hamare Do" (We are two. Our kids are two)
2. "Paanch Ped, Ek Baccha Yahi Desh Ka Nara Sachcha" (Five trees, one kid. This is the country's real slogan) become a reality, then the small farmer's, family labour will also considerably decline. Thus unable to face the challenges of liberalisation and globalisation and unable to benefit by the higher food prices, the small farmers will be forced to "distress sales" of their land and find employment in non-farm activities or join the ranks of the urban poor. Ultimately, many of them may be reduced to wage earners in large Agri-business firms, conducting agricultural operations over extensive areas.

SCENARIO II

Adaptations and Innovations

Globalisation should translate itself into increased prices for agricultural products and increased employment avenues. However, since these cannot benefit the small farmers they will be compelled to adapt their operations to achieve self sufficiency and an export surplus in farm production. They may achieve this in various ways by making new adaptations and innovations.

The struggle to face the severe competition may lead them to invent new institutional innovations leading to mutual cooperation and collective self-reliance among themselves. They may be compelled to adopt group cooperation in areas such as, land and water management, integrated pest management, integrated nutrient supply, improved post-harvest technology and organise themselves into local level groups (poly-centric groups) for the maintenance and management of beneficiary programmes, for planning and for conducting evolution exercises. In extreme cases, they may resort to pooling of their lands into viable large "farming estates", avail of all latest technologies and compete effectively in the global export market. However in this case, they will require a lot of technical financial and managerial assistance from various sources, including the government, to improve the productivity, profitability, stability and sustainability of their farming systems. The nature of the packages of technology, services and public policies, as well as the desirable forms of research effort necessary in this context have to be worked out meticulously by public institutions, like the Indian Council of Agricultural Research (ICAR) and agricultural universities. Information technology, bio-technology and genetic engineering will be the driving forces of change. The information technology will be able to produce a penetrating and clinical mapping of the land, encompassing the physical, chemical and biological features, underground water resources and forecast of climatic conditions in a focussed manner, that even small geographical segments— the small farms—can be benefited through the guidance provided by the ways in which natural and human resources can be optimally combined with appropriate technologies, inputs and options to enhance and diversify production. Technology, in and of itself, is not sufficient to bring about significant changes in the current economic and social situation of small farmers. It is important to recognise that the adoption of technology calls for the convergence at the small-farm level, not only of appropriate and adequate technology, but also of markets, prices, availability of inputs, adequate credit and everything else that fuels the operation and growth of the production system (Pineiro, 1989). Biotechnology and genetic engineering will help develop seeds to meet specific ecological and biological constraints. The positive features of the local environment can be programmed to be exploited. Tissue culture will help in developing new plant varieties to suit the land and climatic factors.

Genetic engineering would make it possible to introduce genes for DNA or to manipulate plant tolerance for salinity (Pineiro, 1989). The emergence of the second green revolution founded on commercialisation of agriculture and introduction of biotechnology for commercial use, the development of transgenic crops that will allow greater flexibility in crop management, reduce dependence on conventional insecticides and herbicides, increase yields and produce cleaner and higher grade of products, are all parts of the grand vision projected for the next millennium by the scientists. Undoubtedly there are some dissenting voices here about the possible scientific risks in genetic engineering. Notwithstanding these dissensions, one development that is sure to occur is greater and more diversified production through commercialisation of agriculture. Small farmers cannot remain aloof when these developments occur. They will be forced to consolidate their holdings, organise themselves into commercial enterprises adopting the most appropriate form of organisation *i.e.* the company or the cooperative. The former *i.e.* the company has worked extremely well in industry and can give similar results in agriculture too. On the working of the cooperatives too, we have the example of the Anand milk cooperatives.

If these possibilities can be realised, and if Indian agriculture can be protected from the negative fallout of globalisation and on the other hand the full advantage of the opportunities afforded by it, then the picture afforded by the new millennium will be truly exciting. The countryside will be buzzing with economic activities, rooted in the soil, with the small farmer sharing the benefits equitably and with a greater degree of technological awareness, with village clusters that are alive and prosperous, but not in splendid isolation but with a greater sense of community and fellowship in an interdependent framework of rural urban settlements constituting a micropolitan environment. The restructuring implied here entails a greater emphasis on multilevel (decentralised) planning, which will be a major departure from the overall top-down planning. With its devolution of political, administrative and fiscal powers, it will bring in its wake a climate conducive to increased economic activity at the local levels away from urban conglomerates. Small farmers empowered and organized into politically active, local and regional associations and fora, will be transmitting their felt needs forcefully:

(i) for technology to research stations;

(ii) for suitable organizational and administrative structures to administrators; and

(iii) for incentives to policy-makers (Sundaram, 1994). The whole chain of transformation will bring about a symbiotic linkage between agriculture and technology and also between agriculture and rural industry, completing a "virtuous circle" of sustainable growth.

Summing Up

In this Chapter, we have attempted to synthesise the various operational ways adopted by governments in the developing countries of Asia to encourage small farmer development. During the course of the discussion, the relevance of as well as the inadequacies in the various strategies adopted by governments have been critically commented upon. They point to certain correctives as well as lessons and provide some directions for change. There is sufficient indication in our analysis to show that the small farmer has not received the just and equitable share of development benefits from the normal development strategies.

As we are entering the new century, some major challenges as well as transformation are in the offing for the small farmers. There are three major driving forces that will determine these changes. They are : liberalisation and economic reforms, globalisation and technological progress. How these will affect the small farmers is depicted through two plausible scenarios of change, which however, are not mutually exclusive. The emphasis that in the wake of these developments, the small farmer development strategy should make full use of frontier technologies, including bio-technology and ensure the linkages between research, technology and production on the one hand, and effectiveness of the delivery system and extension network to carry the fruits of science and technology to the farmer on the other.

REFERENCES

Alexandratos, Nikos (Ed.), (1988). *World Agriculture Toward 2000—An F.A.O. Study*, F.A.O./U.N. 1988, Belhaven Press, London.

Bhalla, G.S. and Chadha, G.K., (1983). *Green Revolution and the Small Peasant:*

A Study of Income Distribution among Punjab Cultivators, Concept Publishing Company, New Delhi.

Bhalla, G.S. (Ed.), (1994). *Economic Liberalization and Indian Agriculture*, Institute for Studies in Industrial Development, New Delhi.

Chadha, G.K., (1986). *Impact of Agricultural Policies and Programmes of Small Farmers in India*, Report prepared for F.A.O. Regional Workshop on "Impact of Policies and Programmes for Increased Agricultural Production, particularly Food by the Small Farmers, RAPA, Bangkok, 11-15 November.

FAO/UN, (1989). Sustainable Agricultural Production-implications for international agricultural research, FAO Research and Technology paper No. 4, Rome.

Hanumantha Rao, C.H. and Gulati Ashok, (1994). *Indian Agriculture: Emerging Perspectives and Policy Issues*, Intrernational/Food Policy Research Institute.

Kessaba Abbas, M., (1989). *Technology Systems for Small Farmers—Issues and Options*, Westview Press, London.

Kirit S. Parikh (Ed.), (1999). *India Development Report, 1999-2000*, Oxford University Press, New Delhi.

Pineiro, M.E., (1989)."Generation and transfer of Technology for poor small farmers", in Kesseba, M. (Ed.), *Technology Systems for Small Farmers — Issues and Options*, Westview Press, London.

Sands, Deborah Merrill, (1986). *The Technology Applications Gap: Overcoming Constraints to Small-Farm Development*, F.A.O. Research and Technology paper 1, F.A.O./U.N. Rome.

Sirirugsa, C., (1986). "Impact of Agricultural Policies and Programmes on Increased Agricultural Production especially Food by Small Farmers, *The Case of Thailand*, F.A.O. Regional Workshop on Impact of Policies and Programmes on Increased Agricultural Production particularly Food by the Small Farmers, RAPA, Bangkok, 11-15 November.

Sundaram, K.V., (1996). *Report on Rural Development in CIRDAP Member Countries, 1994-95*, CIRDAP Series No. 171, CIRDAP, Dhaka.

Sundaram, K.V., (1994). "Decentralised Planning for Agricultural and Rural Development", in Velayudham, T.K. (Ed.), *Challenges Facing Agriculture and Rural Development*, Oxford and IBH Publishing Co. Pvt. Ltd., New Delhi.

Swaminathan, M.S., (1990). *Small Farms and Sustainable Agriculture"*, in *"Sustainable Agriculture in India*, (Ed.) Pradeep Chaturvedi, published by Indian Association for the Advancement of Science and Food and Agriculture Organisation, New Delhi.

Randhawa, N.S. and Sundaram, K.V., (1990). *Small Farmer Development in Asia and the Pacific: Some Lessons for Strategy Formulation and Planning*, FAO Economic and Social Development paper No. 87, FAO/UN Rome.

15

POTENTIAL ZONES FOR GROUNDWATER DEVELOPMENT IN CHINAR BASIN, ASADABAD, HAMEDAN, IRAN

A. TAHERI TIZRO

Introduction

Chinar basin situated between latitudes 34°46" to 34°48" and longitudes 48°7" to 48°13" and is lying in the Asadabad district of Hamedan state (Fig. 15.1). The basin is bounded by schistose rocks of Jurasic age. In West and East it lies in hot and cold climatic zone with a temperature varying from 15 to 30°C (hottest months June-July and coldest months January and February) and receives average annual rainfall of about 300 to 400 mm. The public depends on groundwater resources wich are exploited through deep wells and partly on springs running on fractured zone rocks, these springs discharging throughout the year and are perennial springs.

An attempt has been made in this study to delineate the groundwater potenital zones through the applicaton of geoelectrical soundings.

Hydrogeology

The entire basin is occupied by

1. Jurassic schistose formations, limestone volcanised-Limestone).
2. Jurassic-creataceous formations.

3. Recent and sub to sub-recent sediments (clay and alluvium) (Fig. 15.2).

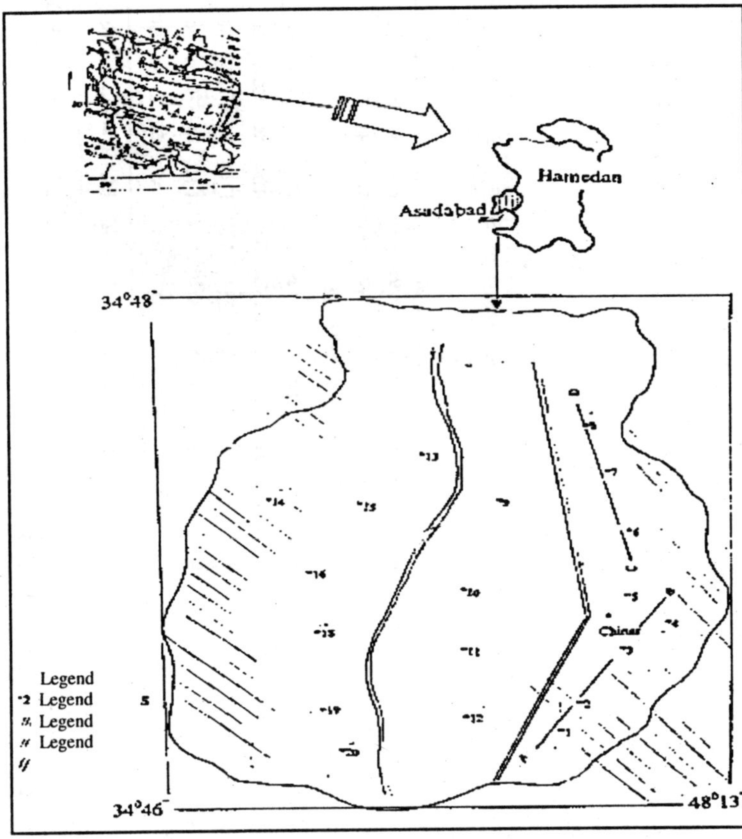

Fig. 15.1 : Index Map Study Area and Location of Vertical Electrical Soundings

Groundwater occurs under watertable conditions in the alluvial, weathered, jointed, fissured and fractured zones. Detailed study of borehole lithology data and geophysical data reveals that the depth to the basement topography ranges from 20 to more than 130 m bgl and is deep along river course (Fig. 15.3). The aquifer thickness varies from 30 to 100 metres. The depth to watertable ranges from 5 to 30 m bgl. Watertable is shallow condition in fractured zone wells with high yield.

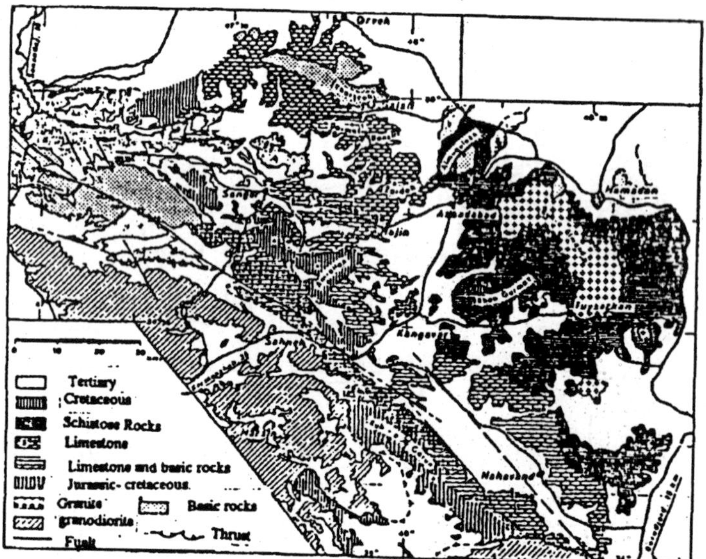

Fig. 15.2: Regional Geological Map of Area (After GSI of IR, Iran, 1990)

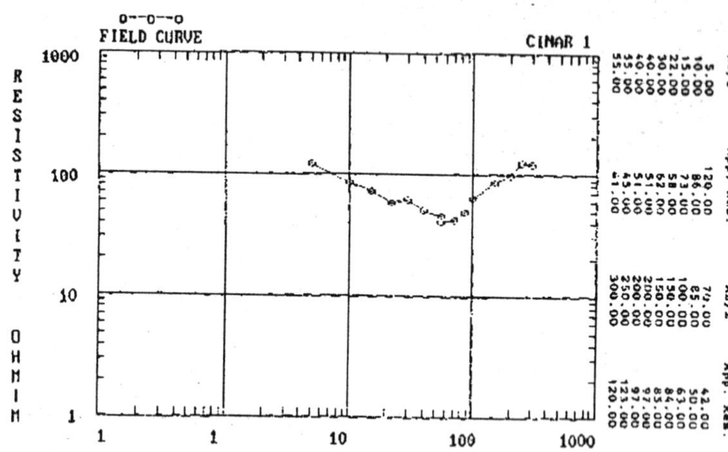

Electrode Spacing (AB/2) in Meters

Fig. 15.3: A Selected VES Field Curve

Vertical Electrical Soundings

Aplication of surface geophysics to groundwater investigations have been given by Zohdy *et al.*, (1974). Geophysical method is an effective tool for ascertaining sub-surface geologic framework of an area (Keller and Frisckneehdt, 1966; Griffiths and King, 1965; Zohdy *et al.*, *op. cit.*). One of the most widely used methods of geoelectric exploration is known as the resistivity methods. Modern development in resistivity data interpretation have considerably increased the depth of investigations with reasonable accuracy upto depth of about 500 m. There are several electrode configurations (Telyord *et al.*, 1979), but the Schlumberger arrangement is found most suitable and is widely used for groundwater investigations.

Vertical Electrical Soundings (VES) data conducted with Schlumberger configuration at 20 locations of this basin (Fig 15.1). A selected VES field curve is presented in Fig. 15.3. The maximum current electrode was kept between 400-500 m. The field apparent

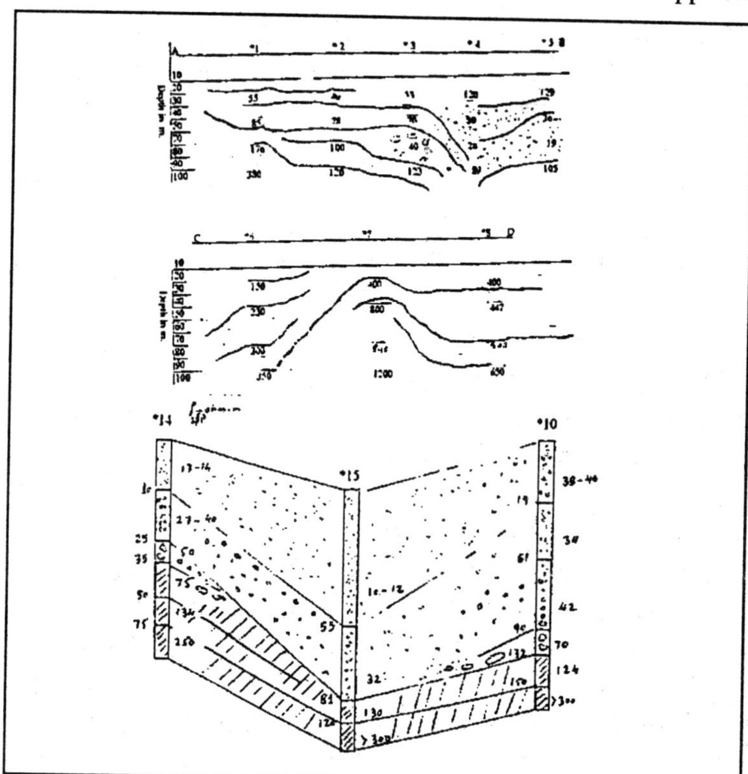

Fig. 15.4 : Shematic Geoelectrical Sections

resistivity data for different values of AB/2 have been processed and sounding data have been interpreted quantitatively and geologically as described by (Kunetz, 1966, Bhattacharya and Patra, 1988; Zohdy, 1989). The interpreted quantitative resistivity data of VES sounding has been carefully compared with the available lithologs and resistivity ranges adopted for different geological formations. The geoelectrical section from the interpreted earth model data was prepared (Fig. 15.4). It was found that potential zones of groundwater lies in hard rocks which are, fractured and jointed in nature. This was followed by drilling of few wells in the area and found to have high yield.

Conclusions

The Chinar basin is a typical hard rock terrain, where the occurrence of groundwater with high potential is confined to the highly weathered, fractured and jointed rocks. The groundwater potential zones of this basin, have been delineated using the details of geological field studies, vertical electrical soundings (VES) and followed by field reconnaissances.

REFERENCES

Bhattacharya, P.K. and Patra, H.P., (1968) *Direct Current Geoelectric Sounding*, Elsevier Company, Amsterdam/London/New York, pp. 12-1282.

Dobrin, M.B., (1985) *Introduction to Geophysical Prospecting*, McGraw-Hill Pub.

Griffiths and King, (1965) *Applied Geophysics for Engineers and Geologists*: Pergamon Press, London, p. 223.

Keller, G.K. and Frischrnecht, F.C. (1966) *Electrical Methods in Geophysical Prospecting*, Pergamon Press, Oxford, p. 517.

Kunetz, G., (1966) *Principles of Direct Current Resistivity.*

Telford, W.N. Geldart, L.P. *et. al.*, (1979) *Applied Geophysics Prospecting*, Cambridge University Press, New York, pp. 632-693.

Zohdy, A.A.R., Eaton, G.P. *et.al.*, (1974) *Application of surface geophysics to groundwater investigations, Techniques of water resource investigation of the USGS*, pp. 116.

Zohdy, A.A.R., (1989) A new method for automatic interpretation of Schlumberger and Wenner sounding curves, *Geophysics*, Vol. 54, No. 2, pp. 245-253.

16

NEW TECHNOLOGY AND AGRICULTURAL DEVELOPMENT

A Study of Dhaulpur (Rajasthan), India

S. HASAN ANSARI

It is well known fact that a new technology has been started in the sixties of the past twentieth century in our country. During 1960 the economic conditions and low income per head was on very high. The bigger industries were running in loss and the unemployment was on the alarming state. All these problems were forced to adopt new techniques for seeding, sowing, ploughing and harvesting which were based on new agricultural policies which covers wide agricultural land. Institutional development, community development works, innovations and new agricultural implements pave the way of transformation of traditional ploughing, sowing, harvesting, thrashing, storing and the use of chemical fertilizer and insecticides and pesticides.

With an increasing population, shrinking farm size and environmental degradation, the challenge is grow more. The land resource constraints and potential, technology transfer in land-use, environmental policy, legal and institutional framework for environmental protection is critical importance for several reasons. It has also been noted that Indian farms were producing one tonne per hectare, and the challenge was to get another tonne of foodgrains from the same farms, while land was shrinking and water and soil quality degraded. Today without new technology, the production could not be doubled. Hence to get maximum benefit of new agricultural policies, assured irrigation facilities, timely availability of chemical fertilizers, use of HYV seeds of wheat and rice and pesticides and insecticides at the time of maturity are necessary. An attempt has been made to see the role of innovation of new technology for agricultural development in a district of Rajasthan as a case study.

ABOUT STUDY AREA

The study area 'Dhaulpur district' of Rajasthan State is located from 26°22′ N to 26°57′ N latitudes and 77°14′ E to 78°16′ E longitudes which is surrounded by district of Bharatpur in north-west, Karauli in west, U. P. in north-east and Chambal river in the south. It accommodates 3 towns, 4 tehsils, 4 Panchayat Samiti and 551 villages spread over an area of 3008.98 km^2. According to 1991 census, there is 7,49,479 persons inhabited in rural and urban settlements and shared 82.81 and 17.19 per cent respectively out of 29.58 per cent workable population, there are 71.25 per cent agriculturist, 6.53 per cent agricultural labour, 4.20 per cent trade and commerce and 8.12 per cent engaged in other services.

OBSERVATION AND DATA ANALYSIS

The observation and analysis of data of the area under study are given as follows :

Changing Nature of Agricultural Implements

Table 16.1 records that the number of agricultural implements and instruments had been decreased from 68,108 to 47,716 during 1986-87 and 1995-96 in which the number of 'Hal' , 'Kolhu' and 'Carts' drastically decreased from 50,942 to 21,510, 8,435 to 49 and 7,077 to 4,160 shared—57.78 per cent, 99.42 per cent and 41.22 per cent respectively during 1986-87 to 1995-96. While in the case of 'Tractor', Diesel Engine and Electric pump increased from 665 to 2,971, 306 to 17,856 and 683 to 1,221 for which ratio comes +346.77 per cent, +5718.62 per cent and +78.77 per cent respectively during the above mentioned years.

Assessment of Energy

The assessment of energy (in horse power) per unit per hectare of Dhaulpur district may be seen in Table 16.2. It shows that the lowest energy recorded in human being *i.e.*0.20 horse power and the highest of camel *i.e.* 2.00 horse power. While in mechanical energy, the highest power noted for tractor and the lowest in diesel and electric pump.

Table 16.1: Agricultural Implements in Dhaulpur District

Name of implements	*Years*		*Increase in number*	*Increase/ decrease in percentage*
	1986-87	*1995-96*		
Plough	50942	21510	–29432	–57.78
Carts	77	4160	–2917	–41.22
Diesel Engine	306	17805	+17499	+5718.62
Electric Pump Set	683	1221	+538	+78.77
Tractor	665	2971	+2306	+346.77
Total	68108	47716	–20392	—

Source: Census of Animals, Dhaulpur, 1999.

Table 16.2 : Assessment of Energy (in Horse Power) per unit per Hectare in Dhaulpur District

Name of source	*Energy (in horse power)*
Human	0.20
Bullock	0.50
Buffalo	0.90
Camel	2.00
Diesel and Electric Pump Set	5.00
Crusher	7.50
Tractor	50.0

Source : *Annals of the Rajasthan Geographical Association*, Vol. III.

Changing Use of Power for Agricultural Activities

The power of agricultural labour, animal and agricultural implements has been calculated in the form of energy noted in Table 16.3. The total energy of the district has been increased from 1,70,989.9 horse power to 2,98,445 horse power during 1986-87 and 1995-96 . It may be also clear that biological energy is replaced by mechanical energy as records shows that human and animal energy decreased by +4.81 per cent and –19.63 per cent and mechanical energy has been increased by +22.44 per cent during 1986-87 and 1995-96. Hence it is clear that biotic energy (489.32 horse power) used in per thousand agricultural area in 1987 has been decreased (379.86 horse power) in 1996 *i.e.* –109.46 per cent in 10 years. At the district, the maximum amount decrease in Dhaulpur (–280.86 per cent) while use of biotic energy increased by +132.43 horse power in Rajakhera tehsil, which shows that more and more people involved in agricultural sector and depend on old agricultural implements and on biotic energy. The more details may be seen in Table 16.4. There is another Table 16.5 indicates that the use of mechanical energy in the form of agricultural implements has been highest in Rajakhera tehsil *i.e.* +2325.64 horse power and the lowest in Dhaulpur tehsil *i.e.* 986.78 horse power per thousand net agricultural area during 10 years. Which again shows that the use of agricultural implements energy has been increased day by day due to saving time, labours and less cost involved in long-term.

Table 16.3 : Use of Energy in Agricultural Activity in Dhaulpur District

Classification of Energy	*Use of Energy*				*Change in 10 years*
	1986-1987		*1995-96*		
	In Horse Power	*In percentage*	*In Horse Power*	*In percentage*	
Human	26587.80	15.55	32039.00	—	–14.81
Animal	42944.60	25.12	22358.50	7.49	–19.63
Machine	101457.50	59.33	244047.50	81.77	+22.44
Total	170989.90	100.00	298445.00	100.00	

Source : Based on Human and Animal Census Report, Rajasthan.

Table 16.4 : Use of Biotic Energy in Gross Cropped Area ('000 ha) in Dhaulpur District

Tehsil	*Years*		*Changes/Transformation*
	1986-87	*1995-96*	
Baseri	527.93	499.68	–28.25
Bari	461.61	393.36	–68.25
Dhaulpur	521.38	240.52	–280.86
Rajakhera	402.03	534.46	+132.43
District	489.32	379.86	–109.46

Source: Based on Human and Animal Census Report, Rajasthan.

Table 16.5 : Use of Mechanical Energy in Gross Cropped Area ('000 ha) in Dhaulpur District (Rajasthan)

Tehsil	*Years*		*Transformation*
	1986-87	*1995-96*	
Baseri	131.15	1786.40	+1655.25
Bari	199.28	1514.68	+1315.40
Dhaulpur	374.22	1361.00	+986.78
Rajakhera	277.79	2603.43	+2325.64
District	268.75	1697.96	+1429.21

Source : Based on Census Report of Animals, Rajasthan.

Cultivated Area under HYV Seeds

There is another parameter to assess the impact of transformation in agriculture *i.e.* HYV seeds. It becomes possible due to high yield, chemical fertilizer, natural adoptive character of HYV seeds, lesser height of plants and more and more fruits led a new hope in high production of food crops. An early ripe, use of chemical fertilizers in required ratio and irrigation facilities have paved the way to use HYV seeds in the desert region. Table 16.6 shows area of HYV seeds in Dhaulpur district. It indicates that the area under cultivation HYV seeds has been increased by 79.14 per cent, 6.61 per cent and 1.00 per cent in respect of sarson, wheat and bajra in 10 years. The dry crops such as bajra and sarson produced in more than 98 per cent of the area under study and where irrigation facilities available, wheat crop is also produced. Some important HYV seeds of sarson are Pusa Kalyani, Durgamani, T-59, Kranti, Pusa Bold, R.H.30 and Krishna. Due to HYV of sarson, the cultivated area has been increased from 2.17 per cent in 1986-87 to 81.31 per cent in 1995-96.While in the

Table 16.6 : Area of HYV Crops in Dhaulpur (Percentage of Crops over Total Agricultural Area)

Crops	*Years*		*Decadal change*
	1986-87	*1995-96*	
Wheat	93.39	100.00	+6.61
Bajra	99.00	100.00	+1.00
Sarson	2.17	81.31	+79.14

Source : District Agriculture Officer, Dhaulpur, Rajasthan.

case of bajra, the HYV seeds are RCCWC 75, MH 179 and C.M.46. it is widespread crops since it covers entire cultivated area during 1995-96. The HYV of wheat are Kalyan Sona, Sonalika, Raj 1482, Raj 1972, Raj 2184, Raj 3077, HD 2329, HD 2204, WH 147, Durgapura 65, Sujata and Mukta.

Change in the Use of HYV Seeds

The change in the use of HYV seeds has been recorded in Table 16.7 which recorded that 5,20,900 kg was use in 1986-87 but after 10 years it increased to 6,89,500 kg during 1995-96 in the district. On an average per hectare comes 2.92 kg and 3.88 kg of HYV seeds used which again indices an increase of 0.96 kg per hectare in 10 years. Hence the farmers of the study area have taken interest to use more and more HYV seeds to increase production of crops based on modern agriculture implements of plough, threshing; chemical fertilizers, insecticide and pesticide and the subsidy provided by Central Government.

Table 16.7: Use of HYV of Seeds in Dhaulpur District (Rajasthan) during 1986-87 and 1995-96

Name of Tehsil	*Use of All HYV Seeds*				*Change (in kg)*
	1986-87		*1995-96*		
	in kg	*kg/ha*	*in kg.*	*kg/ha*	
Baderi	131246	3.37	158263	4.20	+0.83
Bari	98142	2.54	104502	2.77	+0.23
Dhaulpur	216612	3.02	321613	4.50	+1.48
Rajakhera	74900	2.57	105122	3.41	+0.84
District	520900	2.92	689500	3.88	+0.96

Source: Office of the District Agriculture, Dhaulpur, Rajasthan.

Changing Nature and Use of Chemical Fertilizers

It is recorded in Table 16.8 that the use of chemical fertilizer accounted for 24.87 kg per hectare during 1986-87 which has been increased as 44.50 kg per hectare during 1995-96. It means there is an increase of 19.63 kg of chemical fertilizer in 10 years. Hence there is no significant change in the use of chemical fertilizer in an individual year due to unassured irrigation facilities in the area under study. There are three types of chemical fertilizers used in Dhaulpur district *viz.* Nitrogen, Potassium and Phosphate. As per records 4,433 metric tonnes of ammonium sulphate, 712 metric tonnes of superphosphste and 46 metric tonnes of potassium phosphate used in 1986-87 which has been increased to 7,907, 1,279 and 82 metric tonnes respectively during1995-96. It means an increase of 3474 metric tonnes of ammonium sulphate, 567 metric tonnes of superphosphate and 36 metric tonnes of potassium sulphate in 10 years . Due to more use of chemical fertilizer, of course, production of crops has been increased but more and more ecological problem arises, such as, soil pollution, water logging, cost of agriculture and nutritious food crops etc.

Table 16.8: Use of Chemical Fertilizer in Dhaulpur District (Rajasthan) during 1986-87 and 1995-96

Name of Tehsil	*Use of Chemical Fertilizer*				*Increase per ha (in kg)*
	1986-87		*1995-96*		
	in kg	*kg/ha*	*in kg.*	*kg/ha*	
Baseri	953	24.53	1524	40.45	+15.92
Bari	946	24.50	1449	38.39	+13.89
Dhaulpur	1952	27.36	3754	52.49	+25.13
Rajakhera	582	19.93	1180	38.38	+18.45
District	4433	24.87	7909	44.50	+19.63

Source : Office of the District Agriculture, Dhaulpur, Rajasthan.

Size of Land Holdings

Table 16.9 shows size of landholding of Dhaulpur district during 1986-87 and 1995-96 which is divided into 6 categories *viz (i)* Marginal (<1 ha), *(ii)* very small (1-2 ha), *(iii)* Small (2-4 ha) *(iv)* Medium (4-10 ha), *(v)* Large (10-20 ha), and *(vi)* very large (>20 ha). It was recorded that total number of landholdings was 1,02,436 spread over an area of 184154.63 ha in which the highest number of landholding was 47,156 of marginal farmers and the lowest was 71 of very large size farmers during 1986-87 and the same trend noticed in 1995-96 also, but number of landholding of very large size farmers has been increased by +0.64 per cent and the number of medium and large size landholding decreased by –1.05 per cent and –0.89 per cent respectively in 10 years. There is also a significant point to note that the number of size of landholding of lower category of farmers has been increased due to either further division of land in the family or the medium and large size of landholding farmers sold their land. It indicates that medium size landholding farmers did not take much interest in green revolution due to uncertainty of irrigation, higher cost of fertilizer etc. and they sold their land to the very large size farmers who used HYV seeds, chemical fertilizers, tractor, insecticide and pesticide for taking high production of wheat, bajra and sarson as Table 16.6 also proved it.

Table 16.9: Size of Landhgolding in Dhaulpur district during 1986-87 and 1995-96

Sl. No.	*Classification of land holding ha*	*1986-87*			*1995-96*			*Change %age*
		No.	*Area in ha*	*%age*	*No.*	*Area in ha*	*%age*	
1.	Marginal (<1)	47156	24154.83	13.12	48778	23322.62	14.02	+0.90
2.	Very small (1-2)	25629	36596.70	19.87	23850	33708.54	20.26	+0.39
3.	Small (2-4)	19149	53790.42	29.21	17758	49661.65	29.85	+0.64
4.	Medium (4-10)	9485	54890.21	29.81	8246	46793.06	28.13	–1.05
5.	Large (10-20)	946	12364.84	6.71	761	9676.94	5.82	–.89
6.	Very large (>20)	71	2348.63	1.28	114	3198.21	1.92	+0.64
	Total	102436	184154.63	100.00	99507	166360.66	100.00	

Source: Office of the Agriculture Officer, Dhaulpur, Rajasthan.

CONCLUDING REMARKS

Thus above findings show that the impact of new technology in the field of agriculture has been clearly noticed as the number of traditional agricultural implements decreased and the mechanical energy increased

Table 16.10 : Average Size of Landholding in Dhaulpur District during 1986-87 and 1995-96

Sl. No.	*Size of land-holding (in ha)*	*Average size (1986-87)*	*Average of landholding size (1995-96)*	*Change in Percentage*
1.	Marginal (<1)	0.51	0.48	–0.03
2.	Very small (1-2)	1.43	1.41	–0.02
3.	Small (2-4)	2.81	2.80	–0.01
4.	Medium (4-10)	5.79	5.67	–0.12
5.	Large (10-20)	13.07	12.72	–0.35
6.	Very large (>20)	33.08	28.05	–5.03

Source : Office of the Agriculture Officer, Dhaulpur, Rajasthan.

which led the farmers to use HYV seeds and the chemical fertilizers on ensured facilities of means of irrigation.

REFERENCES

Ali, Mohammad (1978). *Studies in Agriculture Geography*, Rajesh Publication, New Delhi.

Boserup, E. (1965). *The Condition of Agricultural Growth,* Allen & Unwin, London.

Chauhan, D.S. (1966), *Studies in the Utilization of Agricultural Land*, Ist edition.

District Census Hand Book, Census of India, 1991, series 21, Part XII A&B, Rajasthan.

Dube, R.S. (1990). *Population Pressure and Agrarian Change*, Rawat Publications, Jaipur.

Gupta, N.L. (1979). *Rajasthan Me Krishi Vikash*, Rajasthan Hindi Granth Academy, Jaipur, pp. 45-46.

Gurjar, R.K. (1987). *Irrigation for Agriculture Modernization*, Scientific Publishers, Jodhpur, pp. 130-137.

Jain, M.B. (1985). *Rajasthan Me Krishi Utpadan*, Rajasthan Hindi Granth Academy, Jaipur.

Kharif Ki Pramukh Phasalen (1999), Information of Agriculture, Department of Agriculture, Rajasthan, Jaipur.

Shafi, M. (1984). *Agricultural Productivity and Regional Imbalances : A Study of U.P.,* Concept Publishing Company, New Delhi.

Shaikh, N.A. (1983). 'Use of Energy in Agriculture', *Annals of the Rajasthan Geographers*, Vol. III, pp. 41-48.

Sharma, A. N. (1984). *Economic Structure of India — Agriculture*, Himalayan Pub. House, Bombay.

Singh, Srinath (1976). *Modernization of Agriculture — A Case Study in Eastern U.P.,* Heritage Publishers, New Delhi.

Singh, B.B. (1996). *Krishi Bhoogol*, Gyanodaya Prakashan, Gorakphur.

Singh, C. (1984), *Rabi Phaslon Ki Vaigyanic Kheti Aur Shasya Vigyan Ke Sindhant*, Bharati Prakashan, Meerut.

Singh, J. (1974-75). *An Agricultural Atlas of India — A Geographical Analysis*, Vishal Publication, Kurukshetra.

Singh, J. (1976), *An Agricultural Geography of Haryana*, Vishal Publication, Kurukshetra, Haryana.

Tiwari, R.C. & B.N. Singh (2000). *Krishi Bhoogol*, Prayag Pustak Bhawan, Allahabad, pp. 392-97.

Weaver, J.C. (1954). Crop Combination Regions in the Middle West, *The Geographical Review*, Vol. 48, pp. 175-200.

Zobler, L, (1962). The Economic Historical View of Natural Resources Use and Conservation, *Economic Geography*, Vol. 38, p. 89.

परिशिष्ट -16.1 : प्रति हजार शुद्ध काश्त क्षेत्र पर प्रमुख उपकरणों की संख्या

	हल			ट्रैक्टर			विद्युत पम्प			डीजल पम्प			गाड़िया			गन्ने के कोल्हू		
तहसील/ जिला	1986-87	1995-96	परिवर्तन	1986-87	1995-96	परिवर्तन	1986-87	1995-96	परिवर्तन	1986-87	1995-96	परिवर्तन	1986-87	1995-96	परिवर्तन	1986-87	1995-96	परिवर्तन
बसेड़ी	383.94	228.98	−154.96	2.28	20.31	+18.03	0.40	0.73	+0.33	3.07	152.66	+149.59	79.35	55.65	−23.70	62.29	0.48	−61.81
बाडी	351.88	164.76	−187.12	3.40	18.13	+14.73	1.01	1.70	+0.69	4.92	118.51	−113.59	45.95	34.91	−11.04	55.08	0.93	−54.15
धौलपुर	356.85	65.32	−291.53	7.13	15.81	+8.68	2.88	3.49	+0.61	0.68	112.19	+111.51	41.83	14.75	−27.08	78.99	0.05	−78.94
राजाखेड़ा	339.27	226.65	−112.62	3.52	35.89	+32.37	19.42	38.71	+19.29	0.98	122.89	+121.91	37.07	20.56	−16.51	15.53	0.08	−15.45
जिला धौलपुर	358.66	149.66	−208.79	4.68	20.67	+15.99	4.80	8.50	+3.70	2.15	123.88	+121.73	49.80	28.94	−20.86	59.35	0.34	−59.01

17

IRRIGATION EFFICIENCY AND SUSTAINABLE AGRICULTURAL PRODUCTIVITY IN CENTRAL ASIAN STATES

G.M. MIR

ISSUE AREA

Central Asian States are passing through a transitional phase and have very recently started addressing to their priorities. One such thrust area concerns the lower irrigation efficiency. The main reasons of inefficient irrigation system are multi-dimensional *e.g.*, the operational procedure of irrigation projects lack sophistication resulting too much or too little water being applied to the crops. In former case, coupled with inadequate drainage system, it has led to life threatening salinity problem. Water mismanagement has further worsen the situation. When the crops do not get the required quantity of water at required time with required quality, it adversely affects the agricultural productivity. Here the benefits of irrigation are being balanced by the losses rendered by poor drainage system. During the early stage of *Perestroika* and *Glasnost* it was estimated that the area going out of production due to widespread waterlogging, salt buildups and anaerobiosis exceeded the area of new land which was coming under irrigation. The other dimension of the problem is that only about 50 per cent of water withdrawals for irrigation actually reaches the field. Moreover, the faults in water management leading to lower irrigation efficiency have adversely affected the productivity pattern in the region owing to the fact that there exists a positive correlation between the two variables (Mir, 1999). Misuse of water has further aggravated the water drying up disaster in the region. The Aral sea has lost 60 per cent of its volume since 1960 dropping its level by 50 feet and now exists only on 55 per cent of its original area and 31 per cent of the original volume (Beaumont, 1989). If the water continues to be drawn ruthlessly from Amu and Syrdarya for mismanaged irrigation, the fears of the dessication of the Aral sea will be there and it is apprehended that this might led to drastic drop in watertable of the surrounding areas, a significant climatic change, and widespread wind erosion of the exposed salt flats would produce further salinization of huge territory of which the basin is surrounded, resulting in the sprawl of huge wastelands and a drastic decline in agricultural productivity.

OBJECTIVES

The data pertaining to the latter part of the transitional phase (1994-95 to 1997-98), however, shows some signs of improvement as a result of the conscious efforts of the respective governments of CAS and the co-operation and assistance extended by international financial institutions and other world bodies in preserving, developing and monitoring water resources of the region through some recently emerged institutions. Hopefully, this will go a long way in solving some grave agro-hydel problems of the region. These water management institutions have to keep a strong vigil from source to recycling process as the study area is leading in water conveyance losses. Moreover, the highly imbalanced water network evident from the facts that four-fifths of the regional hydro network is located in Kyrgyzstan and Tajikistan which have a small land area. Uzbekistan and Turkmenistan which have approximately three-fourths of the region's area and most of its arable land, have only one-fifth of the water network. In Kazakhstan the situation is even worse in this relationship. The riparian states share only 38.45 per cent of the regional water flow through irrigation whereas the non-riparian states share more than 50 per cent (Weinthal, 1993). Furthermore, water rich states are energy poor ones. So the expected inter-state as well as riparian conflicts will have to be changed into potential for co-operation. The basic objective of this Chapter, therefore, is to examine that how far the region has succeeded in improving water use efficiency, how and with which intensity the productivity is responding in the post-independent phase, and finally, to explore the possibilities for further improvement in water-use efficiency and agricultural productivity.

METHODOLOGY

In order to have the desisred results with a score of other relevant combinations, the techniques applied in the present study follows:

Agricultural Productivity

The three best known partial productivity measures are productivity of land, labour and capital. Productivity of land is the most permanent and fixed among the three categories of inputs, hence, being considered the most important productivity measure. Productivity of labour is mainly significant as determinant of income and living standards of population engaged in agriculture. So far the productivity of capital is concerned due to the nature of the problem *i.e.* homogeneous character of capital input due to prolonged socialist structure as well as owing to other allied issues its inter-State variation is negligible. So only the first two variables will form the basis of the study.

Productivity of Land

For the assessment of unit area productivity the formula given by Shafi (1972) has been used which is expressed as under :

$$\left(\frac{yw}{t}+\frac{yr}{t}+\frac{ymi}{t}.....n\right):\left(\frac{Yw}{T}+\frac{Yr}{T}+\frac{Ymi}{T}.....n\right)$$

where

yw, yr, etc.	= total yield of the respective crop in the unit;
Yw, Yr, etc.	= total yield of the respective crop in the region;
t	= area planted under the crop in the unit; and
T	= area planted under the crop in the region.

Productivity of Labour

So far as the labour productivity is concerned, its measurement is based on the technique followed by Singh and Chauhan (1977). The formula would be read thus :

$$Iwi = \frac{wi}{wr} \times 100$$

where

Iwi = agricultural worker index;

wi = per cent agricultural worker/hectare sown area in the unit; and

wr = corresponding data of the region.

After calculating agricultural worker index (Iwi), the composite index of agricultural worker output has been worked out by the following equation :

$$Iewi = \frac{Iei}{Iwi} \times 100$$

where

$Iewi$ = agricultural productivity/agricultural worker;

Iei = agricultural productivity/hectare cropped area; and

Iwi = agricultural worker index.

Irrigation Efficiency

Measures of irrigation efficiency are used to determine what crops can be grown in any area to yield the maximum with a given water supply system. The ratio between the amount of water applied to the crop and the amount of water delivered at the turnout or source, is termed as the farm conduit efficiency. To work out such efficiency there has been little standardisation of measurement of water losses, however, some comon methods are outlined hereunder :

Field Irrigation Efficiency

It is the ratio of volume of water actually consumed by the crops to the volume of water applied to the field, and is usually expressed as a fraction or in percentage. Finkel (1982) determined it by dividing the amount of water required by a crop in the field by the amount of water applied to the field. Another frequently used determination is to take the ratio of average depths of irrigation water which is beneficially used to the average depth of irrigation water applied.

Field Application Efficiency

It is defined as the ratio of the average depth of water reaching and remaining in the route zone to the average depth of water applied to the field. Wade (1982) uses such efficiency to indicate a project's technical efficiency and expresses it as the ratio of volume of water used for crop production to the volume of water diverted from or stored at source.

Water Productivity

This is the ratio of crop yield to quantity of water used.

However, every method has thoroughly been analysed, tested for relevance and treated with caution in order to give the desired results.

RESULTS

There is a positive correlation and in some cases near to linear relationship between the irrigation efficiency and agricultural productivity in CAS and any chage in one variable is responded by almost a proportionate change in other variable (Table 17.1). An insight in the table reveals that there are two phases well demarcated by efficiency-productivity variations. In the first phase (1991-92 to 1993-94) both the variables have shown a downward trend. The main reasons behind this seems, firstly, the abrupt disintegration of erstwhile Soviet Union leading to independence of these five states. Secondly, the changing priorities of the respective political leaderships. Thirdly, changing cropping pattern in favour of cereals. Fourthly, the break-up of integrated watersheds, and other allied problems. But in the second phase (1994-95 to 1997-98) as evident from the table agricultural sector has established its momentum. Irrigation efficiency has improved in all States, albeit, with varying magnitude and the agricultural productivity has started to increase.

The regional data shows that during the first phase irrigation efficiency decreased from 33.74 per cent in 1991-92 to 33.06 per cent in 1993-94. Consequently, the agricultural productivity also diminished from an index value of 1.087 to 0.979 in case of cereals and from 0.997 to 0.916 with respect to commercial crops. During the second phase the average irrigation efficiency rate showed an abrupt increasing trend from 33.82 per cent in 1994-95 to 38.92 per cent in 1997-98.

Table 17.1: Irrigation Efficiency and Agricultural Productivity Responses

State	*1991-92*		*1992-93*		*1993-94*		*1994-95*		*1995-96*		*1996-97*		*1997-98*	
	IE(%)	*API*	*IE(%)*	*API*	*IE(%)*	*API*	*IE(%)*	*API*	*IE(%)*	*API*	*IE(%)*	*API*	*IE(%)*	*API*
Turkmenistan	34.30	2.189	33.50	2.011	33.80	2.066	34.90	2.396	36.80	2.860	38.51	3.627	43.26	4.416
Kyrgyzstan	34.50	2.135	33.30	1.921	33.90	1.983	34.70	2.119	35.60	2.634	36.98	3.113	39.92	3.646
Uzbekistan	33.90	2.099	33.00	1.889	33.20	1.852	33.60	2.015	34.80	2.227	35.07	2.902	37.54	3.503
Kazakhstan	33.20	1.960	32.70	1.732	32.10	1.696	32.80	1.994	33.10	2.246	34.89	2.764	36.87	3.234
Tajikistan	32.80	2.043	32.60	1.999	32.30	1.883	33.10	2.061	33.90	2.383	34.93	2.946	37.01	3.253

Source : Compiled and computed by the author on the basis of:

- *(i)* FAO Production Yearbooks;
- *(ii)* IMF Economic Reviews;
- *(iii)* World Bank Publications;
- *(iv)* CIS Progress Reports;
- *(v)* UNESCO Newsletters;
- *(vi)* Post-Independent International Irrigation Surveys.

IE — Irrigation Efficiency.

API — Agricultural Productivity Index.

This induced the agricultural productivity to increase from an index of 1.069 in 1994-95 to 1.728 in 1997-98 with respect to cereal crops while the commercial crops showed better response as compared to cereals with respective figures of 1.048 and 1.881. If we compare the two phases we come to know that the efficiency level of water-use has shown an increase of 7.73 per cent over the base phase while the agricultural productivity has multiplied by over one and half times. The commercial crops, however, have shown a higher intensity during the period under study.

Kyrgyzstan, Turkmenistan and Uzbekistan fall above the regional average so far their water-use efficiency in first phase is concerned. In case of agricultural productivity Turkmenistan and Kyrgyzstan are above the regional average but Tajikistan has shown good results particularly in cereal productivity. In the second phase interestingly Turkmenistan and Kyrgyzstan are above the regional averages in both the variables.

The trend of agricultural productivity *vis-a-vis* irrigation efficiency in the region as evident from Table 17.2, shows that if we compare the productivity of cereals and commercial crops at two irrigation efficiency levels *i.e.* 33.26 per cent and 35.83 per cent, we observe that all States have shown a marked increasing trend in both cereals and commercial crops while improving their irrigation efficiency. Turkmenistan is leading in commercial crops while Kyrgyzstan in creals followed by Turkmenistan, Tajikistan, Kazakhstan and Uzbekistan in cereals while Kyrgyzstan, Kazakhstan, Uzbekistan and Tajikistan in commercial crops.

The irrigation efficiency rate which has increased from 33.26 per cent during the first phase to 35.83 per cent in the second phase, can be raised upto 75 per cent which is possible if some measures are taken, and that can revolutionise the regional productivity pattern. When the existing irrigation efficiency will be raised to 75 per cent then, on the basis of trend values, Kyrgyzstan whose yields are responding fast to irrigation efficiency can increase its productivity upto an index of 3.942 against 2.877 at the efficiency rate of 35.83 per cent. This will be followed by Turkmenistan which can raise its productivity level upto 3.640 against 3.324 at 35.83 per cent efficiency rate. Uzbekistan which has a productivity level of 2.660 at existing rate of irrigation efficiency can have a productivity index of 3.536. Similarly if the water-use efficiency rate will be raised upto 75 per cent in the study area the productivity of Tajikistan and

Table 17.2: Agricultural Productivity at Various Irrigation Efficiency Levels

State	*Irrigation Efficiency Rate : 33.26%*		*Irrigation Efficiency Rate : 35.83%*		*Irrigation Efficiency Rate : 75%*	
	Productivity Index		*Productivity Index*		*Productivity Index*	
	Cereals	*Commercial Crops*	*Cereals*	*Commercial Crops*	*Cereals*	*Commercial Crops*
Turkmenistan	1.124	0.964	1.550	1.774	1.645	1.995
Kyrgyzstan	1.005	1.007	1.453	1.424	1.731	2.211
Uzbekistan	0.958	0.988	1.307	1.353	1.671	1.865
Kazakhstan	0.958	0.837	1.320	1.238	1.562	1.752
Tajikistan	1.013	0.962	1.386	1.274	1.665	1.689

Source : Based upon :

(*i*) Table 17.1.

(*ii*) Lewis, R.A., "The Irrigation Potential of Soviet Central Asia", *Annals of the Association of American Geographers*, Vol. 52, 1962, pp. 99-113.

Kazakhstan will also respond as desired, the former increasing its level by 0.694 and the latter by 0.756.

An examination of the Table 17.3 showing a composite picture of productivity at three efficiency levels leads to conclude that on an aggregate the relationship between the two variables becomes more evident and it is the better water-use which has improved the productivity from first to second phase and will improve it further at a higher efficiency degree. Turkmenistan having a composite score of 7.64 in irrigation efficiency corresponds with the highest level of productivity index. It is followed by Kyrgyzstan encompassing an efficiency rate of 4.58 and productivity index of 4.69. Uzbekistan which has a long irrigation history within CAS accumulating all drawbacks of high irrigation intensity with poor drainage system, stands at third position so far as the degree of both the variables are concerned. Tajikistan is showing better performance and has been able to leave behind Kazakhstan in this development game .

MEASURES

The logic behind the possibilities for improving the existing irrigation efficiency upto a rate of 75 per cent is directly or indirectly connected with irrigation management in the region. For desired-cum-sustainable productivity the three sub-systems of irrigation *i.e.* water delivery sub-systems, farm sub-systems and water removal sub-system must function satisfactorily. The relevant suggestions are summarized as under:

1. From source of water to field water supplies conveyance losses through seepage, spillage, evapo-transpiration and theft etc. It has been estimated that in CAS the conveyance losses are in excess of 50 per cent of water withdrawals (Beaumont, 1989). Further, it has been calculated that only 15 to 30 per cent of the regional water diverted for irrigation takes the form of return flow as groundwater (Shul'ts). The most obvious way of reducing conveyance losses from canal supply system is to route canals to take the advantage of topography and line any which leak at unacceptable levels. Conservation of water by lining of canals etc. is possible with plastic film. Also there could be conveyance of water to far-flung distances through plastic pipes leaving least

Table 17.3 : Efficiency-Productivity Matrix with Mean Deviation

State	*Deviations from Mean at Different Efficiency Levels*						*Composite Index*		*Levels of Development (IE+AP)*
	Low (1st Phase)		*Existing (IInd Phase)*		*Highest (Potential)*				
	IER (%)	*API*	*IER (%)*	*API*	*IER (%)*	*API*	*IER (%)*	*API*	
Turkmenistan	$\bar{X}$+0.60	$\bar{X}$+0.125	$\bar{X}$+2.53	$\bar{X}$+0.509	$\bar{X}$+4.51	$\bar{X}$+0.083	+7.64	+0.717	Very High
Kyrgyzstan	$\bar{X}$+0.64	$\bar{X}$+0.049	$\bar{X}$+0.97	$\bar{X}$+0.062	$\bar{X}$+2.97	$\bar{X}$+0.385	+4.58	+0.496	High
Uzbekistan	$\bar{X}$+0.10	$\bar{X}$-0.017	$\bar{X}$-0.58	$\bar{X}$-0.154	$\bar{X}$-0.21	$\bar{X}$-0.021	-0.69	-0.192	Medium
Tajikistan	$\bar{X}$-0.60	$\bar{X}$-0.168	$\bar{X}$-1.42	$\bar{X}$-0.257	$\bar{X}$-4.18	$\bar{X}$-0.243	-6.21	-0.668	Low
Kazakhstan	$\bar{X}$-0.70	$\bar{X}$+0.012	$\bar{X}$-1.46	$\bar{X}$-0.155	$\bar{X}$-3.04	$\bar{X}$-0.203	-5.20	-0.370	Very Low

Source : Computed on the basis of Tables 17.1 and 17.2.
IER = Irrigation Efficiency Rate.
API = Agricultural Productivity Index.

chance of water loss through leakage, evapo-transpiration etc. This technique has the added advantage of protecting the local environment by using plastic instead of wood for packaging (Agarwal, 1997).

2. Groundwater is being considered an important source in raising the efficiency level in CAS. Recent investigations comparing yields with the use of ground and canal water suggest that with the use of groundwater no significant deduction in yields generally can be expected. Thus, by using the available groundwater the water supply of the area could be increased by about 10 per cent. This could obviate raising the efficiency to the level nearer to 70 per cent.
3. The irrigation technology available today offer great potentials for improving the irrigation systems by reducing the chance of water losses during conveyance from source to field and decreasing the acreage under wastelands through lesser possibilities of salt buildups and anaerobiosis, given better irrigation management policy. Moreover, such techniques have a capacity of considerably reducing crop failure, and have the added ability to offer considerable indirect benefits like improving the sustainability of productivity. On the basis of method of water application, there are basically two types of irrigation *e.g.* sub-irrigation and surface irrigation but due to the higher efficiency, the overhead irrigation method has been added. Overhead irrigation methods are comparatively recent innovation, and in future their costs and disadvantages are likely to decline and have capacity to offer higher water-use efficiency and control over water application. By comparing the field irrigation efficiency methods, the sprinkler (over-head) has the highest efficiency (60-80 per cent) followed by surface-basin (except rice), border-furrow corrugation (40-60 per cent) and flooded rice (30 per cent).
4. The system of underground water channels called *karez*, horizontal wells (*quanats*) and other sub-ground techniques which draw substantial water to the fields without loss through evapo-transpiration etc. and which are in operation in CAS, should have denser network.
5. The efficiency of the irrigation system could also be raised by the improvement in diversion and conveyance facilities

and more exact water budgeting. The "4R" motive *e.g.*, required water at required time with required quantity and quality and for the required crops, should be the focus of irrigation management. This would be relatively inexpensive and more effective. About 10 per cent of the plan increase in efficiency was attributed by Mirkin to these measures.

6. At present CAS have made few changes in the way that water is used and have not yet defined yet a comprehen-sive plan to conserve water resources. Some States continue to pursue inefficient water policies because it remains unclear to whom water belongs. Neither prices are charged for water used in irrigation nor there is any check to water quantity applied. It appears that payment for water will be one of the first areas addressed to deal with improving water management and economic restructuring.
7. A high degree of co-operation among the CAS that would coordinate research and planning is required to fill the institutional vacuum left behind by the collapse of the former Soviet Union. Cooperation among States is essential also because for planning water sources a watershed/river basin is an ideal ecological unit (eco-system) for water management.

MAJOR FINDINGS

The analysis of data in the preceding study leads to some of the main observations as follows :

— There exists a positive correlation between the independent variable of irrigation efficiency and the dependent variable of agricultural productivity in CAS and any change in water-use efficiency is responded by a change almost of the similar magnitude in productivity. However, in first phase of the study (1991-92 to 1993-94) because of the turmoil of early phase of transition, both the variables have shown a downward temporal trend. The efficiency of irrigation system has improved in second phase with marked impact on agricultural productivity showing an increase of 7.73 per cent in case of former and a multiplication of one and a half times in case of latter.

— The agricultural productivity increased from an index value of 1.963 at an efficiency rate of 33.26 to 2.815 when the water-use efficiency was raised to 35.83 per cent in second phase of the transitional period. The inter-State analysis shows higher figures of productivity with increasing irrigation efficiency. Turkmenistan and Kyrgyzstan have been able to increase their irrigation efficiency from first to second phase which is higher than other three States corresponding to productivity levels much above the regional average. It is worthy to note that, on aggregate, about one per cent increase in the rate of irrigation efficiency will increase a productivity level by an index value of 0.155 and if we compare the expenditure on the investment of one per cent increase in irrigation efficiency with the corresponding output increase by 0.155 in productivity level, we will come to know that the former is almost negligible and this can revolutionize the rural economy in particular and the region in general.

— Central Asian States have much potential for improving their irrigation efficiency and have the capacity of improving their existing rate of efficiency upto 75 per cent. This is possible with a rational water management policy. The conveyance losses in which the region is leading, can be minimised through routing canals to follow topography, lining of canals with plastic film and using plastic pipes for far-flung distances to reduce costs and control leakage as well as loss of water through evapo-transpiration. The efficiency of irrigation system could also be raised by the improvement of diversion facilities and more exact water budgeting. The water supply of the region which could be increased by 10 per cent through tapping the available groundwater which could obviate raising the efficiency to a level nearer to 70 per cent. The system of underground channels and horizontal wells (*karez* and *quanats*) existing in CAS should have a denser network and be encouraged in far-flung rural areas as well because these have the capacity to deliver goods. The relevant irrigation technology is another area of concern. In addition to some surface and sub-surface techniques, overhead irrigation methods which are recent innovation, should be applied. A comparative study reveals that sprinkler method has over 73 per cent of water application efficiency—the highest as compared to other irrigation techniques.

The study leads to believe that Central Asian States have much potentials for the improvement of sustainable agricultural productivity by way of increasing their existing irrigation efficiency which is quite possible but requires a rational irrigation management policy.

18

MECHANICAL TECHNOLOGY IN FARMING BY TRIBES

A Study of the Betul-Chhindwara Plateau, (Madhya Pradesh), India

D.D. VISHWAKARMA

Introduction

Machines and improved implements have become integral components of the developed agriculture. Machines, as substitute of animate power, are useful in areas of scarcity of labour because of low density of population. There is another group of machines and implements which increases the efficiency of man, helps in proper utilisation of inputs like water, fertilizers, insecticides etc. and enhances the quality of produces. The plough has been main tillage implement since time immemorial. Other machinery operations assist in cultivating, sowing, harvesting, drainage and irrigation, but ploughing is the fundamental operation. Now, the tractor is replacing the draft animal so the design and complexity of the plough has increased correspondingly. Along with tractor, use of harvester is increasing, particularly in level areas. Thresher is also added in the list of labour and time saving implements. Other commonly used machines are those used for lifting water from wells, rivers etc. Now-a-days, tubewell can irrigate larger area more efficiently. With the modernization of agriculture, use of different types of chemicals for spraying on crops is increasing. For spraying agrochemical different types of spraying machines and dusters are in vogue. It is proposed to analyze the influence of socio-economic condition of the farmers on use of these improved implements and technology.

Emphasising the significance of the mechanical inputs, Jasbir Singh and Sharma proclaimed that, 'it is the input of inputs' (Singh and Sharma, 1985 : 436). The terrain, edaphic and climatic conditions, determines actual use of mechanical power. Even in physically suitable areas, adoption of new and improved techniques, tools and practices. called innovations, depends, with other things, on characteristics of innovations themselves and characteristics of adopters. Among the characteristics of innovations, utility, cost, compatibility, complexity, communicability, and divisibility are important (Dasgupta, 1989 : 13). Most of these characteristics are not the inherent characteristics of the innovation itself but the reflection of the perception of farmers. There are a large number of variables related to the farmer's adoption behaviour. These characteristics are grouped as personal, situational and social characteristics. These traits of adopters are associated with the behaviour and innovativeness of the farmer. Out of them, tenure and size of holdings and caste structure function as cornerstone (Sharma, 1988 : 27-39). It is because most other social, cultural and economic characteristics are closely associated with them. They play a major role in technological change and adoption of agricultural innovations. The value orientation and behavioural pattern of the village people belonging to higher castes are more progressive than middle and low caste people of the society. As a result, they are more progressive and adoptive (Mohammed, 1992 :150). With this perspective, this chapter envisages to present the comparative picture of the adoption of the mechanical technology by farmers of different communities in general and by farmers of the scheduled tribes in particular, in the Betul-Chhindwara plateau located in the south-central part of Madhya Pradesh.

Methodology

The present analysis is entirely based on data generated by sample survey of 475 cultivating households of different communities randomly distributed in 20 villages scattered in different parts of the region. From each village, about 20 per cent households are selected using sampling techniques and they are thoroughly interviewed, using a well-structured questionnaire. For the convenience of analysis, farmers of different castes have been grouped into well accepted four classes, *viz.* upper castes, backward communties, scheduled castes and scheduled tribes.

Community-wise Ownership of Agricultural Implements and Machines

Since purchase and even use of machines and improved implements require some capital, their ownership and use depend on the capacity of farmers. Consequently, only, well-to-do farmers afford to have and use them. But the socio-economic-cultural backwardness of farmers belonging to lagging communities such as scheduled castes and scheduled tribes presents insurmountable barrier in the way of their utilisation (Sharma, 1989 : 71-80). Even very common implements and machines are beyond their capacity. It is evident from the information about the ownership of different types of these implements collected through the sample survey of 475 households, as mentioned earlier. They present realistic picture than those conveyed by the secondary data. These results of the survey are tabulated in Table 18.1.

Tractor Owners

Tractor is almost unknown here. Only five households own them. Three of them belong to upper communities and two belong to backward communities. They constitute 7.32 per cent and 0.91 per cent of the farmers of their respective classes. None of the households belonging to scheduled tribes own tractor. Probably they cannot afford for it. Overall, one out of 100 farmers has tractor. Density of tractors works out to be 2.8 per thousand hectares of cropped area. Owners of marginal and semi-medium holdings and scheduled caste farmers do not have tractor at all. It is pertinent to mention that holdings of more than half (56 per cent) of the tribal households are either marginal or small. Relationship between the size of holdings and ownership of mechanical technology is evident from Table 18.2.

Thus, it is clear that such capital-intensive mechanical technology is within the reach of big landlords and farmers of higher castes and communities. Scheduled Castes population has poor socio-economic conditions to use such machines in this plateau. Under such conditions, traditional method of ploughing is practiced by all sections of the farmers. Mechanized farming requires managerial skill and close supervision. Machines require careful maintenance because of the rapid deterioration of steel when used in silty loam, silty clay and similar soils. It depends upon the availability of spare parts. At the

Table 18.1: Number and Percentage of Farmers owning Modern Agricultural Implements and Machines by Communities.

(Number of households)

Community	*Bullock Pairs*	*Tractor*	*Trolley*	*Irrigation Well*	*Elect. Pump*	*Diesel Pump*	*Thresher*	*Duster*	*Sprayer*
ST 171	152	0	0	60	13	15	2	2	3
SC 45	37	0	0	10	4	2	1	0	5
BC 218	175	2	1	130	105	11	10	7	38
UC 41	35	3	3	23	22	3	6	2	6
All 475	399	5	4	223	144	31	19	11	52
			Percentage of total households of the class						
ST	88.9	0.0	0.0	35.1	7.6	8.8	1.2	1.1	1.8
SC	82.2	0.0	0.0	22.2	8.9	4.4	2.2	0.0	11.1
BC	80.3	0.9	0.5	59.6	48.2	5.1	4.6	3.2	17.4
UC	85.4	7.3	7.3	56.1	53.7	7.3	14.6	4.9	14.6
All	84.0	1.1	0.8	47.0	30.3	6.5	4.0	2.3	11.0

Table 18.2: Number and Percentage of Farmers Owning Modern Agricultural Implements and Machines by Size of Holdings

(Number of Households)

Size of Holding	*Bullock Pairs*	*Tractor*	*Trolley*	*Irrigation Pump*	*Elect. Pump*	*Diesel Pump*	*Thresher*	*Duster*	*Sprayer*
Marginal	87	0	0	36	23	5	1	2	8
Small	97	1	0	53	25	3	1	1	9
Semi-medium	124	0	0	79	53	10	6	0	13
Medium	75	1	1	43	34	7	7	6	20
Large	16	3	3	12	9	6	4	2	2
All	399	5	4	223	144	31	19	11	52
				Percentage of total households of the class					
Marginal	70.2	0.0	0.0	27.0	18.6	4.0	0.8	1.6	6.5
Small	80.8	0.8	0.0	44.2	20.8	2.5	0.8	0.8	7.5
Semi-medium	91.1	0.0	0.0	58.5	37.3	7.4	4.4	0.0	9.6
Medium	93.8	1.3	1.3	53.8	42.5	8.8	8.8	7.5	25.0
Large	100.0	18.8	18.5	75.0	56.3	37.3	25.0	12.5	12.5
All	84.0	1.1	0.8	47.0	30.3	6.5	4.0	2.3	11.0

same time, machines cannot be used on small and tiny farms by the poor peasantry. These factors are not favourable for the adoption and use of modern power technology in this region.

Irrigation Wells

Out of sample farmers, less than half (47 per cent) have their own irrigation wells. Among scheduled communities, proportion of well owners is lowest (22.2 per cent) among scheduled castes and 35.1 per cent among scheduled tribes. In spite of the fact that the well is not a very costly affair and also government provides subsidy for their construction, STs are far behind the regional average. On the other hand, farmers of backward communities stand at top (59.6 per cent) followed by the farmers of upper communities. At the same time, proportion of farmers possessing irrigation wells increases from 29 per cent among marginal landholders to 75 per cent among large landholders.

Irrigation Pumping Sets

Only few people own irrigation pumping sets. Out of 223 farmers having their own irrigation wells, only 175 have pumping sets for lifting water. It means one-fourth of well owners use traditional devices for lifting water from wells. Consequently, their efficiency is considerably reduced. Out of 60 tribal farmers, having their own irrigation wells, only 28 have irrigation pumps and more than half (32) don't have pumping sets. They get heavy subsidy for this purpose from the government. Scheduled castes are at bottom in ownership of irrigation pumps, however, 60 per cent of well owner SC have pumping sets. It supports the view that farmers of these communities either hesitate in practicing irrigation or depend on public sources of irrigation. As a result, they have smaller proportion of irrigated area. Out of total sample households, only 6.6 per cent have diesel pumps and nearly one-third (30.3 per cent) own electric pumps. In fact, where there is electricity, electric pump is preferred because of the easiness of its operation. Even their distribution is directly related with the size of holdings. Marginal and small farmers are always far behind the average.

Thresher

Like tractor, thresher is rare commodity in this region. Only 4.0 per cent farmers reported the ownership of this. Proportion of thresher owners varies from 0.8 per cent among marginal and small farmers to 8.8 per cent among medium and 25.0 per cent of big landlords. Most of them belong to backward communities and upper castes.

Dusters and Sprayers

Dusters and sprayers are comparatively much cheaper and, therefore, more popular among farmers. These implements are used for spraying chemicals and medicines on standing crop. Nearly 2.3 per cent of farmers have dusters and one-tenth (11.0 per cent) of farmers have sprayers. This proportion varies in different communities. Proportion of duster-owners ranges from nil in scheduled castes and 1.1 per cent in scheduled tribes to 3.2 per cent in farmers of backward communities and 4.9 per cent of upper castes. Their ownership is also associated with the size of holdings. About 1.6 per cent of marginal and 0.8 per cent of small farmers own duster. About 7.5 per cent of farmers possessing medium and 12.5 per cent of big landlords have duster.

Proportion of farmers having sprayer is highest for medium holdings as well as for backward communities. About one-fourth of medium farmers and 17.4 per cent of farmers of backward communities have sprayers. On the other hand, very few of the scheduled tribes (1.8 per cent) own them. It can be concluded that farmers of higher and backward communities as well as medium to big farmers are above the regional averages in respect of the ownership of the improved implements and machines. And thus, they can go for capital intensive agriculture by using mechanical as well bio-chemical inputs on account of their better resource position, greater risk bearing capacity and credit worthiness.

Use of Modern Agricultural Implements

It is worth-mentioning that farmers who own improved implements and machinery they use them. Besides them non-owners also hire them for use from time to time. Because of this fact, proportion of users of improved implements etc. is always higher than the proportion of their owners. Still size of holding and social status are forceful

Table 18.3: Proportion of Farmers Using Modern Agricultural Implements and Machines by Communities

Community	*Bullock Pair*	*Tractor*	*Trolley*	*Irrigation Well*	*Elect. Pump*	*Diesel Pump*	*Thresher*	*Duster*	*Sprayer*
ST	88.9	0.6	0.0	35.7	9.4	16.4	7.6	2.3	4.1
SC	82.2	8.9	0.0	22.2	8.9	6.7	35.6	8.9	20.0
BC	80.3	21.6	0.9	59.6	51.4	6.4	67.4	19.3	39.5
UC	85.4	29.3	7.3	56.1	53.7	7.3	70.7	19.6	51.2
All	84.0	13.4	1.1	47.2	32.4	10.1	43.2	12.2	25.9

Table 18.4 : Density of Modern Agricultural Implements and Machines on per 100 hectares of Cropped Area by Community

Community	*Bullock Pair*	*Tractor*	*Trolley*	*Irrigation Well*	*Elect. Pump*	*Diesel Pump*	*Thresher*	*Duster*	*Sprayer*
ST	39.7	0.0	0.0	11.9	2.0	2.0	0.4	0.4	0.4
SC	19.6	0.0	0.0	5.5	2.5	1.0	0.5	0.0	2.5
BC	32.1	0.3	0.1	24.9	19.6	2.1	1.6	1.1	6.1
UC	16.5	0.9	0.9	11.5	11.6	1.6	1.9	0.9	18.9
All	30.8	0.3	0.2	16.3	10.8	1.9	1.1	0.7	3.2

determinants of their use. Proportion of farmers using modern implements and machines presented in Table 18.3.

It is evident from Table 18.3 that the use of improved implements is associated with the social status of the farmers, excepting the use of bullock-pair and irrigation wells. Tribal farmers are at the bottom in every case except in irrigation pumps. While proportion of users is highest either amongst backward communities or among upper caste farmers. Only one farmer of ST community reported the use of tractor while none owns it. Proportion of tribal farmers using pumping sets is also higher than the proportion of their owners. However, electric pumps are less popular among tribals due to the lack of electrification in the tribal areas. Therefore traditional methods of irrigation is popular among them. Tribal farmers also reported difficulty in obtaining diesel for running pumps because of inaccessibility.

Density of Agricultural Implements and Machines

From the preceding analysis it is undoubtedly proved that proportion of owners and users of modern agricultural implements and machines is directly related with the social status of the farmers connoted by the community. It would be more clear if density of these implements on cropped area is seen. Table 18.4 furnishes these informations by community.

It is evident from Table 18.4 that the tribal farmers are at the bottom in density of five out of eight implements and machines under consideration and are at second place in case of three. On the other hand, highest density is either on farms of backward communities or of upper castes.

It can be summed up that efforts have been made for improving the economic conditions of the tribes and tribal blocks. In this sequence, agriculture has been given top priority, even then it is poorly developed. It is found that the measures being propagated in these areas do not fit in the frame of socio-cultural system of the tribes. Therefore they find it difficult to adopt these measures. It is, therefore, suggested that these measures should be in accordance to their needs and capabilities.

Impact of Adoption of Mechanical Technology

Modern agricultural innovations are used as a package. For instance,

Table 18.5: Use of Modern Agricultural Inputs among Sample Farmers by Communities

Community	*HYVs % HH*	*HYVs % area*	*Fertilizer % of HH*	*Fertilizer % Area*	*Chemical spray % HH*	*Pesticide % HH*	*Land treatment % HH*	*Seed treatment % HH*
ST	60.8	36.0	48.0	33.4	11.1	10.5	2.3	5.0
SC	66.6	39.1	51.1	39.3	11.1	15.6	6.7	8.9
BC	93.6	79.3	83.5	79.6	30.3	30.3	23.4	28.0
UC	100.0	71.3	85.4	82.6	43.9	5.7	29.3	41.5
All	79.8	60.1	67.6	61.6	22.7	23.8	15.0	18.7

optimum benefit from the high yielding variety of seeds is conditioned by the use of irrigation water and fertilizers. These are accompanied by the protectional measures. Performance of all these practices depends on the use of improved implements and machines. Since tribes are yet to start the use of improved mechanical technology, employment of other modern practices is low. For example, impact of the ownership and use of irrigation wells and pumping sets is reflected by the proportion of households practicing irrigation and proportion of cropped area under irrigation. On average, 63.2 per cent of the sample farmers practice irrigation. This proportion is only 49.7 per cent of scheduled tribes and 44.4 per cent of scheduled castes. On the other hand, 82.9 per cent of farmers of upper castes and 73.9 per cent of backward communities irrigate their crops to some extent. Similar is the trend of the proportion of irrigated area. Only 24.9 per cent of the cropped area of sample farmers could be irrigated. This proportion varies from 14.9 per cent for ST and 20.3 per cent for SC to 29.5 per cent for BC and 34.4 per cent for other upper castes. Not only in case of irrigation but also in the use of HYV, fertilizers, pesticides and chemicals for spray and land treatment and seed treatment scheduled tribes are far behind the farmers of other communities. It would be very dear from the information furnished in Table 18.5.

Low adoption of agricultural innovations manifests itself in low productivity. Though the productivity of agriculture is the function of several phyico-socio-economic factors (Sharma and Jain, 1985 : 34), use of inputs is major determinant. Mechanical technology is supposed to be labour saving and they multiply the efficiency of human labour. At the same time, some of them directly influence productivity. For instance, pumping sets raise productivity by providing irrigation water and irrigation, in turn facilitates the use of high yielding variety of seeds and fertilizers, which are yield-raising technology. Others help in protection of crops ultimately leading to increased productivity. Since adoption of agricultural innovations as a whole (Sharma and Vishwakarma, 1998 : 53) in general and of mechanical technology in particular is low among tribal farmers, yield of crops on their farms is very low in comparison to other communities. It is supported from the data presented in Table 18.6.

Conclusions

Although it will be rather over simplification and unrealistic to attempt

Table 18.6 : Yield of Major Crops by Community of Sample Farmers (Yield in kg/ha)

Crops	*ST*	*SC*	*BC*	*UC*	*All*
Wheat	649	888	1233	840	993
Rice	568	593	624	333	584
Maize	496	779	1264	1210	839
Jowar	416	414	940	1157	705
Gram	297	471	653	502	468
Tur	191	505	423	516	363
Soyabean	656	799	916	713	821
Niger	115	173	202	132	143

sweeping generalization on the basis of this discussion, however, certain remarks can be made. It is found that measures propagated in other areas do not fit in the frame of socio-cultural system of the tribals. Therefore, they find it difficult to adopt these measures. It is, therefore, suggested that inputs and technologies should be in accordance of their needs and capabilities. It is found that inputs not fitting in their cultural belief have rare chances of success in tribal areas. It is remarkable that these people get heavy subsidy for purchasing these mechanical technologies. But farmers of upper castes utilise this facility on their names. Thus finance only is not hurdle. In fact operational part is rather more difficult which requires some training and care for maintenance and tribal farmers fail at this count. Unless such problems are not taken care of adoption in general and of mechanical technology in particular is bound to be low among tribes.

Acknowledgements

The author expresses deep sense of gratitude to Dr S.K. Sharma, Professor, Department of Geography; and Dr H.S. Gour University Sagar, M.P. for going through the manuscript and forgiving valuable suggestions.

REFERENCES

Dasgupta, Satadal, (1999). *Diffusion of Agricultural Innovations in Village India*. New Delhi: Wiley Eastern Limited.

Mohammad, Noor, (1992). Anthropogenic correlates and determinants of agricultural Productivity, pp. 139-1184, in Noor Mohammad, ed, *New Dimensions in Agricultural Geography,* Vol. 8. New Delhi: Concept Publishing Co.

Sharma, S.K., (1988). Adoption of agricultural development measures in relation to size and tenancy of operational holdings in Madhya Pradesh, *Journal of Indian Regional Science*. Vol. XX, No. 2, pp. 27-39.

Sharma, S.K., (1989). Use of yield-raising technologies in Tribal blocks of Madhya Pradesh, pp. 71-80, in P.D. Tiwari and A.N. Sharma, eds. *Tribal Ecosystem and Malnutrition in India.* New Delhi: Northern Book Centre.

Sharma, S.K. and C.K. Jain, (1985). Changes in agricultural productivity in Madhya Pradesh, *The Geographer*, Vol. XXXII, No. 2, pp. 34-55.

Sharma, S.K. and D.D. Vishwakarma, (1998). Evaluation of knowledge and use of high variety of seeds by tribals of Betul-Chhindwara plateau. M.P., *The Geographer,* Vol.XLV, No.1, pp. 44-54.

Singh, Jasbir and B. K. Shrama, (1985), "Operational determinants of agricultural productivity", pp. 422-58, *Determinants of Agricultural Productivity: A Sample Survey of Operational Holding for Landuse Planning.* Kurukshetra, Vishal Publications.

19

HUMAN RESOURCE AND AGRICULTURAL DEVELOPMENT OF LUMBINI ZONE, NEPAL

S.B. SINGH and B.R. KARERIYA

INTRODUCTION

The natural endowment comes under resources category when the man thought his mental and physical ability gets benefit through proper use of available materials in that region. The physical factor provide platform upon which the development process proceeds ahead in the condition with technological advancement of man. From this, it is obvious that development hinges much on the response of people. Efficient people may make best out of given resources, while inefficient people may waste them. Hence, economic development under given resources is determined largely by human behaviour, human quality and human institutions (Shrestha, 1998). Joseph Fisher (1962) have rightly remarked that natural resources are in themselves passive in their influence on economic growth. They remain dormant and largely unknown till such time as the general including technology recalls their use value. They acquire a dynamic character largely through changes introduced by men. Further, it is related to sustainable development of both natural as well as human resources. So, the concept of sustainable development, involves rational and successful management of resources either natural or human through technology and efficient energy systems to satisfy human needs, by maintaining or enhancing the quality of environment. The production of wealth and their equitable distribution and protection of environment are equally relevant for improving the quality of human life and fragile ecosystem (Singh, 1995).There is a growing awareness about the need to substitute current production system with a more sustainable and integrated system of agricultural production. The gravity of the environmental degradation question has led to the scientists, planners and geographers to examine ecologically sound, economically viable and socially sustainable farming system for different agro-climatic regions.

PROFILE OF THE STUDY AREA

The study area, Lumbini zone lies at the southern part of the Western Development Region of the Kingdom of Nepal between 27°17' N to 28° 17' N and 82°43' E to 84°23' E with 8,975 km^2 area and 20,13,673 population (Census, 1991). It consists of six districts, namely Arghakhanchi, Gulmi, Palpa, Nawal Parasi, Rupandehi and Kapilvastu. Of which, former three districts fall in the hilly region while latter three ones are in the tarai region. Thus, apparently it can be said that this zone is extended from hilly region, the north to Indian border, the south, she makes her zonal boundary from three sides to the east Narayani, to the west Rapti, and to the north Dhaulagiri and to the north-east direction there lies Gandaki zone while southern boundary of this zone is separated internationally with India (Fig. 19.1). Geographically, it is located at the central part of Nepal which has a considerable bearing on the geology of Nepal as a whole, since the numerous southward thrusts, geographically and chronologically originate here. North of Pokhara lies the dominating heights of Annnapurna Himal/mountain the base of which represents the oldest root-zone of uplift in Nepal Himalaya (Hagen,1960). The paraautoch-thonus zone of Pokhara or so-called Pokhara-Zone is considered the lowest unit in Nepal (Gurung, 1965). But the area, Lumbini Zone is located to the south of Pokhara-Zone extending from hilly region, Mahabharat range (3,000 m) to tarai region (150 m). Between these two unequal topographical parts there exists the Churia range (900 m) running just south of the Mahabharat mountain are analogous to, and contemporary of the Siwalik ranges of north India.

Climatically, this zone is much varied as altitude increases from south to north, the summer and winter temperature are also different in their nature. During summer, temperature ranges from 30° to 20°C and average rainfall is about 200 cm. Tarai is normally hotter than hill as it receives *loo* (heat wave) while in winter temperature decreases from 15° C to 10° C from north to south and it becomes moderately cool. During this period, cold waves locally known as *sirato* descend

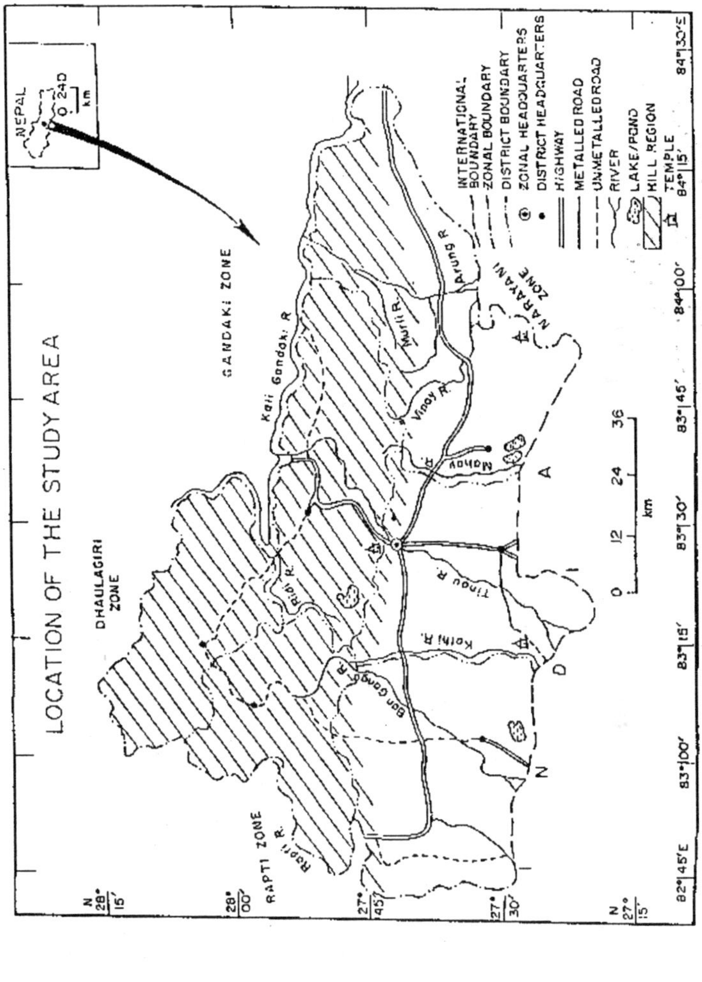

Fig. 19.1 : Location of the Study Area

down from there and make weather very cold in most parts of mid land region. Rainfall is negligible. Sometimes south westerly cyclones approaching from the mediterranean sea cause rainfall which amounts about 40 cm in hilly region and 25 cm in the parts of Lumbini Zone.

STATEMENT OF THE PROBLEM

Natural resources like flora and fauna, minerals etc., were almost unknown and unexploited in the past because of lack of modern technological development and sparse population, but as the population and new technologies began to increase rapidly and simultaneously as well as poverty and regional disparities generally started to be seen. This has been occurred dramatically in Nepal. Nepal, as being one of the developing countries of the world, estimated per capita GDP for the year 1995/96 was US $220. About 90 per cent of 21 million people of the country live in rural areas. Population is growing at the rate of 2.1 per cent per annum, has produced a broad base, increased dependency ratio. Unemployment rate is 4.9 per cent estimated for 1996. Literacy is however increasing but still much lower (40%) to upgrade pace of development by making optimum use of human resources. There is a great challenge to the nation to estimate the massive poverty in the country and provide basic needs to the people. Keeping the view over these challenges, efforts are under way to improve quality of life of the Nepalese people.

Realising the fact, it can be well justified that the study area, Lumbini Zone is also not free from the problems/challenges as the population is growing at the rate of 2.84 per cent per annum. It has been estimated and projected that the population would be 26,63,790 in 2001 which was 20,13,673 in 1991. But if the growth rate follows the same and base year population of 2001 is also the same, the projected population it would be estimated 35,10,750 for the another ten years *i.e.* 2011. Thus, the pressure of population on land in the study area may be much problematic and challenging. To overcome upon these problems and constraints the proper use of the lands giving priorities on agricultural development becomes a must. For this, better land-use planning in the study area should be implemented which requires better policies and management.

MATERIAL AND METHODS

Methodology adopted for information collection to this study is mainly secondary source which has been obtained from different governmental/non-governmental offices. Besides, other books/ journals and statistical year books have also been used for data input. In order to make enriched this Chapter prominent cartographic presentation has also been enveloped which highlights the clear picture of the study area.

RESULT AND DISCUSSIONS

Human Resource

Human resource is the basic resource for the overall development of any region. The human activities are closely related to the geographical environment of the particular region. In other words, it can be stated that man is the essential resource as he utilizes resources and changes the natural environment. He is not only the beneficiary of the entire process of resource development and utilisation (Zimmerman,1951) but also is the most potent and dynamic agent of the production. This resource base is basically useful to mobilise and utilise the natural resources properly. Problems of unemployment and low productivity area, while literacy, martial status etc. are the factors that influence the quality of human resource. It is, therefore, needed precisely planned development policy that these problems should be well tackled and solved so that human resource would be changed into active force for growth and development (Singh, 1988). Human resource and agricultural development of any particular region are primarily based on the socio-economic facilities with their location at appropriate places serve the people and play an important role in the development of a particular geographical area (Singh and Singh,1987).

Keeping the above views, this Chapter focuses on distributional pattern of population, man-land ratio/density such as physiographic, agricultural, caloric and nutrition and literacy etc. It is also examined that how far human resource's impact is on agricultural development of Lumbini Zone, Nepal.

Population Distribution and Man-Land Ratio

Man-land ratio is mainly affected by geographical factors such as terrain, climatic conditions, availability of water, availability of arable land and other socio-economic factors. The hilly districts of Lumbini

Zone are sparsely populated which constitutes 1,80,884 population in Arghakhanchi, 2,66,331 population in Gulmi, 2,36,313 population in Palpa, while the tarai districts such as Nawal Parasi possesses 4,36,217 population, Rupandehi and Kapilvastu possess 5,22,150 population and 3,17,778 population respectively.Tarai districts as being feasible from every type of economic point of view mainly with agricultural activities is economically viable. So, the concentration of the people here is higher as campared to the hilly region. Furthermore, this can be proved with the help of densities of existing districts of the region. Thus man-land ratio is the ratio of total population to the total area of the zone. In other words, it can be well justified from densities of that particular region.

The 1981 Census data reveals that the people residing per km^2 in Arghakhanchi, Gulmi, Palpa, Nawal Parasi. Rupandehi and Kapilvastu was 131.8 persons/km^2, 207.2 persons/km^2, 156.18 persons/km^2, 142.8 persons/km^2, 278.2 persons/km^2 and 155.3 persons/km^2 respectively, while in 1991 the man-land ratio is higher as compared to1981 and the above mentioned districts possess 151.6 persons, 231.7 persons, 172.1 persons, 201.9 persons, 383.9 persons and 213.9 persons per km^2 respectively. The density of Lumbini Zone was 174.6 persons/ km^2 in 1981 while it increased to 224.3 persons/ km^2 in 1991 (Table 19.1).

Table 19.1: Population Density in Lumbini Zone

Area/ District	*Density*	
	1981	*1991*
Arghakhanchi	131.8	151.6
Gulmi	207.2	231.7
Palpa	156.18	172.1
Nawal Parasi	142.8	201.7
Rupandehi	278.7	383.9
Kapilvastu	155.3	213.9
Lumbini Zone	174.6	224.0

Source : Population Census, 1981 and 1991, Kathmandu.

Similarly, other types of densities such as physiographic, agricultural, caloric and nutrition have also been computed.

Physiographic density is the ratio of total population to the total cultivated area, and densities 609.1 persons/km^2. Agricultural density is the ratio of total rural population to the total cultivated area and density is 568.7 persons/km^2 while caloric and nutrition densities are the ratio of total rural population and total population to the total food cropped area and densities are 500.3 persons / km^2 and 535.7 persons/ km^2 respectively (Table 19.2).

Table 19.2 : Density Types in Lumbini Zone

Types	*Density (persons/km^2)*
Physiographic	609.1
Agricultural	568.7
Caloric	500.3
Nutrition	535.7

Source : Population Census, 1991.

Sex-Ratio

Sex-ratio is an essential element to analyse thc human resources as it determines the quantum of labour force supply and influence the socio-economic structure. In the study area of hilly districts: Arghakhanchi, Gulmi and Palpa the sex- ratio is 1148, 1204 and 1141 respectively as per 1000 males while in the districts of tarai such as Nawal Parasi, Rupandehi and Kapilvastu, the sex-ratio is 1003, 973 and 941 respectively as per 1000 males. But as a whole in Lumbini Zone there are 1036 females residing as per 1000 male numbers (Table 19.3).

Thus, the sex-ratio is higher in the hilly districts of the study area than the tarai districts. It is because more male numbers remaining out of their homes for long periods for employment as well as job opportunities.

Agriculture Resource

Agriculture is the most important constituent of man's primary occupations and forms the first vital link of human interaction with his environment (Singh, 1995). Under the land-use pattern agriculture is the life blood of any region. This is borne by the fact that out of a total reported area of 8,97,500 ha, 4,66,230 ha (51.95%) lands have been devoted to agricultural arable land. The whole study area covers the total geographical area, 8,97,500 ha of which, 51.95 per cent is under agriculture. Besides, 3,54,171 ha (39.46%) forest,1806 ha (2.01%) grazing lands/pastures, 42,178 ha (4.70%) roads and settlements and rest 16,855 ha (1.88%) is not available for agricultural purposes.

Table 19.3 : Sex-Ratio in Lumbini Zone by District-wise

District	*Male*	*Female*	*Total*	*Sex-Ratio*
Arghakhanchi	84172	96712	180884	1148
Gulmi	1210795	145536	266331	1204
Palpa	110325	125988	236313	1141
Nawal Parasi	217731	218486	436217	1003
Rupandehi	264544	257606	522150	973
Kapilvastu	191444	180334	371778	941
Lumbini Zone	98901	1024662	2013673	1036

Source : Nepal Population, Census, 1991.

Agricultural Intensity

It has been observed that all districts situated in the same ecological and agro-economic environment, records diversity in agriculture. The rational use of land at different levels shows three types of intensities *i.e.* cropping intensity (ratio of harvested area to arable area) land-use intensity (ratio of gross sown area to net sown area) and irrigation intensity (ratio of gross irrigated area to net irrigated area) have been distinguished. The density of population is much greater and land resources available for extension of cultivation are much smaller. In the study area, cropping intensity varies from 105.29 (Rupandehi) to 149.14 per cent (Nawal Parasi). But cropping intensity for whole Lumbini Zone is 119.16 while land-use intensity varies from 102.77 in Rupandehi to 144.89 in Arghakhanchi and for whole Lumbini Zone is 112.54. Similarly, there is also variation in irrigation intensity as it is differentiated from 139.98 in Arghakhanchi to 249.60 in Kapilvastu. But for whole study area the irrigation intensity is 201.99 per cent (Table 19.4).

Table 19.4 : Land-use Character in Lumbini Zone

Area/District	*Cropping Intensity*	*Land-use Intensity*	*Irrigation Intensity*
Arghakhanchi	125.10	144.89	139.98
Gulmi	142.65	108.05	190.74
Palpa	112.15	129.95	210.43
Nawal Parasi	149.14	118.71	165.39
Rupandehi	105.29	102.77	195.14
Kapilvastu	108.19	103.39	249.60
Lumbini Zone	119.16	112.54	201.99

Irrigation System

It has been well recognised that irrigation plays a vital role to farmers' prosperity and national agricultural production. The water supply is the need for a well-controlled year around without which it is virtually impossible to accelerate agricultural growth rates to a level of bringing rural prosperity and economic transformation of the country. Although irrigation has been the centrepiece of investment programmes in Nepal for more than forty years; the emphasis has been largely on expanding supplementary monsoon irrigation. Only 18 per cent of the arable area is commanded by well-controlled year-round water supplies (Nepal/APP, 1995) because irrigation has to be dependent on the seasonal/monsoonal character of rainfall which generally occurs between 15th June to 15th October, except for a few days during the winter season. The availability of rainfall is the main cause for the development of irrigation facilities in the area.

Thus, keeping the view, it can be well conceived that in the study area, the irrigation system is quite peculiar as tarai region is drained by both groundwater system and river flowing from the north while hilly region is basically irrigated with terraced method and the perennial and seasonal rivers are the main source for these purposes from where streams/drains locally known as *Kulos* are made.

Agricultural Productivity

The term production has been used to denote the per ha yield of crops. It has been well understood that the tarai districts of Lumbini Zone are more productive in both food crops as well as in cash crops as compared to hilly districts of the study area. It is because of the role played by some major factors *viz.*, plain area, accessibility, irrigation facilities, modernized implements used in farms, availability of transportation and marketing facilities and so on. In Lumbini Zone the overall production of food crops is 2,024.5 kg per ha while per capita production is 377.8 kg. Similarly the cash crop production per ha in the study area is 13,799.6 kg while per capita production is 234.9 kg. In the whole study area, Nawal Parasi district has the highest cash crop production (18,053.1 kg/ha) while 579.4 kg is the per capita production and the lowest is possessed by Arghakhanchi district (2,229.7 kg/ha and 18.2 kg/capita production). Some cash crop such as oilseeds and sugarcane are also grown. Potatoes and other vegetables are grown mainly for household consumption. By substantially increasing agricultural and food production, these improvements would improve the food security situation in the region (Hirmiz and Karim, 1996). Hence to analyse the general agricultural productivity level in the study area at two points of time period *i.e.* (1986/87 and 1995/96) production of respective crops has been taken into accounts which has further been depicted on the basis of the formula (Shafi, 1972):

$$\left(\frac{yw}{t}+\frac{yr}{t}+\frac{ymi}{t}.....n\right):\left(\frac{Yw}{T}+\frac{Yr}{T}+\frac{Ymi}{T}.....n\right)$$

$$=\frac{y\sum w}{t}:\frac{n\sum Y}{T}$$

where

yw, yr etc. = total yield of the respective crops in the district,
Yw, Yr etc. = total yield of the respective crops at the national level.
t = area planted under the crop in the district, and
T = area planted under the crop at the national level.

The result exhibits that five types of agricultural productivity level which are high level of agricultural productivity (index value, 1.75-2.00), medium level of productivity (index value, 1.00-1.75), low level of productivity (index value, 0.75-1.00), very low level of productivity (index value, 0.50-0.75) and very very low level of agricultural productivity (index value <0.50). Accordingly, the calculation justifies that during 1986-87, Nawal Parasi having index value 1.40, and Rupandehi having index value 1.13, were under medium level of productivity while Gulmi, Palpa, Kapilvastu and Arghakhanchi possess 0.80, 0.80. 0.87 and 0.87 index value respectively and are put under the low level of productivity in reference to the food production. But this figure is varied from 1995-96. The all tarai districts have fallen in medium level while the hilly districts are under low level of productivity followed by Gulmi (index value 0.70), Palpa (index value 0.84), Nawal Parasi (index value 1.26) and Kapilvastu (index value 1.16) respectively (Table 19.5).

Table 19.5 : Food Crop Productivity Level in Lumbini Zone

		1986-87		*1995-96*	
Rank	*Index Value*	*Name of District*	*Per cent*	*Name of District*	*Per cent*
I	1.75-2.00	—	—	—	—
II	1.00-1.75	Nawal Parasi, Rupandehi	33.33	Nawat Parasi, Rupandehi Kapilvastu	50.00
III	0.75-1.00	Gulmi, Palpa, Kapilvastu, Arghakhanchi	66.67	Gulmi, Palpa, Arghakhanchi	50.00
IV	0.50-0.75	—	—	—	—
V	< 0.50	—	—	—	—

Source : Statistical Year Book of Nepal, 1997.

Besides, in the context of cash crops production in 1986-87, three districts—Nawal Parasi (index value 2.41), Rupandehi (index value 1.79), Kapilvastu (index value 2.30) were in high level of productivity. Out of these, only one district *i.e.* Gulmi, possessing the index value 0.98 was under the low level category while two districts Palpa and Arghakhanchi were categorised under very low level of productivity which possess 0.68 and 0.57 index value respectively. But this figure is completely varied as compared to 1995-96. During this period, two districts Nawal Parasi and Kapilvastu having 2.38 and 1.79 index value respectively are under high level of productivity. Rupandehi and Palpa have 1.66 and 0.58 index value are under medium level of productivity and very low level of productivity respectively. But two districts Gulmi and Arghakhanchi are categorised into very, very low level of agricultural productivity which possess <0.50 index value *i.e.* 0.45 and 0.29 respectively (Table 19.6 and Fig. 19.2)

Table 19.6 : Cash Crop Productivity Level in Lumbini Zone

		1986-87		*1995-96*	
Rank	*Index Value*	*Name of District*	*Per Cent*	*Name of District*	*Per Cent*
I	1.75-2.00	Nawal Parasi, Rupandehi, Kapilvastu	50.00	Nawal Parasi, Kapilvastu	33.33
II	1.00-1.75		—	Rupandehi	16.67
III	0.75-1.00	Gulmi	16.67		–
IV	0.50- 0.75	Arghakhanchi, Palpa	33.33	Palpa	16.67
V	<0.50		—	Gulmi, Arghakhanchi	33.33

Source : *Statistical Year Book of Nepal*, 1997.

EPILOGUE

It is, therefore, concluded that there is an urgent need of proper human resource development providing them different types of opportunities for employment on the one hand, and suitable land-use planning accentuating in agricultural development on the other hand. For this, the following steps should be taken into consideration:

1. Land-use should be done in planned way but not haphazardly so that the interaction between rural and urban may be at the maximum level.
2. The empowering of agricultural development with special references to irrigation system, inputs and other innovative technologies should be done both in tarai and hilly region of the study area.
3. Lumbini zone, as being an agricultural region, there should be another green revolution, white revolution (milk), blue revolution (fish) and yellow revolution (oilseeds), so that

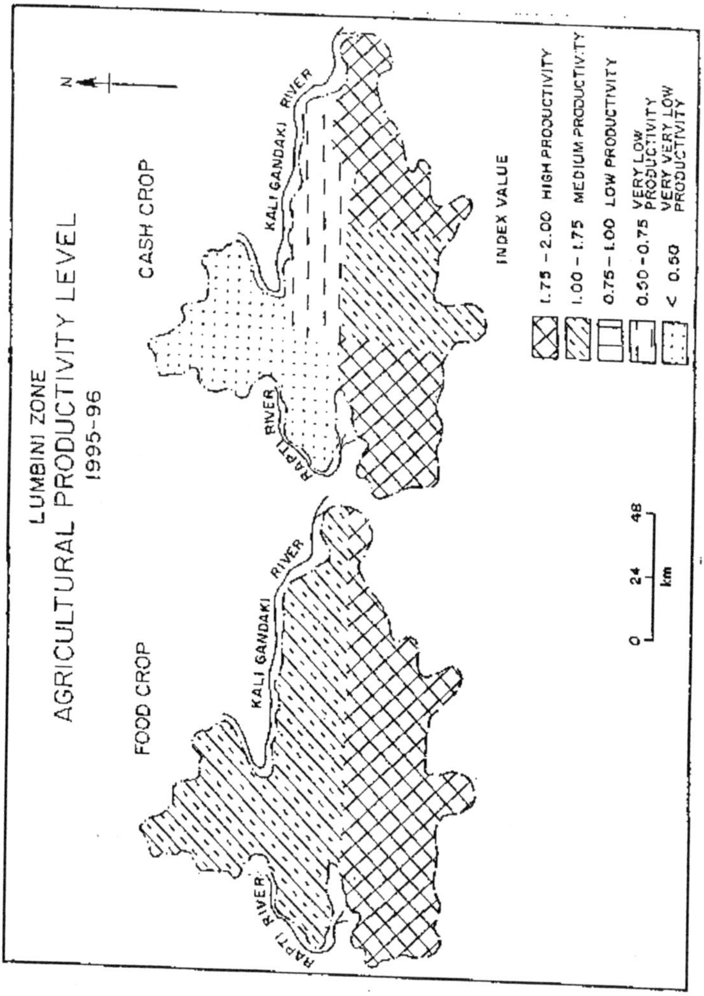

Fig. 19.2 : Lumbini Zone—Agricultural Productivity Level, 1995-96

human resource and agricultural development may be possible further leading to enhancing the employment opportunities *vis-a-vis* poverty alleviation.

REFERENCES

Fisher, J.L. (1962), *Role of Natural Resources in Economic Development, Principles and Problems*, London, p. 32

Gurung, Hark (1965), *Pokhara Valley, Nepal Himalaya :A Study in Regional Geography*, unpublished Ph.D. Thesis, University of Edinburgh, U.K. p. 17.

Hagen, Tony (1960), *Reports on Geological Survey of Nepal*, Zurich.

Hirmiz, Toma and A. Karim (1996), *Agriculture and Development in Western Asia*, Joint Publication of United Nations Economic and Social Food and Agriculture Organization of the United Nations, No. 18, Baghdad, p. 32.

HMG Nepal and Asian Development Bank (1995), *Nepal Agriculture Perspective Plan*, Inc. Washington, D.C., p. 69.

Shafi, M. (1972), Measurement of Agricultural Productivity of the Great Indian Plains, *The Geographer*, Vol. 19(1), pp. 4-3.

Shrestha, S.H. (1998), *Natural Resources, Economic Geography of Nepal*, Educational Enterprises Pvt. Ltd., Kathmandu, p. 73.

Singh, B.N. (1988), *Integrated Rural Area Development and Planning*, Anupama Publication, Delhi.

Singh, S.B. (1995), Sustainable Development of Rural Environment: A Case Study of Suithakalan Block, U.P., *National Geographer*, Vol. 30(2), pp.123-130.

Singh, S.B. and Singh, B.N. (1987), Planning for Socio-Economic Facilities in Ghazipur Tehsil, U.P., *Nat. Geog. Jl. of India*, Vol. 33(1), p. 89.

Zimmerman, W. (1951), *World Resource and Industries*, Harper and Row Publisher, New York, p. 91.

20

RESOURCE POTENTIAL IN CENTRAL ASIA

ABDUL RAUF SHAH

Studies of alternative allocation of resources (spatial temporal and functional) indicate a role for the geographer as resource analyst. They indicate a predisposition towards the compilation and manipulation of data about natural resources, or the inquiry into the manner in which resources are allocated among competing uses and users. As resource analyst geographer seeks to understand the fundamental characteristic of natural resources and the process through which they are, could be and should be allocated and utilised.

Existing resources of Central Asia are vast comprising oil, natural gas, non-ferrous minerals, precious mineral such as gold and silver, rare mineral such as uranium etc. Exploration of this natural wealth can play a major role in the overall development of these recently emerged Central Asia States (CAS). The study undertaken with the help in the formulation of programmes for more balanced regional development in future. However, this Chapter broadly aims at the following objectives :

(i) To analyze spatially potential resources of all five republics of Central Asia.

(ii) To figure out annual variations in all resources of Central Asian States.

(iii) To study variations in trends of all resources of Central Asian States.

(iv) To develop a resource base model of all republics of Central Asia in order to understand the composite picture as well as variations in resource potential.

In this Chapter resource potential data from 1987 to 1997 have been collected to find out the annual change under each potential. The average of these 11 years have been taken into account as a base showing resource potential variations of all these Central Asian republics.

Central Asia comprises the five republics of Turkeministan, Tajiskistan, Kyrgygstan, Uzbekistan and Kazakhstan. It is a vast area of deserts and high mountains though with some densely populated oases. Kazakhstan is characterised by a series of sub-regional industrial clusters, namely heavy industry and mining in centre, extensive grain farming and industrial clusters in the north, energy intensive and non-ferrous metals in the north-east, irrigated agriculture, non-ferrous chemicals in the south and petro chemicals in the west.

Within manpower resources, secondary sector has recorded a remarkable growth. From 1977 to 1998 the growth rate is in positive order. Within these years the year 1987, 1990 and 1992 have shown high trend as compared to the remaining years. Within tertiary sector the year 1987 and 1995 have gone in positive direction. Primary sector on the whole has maintained low momentum as compared to secondary sector (Table 20.1). From 1987 to 1993 an increasing growth rate order has been observed in total employed persons within the Kazakhstan republic as a whole.

Kazakhstan is an important producer of gold, iron ore, coal, copper, chrome and zinc. Other mineral resources found in exploitable amounts are oil, mineral gas, nickel, titanium, tungsten, molybdenum, lead, zinc, manganese and aluminium. From 1987 to 1997 iron bearing ores have recorded a positive growth. Out of these 11 years the starting four years (1987 to 1991) are in high positive order, while as the remaining years comparatively have not maintained the same momentum. But on the whole the iron ore industrial product is a leading product as compared to other industrial production resource (Table 20.2). Next to iron bearing ore, crude petroleum and natural gas are also leading industrial production resources. Crude petroleum product has almost recorded an increasing growth trend. High production was observed in the year 1987, 1988, 1991 and 1994. Natural gas on the other side has shown a slight decreasing growth as compared to crude petroleum. But in contrary to other industrial products such as mining, lignite wheat flour cement, the natural gas product is a leading industrial sector. Among the remaining industrial products wheat flour has comparatively gone in higher order with

Table 20.1: Kazakhstan : Manpower Resources : Sectorwise Employment and Changing Pattern *(Annual)*

Sectors	*1987*	*1988*	*% Change*	*1989*	*% Change*	*1990*	*% Change*	*1991*	*% Change*	*1992*	*% Change*	*1993*	*% Change*	*1994*	*% Change*	*1995*	*% Change*	*1996*	*% Change*	*1997*	*% Change*
Employed Persons	7295	7400	+1.44	7467	+0.90	7563	+1.28	7716	+2.02	7572	– 1.86	6926	– 8.53	5682	–17.96	6552	+15.31	6117	– 6.63	6334	3.54
Primary Sector (Agriculture, Mining Factory)	1883	1921	+2.02	1867	– 2.81	1921	+2.83	1963	+2.18	2011	+2.44	1969	– 2.08	1615	–17.97	1632	+1.05	1623	– 0.55	1627	0.24
Secondary Sector (Manufacturing Electricity, Gas), Construction)	1977	2031	+2.73	2118	+4.28	2146	+1.32	2377	+10.76	2313	– 2.69	2010	– 13.09	1784	–11.24	1446	–18.94	1615	11.68	1699	5.20
Tertiary Sector	1308	1281	– 2.06	1218	– 0	1214	– 0.32	1293	+6.50	1244	– 3.78	1119	– 10.04	1447	+29.11	1592	+10.02	1519	– 4.58	1555	2.36

Table 20.2: Kazakhstan: Variations in Industrial Production Resources *(Annual)*

(Thousand Metric Tonnes)

Products	*1987*	*1988*	*% C*	*1989*	*% C*	*1990*	*% C*	*1991*	*% C*	*1992*	*% C*	*1993*	*% C*	*1994*	*% C*	*1995*	*% C*	*1996*	*% C*	*1997*	*% C*
Mining & Quarrying	142	143	0.70	138	3.49	131	–5.07	130	–0.76	127	–2.30	112	–11.81	105	–6.25	83	-–20.95	77	–7.22	73	5.19
Petroleum	21814	22248	1.98	21953	1.32	21676	–1.27	22036	1.66	21934	–0.46	19572	–10.76	18544	–5.25	17032	–8.15	10123	–40.56	13577	34.12
Natural Gas	6311	7134	13.04	6710	–5.94	7114	6.02	7885	–10.83	8113	2.89	4488	–44.68	5916	32.40	6524	10.27	8114	24.37	8319	2.54
Iron Bearing	24224	24342	0.48	23764	2.37	23846	0.34	21993	7.77	17671	–19.65	10521	40.46	14902	41.64	12975	–13.65	13135	1.23	13055	–0.60
Manufacturing	632	689	9.01	727	5.51	710	2.33	724	3.70	596	–17.67	662	11.07	642	3.02	548	–14.64	463	15.41	506	9.17
Milton Lamb	258	279	8.13	289	3.58	270	–6.57	260	2.59	243	–6.53	275	13.16	258	–6.52	212	–17.82	172	18.86	192	11.62
Wheat flour	1935	1909	1.34	1968	3.09	1962	–0.30	2003	–2.59	1970	–1.64	1940	–1.52	1960	1.03	1575	–19.64	1593	1.14	1236	22.41
Cement	8349	8446	1.16	8650	2.41	8301	4.03	–8.74	–15.03	6436	–68.41	2033	–12.83	1772	–37.07	1115	29.46	1443	–11.36	1297	–11.36

respect to its growth in production (Table 20.2). Production growth rate of cereal crops has gone higher as compared to remaining one. Wheat crop occupies higher place (Table 20.3). The growth in production of wheat crop is remarkably higher as compared to remaining crops. The growth trend is almost in positive direction. The high production recording growth year includes 1987, 1988, 1989 and 1992. Within these years the production rate is comparatively high as compared to remaining years. Wheat production growth is followed by barley, potato and sugar beat. Among these crops barley has shown high growth in production. As compared to wheat the growth in barley production has also maintained a parallel positive trend. The year 1987, 1988 and 1990 have recorded high growth in production. While as in remaining periods the growth in production has slightly gone down. Among the remaining crops *i.e.* rice, maize, oats millets, soybean and seed the maize is a leading crop followed by rice. Yield kg/hectare within cereal crops has gone high in rice and maize (Table 20.4). Wheat, barley and oats have gone down. Among the remaining crops highest yield kg/hectare was recorded in sugar beat. The years 1987, 1988, 1989, 1990 and 1991 have recorded high growth in yield kg/hectare (Table 20.4).

Next to Kazakhstan, Uzbekistan republic is also rich in natural resources. The republic lies along the ancient Silk Road between Europe and the Far East. It has an area of 447,000 square kilometres and is slightly smaller than France. The republic is land locked and almost three-fifth of its area consists of steppe, desert and semi-desert terrain; the remainder constitute fertile valleys.

Uzbekistan with respect to manpower resources has recorded better growth in primary sector as compared to secondary one. Within the primary sector almost all years have gone in positive order from the year 1989 to 1995. However the year 1989, 1992, 1993 and 1995 have shown high trend as compared to previous years.The tertiary sector remarkably has gone down as compared to primary and secondary one. Within the secondary sector the year 1989, 1990 and 1991 have almost recorded high growth. The remaining years comparatively have not shown much growth. Employed persons are slightly more than that of Kazakhstan. In all the years the trend in employed persons has gone in positive order (Table 20.5).

In case of Industrial production Natural gas mining crude petroleum and crude steel are the leading industrial products. Their production as compared to the remaining one is at higher momentum.

Table 20.3 : Kazakhstan: Annual Variation in Production of Principal Crops

(Thousand Metric Tonnes)

Crops	*1987*	*1988*	*% A*	*1989*	*% A*	*1990*	*% A*	*1991*	*% A*	*1992*	*% A*	*1993*	*% A*	*1994*	*% A*	*1995*	*% A*	*1996*	*% A*	*1997*	*% A*
Rice	606	626	2.96	555	–11.34	579	4.32	515	– 11.05	467	– 9.32	403	– 13.88	283	– 29.77	184	– 34.98	226	22.82	255	12.83
Wheat	16108	12162	–24.49	10784	–11.33	16197	50.19	6889	– 57.46	18285	165.42	11585	– 36.64	9052	– 21.86	6490	– 28.30	7678	18.30	8955	16.63
Maize	477	561	17.61	479	–14.61	442	– 7.72	330	– 25.33	370	12.12	355	– 4.05	234	– 34.08	136	– 41.88	122	– 10.29	111	– 9.01
Barley	6929	5850	–15.57	5310	–9.23	8500	60.07	3085	– 63.70	8511	157.88	7149	– 16.00	5497	– 23.10	2208	– 59.83	2696	22.10	2583	– 4.19
Oats	459	345	–24.83	259	–24.92	611	135.9	231	– 62.19	727	214.7	802	10.31	822	2.49	250	– 69.58	359	43.6	286	– 20.33
Millet	549	577	5.10	459	–20.45	940	104.79	235	– 75.00	447	90.21	233	– 47.87	130	– 44.20	39	70.00	30	– 23.07	64	113.3
Potato	2066	2260	9.39	1784	–12.06	2324	30.26	2143	– 7.78	2570	19.92	2296	– 10.66	2040	– 11.14	1720	– 15.68	1656	– 3.72	1472	11.11
Soyabean	40	38	5.0	30	–21.05	25	– 16.6	14	– 44.0	12	– 14.28	6	– 50.0	6	10.00	4	33.33	3	25.00	3	0.00
Sugar Beat	1804	1321	– 26.77	1188	–10.06	1134	– 4.54	726	– 35.97	1276	67.03	925	– 27.50	433	– 53.18	371	– 14.31	341	– 8.08	341	0.00
Seed Cotton	312	325	4.16	315	–3.07	324	2.85	291	– 10.18	246	– 15.46	198	– 19.5	208	5.05	223	7.21	183	– 17.93	198	8.19
Grapes	141	94	33.33	48	–48.93	139	189.5	66.	– 52.51	52	– 21.21	40	– 23.7	37	– 7.5	68	83.78	37	– 45.48	36	2.70

Table 20.4 : Kazakhstan: Yield kg./Hectare of Principal Crops

Crop	*1987*	*1988*	*1989*	*1990*	*1991*	*1992*	*1993*	*1994*	*1995*	*1996*	*1997*
Rice	4.55	4.63	4.17	6.63	4.36	3.85	3.54	2.77	1.93	2.53	3.00
Wheat	1.05	0.81	0.74	1.15	1.35	1.31	0.90	0.71	0.51	0.62	0.77
Maize	4 00	4.09	3.54	3.42	3.65	2.93	3.03	1.69	1.56	1.41	1.60
Barley	1.00	0.82	0.78	1.27	1.28	1.48	1.02	0.90	0.47	0.47	0.81
Oats	0.95	0.98	0.61	1.59	1.19	1.48	1.47	1.27	0.50	0.79	0.73
Millet	0.81	0.82	0.59	1.20	1.27	1.59	0.44	0.35	0.16	0.11	0.41
Potato	10.81	3.22	8.57	11.26	10.70	10.30	9.40	9.35	8.34	8.76	8.36
Soybeans	1.05	1.26	1.20	0.96	1.47	0.92	0.85	1.00	0.8	0.75	0.09
Sugar Beat	2.43	2.46	2.64	2.70	2.48	2.19	1.8	1.87	2.02	1.72	18.00
Grapes	7.42	5.22	2.66	8.17	3.88	3.05	2.5	1.85	3.57 .	2.31	3.27

Table 20.5 : Uzbekistan: Manpower Resources: Sector-wise Employment and Changing Pattern *(Annual)*

Sectors	*1987*	*1988*	*%C*	*1989*	*%C*	*1990*	*%C*	*1991*	*%C*	*1992*	*%C*	*1993*	*%C*	*1994*	*%C*	*1995*	*%C*	*1996*	*%C*	*1997*	*%C*
Employed Persons	7716	7791	0.97	7642	– 1.91	7941	3.91	8111	2.14	8271	1.97	8259	– 0.14	8150	– 1.31	8158	0.09	8154	0.02	8156	0.02
Primary Sector (Agricultural, Mining and Factory)	2983	3029	1.54	2938	– 3.00	3120	6.19	3291	5.48	3612	9.75	3667	1.52	3603	1.34	3755	4.21	3679	2.50	3717	1.03
Secondary Sector (Manufacturing Electricity, Gas & Construction)	1882	1892	0.53	1873	– 1.00	1911	2.02	1867	– 2.30	1147	–38.59	1166	1.65	1067	– 8.49	1010	– 5.34	1038	– 1.34	1024	– 1.34
Tertiary Sector	840	844	0.47	836	– 0.71	852	1.91	542	–36.52	828	52.76	783.00	– 8.09	736	– 6.00	784	6.52	760	– 1.57	772	1.57

Highest production of 48,980 thousand metric tonnes of natural gas was recorded in the year 1996 (Table 20.6). The trend in natural gas production from 1987 to 1996 is almost in positive order. Mining industrial product on the other side has not gone parallel with respect to growth in production from 1987 to 1996 as compared to natural gas. High growth in production of 6,240 thousand metric tonnes was observed in the year 1989. The year 1987 and 1988 have slightly shown a low growth as compared to the year 1987 and 1988 (Table 20.6). From 1989 to 1996 the production has gone down. Crude petroleum production from 1990 to 1996 has recorded high growth in production as compared to previous years. The year 1992, 1993, 1994 and 1996 have shown positive trend. Comparatively the previous years have recorded low growth. The remaining industrial products on the whole have recorded low growth trend as compared to the industrial products of Kazakhstan.

With respect to cereal crops the rice and wheat are leading crops so far as their production trend is concerned. In case of rice the years 1987, 1988, 1990, 1991 and 1993 have shown high growth in production, while as the remaining years have slightly gone down. Wheat crop also have shown a positive trend from 1987 to 1996, but the growth in production of the crop from 1993 onwards upto 1996 has recorded remarkably a high growth in production as compared to all remaining crops. Within the remaining crops the seed cotton has gone very high with respect to its production growth rate (Table 20.7). Yield kg/hectare within cereal crops is high in case of maize. The crop has almost gone in positive order from 1987 to 1996. The growth trend in all these years has not shown much fluctuation. Within the remaining crops yields kg/hectare of potato has gone high (Table 20.8).

Turkeministan republic on the other hand has not shown much variation in manpower resources. Primary sector as usual is leading as compared to secondary sector. But tertiary sector on the whole has remarkably gone very high (Table 20.9). The republic has substantial energy resources primarily natural gas and oil. Natural gas production exceeds the production of mining and quarrying. From 1987 to 1991 the production has shown as increasing trend. Within these years the year 1989 has shown high production. Onwards from 1990 to 1997 the mining and quarrying production has not gone so high (Table 20.10). Natural gas production is followed by mining and quarrying. The high growth in mining production was recorded in the year 1987, 1989, 1990 and 1992 (Table 20.10). The remaining years have not

Table 20.6: Uzbekistan: Variations in Industrial Production Resources *(Annual)*

Industrial Products	*1987*	*%C*	*1988*	*%C*	*1989*	*%C*	*1990*	*%C*	*1991*	*%C*	*1992*	*%C*	*1993*	*%C*	*1994*	*%C*	*1995*	*%C*	*1996*	*%C*	*1997*	*%C*
Mining	5030	8.74	5470	14.07	6240	3.79	6477	−8.16	5948	−21.31	4680	−18.65	38.07	−1.97	3732	−20.57	2964	−4.28	2837	1.93	2892	1.93
Crude Petroleum	1027	3.31	1061	13.19	1201	7.07	1286	7.62	1384	26.30	1748	37.64	2406	45.46	3500	−57.8	1477	236.96	4977	58.64	7896	58.64
Natural Gas	39777	0.41	39943	2.87	41090	−0.79	40762	2.74	41881	2.20	42805	5.21	45036	4.98	47283	1.51	48000	2.04	48980	2.04	48490	−1.00
Manufacturing	246	10.56	272	7.35	292	1.02	295	9.49	323	0	323	17.02	378	3.17	390	3.07	402	−5.47	380	−5.47	391	2.89
Meat	56	−1.78	55	10.90	61	9.83	67	−7.46	62	9.67	68	8.82	74	−1.35	73	13.69	83	−6.02	78	−6.02	80	3.20
Mutton & Lamb	45	8.88	49	4.08	51	0	51	−71.37	14.6	16.43	17	5.88	18	−22.22	14	−25.0	10.5	−52.38	5	−49.52	7.75	51.0
Cement	5512	1.28	5583	10.94	6192	3.11	6385	−3.0	6191	−4.15	5934	−12.75	5177	−7.66	4780	−28.47	3419	−4.15	3277	−4.15	3348	2.16
Crude Steel	1044	−2.68	1016	−0.29	1013	−15.20	859	−20.37	684	−10.81	610	−3.78	622	1.96	616	−0.96	619	0.48	466	−24.71	542	16.41

Table 20.7: Uzbekistan: Annual Variation in Production of Principal Crops

(Thousand Metric Tonnes)

Crops	*1987*	*1988*	*%C*	*1989*	*%C*	*1990*	*%C*	*1991*	*%C*	*1992*	*%C*	*1993*	*%C*	*1994*	*%C*	*1995*	*%C*	*1996*	*%C*	*1997*	*%C*
Wheat	495	563	13.73	342	−39.25	553	61.69	610	10.30	964	58.03	876	−9.12	1362	55.47	2347	72.32	2535	8.05	1700	−32.93
Maize	421	520	23.51	460	−11.53	431	−6.30	431	0.0	367	−14.84	404	10.08	276	−31.68	186	−32.60	250	34.40	220	12.0
Barley	289	391	35.29	246	−37.08	385	56.50	324	−15.84	361	11.41	292	−19.11	300	2.73	321	7.00	350	9.03	250	−28.57
Rice	506	581	14.82	284	−51.11	503	77.11	515	23.85	539	4.66	545	1.11	498	−8.62	328	−34.13	450	37.19	294	−34.6
Potato	261	308	18.00	325	5.51	336	3.38	351	4.46	365	3.98	472	29.31	567	20.12	440	−22.39	490	11.36	510	4.08
Seed cotton	4858	5365	10.43	5292	−1.32	5058	−4.42	4646	−8.14	4128	−11.14	4235	2.59	3938	−70.12	3994	1.42	3300	17.37	3400	3.03
Cotton	1505	1732	15.08	1656	−4.38	1593	−3.80	1460	−8.34	1274	−12.73	1321	3.68	1258	−4.76	1265	0.55	1081	14.54	1080	0.09
Grapes	652	655	0.46	416	−36.48	745	79.0	480	−35.57	439	−8.54	381	−13.21	353	−7.34	577	63.45	580	0.51	580	0.00

Table 20.8: Uzbekistan: Yield kg./Hectare of Principal Crops

Crops	*1987*	*1988*	*1989*	*1990*	*1991*	*1992*	*1993*	*1994*	*1995*	*1996*	*1997*
Wheat	1.06	1.23	0.98	1.27	1.25	1.53	1.25	1.41	2.01	2.41	1.7
Maize	3.56	4.48	4.14	3.95	3.99	3.70	3.84	3.10	3.00	4.16	3.66
Barley	1.20	1.38	1.04	1.32	1.09	1.28	1.05	1.02	1.28	1.55	1.25
Rice	3.26	3.47	1.76	3.42	3.21	2.96	3.01	2.98	1.95	2.43	1.68
Potato	8.41	9.93	9.28	8.00	8.77	8.48	10.48	10.90	9.36	10.88	10.20
Grapes	7.24	7.11	1.91	7.84	5.05	5.28	4.01	3.71	6.07	6.82	6.82

Table 20.9: Turkeministan: Manpower Resources: Sector-wise Employment and Changing Pattern *(Annual)*

Sectors	*1987*	*1988*	*% Change*	*1989*	*% Change*	*1990*	*% Change*	*1991*	*% Change*	*1992*	*% Change*	*1993*	*% Change*	*1994*	*% Change*	*1995*	*% Change*	*1996*	*% Change*	*1997*	*% Change*
Employed Persons	1411	1447	+2.55	1494	+3.48	1546	3.48	1585	2.52	1584	—	1520	–4.04	1552	2.10	1536	–1.03	1540	–0.12	1538	–0.12
Primary Sector (Agricultural, Mining Factory)	574	594	+3.48	625	5.21	651	4.16	675	—	663	6.22	669	0.90	666	0.44	667	0.15	668	0.14	665	–0.44
Secondary Sector (Manufacturing Electricity, Gas and Construction)	293	289	–1.36	309	6.92	320	3.55	319	–0.31	304	4.70	311	2.30	315	1.28	317	0.63	316	–0.31	318	0.63
Tertiary Sector	580	602	+3.79	598	–0.66	613	2.50	630	2.77	621	1.42	625	0.64	623	–0.32	624	0.16	199	–68.1	131	–34.17

Table 20.10: Turkeministan: Variations in Industrial Production Resources *(Annual)*

(Thousand Metric Tonnes)

Products	*1987*	*1988*	*% C*	*1989*	*% C*	*1990*	*% C*	*1991*	*% C*	*1992*	*% C*	*1993*	*% C*	*1994*	*% C*	*1995*	*% C*	*1996*	*% C*	*1997*	*% C*
Mining & Quarrying	5101	5033	–1.33	5045	0.23	4996	–0.97	4828	–3.36	5200	7.70	4301	–17.28	4083	–5.06	4394	7.61	4424	0.68	4409	–0.33
Natural Gas Production	88135	88303	0.191	89921	1.83	87763	0.07	84348	–3.89	60102	–28.74	65154	+8.40	43513	–33.21	35722	17.90	32264	–9.68	35182	0.90
Beef & Veal	44	44	0.00	45	2.27	47	+4.44	44	–6.38	46	+4.54	18	60.8	51	183.0	51	0	52	1.96	51.5	–0.96
Mutton	32	32	0.00	34	6.25	27	–20.58	30	11.11	35	16.66	45	28.57	45	0.0	50	11.11	43	14.0	46	6.97
Cement	1066	1110	4.12	1057	4.47	1085	2.64	904	–16.68	1050	16.15	1118	6.47	690	–38.28	437	–36.6	563	28.83	500	–11.19

shown such a trend in production. Among other industrial products distilled oil and fuel figures out in terms of output growth. Hydropower production is more than thermal and geo-thermal. Production and consumption in 1992 was 13.18 billion kilowatt and has gone down in 1995 (Table 20.11).

The main principal crops include wheat, barley, rice, maize, potatoes, seed cotton and grapes. Within cereal crops maize has recorded high growth in production. From 1987 to 1996 the trend is almost in positive order. Highest growth in production was recorded in the year 1993. Among remaining years the year 1988, 1989 and 1990 have also shown an increase in their production record. While as the remaining years are in fluctuating order. Within the remaining cereal crops including barley and rice, the barley crop has shown high growth in production. From 1987 to 1992 the growth in production of the crop has recorded an increase while as in the remaining years there is an decrease in the production record (Table 20.12). Yield kg./hectare within cereals has gone high in rice. In this crop in all the years there is an increase in the yield kg./hectare of the crop. The yield record is followed by barely and wheat. Among other crops the high yield kg./hectare of the crop was recorded in seed cotton and potatoes (Table 20.13). Production of cereal crops have remarkably low growth rate as compared to Uzbekistan and Kazakhstan.

Next to Turkeministan, manpower resources in Kyrgyzstan have recorded better growth within this sector of resource potential. Number of employed persons from 1987 to 1992 have shown an increasing trend. Out of these four years high record was observed in the year 1992 (Table 20.14). Onwards from 1992 to 1997 there has been a decreasing growth tendency. Primary sector comparatively has a leading edge as compared to secondary one. Annual variation in primary sector from 1987 to 1997 have almost recorded a positive growth. From 1992 onwards to 1996 the resource sector has gone in an increasing order. Within these years the year 1995, 1996 and 1997 as compared to previous years are in high growth order. Within negative growth recording years the year 1990 has a low negative growth rate. Secondary resource sector in Kyrgyzstan is leading as compared to tertiary sector. Within secondary sector the year 1989, 1990, 1991 and 1992 have shown high growth as compared to the years from 1993 onwards. The tertiary sector maintained low momentum comparatively as compared to primary and secondary resource sector. Annual variation in tertiary sector from 1987 to 1997 has on the whole fluctuated. Most of the years have shown negative growth.

Table 20.11: Turkeministan: Annual Variations in Energy Resources

(Billion Kilowatt Hours)

Products	*1987*	*1988*	*%C*	*1989*	*%C*	*1990*	*%C*	*1991*	*%C*	*1992*	*%C*	*1993*	*%C*	*1994*	*%C*	*1995*	*%C*	*1996*	*%C*	*1997*	*%C*
Production	13.27	12.89	– 2.86	14.51	12.56	14.61	0.68	14.90	1.98	13.18	– 11.54	12.64	– 4.09	– 4.09	0.16	9.8	– 6.66	9.99	1.93	9.85	– 1.40
Thermal	15.68	14.80	– 6.68	16.93	+14.39	12.68	– 25.10	12.19	– 3.86	13.18	8.12	12.63	– 4.17	10.49	– 16.94	9.8	6.63	10.14	3.46	9.97	– 1.67
Imports	0.94	0.98	4.25	1.06	8.16	1.11	+4.71	1.13	1.80	1.15	1.76	1.25	8.69	1.04	– 16.8	0.98	5.76	1.01	3.06	0.99	– 1.98
Exports	5.88	5.34	– 9.18	6.30	17.97	6.07	– 3.65	6.30	+3.78	5.45	– 13.49	4.44	– 18.53	3.69	– 16.89	3.00	18.64	3.34	10.0	3.17	– 5.08
Consumption	14.21	13.87	– 2.39	15.57	12.25	15.72	0.96	16.03	1.97	8.88	– 44.6	9.45	+6.41	7.85	– 16.93	7.78	6.89	7.81	0.38	7.79	– 0.25

Table 20.12 : Turkeministan: Annual Variations in Production of Principal Crops

(Thousand Metric Tonnes)

Crops	*1987*	*1988*	*% C*	*1989*	*% C*	*1990*	*% C*	*1991*	*% C*	*1992*	*% C*	*1993*	*% C*	*1994*	*% C*	*1995*	*% C*	*1996*	*% C*	*1997*	*% C*
Wheat	93	92	– 10.75	86	– 6.52	130	51.16	205	57.69	368	79.51	502	36.41	1063	111.75	878	– 17.40	430	– 51.02	430	0.0
Barley	89	105	17.97	106	0.95	139	31.13	128	– 7.91	124	– 3.12	197	58.87	106	–46.19	62	– 41.50	50	– 19.35	50	0.0
Rice	45	50	11.11	45	+10.0	42	6.66	54	28.57	64	18.51	88	37.5	109	23.86	89	– 18.34	60	– 32.58	55	– 8.33
Maize	95	159	67.36	140	– 11.94	132	– 5.71	124	– 6.06	147	18.54	202	37.41	136	–32.67	80	– 41.17	60	25.0	60	0.00
Potatoes	34	38	11.76	37	– 2.63	10	– 72.97	30	200.00	35	16.66	30	– 14.28	30	0.0	21	– 30.00	36	71.42	28	– 22.22
Seed Cotton	1772	1341	– 24.32	1382	3.05	1457	5.42	1433	– 1.64	1290	– 9.97	1341	3.95	1283	– 4.32	1293	0.77	550	– 57.46	620	12.72
Cotton	1380	410	– 70.28	397	– 3.17	437	10.07	430	– 1.60	390	– 9.30	402	3.07	385	– 4.22	379	– 1.55	131	– 65.43	190	45.03
Grapes	154	165	7.14	124	– 24.84	169	36.29	110	– 34.91	125	13.63	79	36.8	147	86.07	163	10.88	165	1.22	165	0.0

Table 20.13: Turkeministan: Yield kg./Hectare of Principal Crops

Crop	*1987*	*1988*	*1989*	*1990*	*1991*	*1992*	*1993*	*1994*	*1995*	*1996*	*1997*
Wheat	1.45	1.53	1.53	2.16	1.38	1.91	1.96	6.18	4.62	2.15	2.15
Barley	1.48	1.54	1.70	2.10	1.96	2.03	2.34	1.25	0.42	0.5	0.5
Rice	2.5	2.77	2.64	2.62	2.84	2.56	2.25	2.94	3.06	1.5	1.44
Maize	2.1	3.31	3.25	3.21	3.26	3.76	4.29	3.16	2.28	1.33	1.33
Potatoes	8.5	9.5	9.25	10.0	10.0	11.66	7.5	2.30	5.25	7.5	7.0
Seed Cotton	2.00	2.10	2.18	2.33	2.38	2.27	2.31	2.29	2.13	1.05	1.12
Grapes	8.10	7.85	7.29	9.38	6.87	6.94	4.38	7.73	7.76	7.85	7.85

Table 20.14 : Kygryzstan: Manpower Resources: Sector-wise Employment and Changing Pattern *(Annual)*

Sectors	*1988*	*1989*	*% C*	*1990*	*% C*	*1991*	*% C*	*1992*	*% C*	*1993*	*% C*	*1994*	*% C*	*1995*	*% C*	*1996*	*% C*	*1997*	*% C*
Employed Persons	1716.1	1738.0	1.28	1747.9	+0.51	1754.1	+0.35	1835.0	+4.66	1680.6	– 8.45	1645.4	– 2.00	1641.7	– 0.22	1651.5	0.59	1689.3	2.28
Primary Sector (Agricultural, Mining, Factory)	599.9	600	+0.01	595.4	– 0.76	638.1	+7.17	716.4	+12.27	670.3	– 6.43	704.3	+5.07	786.6	+11.68	781.5	– 0.64	818.3	5.12
Secondary Sector (Manufacturing Electricity, Gas, Construction)	396.0	415.4	+4.89	416.8	+0.33	374.0	– 10.26	331.4	– 11.39	282	– 14.95	249.6	– 11.48	209.3	– 16.14	180.3	– 13.85	160.9	– 10.75
Tertiary Sector	111	208.2	+87.56	209.0	+0.38	203.4	– 2.67	201.0	– 1.17	205.3	+2.13	198.3	– 3.40	179.1	– 9.68	22.2	13.27	251.7	13.37

Annual pattern in Industrial production resource of the republic on the other side has revealed somewhat a different situation. Among all industrial products the mining and quarrying industrial sector has shown a remarkable growth. The annual variation in this industrial sector from 1987 to 1995 has almost recorded a negative growth. Within these negative growth recording years high negative growth was observed in the year 1992. Among positive growth recording years the year 1988 has gone high as compared to the year 1987. Next to this industrial sector crude petroleum and natural gas comparatively have recorded better growth as compared to remaining industrial products. Annual variation in crude petroleum from 1987 to 1996 has almost shown negative growth, but their overall growth is higher than that of natural gas. Among negative growth recording years the lowest negative growth in crude petroleum industrial product was observed in the year 1993. In case of natural gas the year 1987 and 1988 have recorded high growth, while as onwards the trend is almost in decreasing order. Among the lowest growth recording years the lowest negative growth has been recorded in the year 1993 (Table 20.15). The remaining industrial products have not comparatively gone so high as compared to the industrial products of Turkeministan and Uzbekistan.

With respect to production of principal crops the wheat crop has shown better growth as compared to remaining cereal and non-cereal crops. Annual variation in the production of the crop has almost recorded a positive growth, but comparatively has not gone to the extent of the production of principal crops in Kazakhstan. Among maize and barley, the barley production is higher than that of maize crop (Table 20.16). Most of the years have recorded negative growth. The annual variation in the production of the crop has not fluctuated to the extent of maize crop. Among remaining crops, potatoes production growth is more than that of seed cotton and grapes. Productivity per hectare has gone high in maize as compared to wheat and barley. The annual variation in the productivity of the crop has on the whole recorded a positive growth. In case of wheat and barley the wheat crop has shown a better profile with respect to its productivity per hectare. But on the whole the maize crop has highly figured out in output category of yield kg./hectare (Table 20.17).

The remaining republic *i.e.,* Tajkhistan has not shown much variation in primary sector from 1987 to 1996. Almost in all the years the primary sector has gone in positive direction. However the high

Table 20.15 : Turkeministan: Annual Variations in Industrial Production Resources

Products	*1987*	*1988*	*% C*	*1989*	*% C*	*1990*	*% C*	*1991*	*% C*	*1992*	*% C*	*1993*	*% C*	*1994*	*% C*	*1995*	*% C*	*1996*	*% C*	*1997*	*% C*
Mining & Quarrying	3875	4030	4.00	3997	−0.81	3742	−6.37	3473	−7.18	1040	−70.05	1721	65.48	793	−53.92	413	−47.91	432	4.60	422	−2.31
Crude Petroleum	172	166	−3.48	158	−4.81	151	4.43	138	−8.60	113	−18.11	88	−22.12	88	0.00	89	1.13	84	−5.61	86.5	2.97
Natural Gas	112	105	−6.25	105	0.00	96	−8.57	83	−13.25	72	−13.25	42	−41.66	39	−7.14	36	−7.69	26	−27.7	31	19.22
Beef & Veal	66	72	15.15	85	18.05	7.05	−7.05	87	4.39	88	1.14	88	0.0	82	−6.81	85	3.69	90	5.88	87.5	−2.77
Mutton & Lamb	67	72	11.94	75	−4.16	70	−6.66	64	−8.57	70	−9.37	82	−17.14	76	−7.31	64	−15.7	49	−23.21	56.5	33.67
Wheat flour	426.8	417.2	−2.11	436.5	−4.67	398	8.82	475	19.3	321	−32.4	426	32.7	320	−24.8	267	−16.56	252	−5.61	259	2.77
Woven Cotton Fabric	116.16	115.6	−0.48	116.7	0.95	114.5	−1.88	119	3.93	119	0.0	65	−47.37	49	−24.61	21	−57.14	25	19.04	23	−8.0
Cement	1311	1380	5.26	1408	2.02	1387	−1.49	1320	−4.83	1096	−16.96	692	−36.86	426	−38.43	310	−27.23	546	76.12	428	−21.61

Table 20.16: Kyrgzstan: Annual Variations in Production of Principal Crops

(Thousand Metric Tonnes)

Crops	*1987*	*1988*	*% C*	*1989*	*% C*	*1990*	*% C*	*1991*	*% C*	*1992*	*% C*	*1993*	*% C*	*1994*	*% C*	*1995*	*% C*	*1996*	*% C*	*1997*	*% C*
Wheat	709	548	– 22.7	581	6.02	482	– 17.03	434	–9.77	679	56.45	885	30.33	608	– 31.29	677	11.34	964	42.39	965	0.10
Maize	460	497	8.04	452	– 9.05	406	– 10.17	365	– 10.09	281	– 23.01	184	– 34.51	129	– 29.89	116	– 16.07	182	56.89	184	1.09
Barley	628	602	– 4.14	544	– 9.63	592	+8.82	556	– 6.08	621	+11.69	510	– 17.87	310	– 39.21	173	44.1	164	– 5.20	185	12.80
Potatoes	287	333	16.02	324	– 2.70	365	+12.65	303	– 16.98	362	19.47	308	– 14.91	311	0.97	432	38.90	562	30.09	558	0.71
Grapes	36	28	– 22.2	33	+17.8	43	+30.30	29	– 32.55	31	6.89	9	– 70.96	18	100.0	20	11.11	14	– 30.0	16	14.28

Table 20.17: Kyrgystan: Yield kg./Hectare of Principal Crops

Crops	*1987*	*1988*	*1989*	*1990*	*1991*	*1992*	*1993*	*1994*	*1995*	*1996*	*1997*
Wheat	3.15	2.78	2.94	2.48	2.23	2.73	2.61	1.82	1.85	2.13	2.13
Maize	6.86	7.0	6.84	6.15	5.88	5.10	4.48	3.48	3.31	3.87	4.0
Barley	9.64	3.64	8.17	3.33							
Potatoes	10.00	16.86	14.08	14.0	13.17	13.40	11.40	6.11	8.81	6.67	11.16
Seed Cotton	2.35	3.76	2.74	2.1	2.38	2.13	2.1	2.07	2.27	2.28	2.03
Grapes	6.0	4.66	4.71	6.14	4.14	5.16	1.5	2.57	2.85	2.0	2.28

positive growth recording years are 1989, 1990, 1992 and 1996. Among these years the year 1996 has shown high growth (Table 20.18). Annual variation in employed persons from 1987 to 1990 is almost in increasing order. Among these years the highest number of employed persons are recorded in the year 1991. Onwards from 1992 to 1997 there has been a random fluctuation. Lowest number of employed persons were recorded in the year 1996. Secondary sector comparatively has gone ahead as compared to tertiary sector (Table 20.18).

Within Industrial production resources mining quarrying sector has gone ahead as compared to other products. But Tajik republic in overall has not gone high as compared to remaining republics with regard to growth in Industrial production resources. Next to mining and quarrying petroleum product is slightly showing an increasing momentum. But their annual variation from 1987 to 1997 has fluctuated to a wide extent. High positive growth was recorded in 1987 and 1988, while as remaining years are recording a negative growth (Table 20.19). On the other side within the agricultural sector wheat crop has shown a better growth as compared to other crops. Annual variation in the crop has on the whole recorded a positive growth. Low growth was recorded in the year 1989. The remaining years are almost in positive direction. Next to wheat, maize crop has gone high as compared to barley and rice (Table 20.20). Productivity per hectare among cereal crops has gone high in maize followed by rice and wheat (Table 20.21).

Above analysis reveal that Uzbekistan republic is leading with regard to primary sector (Agricultural forestry mining and fishing). Next to Uzbekistan comes Kazakhstan and afterwards Tajikistan (Table 20.22). Among the remaining republics (Kyrgyzstan and Turkeministan) Kyrgyzstan has gone ahead as compared to Turkeministan. But the situation is different in Industrial sector. Within this resource potential Kazakhstan gone forward. The republic has shown the highest growth in this sector and is followed by Uzbekistan, while as among the remaining three states Turkeministan is ahead of Kyrgyzstan and Tajikistan. Lastly in this industrial sector Tajikistan has an edge over Kyrgyzstan. With regard to manpower Uzbekistan has figured out at the top. The republic is followed by Kazakhstan. While as among the remaining republics Tajikistan has gone ahead followed by Kyrgyzstan aad Turkeministan (Table 20.22).

Table 20.18 : Tajikistan: Manpower Resources: Sector-wise Employment and Changing Pattern

Sectors	*1987*	*1988*	*% C*	*1989*	*% C*	*1990*	*% C*	*1991*	*% C*	*1992*	*% C*	*1993*	*% C*	*1994*	*% C*	*1995*	*% C*	*1996*	*% C*	*1997*	*% C*
Employed Persons	1775	1820	+2.53	1879	+3.24	1934	+2.92	1969	1.80	1908	+3.09	1854	– 2.83	1855	+0.05	1853	– 0.10	1731	6.58	1792	3.52
Primary Sector (Agricultural, Mining, Factory)	745	765	+2.68	792	3.52	833	5.17	881	+5.76	892	1.24	954	6.95	1002	5.03	1095	+9.28	1026	+6.30	1060	3.31
Secondary Sector (Manufacturing, Electricity, Gas, Construction)	382	384	+0.52	417	8.59	422	+1.19	405	– 4.02	382	– 5.67	335	12.30	315	– 5.97	264	– 16.19	249	– 5.68	143	– 42.57
Tertiary	223	223	0.00	201	– 9.86	200	– 0.49	201	+0.5	119	– 40.79	159	33.61	151	– 5.03	145	– 397	115	– 20.68	147	27.8

Table 20.19: Tajikistan: Annual Variations in Industrial Production Resources

Products	*1987*	*1988*	*% C*	*1989*	*% C*	*1990*	*% C*	*1991*	*% C*	*1992*	*% C*	*1993*	*% C*	*1994*	*% C*	*1995*	*% C*	*1996*	*% C*	*1997*	*% C*
Mining & Quarrying	593	673	+13.44	515	– 23.47	475	– 7.76	313	– 34.10	216	– 30.99	200	– 7.40	140	– 30.00	30	– 78.57	85	183.3	58	– 31.76
Hard Coal	282	242	– 14.18	176	– 27.27	136	22.72	99	– 27.20	57	– 42.42	39	– 31.57	32	– 17.94	64	100.0	48	– 25.0	56	16.66
Petroleum	280	235	– 16.07	195	– 17.02	111	– 43.07	93	– 16.21	72	– 22.58	50	– 30.55	30	– 40.0	40	33.33	35	12.5	36	– 97.14
Natural Gas	68	62.5	– 8.08	66	5.6	59	– 10.60	73	23.72	45	– 38.35	63	40.0	73	15.87	65	– 10.95	69	3.07	67	– 2.89
Beetle leaf	57	58	1.75	58	0.0	55	– 5.17	48	– 12.72	41	– 14.58	34	– 17.07	36	5.88	33	– 8.33	30	– 9.09	31	3.33
Mutton Lamb	25	26	4	24	– 7.69	25	4.16	21	– 16.	20	– 4.76	20	0.0	21	5.0	20	– 4.76	18	– 10.0	19	5.55
Wheat flour	716	708	– 1.11	724	2.25	692	– 4.41	756	9.24	628	– 16.93	667	6.54	360	– 46.02	283	21.38	746	163.60	514	– 31.09
Cement	1094	1109	1.37	1110	0.09	1067	– 3.87	1013	– 5.06	447	– 55.87	262	– 41.38	178	– 32.06	78	– 56.17	128	64.10	103	– 19.53

Table 20.20: Tajikistan: Annual Variations in Production of Principal Crops

(Thousand Metric Tonnes)

Crops	*1987*	*1988*	*% C*	*1989*	*% C*	*1990*	*% C*	*1991*	*% C*	*1992*	*% C*	*1993*	*% C*	*1994*	*% C*	*1995*	*% C*	*1996*	*% C*	*1997*	*% C*
Wheat	175	191	9.14	109	42.93	143	31.19	513	-61.4	170	-66.86	175	2.94	148	-18.28	188	31.46	326	73.40	326	0.30
Maize	85	88	3.52	109	23.86	85	-22.41	60	-29.41	62	3.33	34	-45.16	18	-47.05	20	11.11	15	-25.0	16	6.66
Barley	43	43	0.00	34	20.93	48	41.17	51	-6.25	42	17.64	32	-23.80	23	-28.12	20	-13.04	23	15.0	25	8.69
Rice	27	30	11.11	28	6.66	29	3.57	26	-10.34	20	-23.07	23	15.0	20	-13.04	20	0.0	24	20.0	20	-16.66
Potatoes	192	183	4.68	217	18.57	207	4.60	181	-12.56	167	-7.73	147	-11.97	140	-4.76	110	-21.42	112	1.81	115	2.67
Seed Cotton	872	964	10.55	921	-4.46	842	-8.57	826	-1.90	442	-46.48	524	+18.55	529	0.95	417	-21.17	312	-25.17	350	14.74
Cotton	276	294	6.52	290	-1.36	256	-11.72	247	-3.51	126	-48.98	150	19.04	151	0.66	119	-21.19	49	-58.8	102	14.60
Grapes	132	178	34.84	174	2.24	190	9.19	121	-36.31	100	-17.35	88	-12.0	80	-9.09	85	6.25	87	2.35	87	0.0

Table 20.21: Tajikistan: Yield kg./Hectare of Principal Crops

Crops	*1987*	*1988*	*1989*	*1990*	*1991*	*1992*	*1993*	*1994*	*1995*	*1996*	*1997*
Wheat	1.21	1.22	1.00	1.00	1.06	6.92	0.87	0.86	0.98	1.91	1.91
Maize	4.47	4.63	6.05	5.31	4.00	2.90	3.09	1.63	1.81	1.25	1.33
Barley	1.13	1.13	1.06	1.09	1.10	0.76	0.84	0.57	6.55	0.57	6.92
Rice	3.0	3.33	2.8	2.9	2.88	1.81	1.91	1.53	1.66	1.71	0.50
Potatoes	17.45	16.63	21.7	14.78	13.92	12.84	12.25	11.00	27.5	28.0	28.75
Seed Cotton	2.69	3.01	2.87	2.76	2.76	1.55	1.90	1.86	1.54	1.36	1.64
Grapes	5.73	7.12	6.96	7.30	4.48	3.44	3.03	2.96	2.93	2.9	2.9

Table 20.22 : Central Asian Republics showing Resource Potential Variations

(Average of 1987-1997)

Central Asian Republics	*Population Resources*	*Manpower*	*Agriculture, Forestry, Mining & Fishing*	*Industry*
Kazakhstan	16.22	7131.01	1864	3324.22
Uzbekistan	21.46	8075.14	3423.71	2200.28
Turkeministan	44.44	1496.6	623.8	912.4
Kyrgyzstan	4.43	1710.14	691.08	486.99
Tajikistan	5.51	1857.8	888.8	528.0

REFERENCES

Conolly, V., *Economic Development in Soviet Union*, Oxford.

Cole, J.P. and German, F.C., (1967), *A Geography of the USSR* : *The Background to a Planned Economy,* London.

Dewney, J.C., (1980*), A Geography of Soviet Union*, Oxford.

FAO Production Year Books (1987-1997).

Gerasimove, D.B., (1980), *The Natural Resources of Soviet Union*, Francisco.

German, F.C., (1961), *A Geography of USSR*, Agrarian Change Under Population Pressure, London

IMF, (1993), *Kazakhstan Economic Review.*

IMF, (1995), *Kyrgyzstan, Economic of Kyrgyzs Republic.*

IMF, 1994), *Uzbekistan, Economic Review.*

IMF, (1993), *Tajikistan, Economic Review.*

Statistical Year Book for the Asia and the Pacific, (1998), Statistical Division of the Economic and Social Commission for Asia.

UNESCO, (1994), *Statistical Year Book.*

UN, (1994), *Statistical Year Book.*

UN, (1993), *Statistical Year Book.*

21

SLOPE CULTIVATION IN THE POKHARA VALLEY, NEPAL

M.P. TRIPATHI

The location of the valley under the shadow of the Annapurna and Machapuchre Himalaya has made the region one of the most beautiful spots in the country. Unique in the natural scenery.

The Pokhara valley lies in the Western Development Region of Nepal between 25° to 28° N. and 83° to 84° EL. It covers an area of 245 sq.km. and about 3,00,000 population at present.

It has remained cut off from the rest part of the country till recently (1953) and its habitats, economy and social structure have grown more or less independently, having better influenced of the outside world. But the valley has experienced much change in all aspects of life. Within last two and half decades (2.5 decades). It is now linked with Kathmandu and other part of the country and abroad by land and air transport. The Pokhara valley has been divided into two parts—upland and lowland.

Pokhara valley is located in the central part of Nepal. It is the important tourist centre. Pokhara city is the capital of this region (Western Development Region). This valley is important from Geographological and Geological point of view. The valley has a slopes of different gradient in north. The cultivation is done along the slopes (especially, south facing) successfully.

The valley and its adjoining regions are composed of sedimentary and transported rocks (lowland valley region). The region has multicyclic landscape and terraces as evidently seen clearly of varying heights and slopes and V-shaped valleys. It presents a thick strata of folding and faulting nature. The valley has a sub-tropical type climate in the lowland and temperate climate in the upper part of the slopes.

The mean temperature in summer is high (39°C). But the winter temperature is, as low as – 3 or 4° sometimes. The annual average rainfall is about 4500 mm. Actually this valley gets maximum rain in the country (5600 mm).

Broadly speaking the temperature condition of the valley are so variable that the average monthly figures give quite an erroneous picture of the temperature conditions. An analysis of the mean daily maximum and minimum of different months of the years would give clear picture of the temperature condition given in Table 21.1.

Table 21.1 : Monthly Maximum and Minimum Temperature in the Pokhara Valley

Average Maximum and Minimum in °C

Years	*Temp.*	
1957	26.5	N.A.
1958	27.0	26.4
1959	26.9	26
1960	26.3	26
1961	25.2	26
1962	25.8	24
1963	15.7	16
1964	15.2	15
1965	14.2	16
1966	15.2	16
1967	15.6	24
1968	15.1	
1969	15.6	
1970	15.9	

RAINFALL CONDITION

Pokhara valley gets on an average annual rainfall of 3247 to 5610.4 mm of which 87.4 per cent falls during four months (from June to September). This is the rainiest period of the valley.

Rainfall (mm)

Years	*mm*
1957	3227.9
1958	3319.2
1959	3263.8
1960	3936.4
1961	4006.4
1962	3234.6
1963	3616.4
1964	3345.5
1965	3888.3
1966	3288.0
1967	5610.8
1968	
1969	
1970	
1971	
1972	
1973	
1974	

In this Chapter the author has prepared slope profiles of some of the important hills of the valley which are under agricultural use. The sample of soils are collected from different heights of different hills. The different soils have been measured at different points along the hill slopes. It is remarkable feature that along most of the hill of the valley, the local people are carrying agriculture practices along steep slope (even up to 70°) where maize and potatoes are grown in abundance.

LAND-USE PATTERN

The slope profile of Pokhara valley is similar to the classification of Wood (1942), King (1967) and other scholars of geomorphology. As they have broadly categorised into three stages:

1. The Crest or convex slope,
2. Debris slope, and
3. Pediment slope.

The nature of slope profiles of the valley is quite different from the vidhyan or other ancient hills (Rai, 1969). The scarp free is missing and the whole surface of the valley is covered with debris slope especially the lower part of the valley. Although the angle of the slope reaches even up to 70°. In the beginning stage the scarp free face might have been present but gradually it has been covered with debris and free face, is totally now missing. When all the element of these hill slopes (six slopes) are analyzed carefully one can find some interesting features which are practically prevailing in this area.

The Crest Slope

This part of the slope is covered crest of the hill usually related to the zone of weathering commonly longer where the depth of surfacial detritus is greatest *i.e.* where streams have not yet fully developed. The profiles show slightly covered crest with top surface exhibiting a mild slope towards its edge. The slope is covered with thick weathers reddish laterite and yellow soil where thickness is more than a meter in profile No. 1A, but less than 1 metre in profile No. 1B. Moreover it is found that the thickness of the soil within the same hill varies from place to place, *i.e.* from one slope element to an other. The crest of profile No. 1C is covered with thick forest at several places while in other areas the vegetation is neglected and destroyed by the people of the villages. Due to loose surface soil at several points and heavy rainfall, the rills coming down, the slope are flooded with weathered materials in the form of solution or suspension. In the hill slope C I the local people have cut down most of the woods and cleared down lands for cultivation accelerating the process of soil erosion. Heavy rainfall causes intensive floods and transportation of large volumes of debris in the form of transportation rock materials.

Debris Slope

Just below the crest the slope is covered by debris. The angle of slope here varies from 30° to a maximum of 45° to 60°. It is an intensive zone of slope profiles. Around the valley the hill slope profile show that along the hills the debris surface is most active zone from agriculture point of view. Through the slope in profile No. 1A is to maximum of 60° to 70° of the height of 2000-3000 m but in profile No. lB it is nearly 40° at the height of 1500 - 2000 m. In profile No.1C the maximum slope is 45° to 50° at the height of 1200 to 1500 m. In profile 2A to 2C the maximum angle of slope is 60, between 500 to 1500 m respectively. Thus in all profiles the angle of the debris slope is under active agricultural practices, except the profile 1C other profiles are the devoid of vegetation. It is seen that even up to a slope of 60°-70° parallel narrow strips are prepared along the hill slopes. To check soil erosion at a particular distance the position of strip and channel is changed. The most crops grown on these sloped area maize, potatoes, wheat and high altitude and local paddy. The local people have more care to have these crops along these slopes. Though they have the less scientific knowledge about the improved agricultural techniques.

Pediment Slope

In profile No. lA and 1B, the pediment surface is noticed at the height of 1200 to 1600 m, but it is not well marked in other profiles. In most of the profiles the pediment surface in quite important from land-use and settlement point of view. In profile No.1A along pediment surface the land is put under maize, potatoes and wheat cultivation and settlements are seen there. The thickness of soil is poor. The pediment slopes are found developed on the sedimentary and gneiss rock strata these surfaces have been formed by the gradual erosion and recession of debris slope. The topography have been formed in steep slope and deep narrow valleys. Thus the process of recession of slope is not well marked and the developed of pediment surface is found in retarded form.

SUGGESTION AND CONCLUSION

The study of slope in the valley shows an interesting relationship between soils and terrain. The association of particular soils with distinct slope has demonstrated different zone of land-use. It is also noticed that the agricultural practices in the valley show a striking variety which has been developed under the influence of changing soil's economic conditions.

The production of rice, maize, potato and wheat per ha. is very low in comparison to plane areas. But due to limited availability of level land they are forced to carry on farming along dip slopes. These are marked differences in the intensity of farming and crops type. Potatoes and maize cultivation along such slopes are a common practice due to its higher remunerative price per/ha. yield. But the cultivation has created a serious problem such as soil erosion in these areas. As potatoes and maize grown below the surface soil becomes loose and during rainy season huge amount loose surface soil is washed away by rain water in the form of surface run-off. It has been seen practically in the valley that per/ha. yield is decreasing day-by-day instead of increasing use of fertilisers and pesticides. The reason is very clear that is the cultivation of maize and potatoes which help the top soil to be washed and the hill slopes have become more or less a barren surface. To check up soil erosion one has to think to develop some scientific techniques so that soil erosion could be minimized and the yield of crops may be increased.

At present more and more land or the hill slope of the valley put under vegetable cultivation because of increasing demand in Pokhara city and other suburban areas too. Most of the hill slopes from debris to the pediment zone surfaces are seen with vegetable crops—potatoes, cabbage, green leaf, sweetpotato, cauliflower, chillies, onions and many other crops; most of the areas of low land valleys are being used in housing, hoteling and other commercial and residential uses.

The surfacial deposits vary in maize, paddy, potatoes and confined to certain pockets, controlled by topography and geomorphic agents. In this region heavy rain causes in intensive gulling and transportation of large materials is washed away annually. By adopting contouring method conservation of surfaces are maintained to a great extent (with afforestation and other uses). The soil texture varies from course materials. Though the *p*H values of soil is low and poor, in some areas. Though the cultivation of rice, maize and potatoes are done but the production is decreasing. Therefore to meet the consequences of the farmers, maize farming to maintain sustainable development and to minimise effect of soil erosion is advisable.

REFERENCES

King, I.C. (1967), *Morphology of the Earth* (Olive and Bond).

More, R.P. (1962-63), Soils, slopes and land-use in part of South-Western Nigeria. Some implications for the planning of agricultural development, *Inter tropical Africa, Trans, Inot. of Br, Geog,* Vol. 30-32 (London).

Nookargu D., and Vadhyavathan R. (1977), Hill slope elements and surfacials deposits near Visakhapatnam (A.P.) *Journal of the Inct, Geog. Science/ Association,* Vol. 13 pp. 45-57.

Rai, K.K. (1971), The study of hill side slopes in the Sonar Burma Basin *Geographical Outlook,* Vol. 3, pp. 170-74.

Rai, K.K. (1971), The study of hill side slopes and its impact of land-use around Sagar, *Trans Inst. of Ind. Geo.,* Vol. I.

Wood, A. (1942), Development of hill side slope, *Proce. Geol. Asso :* Vol. 43, pp. 128-140.

22

MARKETING SYSTEMS AND ITS IMPACT ON MARKET DEVELOPMENT

A Study of Central Development Region of Nepal

ALPANA CHANDRA SCOTT

Market centres play a vital role in the economic development of the area they serve. In backward and agricultural countries like Nepal, market centres are one of the major sources controlling the economy. Rural population comes into direct contact only with market centres of lower order for exchange of goods and services. The rural population disposes of its produce through these markets and obtains the goods not produced by it. Thus rural market centres play a vital role in the economy of agricultural countries which influences the process of development.

It is a well known fact that the market development is a slow and steady going process.[1] In the region it is governed by human behaviour. Daily needs of the increasing population in the region and the motivation has implied inducement to produce more, at a rate higher than the minimum acceptable level. This process of production and the produce reaching to the consumer gives rise to certain marketing systems. Marketing systems of a particular region also lead to the development of market centres of lower order and higher order both.

This Chapter examines different marketing systems operating in the Central Development Region of Nepal—a country dominated by agrarian economy, and its impact on the development of market centres in the area.

OBJECTIVES

The aim of this study is to examine the internal market systems operating in the ecological zones in Central Development Region and to assess the potential of Market Development.

This study also illustrates the spatial patterns, structures and processes trade between the three ecological zones of the study area, *i.e.* mountain, hill, tarai.[2]

HYPOTHESIS

The following hypothesis is likely to be tested in the present research work:

1. The nature of marketing systems can be explained by the form of major variables of ecological complex, *i.e.*, environment, population, socio-economic organisation and technology.
2. The areas where transport network and other infrastructure facilities are good, an efficient marketing system exists.

METHODOLOGY

The present study is mainly based on primary data which have been collected from consumers and traders visiting the market places and from urban dwellers. The agricultural officer of Janakpur and Kathmandu, the industrial officer of Balaju Industrial State, consumers and traders and some village residents were interviewed with the help of questionnaire. This questionnaire contained questions about the mode of sale of agricultural and animal produce, type of marketing agencies and channels involved, problems faced in the disposal of surplus produce and profit earned by traders.

Thus information collected was tabulated and graphed.

MARKETING SYSTEMS

The systems of marketing in backward agrarian area as the one under investigation differ from a developed area.[3] The marketing chain-linking producer to consumer can be divided into two sections, first bulking or collecting, and second distributing. Bulking consists of the accumulation and transport of successively larger parcels of a commodity until a limiting consignment size, usually a truckload is reached. Distribution is the reverse of bulking, involving the transport and sub-divisions of successively smaller parcels until finally they reach their minimum size at the last transaction, between the retailer and the consumer. The wholesalers with most capital are more likely to deal with the stages of marketing nearest to the limiting consignment, when relatively large loads are transported over long distances and economies of scale are possible in transport and storage, which are too less in number and mainly concentrated in the Kathmandu valley.

Study of the marketing systems of various products in the study area is carried out to illustrate the spatial patterns, structures and processes of trade between the three ecological zones of the study area, *i.e.* mountain, hill and tarai.[4]

The marketing systems of various products in the study area may be divided in four categories :

(a) Marketing system of foodgrains. (Fig. 22.1)
(b) Marketing system of perishable goods. (Fig. 22.2)
(c) Marketing system of manufactured goods. (Fig. 22.3)
(d) Marketing system of agricultural tools. (Fig. 22.4)

Marketing System of Foodgrains

Among the foodgrains rice and wheat are found to be two important foodgrains in the Central Development Region. Rice is the dominating crop of the study area and the staple food of people. Wheat production has recently gained momentum. It is more durable and costly so its marketing is profitable. The marketing system of both rice and wheat prefer more or less similar channels of distribution. Hence only the marketing system of rice has been attempted here.

Marketing System of Rice

The marketing system of rice presents a rather complex structure (Fig. 22.1). Two type of marketing systems have been identified :

1. Private sector or informal sector.
2. Government sector or formal sector.

Private Sector : In open markets (daily and periodic both), farmers bring their produce and are free to dispose off to wholesalers, middlemen or consumers directly.

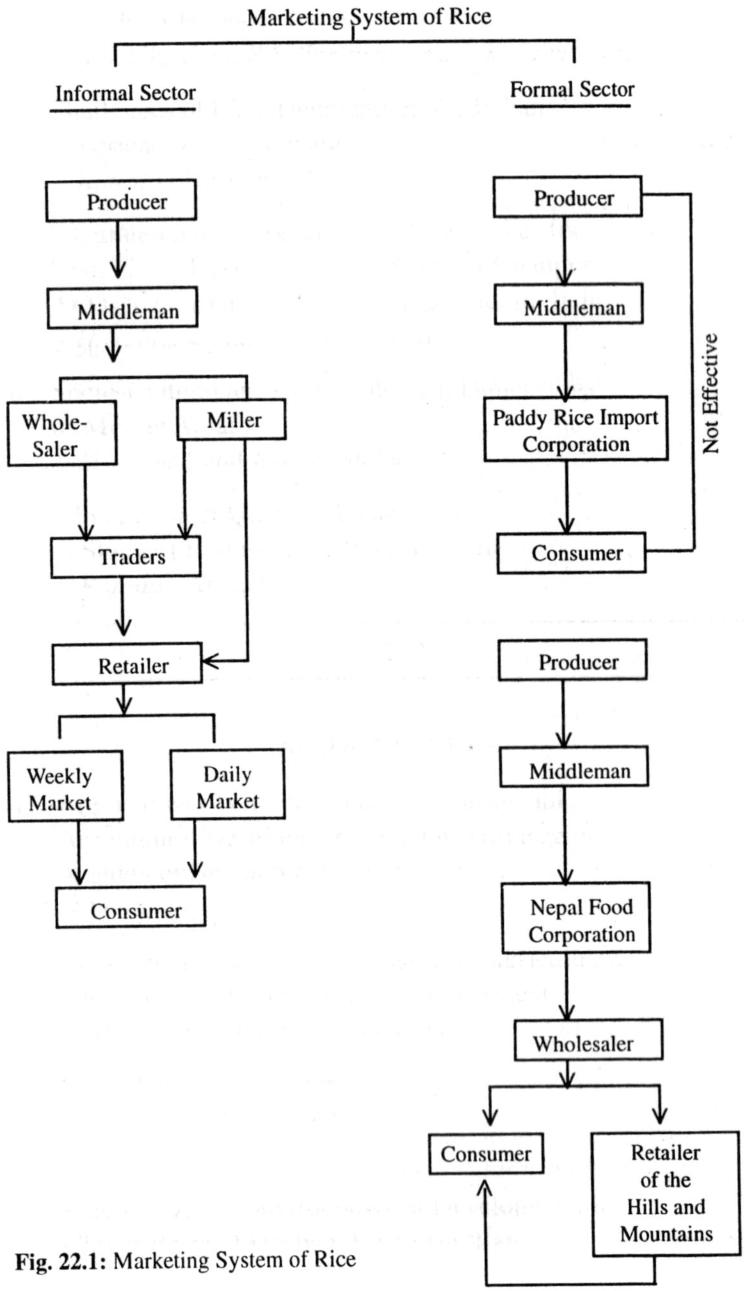

Fig. 22.1: Marketing System of Rice

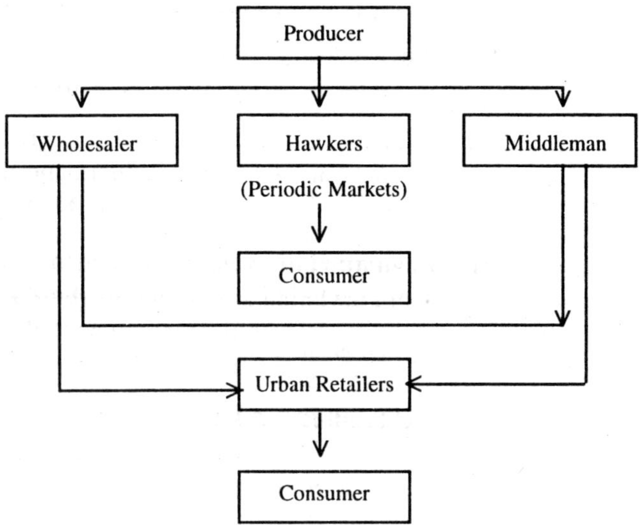

Fig. 22.2: Marketing System of Perishable Goods

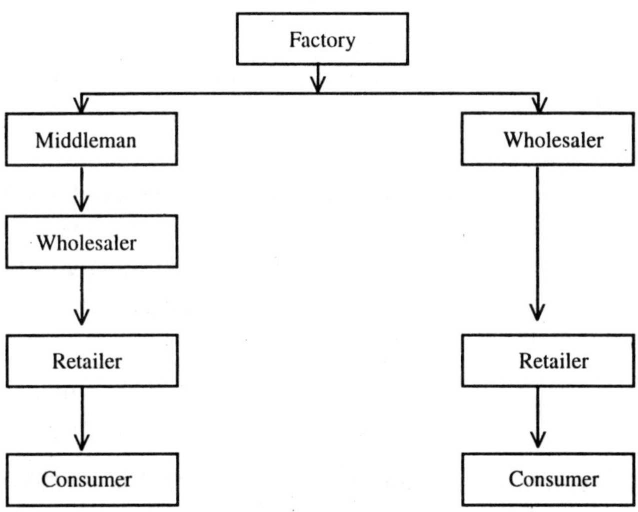

Fig. 22.3 : Marketing System of Manufactured Goods

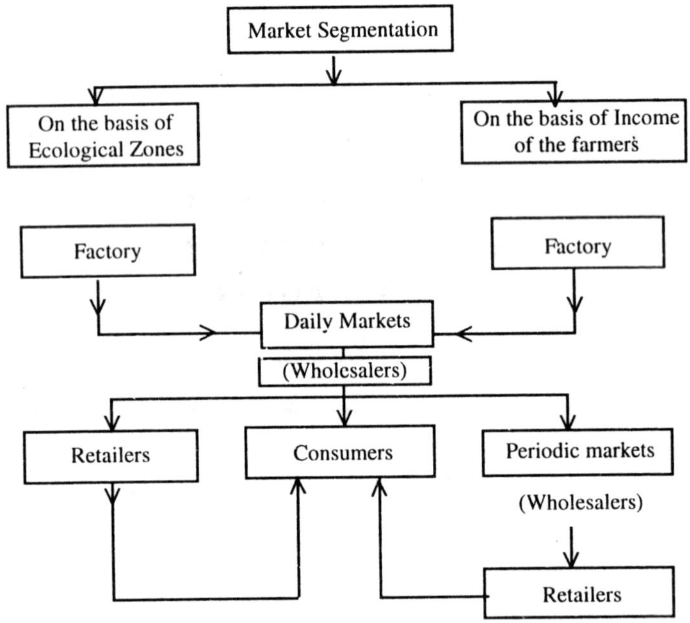

Fig. 22.4 : Marketing System of Agricultural Tools

In this marketing system the middlemen and the millers play important role. A major quality of the marketable surplus of paddy is sold by the farmers or the producers to the middlemen or millers in their respective villages at a much lower price than they can obtain in the market. This type of sale in the study area is very prominent because the farmers' economic condition is very poor and they are often in debt. It was observed during the survey that most of the low-income farmers remain in the clutch of the middlemen, millers or the contractors who pruchase their grains on their respective farms or fields. It was also noticed during the field study that these middlemen, millers or contractors or their agents also move from village to village for collecting agricultural produce from the farmers at comparatively reduced price at the time of the harvest. Sometimes the farmers are unable to buy seeds, pesticides, fertilizers etc. due to lack of money. The agents of rice millers provide them financial assistance in anticipation and thus indirectly compell the farmers to commit the sale of their produce to the millers.

In this system often the millers buy paddy from the middlemen. The millers sell the rice obtained from paddy to the wholesalers as well as to other traders. The millers are also found to sell rice to the retailers directly. These retailers buy from the wholesalers also. The traders, in often cases sell rice at retailing points, *i.e.* daily and weekly markets. The consumers buy from these retailing shops.

Wholesalers from Janakpur take the rice to Kathmandu and sell either again to the wholesaler or retailers.

Rice mills in the Central Development Region (Fig. 22.1) do not have adequate capacity to mill the available quantity of paddy, hence 50 per cent of paddy flows to the Eastern Development Region for milling purpose. Paddy of Birganj headquarters is carried to Etahari and Biratnagar for milling. Thus from producer to middleman-miller-wholesalers-retailer and consumer represent a long chain of distribution.

Government Sector : The second marketing system of rice operates through the Government agencies. The government established "Paddy and Rice Import Corporation". This corporation was expected to buy from the producer or from the middleman, get it processed from the mills and provide rice to the consumer through government fair prices shops. But due to change of policy and physical problems this system collapsed.

Now "Nepal Food Corporation" has been established. With the help of co-operative societies this corporation fixes the rate of paddy in the peak season. Its branches are located in the villages and it is financed by "Krishi Vikas Bank". In some cases the corporation has its own mills. But in most cases it gets its paddy changed into rice from the millers. After obtaining, rice is supplied to the deficit areas *e.g.* hill and mountain regions. This system is existing now-a-days but not running efficiently due to the monopoly of the middlemen and the millers.

Government has proposed another marketing system for the rice. It has tried to establish wholesale markets and ask the producers to sell their produce at these wholsale centres to consumers and traders both.

There are four wholesale markets in the region *i.e.* Trishuli, Nuakot, Surkhet and Janakpur.

This system is yet not fully established and is suffering a hard competition with the millers. Hence it can be said that the small scale producers of the gains apply more labour intensive methods of production than large scale producers. It is the large scale producers, however who are actually best connected to information sources on technical improvements, price variations and alternative marketing strategies, and who have the potential advantages of scale economies in bulk handling, storage and transport. Small scale producers can, of course, obtain many of the advantages of the larger scale producers by forming co-operatives, but although some co-operatives have been formed in the study area, and the levels of farmer participation, mutual confidence and co-operation are generally very low.

Marketing Systems of Agricultural Tools

In this system market segmentation is done in two ways:

(*a*) On the basis of ecological zones *i.e.* mountain, hill and tarai.
(*b*) On the basis of income of the farmers, *i.e.* small land holders of mountain and hills and big land holders of Tarai.

The needs of the consumers directly reflect the category and the nature of the products.

The daily markets are the major centres of marketing of agricultural tools. The farmers buy from these daily markets. The wholesalers of these daily markets sell the tools to the retailers and also directly to the consumers. The consumers also buy it from the retailing shops (Fig. 22.4).

The above discussion reveals the dominance of middleman in the marketing system of various products in the study area. It shows that the middleman takes a big part of profit and plays no important role.

After discussing the important marketing systems of various products operating in the study area, it is relevant to discuss the impact of marketing systems on market development.

SOME CHARACTERISTICS OF MARKET DEVELOPMENT

The important facts of market development are highlighted in this Chapter on the basis of the marketing systems which in turn effects transport technology, environment, population and socio-economic organisation. The following facts emerge out of the present study.

1. The actual structure of the marketing chain may vary

considerably according to the prevailing prices, *e.g.*, during periods of low grain production and high prices, wholesale buyers compete intensity with one another to obtain grain, and the producers have little difficulty selling where they wish at a reasonable price, which directly affects the market expansion, its periodically and development. In constrast, periods of excessive production and low prices mean that the wholesale buyers have little difficulty in obtaining grains, and that the producer may have to accept the financial hardship and inconvenience of selling wherever the wholesalers wish, and at whatever prices the wholesalers may offer. This aspect leads to the development of seasonal markets.

2. Rice is the dominating crop of the study area and the staple food of people. Wheat production has recently gained momentum. It is more durable and costly so its marketing is profitable. Hence the market involved in selling rice must also have storage facility to bulk which is costly, whereas wheat being a durable product, its marketing is easy and wheat markets may be established easily.
3. Influence of middlemen and millers hinders the development of the market. These people move at the farmers' place and buy the produce and hence the farmer is devoid of going to the market centre as well as getting reasonable which in turn affects the development of the market in that particular area.
4. The inadequate capacity to mill the available quantity of paddy in the study area indirectly promotes the origin and development of market centres in the other region towards which the paddy flows.

   Long chain of distribution of produce hinders origin and development of market centres.
5. The interference of the government in the marketing system to an extent shows good signs for the development of markets. As it is evident from the field survey that the government is trying to establish wholesale markets, for example wholesale markets of Trishuli, Nuakot, Surkhet, Janakpur.
6. The importance of the transport to develop an efficient marketing system which in turn affects the market development in the Central Development Region of Nepal, is considered almost exclusively from the point of view of the mechanized world.

Transportation causes local changes for people, goods and information. It is used to satisfy the consumers' requirements and plays a significant role in the division of labour and development of efficient marketing system, *e.g.,* in the markets of Kathmandu, Janakpur and Birganj the consumers can easily reach at the market centre, as well as wholesale supply is observed to be frequent.

Improved transport also expedites the flow of different types of traffic in various parts of the region and is of utmost importance in the intra-regional and inter-regional movement. A network of road transport at Narayanghat market in the study area links the Central Development Region with the other parts of the country.[5] This road network provides facility for intra-regional movements in the country, and the markets are well feeded for regular supply which promotes a healthy and full developed marketing system and the particular market also develops and increases its effective area.

Improved transport has increased the mobility in the region, of the traders and consumers for an efficient marketing system which leads to the establishment of daily market.

CONCLUSION

Hence it can be concluded that the marketing system and its impact on the market development has a distinct marketing pattern related to the following factors :

1. The distribution, scale, manner and labour requirements of production for agricultural products, land tenure and farm size are perhaps the most important influence.
2. The distribution, scale and manner of consumption.
3. The relations between the bulk, value, perishability and handling characteristics of the product.
4. The seasonality of production and consumption.
5. The elasticity of supply, and the availability and profitability of alternative products. In agriculture, the length of the growing season and the time required for the crop to reach maturity are particularly important.
6. The elasticity of demand, and the availability and the prices of alternative products. For food products, the main variations in marketing patterns are between basic food staples, frequently consumed delicacy foods, and rarely consumed luxury or novelty foods.

The above discussion reveals the dominance of middleman in the marketing system of various products in the study area. It shows that the middleman takes a big part of the profit and plays no important role. Hence it is clear that marketing systems affect the development of markets as well as the development of the whole area.

NOTES

1. Symanaski, Richard and Bromley, R.J. (1974), "Market Development and the Ecological Complex". *The Professional Geographer,* Vol. XXVI, No. 4, Nov., pp. 384-386.
2. Srivastava, R.C. (1985), Marketing System and Problems in the Patha Area of Bundelkhand (U.P.) Commercial Activities and Rural Development in South Asia". *Proceedings of the International Conference of the I.G.U.* Study Group on Geography of Commercial Activities, Gorakhpur, pp. 149.
3. Shrivastava, V.K. and Naik, H.P. (1980), विपणन केन्द्र स्थलों का उद्भव एवं विकास, *उत्तर भारत भूगोल पत्रिका,* अंक 16, सं. 2.
4. Stanlon, W.J. (1978), *Fundamentals of Marketing* (5th Edition), McGraw-Hill Kogakusha Ltd., Tokyo.
5. Chitrakan, R.C. (1983), *Marketing System in Private Enterprises—A Case Study of Biscuit Industry,* CEDA Publication, Tribhuvan University, Kirtipur, Kathmandu, pp. 110-112.

23

AGRICULTURAL MARKETING FACILITIES THROUGH REGULATED MARKETS IN POST-INDEPENDENCE INDIA

A Geographical Analysis

NIZAMUDDIN KHAN

India is an agricultural country. About 65 per cent of working force is still engaged in agricultural and allied activities. She has achieved a great stride in agricultural sector through the adoption of new high technological agricultural revolution, called as Green Revolution during the post-Independence era. Production of both food and non-food crops rose up manifold since Independence. The production of foodgrains rose up from 59.14 million tonnes during 1950-51 to 195.30 million tonnes during 1998-99. Production of oilseeds also increased from 5.15 million tonnes to 25.0 million tonnes during 1950-51 to 1997-98. Similarly sugarcane, potato, tea, coffee and so on also recorded a great improvement in amount of production during the same period. The production of sugarcane rose up from 57.05 million tonnes to 277.20 million tonnes, potato from 1.66 million tonnes to 25.10 million tonnes, tea from 278 million kg to 810.6 million kg. and from 0.19 lakh tonne to 2.30 lakh tonnes during 1950-51 to 1997-98.

Thus the country reached to a level of self-sufficiency in terms of production of many crops especially in foodgrains. It is not only the end but the agriculture is very much diversified and changed in the motto or objective of the production.

Moreover, Indian agricultural transformation from peasant or subsistence or traditional forms to commercialisation or market oriented nature has been started during 1980s. Now the production of crops is market oriented. Area under commercial crops are increasing at the cost of hectarage under foodgrains. The surplus of foodgrains have also increased due to bumper production over the decades. In this way, the importance of an efficient marketing system which may ensure a remunerative prices to the growers, is urgently needed. As compared to the growth in production sector, the distributive sector has largely remained traditional. Studies revealed that more than 80 per cent of vegetable and fruits are sold still in traditional markets like *haat* and shandies (Khan, N., 1991). Larger share of marketed surplus of non-perishable agricultural commodities are transacted with the mobile or village traders at the producer's home during the glut seasons especially by marginal and small farmers. Producers are exploited at every level by middlemen in the form of undesirable deductions. Distress sale, have become common features with Indian farmers. This inefficient and non-remunerative agricultural marketing system has created unrest among both the producers and the consumers because the higher proportion of consumer price of food crops goes to the middlemen. With a view to make Indian marketing system more competitive and to get rid of malpractices, the Central Government and various State Governments have revitalized operationalisation and establishment of wholesale assembling markets all over the country since launching of the five year Planning Strategy since 1951 (Viswas, P.K. and Parasher, M.P., 1990). Nomenclature for these wholesale markets varies from State to State but commonly they are known as Regulated Markets. These markets are regulated by certain Acts designed by concerned States. The functions and operations in the markets are supervised by a market committee, constituted of representative of producers, traders and the Government nominees.

Keeping in view the needs of an efficient remunerative marketing systems for agricultural produce in the country in the era of globalisation of trades and increasing trends of commercialisation of the agriculture, author is interested to understand the Government efforts for ensuring the better price to the growers for their products since Independence. So the present study is intended with the following objectives:

1. Temporal Analysis of Regulated Markets,
2. Spatial Pattern of Regulated Markets,
3. Functions of Regulated Markets, and
4. Conclusion and Suggestions.

Origin and Evolution of Ragulated Markets

The beginning of regulation of agricultural marketing can be traced back to the British rule during the second half of the nineteenth century. But the regulation of marketing of agricultural produce was not for the protection of the producers but for the regular supply of raw materials to the cotton industries located at Manchester in Great Britain. Thus for the first time in India Cotton Market named Karenja was brought under regulation in 1886 under Hyderabad Residency order. Subsequently a new special law known as Berar Cotton and Grain Market Law was enacted in 1897. It was the first exclusive statute on the regulation of marketing of Agricultural produce. But this Act failed to represent the growers in the market committee. In 1927 some improvement was made on Berar Act and a new Act named Bombay Cotton Market Act was formulated. Under this Act, the first regulated market was established at Dhulia during 1930-31.

However, the need of regulation of agricultural product markets was stressed at State level by the Royal Commission on Agriculture constituted in 1928. Hyderabad was the first State which passed the Agricultural Produce Market Act in 1930. After this many States like M.P., Bombay, Mysore and Madras also enacted regulation of markets during 1932, 1939, 1939 and 1932 respectively. Thus by the end of 1948 Patiala Agricultural Produce Act was enacted. At the time of independence, there were 236 regulated markets which accounted for only 4.16 per cent of total wholesale assembling markets (5,678) in the country. By the end of 1958-59, most of the States took step to pass special legislation to regulate the marketing of Agricultural produce in the respective States. Table 23.1 indicates that the trends of regulation of agricultural markets improved very fast after the end of third Five Year Plan. The number of regulated markets increased from 1,012 in 1966 to 7,062 in 1997. The percentage of regulation of total wholesale assembling markets also rose up from 18 per cent to 99 per cent during the same period. Progress is not ended but there is target to establish more such markets in the country in future. The National Commission on Agriculture in its fourth report emphasised on the need of establishment of about 30,000 regulated markets upto 2000 AD to provide well adequate facilities to growers for the selling of their surplus at the fair price.

The growth of regulated markets is not uniform in all the States. Date of beginning also varies from year to year. Annual growth of regulated markets during 1950-51 to 1990-91 is given in Table 23.2.

Table 23.1 : Progress of Market Regulation in India (Post-Independence Period)

Years	*No. of Regulated Markets*
1951	236
1956	470
1961	715
1966	1012
1973	2754
1976	3528
1979	4345
1981	4605
1984	5579
1991	6052
1997	7062

Source : 1. Acharya, S.S. & Agarwal, N.L. (1982) Agricultural Marketing in India.
2. *Economic Survey*, 1997.

Table 23.2 : State-wise Growth of Regulation of Markets in India during 1950-51 to 1990-91

States	*Number of Regulated Markets*		*Annual Growth in (%)*
	1950-51	*1990-91*	
Andhra Pradesh	35	568	38.07
Arunachal Pradesh	—	—	—
Assam	14 (1980)	32	12.85
Bihar	144 (1970)	788	23.36
Goa	3 (1980)	5	6.66
Gujarat	236 (1970)	317	1.71
Haryana	150 (1970)	253	3.43
Himachal Pradesh	31 (1980)	44	4.19
J&K	—	—	—
Karnataka	23	360	36.63
Kerala	4 (1960)	4	0
Madhya Pradesh	3	498	412.0
Maharashtra	121	736	12.72
Manipur	—	—	—
Meghalaya	—	—	—
Mizoram	—	—	—
Nagaland	—	—	—
Orissa	15 (1960)	129	25.33
Punjab	92	670	15.70
Rajasthan	152 (1970)	379	7.46
Sikkim	—	—	—
Tamil Nadu	11	272	60.68
Tripura	1 (1970)	21	100.00
Uttar Pradesh	132 (1970)	630	18.86
West Bengal	116 (1980)	321	17.67
India	236	6052	61.61

Table 23.3 : State-wise Distribution of Cropped Area served as per Regulated Market, 1990-91

States	*Number of Regulated Markets*	*Area Thousand hectares*
Andhra Pradesh	568	12.38
Arunachal Pradesh	—	—
Assam	32	19.28
Bihar	798	12.21
Goa	5	12.40
Gujarat	317	13.67
Haryana	253	12.98
Himachal Pradesh	44	12.159
Jammu & Kashmir	—	—
Karnataka	360	34.53
Kerala	4	70.50
Madhya Pradesh	498	45.84
Maharashtra	736	29.33
Manipur	—	—
Meghalaya	—	—
Mizoram	—	—
Orissa	129	7.68
Punjab	670	11.37
Rajasthan	379	50.80
Sikkim	—	—
Tamil Nadu	272	26.31
Tripura	21	21.90
Uttar Pradesh	630	40.54
West Bengal	321	27.04

North Indian States like Rajasthan, Haryana, Punjab and U.P. recorded growth rate less than 20 per cent, Gujarat, Maharashtra and Kerala in South and Assam in NE India come under the 20-40 per cent growth rate. Tamil Nadu and M.P. have shown very high growth above 60 per cent per annum.

Spatial Distribution of Regulated Markets

At present there are 7,062 regulated markets in India during 1997-98, These markets are very unevenly distributed in different States. The highest number of regulated markets was found in Bihar (798) followed by Maharashtra (736), Punjab (670), U.P. (630), A.P. (568), M.P. (498), Rajasthan (379), Karnataka (360), west Bengal (321) and Gujarat (317) during 1990-91 (Table 23.3). The States which accounted for less than 100 regulated markets were Himachal Pradesh (44), Assam (32), Delhi (11), Kerala (4) and Goa (5). Seven States *viz.* J&K, Sikkim, Arunachal Pradesh, Nagaland, Meghalaya, Mizoram, Manipur and Union territories like Andaman & Nicobar, Dadra & Nagar Haveli and Lakshadweep are yet to be covered under regulation of agricultural assembling markets.

The study of regulated markets with reference to population as well as cropped area in different States have also shown great regional variation in the distribution of regulated agricultural marketing facilities and even after the 50 years of the Independence revealed that Tamil Nadu, partly A.P., Punjab, Haryana and Gujarat served less than 2,00,000 population by per regulated markets. The States of U.P., Bihar, West Bengal, Assam, Orissa, Kerala, Western Coast of Maharashtra, and eastern Rajasthan are serving more than 4 lakh population per regulated market, while the remaining States, except where regulation is not yet implemented, serve population between 2 to 4 lakh persons per regulated market. Thus usually the States of high potential of population as well as agricultural production remained very backward in agricultural marketing facilities though in absolute number they represent probably, higher scores as compared to the States of low population and market rates.

Similarly the ratio between cropped area and regular markets have also revealed imbalances in spatial distribution. Table 23.3 indicates the States of Bihar, Orissa and Punjab and higher density of regulated markets with reference to cropped area in the country. They were accounted for less than 20 thousand hactares of cropped area per regulated market. While the States, which represent the medium size of cropped area and R.M. index, were Andhra Pradesh, Gujarat, Haryana, H.P., Karnataka, Maharashtra, Tamil Nadu, Tripura and West Bengal. They had the ratio between 20-40 thousand hactares, per regulated market (Table 23.3). Remaining States *viz.* U.P. (40) Kerala (760), Rajasthan (50) and M.P. (45) showed lowest concentration of regulated markets. Their contribution for cropped area per regulated market was more than 40 thousand hectares.

Functions of Regulated Markets

The main objective of the regulation of agricultural produce markets is to protect the interest of the producers in the markets. Prior to this the producer sellers were severely exploited the monopoly of traders in unregulated markets. Maximum efforts are made to fulfil the following objectives:

(i) To prevent the exploitation of farmers by the traders in the marketing of their products.
(ii) To make the marketing more effective and efficient with a view to provide better prices of the products to the producer seller and to make available to the consumers at reasonable prices.
(iii) To encourage the farmers for better production both quantitatively and qualitatively by ensuring remunerative price incentives to the producers.
(iv) To make an orderly marketing system of agricultural produce through the development of infrastructural facilities like link roads to villages from the regulated market, storage, credit facilities, inputs facilities etc. in the market complex.

For achieving the above objectives, every regulated market is controlled by a market committee consisting of representative of both producers and traders in their due proportion. Market area is also well defined for every market. The commodities are notified for arrival and transactions in such markets. The number of commodities listed varies from State to State. The lowest number of commodities notified for regulated markets is found in Madhya Pradesh.

Fruits and vegetables, the most perishable commodities, are yet to be notified in many States of the country. The commodities coverage under the existing Act, is by and large satisfactory as far as foodgrains and oilseeds are concerned. However, same is not correct in the case of other commodities such as poultry, limestones, products, forest products, fisheries etc. Sales are made by fair open auction and close tender method under the supervision of an official of market committee of the concerned market. Standard weights and scales are used by licensed weighman to reduce the undue malpractices. Unauthorised deductions are removed and only prescribed charges for commission agents, brokers and weighmen are paid by producers. The commodities are graded by expert personnel and more fair prices are given to the producers. Market news service is also provided especially about the prices of different commodities day-by-day in the market, so that the producers could be saved from any kind of loss of price of their commodities due to ignorance of price structure in the market.

Market fees are also levied on the arrival of commodities and the income generated through the fees are used for the development of infrastructural facilities in and out of market to make more convenient for producer sellers to reach the regulated markets from their villages. Besides, storage facilities, and the credit facilities are also available in them. The payment of sale is made immediately after sale without any unfair deductions.

Conclusion and Suggestions

The regulation of markets and marketing operations has been accepted as one of the important measures for improvement of Agricultural Marketing in India. The Agricultural Produce Market Acts have been implemented in almost all the States and Union territories with the exception of few States and Union territories till now. There are 24 Regulated Market Acts in force all over the country at present. The purpose of all the Acts are the same but varies in the notification of the commodities. These markets have played considerable role in improving the agricultural marketing process and the producer sellers were enabled and encouraged to receive a fair and remunerative prices for their surpluses of agricultural commodities. But the many researchers have questioned the performance of the regulated markets for providing benefits to the targetted people *viz.* producer sellers of market area (Harris 1980 : 81, 84). It is found through the case studies that the producer sellers especially small and marginal farmers visit Regulated Market in very low proportion due to their small amount of the surplus. It would be uneconomical for them after incurring the travel and time cost (Khan, N., 1991). It is further investigated that the mobile traders, moneylenders-cum-traders enter the regulated markets as producer and get the benefits of producer sellers in the markets. Their products are recorded as primary arrival by the fee collectors. The farmers, yet remained exploited by these traders in the villages. Producer sellers usually attend the traditional rural markets for disposing of their rather small amount of surplus (Varma 1985, Doshi 1987, Khan, N., 1999). The convenient locations of these markets (*haat*/shandies) makes them more accessible to the market participants and thus became popular.

However, the vegetables, fruits, flowers, and other forest products have not been included in Agricultural Market Produce Act in many States. About 80 per cent marketed surplus of these produce are transacted in unorganised traditional periodic markets·both in urban and rural settlements (Khan, N., 1988).

Livestock and livestock products are not included in the list of regulated markets even in Haryana and Punjab where milch buffalloes are abundant (Bhatia, G.R., 1990). cattle are transacted only in the traditional nearby rural markets due to the problem of transportation. Thus the rural markets are still, the foundation of agricultural marketing network of the country. These markets are rarely connected by the regulated market in concerned region. Infrastructural facilities and different incentive services for the producer sellers remained more theoretical than practical. Corruption, bribery and unfair dealing, in the name of grading are very common practices in almost all the Regulated Markets in the country.

Moreover, the author would like to present some suggestions to make agriculture marketing more attractive and efficient and remunerative to the producer sellers living in distant inaccessible villages.

(i) The coverage of regulated market act should be extended to all the commercial agricultural commodities *viz.* horticulture, viticulture, apiculture, sericulture, pisciculture, poultry, livestocks and livestock products.

(ii) The States, which have not enacted Agricultural Produce Act, till now, should bring the assembling markets under regulation immediately.

(iii) Entry and title of producer-sellers should be checked at the Gate of market campus to avoid the Proxy representation of mobile or village traders as a producer sellers in the market.

(iv) All the village markets like *hats*/shandies should be brought immediately under the regulation of Agricultural Produce Market Act on the priority basis all over the country. Such markets are estimated to 33,000 by the end of twentieth century.

(v) Rural primary assembling centres should be considered as depot centre/sub-mandi to provide the fair dealings to the small and marginal farmers who remained unable to visit main regulated market located at far distance especially in urban centres.

(vi) Infrastructural facilities should be extended to village level. Link roads, storage facilities, shops of inputs, banking credit facilities, drinking water, auction platforms and the cattleshed etc. are urgently needed in the rural markets which attract 90 per cent of producer-sellers in different States.

Regulated markets, however, are the positive decision of government for improving the return of producer-sellers for their products. All the farmers should be brought in contact with such markets. It will require consistent efforts, through extension services for making the market accessible, both physically and economically, to the producer-seller. In the effort of market reforms through the regulation of markets, one should not ignore the fact that the imposing of rule or any strategy will never succeed unless the target population realises the potential benefit from such policies (Bhuyan, S., 1990).

REFERENCES

Acharya, S.S. & Agarwal, N.L. (1982), *Agricultural Marketing in India*, Oxford and IBH Publishing Co. Ltd., New Delhi.

Bhatia, G.R. (1990), Legal and Administrative Framework for Regulation and Management of Agricultural Produce Markets in India, *Agricultural Marketing,* Vol. 35, No. 4, pp. 14-29.

Bhuyan, S. (1990), Some Intervention in Agricultural Marketing: Is It Necessary? *Agricultural Marketing,* Vol. 35, No. 4, pp. 7-13.

Bhide, Shanshanka *et al.* (1981), Structural Changes in an Agricultural Assembling Markets : A Case Study of Arecanut (Betelnut) Market in Mangalore, Karnataka, *Indian Journal of Agricultural Economics,* Vol. 34, No. 2, pp. 25-34.

Doshi, R.R. (1987), Market Regulation: A Lost Game, *Indian Journal of Agricultural Marketing,* Vol. 1, No. 1, pp. 21-30.

Harris, B. (1980), Regulated Foodgrain Markets: A Critique, *Social Scientist,* Vol. 8, No. 8, pp. 22-31.

Harris, B. (1981), Inaction, Interaction and Action: Regulated Agricultural Markets in Tamil Nadu, *Social Secientist,* Vol. 9, No. 4, pp. 96-137.

Harris, B. (1984), *State and Market : State Intervention in Agricultural Exchange in Dry Region of Tamil Nadu, South India,* Concept Publishing Co., New Delhi.

Khan, N. (1991), *Agricultural Development and Marketing,* H.K. Publisher, Delhi.

Khan, N. (1988), Direct Marketing in Agriculture, *The Geographers,* Vol. 35. No. 2, pp. 68-79.

Khan, N. (1999), Role of Periodic Markets in Agricultural Marketing System in Faizabad District in *Commercial Activities in Ganga Basin* (ed.) Srivastava, V.K., Concept Publishing Co., New Delhi.

Singh, L.P. (1983), *Regulated Markets in India,* Capital Publishing House, Delhi.

Viswas, P.K. and Parasher, M.P. (1990), Development of Agricultural Produce Markets in India, *Agricultural Marketing,* Vol. 33, Vol. 2, pp. 31-32.

24

REGULATED AGRICULTURAL MARKET IN LUCKNOW (U.P.), INDIA

R.S. DIXIT

INTRODUCTION

The case studies have both the practical and the theoretical significance in understanding the totality of a theme under spatial analysis. For presenting the spatial analysis of the regulated agricultural markets—RAMs—of any State, it seems proper and even necessary, too, to study from near at least one RAM. Through this, various details come up which throw light on the real state of affairs. To acquaint with the real conditions under which an institution works, the detailed study is done as a case study.

OBJECTIVE

In view of the above, the scholar has made an effort to present the various minute details of one noted RAM of the State. The objective of this endeavour, therefore, primarily, is to present in detail the case of the Regulated Agricultural Market, Lucknow *vis-a-vis* primary data collected through a detailed survey (Dixit, 1998, pp. 302-335).

SELECTION OF THE CASE

The 'Lucknow Mandi' as it is popularly known, has been selected for case study just on random basis.

The Lucknow Mandi is among the important ones of the mandies of Uttar Pradesh. Lucknow is the capital city of the State. Thus, this market is that of the State's capital city. This mandi is of first order of the hierarchical tier. To study the work done by the State government through the establishment of the mandies, this selection has been made.

METHODOLOGY

The methodology adopted for studying the details of this RAM, is the personal interview through survey schedules. The mandi yard has been visited, the officials of the market committee, the sellers/farmers, and the purchasers/traders have been contacted formally as also informally to get more and correct information of the functioning of the mandi. Three visits were paid to the mandi—one during the peak time immediately after the Rabi harvest in April 1997; another one in the month of June 1997. While yet another one took place in the month of August 1998. Thus, besides collecting the secondary data from various govenment departments, the primary data has been collected for developing the case study. This data is based on three types of survey schedules.

The first schedule is about the information related to the regulated agricultural market itself. It included as many as 21 aspects with various parts/sub-classes of the same. The second schedule is on farmer (producer/seller) information, while the third schedule is meant for trader/purchaser information. The second and the third schedules have 18 items each. On the basis of these schedules the primary data have been collected for the real inside story of the market. The present write-up on the case of Lucknow is also basically divided into three major parts : the market, the farmers, and the traders.

THE CASE OF LUCKNOW

The Market—The Mandi

Location

Situated on both sides of the river Gomti, the city of Lucknow is the capital city of the State and has come up as one of the metropolitan cities of the country. It has been the capital city of the Nawabs of Awadh and has a historical past. Presently, it is also an important centre of education and learning, it is linked by national highways, railways with all the important cities of the country. As per the Census 1991, it has the population of 16,69,204 while the estimates are there that the present population of this metropolis is nearly three millions. It is a big urban agglomeration as also an important commercial centre of the State.

The Lucknow RAM, popularly known as Lucknow grain market established on 22nd July, 1972 and is located near Purania on Sitapur Road—the National Highway No. 24, in the trans-Gomti area in Lucknow. The site, at present, has come well within the city and not outside the city. The new yard of Lucknow mandi was constructed in the year 1978. The area of the yard is 31.312 acres. Besides, 3.5 acres of land just adjacent and attached with this main market has also been purchased in 1996 and the boundary wall has also been constructed. However, the construction of buildings etc. has not been started as yet. At the main new yard site, the transaction of 24 specified/notified items of agricultural produce had been transferred from Pandeyganj, and Daliganj localities on 28th August, 1978. Daliganj mandi had shifted to the present site in 1985 onty and to one part of the yard, the fruit market had been shifted from the Kaiserbagh mandi on 26th January, 1986. By first August 1986, the entire Mandi had been shifted to the present site.

Land-Use Analysis

The entire mandi yard spreads over an area of 1,26,755.00 m^2. The total land has been divided into fourteen types of uses as follows (percentages in brackets):

The area under shops-cum-godown is 5,780 m^2 (4.56),while the space for shops without godown is 12,320 m^2 (0.33), and the retail shops have occupied 4,452 m^2 (3.54). Approximately, equal to the area for the retail shops, there is a space for open platforms in front of shops, 7,736 m^2 (3.10). The office complex of the market yard has been built up over an area of 3,800 m^2 (2.99 per cent). The land has also been used for bank and post-office-buildings to the extent of 2,800 m^2 (2.22). One-fifth of the total land of the yard has been devoted to roads, *i.e.* 25,360.74 m^2 (20.00). The parking space has been provided by the mandi as off-street parking as well as the main parking, 12,818.26 m^2 (10.23) and 6,250 m^2 (4.73) respectively. The total area under open space comes to 83,274 m^2 (25.93) which is one-fourth of the total land under the yard. The auction platforms have been given an area equal to 4,416 m^2 (3.48). The space devoted to pavements is 6,350 m^2 (5.00) while the canteen space is 13.80 m^2 (7.08). It is clear from the above that the largest area is under open space (25.08) while second to it is the area under roads and pavements (25.00). The total area under parking is the next (14.96). Approximately one-fourth (22.72) of land has been occupied by auction platform, open platform in front of shops, shops-cum-godowns, shops without godowns, mandi godowns, and retail shops. The land under the canteen, bank, post-office and the office complex is about half (12.29) of the open space. (Fig. 24.1).

Besides, the main site at Sitapur Road, there are two more sites of the Lucknow mandi—one on Hardoi Road and yet another one in Gomtinagar. The present analysis is related to the Sitapur Road campus only as the Dubagga site has started functioning from June 1998 only while Gomtinagar site is still undeveloped.

On Hardoi Road, at Dubagga, the fruit and the vegetable mandi has been constructed and the sale/purchase of fruits and vegetables has also been just transferred here in June 1998 from the Chowk mandi. The down town, from the evils of environmental pollution on the one hand and traffic jams on the other hand, has, thus, been freed. The Dubagga mandi is rather a big scheme being implemented at a fast rate. The entire developmental work is spread over 15 acres of land for which Rs. 1,40,00,000 has been approved by the government. The preparational work on the third campus in Lucknow—the Gomtinagar mandi campus—is also to start soon.

The Nature of Mandi

The agricultural produce market committee—the mandi Samiti—Lucknow has primarily, a primary market in case of arrivals of various cereals, vegetables and fruits, otherwise it is a secondary market. Since Lucknow is an important commercial centre of the State, in addition to the primary arrivals, there are some secondary arrivals of some crops which are brought to this mandi from various market areas of the State as also outside the State such as Orissa, Maharashtra, Andhra Pradesh, Kerala, Gujarat etc. too. It is a A Special class of mandi as classed by the U.P. Mandi Parishad meaning thereby that much more than Rs. 80,00,000 is collected as market fee every year.

The major primary arrivals consist of paddy, rice, potato, wheat, moong and arhar. The secondary arrivals include gram, wheat, urd, and various fruits like apple, grapes, bannana, orange etc. as also the various Kirana items too.

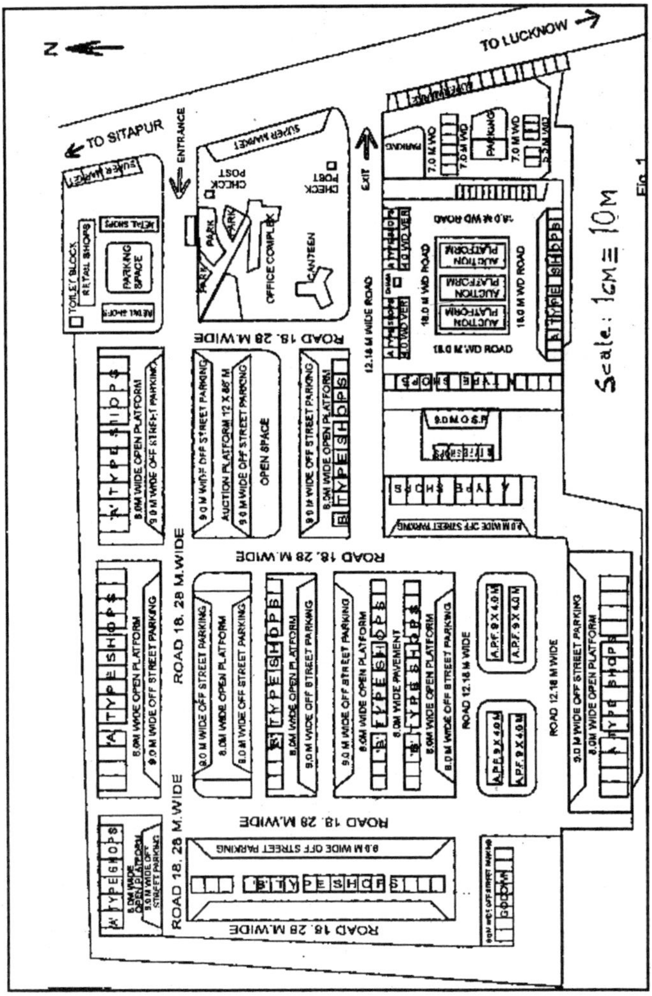

Fig. 24.1: Regulated Agricultural Market: Lucknow—The Mandi Yard

The Dashahari mango variety due to its taste and flavour, is not only widely demanded in this country, rather in other parts of the world as well. This mango variety came out of a small village, named 'Dashahari' near Kakori in Lucknow district. It is widely produced in a large area in and around Malihabad in Lucknow. This variety is in large volume sent to other parts of the country as well as various countries of the world. Ensuring proper production, and sale of this crop a 'mango belt' around Malihabad has been declared by the U.P. Government. This belt is also specially attended to by the governmental and/or non-governmental agencies.

Market Area

The agricultural produce market committee, Lucknow has its mandi area—the notified area—which consists of the area of 57 Nyaya Panchayats, and Gram Sabhas under five development blocks of the district. The area of the various development blocks is as follows : Kakori. 22,594 hectares; Chinhat, 21,446 hectares; Bakshi Ka Talab, 37,782 hectares; Mal 25,382 hectares; Malihabad 21,092 hectares. The largest number of Panchayats and Gram Sabhas are from the Malihabad followed by Kakori, and further followed by Bakshi Ka Talab. All the villages of these 57 Nyaya Panchayats and Gram Sabhas fall under the notified area of the Lucknow Regulated Agricultural Market. Some of the other important villages of the notified area are Kasmandi Kalan, Kasmandi Khurd, Itaunja, Kursi, Bharawan, Gori-Gurwan. Paigaramau, Bidiapur, Asti, Mahona and Behta (besides Kakori, Chinhat, Bakshi Ka Talab, Mal, and Malihabad).These are the major villages from where the primary arrivals of crops take place at this market. The market is well connected with entire market area by metalled/unmetalled roads.

The 10 major crops of this RAM are paddy, wheat, barley, bajra, maize, arhar, gram, urd,. mustard, and groundnut. The percentage of area sown under these crops is as follows : Paddy : 20.50, Wheat: 47.50, Barley : 8.00, Bajra : 3.20, Maize : 4.25, Arhar: 4.20, Gram : 4.30, Urd : 4.25, Mustard: 0.85, Groundnut: 0.95. The percentage production of these major crops in the market area of Lucknow RAM is as follows: Paddy : 12.22, Wheat : 40.21. Barley : 1.00, Bajra : 0.55, Maize : 1.00, Arhar; 1.20, Gram : 1.08. Urd : 0.55. Mustard : 0.03, Groundnut ; 0.06, Potato and others : 42.10.

Some other seasonal items are gur, khandsari, lobia and til.

Sub-Yards

The four sub-yards came into being in 1972 which are attached with Lucknow only. These are Malihabad, Itaunja, Mal, and Bakshi Ka Talab. About Rs. 20,00,000 annual mandi fee is received at each of these markets by the government.

Facilities

The main objectives of the Mandi Samiti would not be fulfilled until and unless the appropriate and adequate arrangement of facilities and conveniences is not made available to the market functionaries specially the farmers at the marketyard for the sale of their produce at the most competitive price. In the new marketyard, a number of basic and modern facilities have been made available for the farmers to a great extent. Amongst these are the facilities of drinking water (with overhead tank as also the India Mark II hand pumps), canteen, bank (State Bank of India), police out-post, post office, rest house, toilets, roads, shades, light, cleanliness etc.

All the basic facilities for market functionaries have been constructed at this new yard (Mandi Vikas, 1990 August-December) but these are not well maintained and rather some of these have been seen under vary poor condition.

Besides the basic facilities, the market committee provides the major modern facilities like existence of Kisan Bazar. This Bazar consists of 50 shops known as input shops near the gate outside the campus. These shops have the items of farmers' needs like good quality seeds, fertilizers, insecticides, pesticides, fodder, etc. In addition, modern machines like weighing scale, moisture meter have also been installed. A very significant major modern facility—the grading unit already installed is however out of order/defective at this time and hence not functioning although the machine is quite costly. There is just one godown (of the mandi with one thousand metric tonnes of capacity) which is centrally air-conditioned. It has been let out to the Central Warehousing Corporation. This godown is spreading over 5-6 acres of land. It includes 10 C type godowns within itself.

A flying squad of the Mandi Samiti also makes surprise inspections from time to time and at place to place in the marketyard especially during the process of sale-purchase.

Shops

The marketyard, besides auction platforms has three types of shops. There are six platforms of 12×72 m size each. No rent or any other charges are taken from traders for the platforms. The shops are A type, B type and C type—old and new. The areas of both the old and new shops are also same, but the rents are different according to the time of construction.

There are 22 old A type shops on Rs. 300 per month rent. The A type shops built afterwards are 26 in number and their monthly rent is Rs. 800 each. The B type shops are let out at the rate of Rs. 180 per month for old ones while Rs. 410 and Rs.440 per month for the new shops. There are 31 C types shops of which rent is Rs. 90 per month each. All these shops are meant for grains and fruits.

The vegetable shops are, however, different in nature. There are 12 A type shops with a monthly rent of Rs. 800; 17 B type shops with Rs. 600 per month rent; and 72 C type shops are on Rs. 480 per month rent.

The Dubagga mandi has 22 B type shops measuring 71.20 m^2 each with Rs. 1000 monthly rent; there are 117 C type shops measuring 33.04 m^2 each on Rs. 180 monthly rent. There are also 96 module shops which have been formed by raising temporary partitions of auction platforms of 12×72 m^2 each, as also 2 other platforms with 9×80 m^2 each, as also 4 of 12×60 m^2 each. All these platforms are covered. Besides, there is a fish mandi also at this site which was transferred to this place on 31.3.1995. The functioning of this mandi started on February 1998 although this transfer was completed by January 2, 1996. It has three types of shops as : A, 2 in number on Rs. 1005 per month; B, 6 in number on Rs. 645 per month; and C, 63 in number on Rs. 420 per month rent.

It is also significant to mention that despite the fact that the Pandeyganj traders have got allotted their shops at this mandi, they have taken their business back within two years to the Pandeyganj mandi only on the pretext of lack of facilities etc. in 1980. The mandi officials keep on watching their business there in Pandeyganj area, and they have sent notices to the traders for Rs. 5,00,000 as mandi fee but despite the High Court judgement (in favour of mandi) dated 5.3.1998 for the recovery, the mandi could not get any payment till now (although, since 1980 till now the fees amount has increased due to accumulation of interest totalling to Rs. 30,00,000).

Attendance

One hundred and forty farmers and about 150 traders were present on the day at the time of visit made by the investigators. However, there are as many as 200 registered traders, besides the village merchants at this market. There are 32 commission agents (*aratiya*), 100-200 millers, and 200-500 traders. The attendance in the market varies from 500-1000 persons. Generally, 150 farmers visit the market daily but during the peak days the number reaches even 300. It has come to the knowledge that a large number of farmers go to the nearby unregulated/ conventional mandis for the sale of their agricultural produce. In case these farmers also bring their produce to this regulated market then the attendance would increase sharply to become even more than 1000.

Income

The mandi Committee has several sources of income but the most important one is that of market fee. The market fees are those charges which are paid by the purchaser-trader to the market committee on the sale of the agricultural produce of the farmer. The market fee had been charged at the rate of just 1.5 per cent upto 14.12.1994. From 15.12.1994 it has been increased to the rate of 2 per cent of the total sold value. It is paid by the purchaser to the mandi. Now 0.5 per cent fee has just been increased with effect from 15 August, 1998 under the head 'development tax'. Thus, 2.5 per cent in all is charged from the trader as market fee. The total amount of market fee is collected during the recent past is much more than Rs. 2,00,00,000 per annum. The present total annual income through this head, thus, is about Rs. 3,00,00,000 to the market committee (despite the fact that a large number of farmers take their agriculture produce to nearby non-regulated agricultural markets which costs a loss of Rs. 1,00,00,000 per annum). It is important to note that the mandi fee from grains is getting down year by year while it is increasing through other items particularly fruits—more particularly apples.

Commodity Structure

The agricultural produce including types and sub-types as mentioned in "Crop arrivals" are brought for sale to this mandi. Thus, the commodities include the farm produce, the horticulture produce, the agriculture animal/livestock produce and forest produce.

Crop Arrivals

Year-wise

The crops are brought by the farmers for sale at mandi. It has been observed that the total arrivals of all the crops during the last ten years, upto the year 1997-1998, have been more or less the same to unusual or less very significantly marked changes in volumes of these crops have been noted. The total arrivals during the period 1988-89 to 1997-98 (i.e. ten years) have been in between 34.43 lakh to 43.24 lakh quintals. It has been 34 lakh quintals during 1992-93 while the volume during 1990-91 was 43 lakh quintals. However, the primary arrival during these years have been varying from 22.15 lakh to 29.48 lakh quintals (during 1997-98 and 1988-89 respectively). The secondary arrivals have been varying during the same period from 10.40 lakh to 17.38 lakh quintals (during 1992-93, 1996-97 respectively). The details of these are shown in Table 24.1 as also in Fig. 24.2.

Commodity-wise

This part includes the arrivals of major groups of produce *i.e.* the five specified groups—the farm produce, horticulture produce, grape agriculture, animal/livestock products, and forest products. The total arrivals of these five types of items during the last four years *i.e.* 1994-95 to 1997-98 have been presented through Table 24.2.

The observations show that the arrivals of cereals varied from 7.38 lakh quintals in 1996-97 to 11.9 lakh quintals in 1994-95. The arrivals of the pulses varied from 1.73 lakh quintals in 1997-98 to 2.33 lakh quintals in 1996-98. The oil-seed arrivals have been varying from 37.47 thousand quintals in 1995-96 to 90.18 thousand quintals in 1997-98. The fibrous crop arrivals have varied from 746 quintals in 1994-95 to 1440 quintals in 1997-98. The inhaler/tobacco arrival has been changing from 8.37 thousand quintals in 1996-97 to 16.10 thousand quintals in 1994-95. The spices-arrivals have been from 36.90 thousand quintals in 1995-96 to 56.50 thousand quintals in 1997-98. The volume of arrivals of the miscellaneous items have been varying from 13.43 thousand quintals in 1997-98 to 24.03 thousand quintals in 1994-95.

The changes in arrivals of the horticultural produce have been as : 10.20 lakh quintals, in 1997-98 to 12.45 lakh quintals, in 1994-95 in case of vegetables, while in case of fruits it has been from 7.88 lakh quintals in 1994-95 to 9.60 lakh quintals in 1997-98.

Table 24.1: Lucknow RAM—Trend of Market Arrivals

Year	*Quintals*		
	Primary	*Secondary*	*Total*
1988-89	29,48,122	12,58,097	42,06,219
1989-90	27,90,270	14,67,205	42,57,475
1990-91	28,16,329	15,08,579	43,24,908
1991-92	28,69,927	12,51,534	41,21,461
1992-93	24,02,319	10,40,923	34,43,242
1993-94	24,53,502	13,97,254	38,50,756
1994-95	26,20,759	15,36,608	41,57,367
1995-96	25,33,361	16,42,875	41,76,236
1996-97	24,96,624	17,38,218	42,34,842
1997-98	22,15,651	17,24,597	39,40,248

Source : Lucknow Mandi Samiti, Lucknow.

As far as it is related to the grape cultivation, the arrivals varied from 15.33 thousand quintals in 1997-98 to 32.59 quintals in 1994-95.

In case of livestock/animal products, the arrivals varied from 7.30 thousand quintals in 1994-95 to 14.45 thousand quintals in 1996-97.

In the case of forest products the arrivals have been changing from 6.74 lakh quintals in 1994-95 to 7.40 lakh quintals in 1996-97.

The total of all these have been varying from 39.40 lakh in 1997-98 to 42.34 lakh in 1996-97. The various figures relating to these items and 4 years as referred have been graphically demonstrated through Fig. 24.3.

Arrivals 1997-98 with Commodities

Efforts have been made to collect the primary data relating to the primary, secondary and the total arrival of all the specified/notified items at the Lucknow Mandi. Some of the salient features of the arrivals are given in Table 24.2 and also in Fig. 24.3.

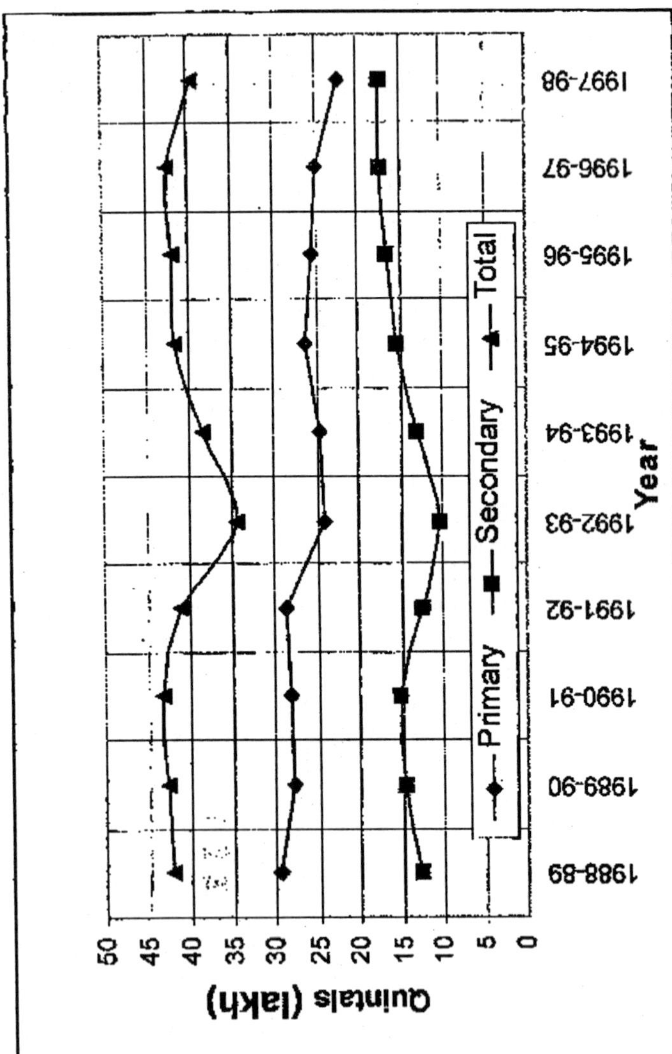

Fig. 24.2: Regulated Agricultural Market: Lucknow—Trend of Crop Arrivals

Table 24.2: Regulated Agricultural Market Lucknow —Arrivals

(Quintals)

Commodity	*Years*			
	1994-95	*1995-96*	*1996-97*	*1997-98*
A. Farm Produce				
Cereals	11,09,310	10,20,907	7,38,589	8,71,781
Pulses	1,74,580	2,10,491	2,33,945	1,73,271
Oilseeds	51,906	37,477	90,180	84,002
Fibrous Crops	764	953	1,423	1,440
Inhailers	16,100	9,057	8,371	9,613
Spices	42,577	36,901	51,864	56,508
Miscellaneous	24,035	15,596	15,492	13,436
B. Horticultural Produce				
Vegetables	12,45,232	12,01,664	18,98,931	10,20,791
Fruits	7,88,073	8,69,685	8,97,926	9,60,403
C. Grape Cultivation	32,495	32,250	25,031	15,332
D. Livestock Produce	7,303	9,736	14,455	12,456
E. Forest Produce	6,74,892	7,16,517	7,40,179	7,25,025
Grand Total	41,57,367	41,76,236	42,34,842	39,40,248

Source: Mandi Samiti. Lucknow.

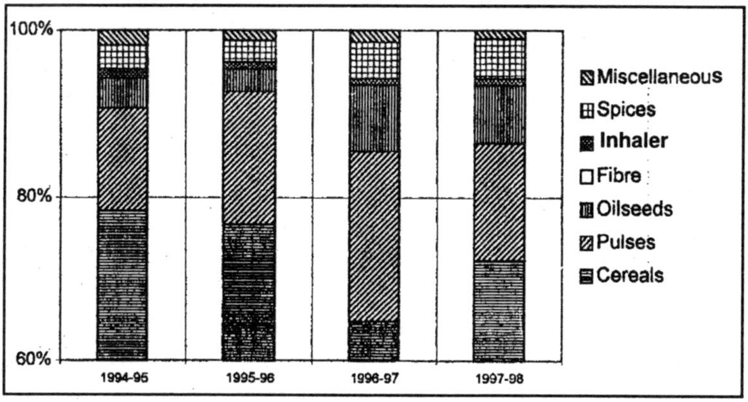

Fig. 24.3 : Regulated Agricultural Market: Lucknow—Farm Produce Arrivals

A. Farm Produce

Cereals

In case of cereals, 8 items—paddy, rice, jowar, bajra, maize, wheat, barley, and jaai—the primary, secondary, and the total arrivals have been as 3.79 lakh quintals, 4.92 lakh quintals and 8.7 lakh quintals respectively showing that there is no striking difference between primary arrivals and secondary arrivals.

Pulses

In case of pulses, the respective figures have been 10,000 quintals, 1.63 lakh quintals and 1.73 lakh quintals indicating striking difference meaning thereby that the secondary arrival is about 16 times than that of primary arrival.

Oil Seeds

In case of oilseeds, the primary arrivals have been 2.50 thousand quintals as against the secondary arrivals of 81.45 thousand quintals. The total of the two has been 84.00 thousand quintals. This shows more than 40 times arrivals of the secondary items than the primary items.

Fibrous Crops

In these crops, there have been absolutely no primary arrivals while the secondary arrivals were of 144 thousand quintals.

Inhaler/Tobacco

In this case also, there were no primary arrival at all while the secondary arrivals were recorded as 961 thousand quintals.

In case of spices, only coriander had the primary arrival of 5 quintals while the secondary arrivals of various other items have been more than 56.50 thousand quintals.

However, the other items considered under miscellaneous items have recorded 21 quintals and 13.40 thousand quintals (primary is gur only) totalling to 13.43 thousand quintals. In this case, except gur, all the arrivals are the secondary arrivals. There has been 3.91

lakh quintals as primary arrival, 8.18 lakh quintals as secondary arrivals, and thus, 12.10 lakh quintals has been the total arrival showing that the secondary arrival is more than double the primary arrival.

B. Horticultural Produce

In case of vegetables, 23 various items like tomato, potato, onion, etc. have been important. The primary, secondary, and the total arrivals of the vegetables have been 7.59 lakh quintals, 2.60 lakh quintals and 10.20 lakh quintals respectively. This shows that the primary arrival is about three times the secondary arrival. In case of fruits (about 20 types) the primary arrival has been 4.77 lakh quintals while the secondary arrival has been 4.83 lakh quintals totalling to 9.6 lakh quintals. This shows that more or less equal figures have been there, in these cases. The totals of the horticultural produce have been as 12.37 lakh quintals of primary arrivals, 7.44 lakh quintals of secondary arrivals, and 19.81 lakh quintals of the total arrival. These figures show that the primary arrival is nearly double than the secondary arrival.

C. Grape Cultivation

Only the secondary arrivals of 15.33 thousand quintals of grapes have been recorded.

D. Animal/Livestock Produce

Only one item *i.e.* ghee is there under this head. The primary arrival of ghee is 11.27 thousand quintals, the secondary arrival is 1.18 thousand quintals, while the total arrival has been recorded as 12.45 thousand quintals. These figures show that the primary arrival is 11 times than that of the secondary arrival.

E. Forest Produce

There are two items, gum (*gond*) and wood in this reference. *Gond* has the secondary while wood has recorded both the primary and secondary arrival, although the primary arrival is more or less 5 times the secondary arrival. The forest produce arrival totals are as 5.75 lakh quintals, 1.49 lakh quintals, and 7.25 lakh quintals. These figures show that the primary arrivals are more or less 4 times the secondary arrivals.

In case of grand total, the figures recorded are 22.15 lakh quintals, 17.24 lakh quintals, and 39.40 lakh quintals. These figures show that the primary arrivals are nearly one and a half-times the secondary arrivals. Thus, the primary arrivals are much more than the secondary arrivals.

Problems and Programmes

Loss of Income

It has been observed that primary arrivals at Lucknow RAM are actually as low as 20 per cent of the expected volume due to the fact that about ten non-regulated/conventional agricultural market places do heavy business in and around Lucknow such as Pandeyganj, Murlinagar, Sahadatganj, Chinhat, Dobagga, Bakshi Ka Talab, Banthra etc. The functioning of these traditional grain markets is neither controlled nor regulated. Hence, it causes a great loss to the State government.

The present Lucknow Mandi is characterised by paddy crop which occupies a large space which is not available at other traditional markets particularly those inside the Lucknow city. Thus, farmers in large numbers bring paddy crop to the RAM otherwise only about 20 per cent of the farmers as well as the volume of crops arrive at Lucknow market which causes a great loss amounting to more than Rs.1,00,00,000 per annum to the government. Therefore, all the grain market are to be shifted to mandi site only. All the arrivals should be thus at one place—the mandiyard. The mandi income, therefore, would increase considerably.

Lack of Adequate Storage Facilities

For proper marketing of perishable items, there is a great need of cold storage for keeping them. These items are fruits, and vegetables. However, there are no adequate cold storage arrangements for keeping enough fruits, and potato, although potato is sent from Lucknow to various markets of the country. There is one godown with 1000 metric tonnes capacity in the mandi campus which is quite insufficient in view of increasing trade in fruits and vegetables in this market. The observations show that the cold storage facilities should be trippled atleast.

Basic Facilities

Some problems are related to the inadequate general/basic infrastructure of physical facilities on the one hand while the functioning of the market itself on the other. Although, there can be no two opinions about the truthfulness of the fact that the RAM of the capital city of the State—Lucknow—is in general far ahead than many others in the State. The existing physical facilities in view of the growing needs of the people with the passage of time are getting inadequate in general. Besides, also there can be no two opinions about the inability and/or slackness on the part of the market committee, in general, to maintain all the installed physical facilities properly as sufficient funds are not available with these bodies. The facilities like open and covered platforms, rest house, parking space, cattle shed, Charhi, canteen, toilets, water huts, and godowns especially are not maintained. The maintenence of the areas in and around the grain market, the fruit market, and the vegetable market is very poor. There are also general complaints about sanitation, electricity, and first aid/medical facilities at the market yard. The fruits, and vegetables being the perishable commodities, if not carefully kept, create the environmental degradation problems having adverse effect on health of the market functionaries.

The unhygienic atmosphere is created by the spoiled fruits and vegetables on the campus, there are no proper facilities for dumping the waste material. There is also just one entrance for the Mandi, which is characterised by a long-queue of the farmers every morning waiting for the entry. This causes a great inconvenience, loss of time and energy for the producer-sellers. This is also one of the reasons that a large number of farmers prefer to go to other markets in the vicinity—non-regulated ones.

The development of infra-structural facilities and maintenance is one of the dire needs of the mandi for which higher budget amount be allotted to the mandi samiti. The mandi officials are of the view that despite heavy income from the mandi, there is a great problem to meet the expenses due to the government orders and the small budgetary allotments. The mandi officials feel, the powers are highly centralised at the Mandi Parishad. Ths market committees have no effective measures to control the traders by effective regulation.

Lack of Proper Marketing Process

It has also been observed that practically at times no auctions take place in the sale process of the agricultural produce of the farmers because of some understanding between the commission agents and the traders. Rather, farmers are, for one reason or the other, indirectly impressed by the purchasers-traders to leave their produce with the commission agents who ultimately pay less remunerative rates to the farmers.

A very surprising fact has come to light that a group of traders also works at the Mandi gate as Mafia traders. The farmers are, generally, threatened by these Mafia traders for taking their agricultural produce to their shops particularly located in Pandeyganj, and Daliganj areas. Thus, it seems that the mandi is not functioning properly in the interest of the farmers to the extent it should.

It has been mentioned by the officials of the market committee itself that for the proper functioning of the mandi, the management has been superseded by a government administrator—the ADM trans-Gomti. This clearly shows that there is a great need for management of the market activities which are not properly functioning at the Lucknow RAM.

In the interest of farmers, the marketing process as such has to be followed without fail by all concerned as also marketing in conventional sites still going on despite a regulated marketyard. This traditional marketing has to be stopped. This would favour both the farmers as also the government besides the general public from the hustle and bustle at the conventional grain market sites.

Programmes

Some new schemes have also been introduced in favour of the farmers by the market committee such as scholarship to farmers' children, Bakhari Scheme, Insurance Policy—Samuhik Bima Yojna known as group insurance.

THE FARMERS—THE SELLERS/THE PRODUCERS

The farmers are the most important part of all the market functionaries as these are the ones for whom the government has brought out the regulation as they were being cheated and not getting the proper benefit for their agricultural produce at the conventional agricultural markets. Keeping in view their significant position, the survey was done about these functionaries of the market through various angles, the results of which are being presented below:

Major Villages

Although, the Lucknow mandi, like every other mandi in the State has its own notified area which includes a number of villages from where the farmers come to sell out their agricultural produce at this market, there are some major villages of which mention is necessary.

About one-third of the farmers surveyed/interviewed came from 20 km distance. Itaunja is one of the most important village in this respect. Next in order, the farmers were from Bakshi Ka Talab, Mehmoodabad, Behta, and Kursi Road. Each of these villages contributed to the extent of 12 per cent meaning thereby that half of the farmers, contacted, came from these villages to the Lucknow mandi. Kamlapur is another important village from where a large number of farmers come to this mandi. The percentage of these farmers is eight. The rest of the farmers came from Kharagpur, Madiaon, and Mohan. Each of these villages contributed four per cent farmers to the market.

Social Structure

Although all the major castes of Hindus have been observed to be present in the market as producer-sellers or the farmers, Muslims also formed an important part of these market functionaries. There were Brahmins, Vaishya among the upper castes; and Blacksmiths, Yadavas, and Pals among the backward castes,while several farmers belonging to scheduled castes also were present as sellers of agricultural produce in the market. Besides, there were Muslims too.

It has been observed that 28 per cent of the interviewed producer-sellers were amongst the upper Hindu castes. Their percentage is second to the backward castes which stands for the highest percentage, 40 per cent. It is quite interesting to note that after these two castes, there were Muslims who contributed 24 per cent meaning thereby that amongst the producer-sellers who were surveyed about one-fourth have been from Islamic origin. Thus, they play an important role in this context. However, the percentage of scheduled castes as producer-sellers has been the lowest—just 8 only. This clearly shows that their position as sellers is not significant in comparison to other Hindu castes and Muslims.

Education

As far as it is related to education of the interviewed farmers, it is worth mention that the percentage of illiterate farmers is the minimum—only 4 per cent. On the other hand also, the percentage of the farmers with higher education is the same *i.e.* four. The highest percentage was of the farmers who had received education upto high school or intermediate standard. The percentage of such farmers is 48. Thus, nearly half of the interviewed farmers had received education upto 10 or 10 + 2 standard. Another significant part is of those who received only basic/primary, and junior high school education. The percentage of such farmers is 44 which is although, second in rank but only little bit less than the first ranker. However, of the total farmers, 4 per cent were illiterates too. On the other hand, there were 4 per cent of the farmers who received higher/university education too. It is significant to mention that those who entered agriculture as their occupation, had not any education and/or training in agriculture. Rather, even after receiving higher education, they entered the agriculture profession due to poverty and unemployment only.

This shows that the most of the farmers had received education and also that the most of such farmers had received education upto high school and/or intermediate standard.

Road Linkages and Mode of Transport

Here, it is also significant to note that all the major villages of the notified area are linked with the Lucknow mandi by *'pucca'* road—metalled road.

In bringing their produce to the mandi from their villages, the farmers take upto two hours. There are 36 per cent farmers who take 1-2 hours. in reaching the market with their produce. Sixteen per cent of the farmers take less than one hour time while 32 per cent farmers take more than two hours to reach the mandi place. This distance is travelled by tractor or tonga in which they carry their produce.

Distance Covered

The most of the farmers who had been interviewed came from a distance of 10-20 km. Their percentage is 72. Thus, nearly three-fourth of the entire sellers came from the villages which are located between 10-20 km. Those who came from less than 10 km. formed 16 per cent while those who came from more than 20 km. formed 12 per cent. This also makes it clear that the most of the major villages which participate in the selling process are located between 10-20 km. of the mandi's notified area.

Reasons of Visits

The farmers come to Lucknow mandi for two major reasons : vicinity and competitive prices. While 45 per cent of the farmers told that the first reason for them to prefer for this mandi was vicinity, the same percentage of farmers was in favour of competitive price too. Ten per cent farmers viewed that they had been coming to the mandi as it was the only big mandi in the entire area of their reach.

The transport cost per quintals is not high. The farmers pay Rs. 5 per quintals for a short distance, and Rs. 10 per quintals for comparatively long ones. There were 48 per cent farmers who paid only Rs. 5 per quintals while 52 per cent of them paid Rs. 10 per quintals.

Yard Visits

The farmers visit the market during both the Rabi and Kharif crop seasons. They have several visits during both the seasons for selling their agricultural produce. However, it has been observed that a large number of farmers visit the market several times during both the seasons.

During the Kharif season 44 per cent of the farmers make as many visits as 13 while 40 per cent of the sellers make only upto 6 visits during the season. The rest of the farmers *i.e.* 16 per cent have their visits during this season in between 6 and 12.

During the Rabi season 48 per cent of the farmers have only upto six visits while 44 per cent of the farmers have as many as 13 visits to the mandi. Only 8 per cent of the farmers have 6 to 12 visits during this season.

This clearly shows that during the Rabi season, the farmers sell out their crops within some visits to the market while during the Kharif season they make more visits to dispose of their produce. The prices play an important role in the number of visits by the farmers to the market.

Weekly Markets

When the farmers were questioned about attending any other market — they responded in a very interesting way. Although, the farmers coming to Lucknow mandi for selling the most of their agricultural produce, sometimes, they also go to sell their produce in the weekly markets too. The most important reason for this is the vicinity of these weekly markets. These markets are located either in the villages of the farmers only or only within 5 km from their villages. For fulfilment of urgent or emergent needs of their families, some farmers sell their produce at these market places. The percentage of the farmers going to various such markets are as follows. Itaunja : 32; Behta : 24; Bakshi Ka Talab: 12; Mehmoodabad : 12; Kamlapur : 8; Mohan : 4; Others : 8. This clearly shows that one-third of the farmers attended the weekly market at Itaunja, while nearly one-fourth of them attended the Behta weekly market. About one-eighth of the farmers go to Bakshi Ka Talab, and Mehmoodabad each while one-sixteenth of them attend the Kamlapur weekly market. It must be noted that the volume of sale in the weekly market is considerably low in comparison to that sold in the regulated market.

Land Holdings

After the survey, it was observed that the farmers who visited the yard had small land holdings. As many as 52 per cent of the farmers had only upto 10 bighas of land each. As against this only 16 per cent of the farmers had more than 20 bighas of land each. Thus, 32 per cent of the farmers had their land holdings between 10 and 20 bighas (each).

Out of these holdings, however, the most of the land is under irrigation. Only 8 per cent of the farmers have more than 20 bighas of their land under irrigation, while as many as 56 per cent of the farmers have only upto 10 bighas of their land under such conditions. Thus, 36 per cent of farmers have their land under irrigation conditions and that their holdings are between 10 and 20 bighas.

It is, however, a point to note that 28 per cent of farmers had absolutely no unirrigated land while only 4 per cent of the farmers had their unirrigated land beyond 5 bighas each. The most of the farmers' unirrigated land falls only under 5 bighas each. The percentage of such farmers is as high as 68.

Crop Structure and Market Surplus

Wheat and paddy have emerged as the major and predominating crops amongst the farmers interviewed for the purpose. It is interesting to note that all the 100 per cent farmers grow both wheat and paddy. Pulses are grown by 44 per cent of the farmers, while potato is grown by 48 per cent of the farmers. Oilseeds and sugarcane are also produced by 4 per cent each. Other crops are just in too low percentages. Thus, cash crops are not the prominent ones. Only the food crops have the key position in the volume of production.

It has also been noted that the marketed surplus of these famers is quite low in relation to the total volume produced. This is because the producer-sellers keep a large volume of the crops for their home consumption. Besides, for seed purpose also quite a significant part is saved. In addition, some farmers also pay wages to the labourers in the form of produce (and not in the form of cash payment due to their financial conditions at that time). And yet as many as 90 per cent of the farmers have mentioned that their annual surplus has been increasing during the last 5 years.

Factors Affecting Sale Decision

When asked about the time for sale, it was observed that the selection decision was affected by several factors. As many as 96 per cent farmers said that their immediate family need has been the most important factor deciding the time of sale which is, generally, immediately after the harvest more particularly in case of Rabi crop due to the general family needs like annual purchases of provision goods, seasonal purchase of cloths, and special purchases relating to marriage or any other ceremonies. Ten per cent of the farmers also had their opinion relating to competitive market rates, lack of storage facilities at home, and family indebtedness, etc. about deciding the time for sale of their produce.

Some suggestions have also been given for betterment of the market functioning to improve the conditions like cattleshed, night stay and parking etc.

A good number of farmers—20 per cent—also disclosed that at the gate of the yard some Mafia traders were also working strongly. They threatened the farmers to bring their produce to particular grain markets and/or traders, otherwise they would have to face the consequences.

Suggestions

The agricultural marketing practices have considerably improved as compared with the past. Now the farmer is getting fair deal and better prices against the village merchant as also several facilities are available to them at the market site under regulation.

Farmers opined about the functioning of the market that they are satisfied—to some extent. However, some farmers viewed that the functioning is not satisfactory as there used to be no true competition in the bidding. In their view, there is some understanding between the traders and commission agents and hence they (the farmers) do not get the maximum possible price of the produce. The traders also may have an understanding amongst themselves to share the volumes of commodities as per their requirements. This sort of understanding reaches in advance and hence the bidding process seems to be a formality only. Since the traders and the commission agents belong to the same occupation. They too have some sort of understanding reached in advance against the farmer's produce. They also have their general meetings at the market as also outside the market. Thus, due to their personal terms, feelings and relations, the commission agents indirectly favour the traders but they pose to be favouring the farmers. The bidding, therefore, becomes a meaningless exercise.

As far as eradication of such practices is concerned, the farmer-producer-seller says that the supervision/inspection in the true sense of the word be done and that strict enforcement of rules and regulation be observed by the market committees. The market committee must practically help the farmers and check if any trader is trying to cheat them. The attitude of the market committee should not be indifferent towards farmers and they should implement all the rules and regulations strictly followed by the surprise checks. Thus, they should take active interest in making the system helpful to the farmers.

Some farmers also express that for want of good number of traders for bidding, government should be an ideal purchaser of the farmer's produce. It has also been observed that some unauthorised labour also entered the market and in process of their work they did some cheating too. It has also been observed that generally the payments are made to the farmer after sale, yet the attitude of the commission agents always have some favour to the traders rather than to the farmers. Such happenings should be carefully watched and checked.

The market committee and the government have provided the bye-laws insisting on payment to the farmer on the same day but in many cases, it has been observed that it is not the case. The commission agents make advance payments to the farmers at the time of their (farmers') urgent needs hence, generally, the commission agents also have the tendency of undue profit making rather than just rendering a service to the farmer as is generally posed by them. It seems that the trade is controlled by the commission agents and the traders rather than by the rules and the regulations. Sometimes, a farmer has to visit the market to meet the commission agent only to collect his payment as it is not done immediately after sale every time.

By and large, the farmers are of the view that there is no alternative to regulated marketing and such markets should continue but their functioning should be made more beneficial to the farmers in practice—although there is almost everything in the favour on paper—in black and white. The drawbacks/problems be removed, and the regulation is needed in practice. There should be no political interventions as also no mafia functioning or parallel (unregulated) trading of agricultural produce.

The Traders—The Purchasers

As has already been mentioned, one survey schedule was prepared for traders' responses too. Thus, these responses have also been recorded. Various interesting features have come up after analysing the records. Some of these are given below :

Place of Origin

The nearby area includes some important localities/villages like Madiaon, Daliganj, and Khadra. The Khadra locality situated on the same road towards Lucknow is only about 3 km, while Daliganj is also about 2½ km. Madiaon is also on the same road and it is located within 4 km from the mandi. There are 40 per cent traders who came from Khadra, while 30 per cent from Daliganj locality. Dobagga contributed to 10 per cent while others, not more than 5 per cent each. This clearly shows that nearly 75 per cent of the traders reach the marketyard from within 5 km distance only.

Social Structure

The investigations about social structure has brought out interesting results; while the sellers were mostly from the Hindu backward castes, the purchasers are mostly from the Hindu upper castes. Against, the 40 per cent sellers from the backward castes, there were as high as 60 per cent purchasers from the upper castes. The most of the upper castes people included Brahmins followed by Vaishyas. There were only 25 per cent purchasers who came from the backward class community. The Muslim community also contributed to the extent of 10 per cent while the scheduled castes just 5 per cent.

Education

When the investigation was done about the educational standard of the traders, it was observed that there were absolutely no traders with higher education or with any university degree. While amongst the sellers, there was an existence of such highly education people. It has already been mentioned that this was particularly because of the unemployment and poverty conditions that the persons with higher education entered the farming occupation. It must be noted here that the highly educated people had absolutely no education or training related to agriculture—rather, it was only the non-agricultural area which they had studied. It was also, specifically noted that this group of purchasers was all literate while there were some illiterate persons also in the selling unit—*i.e.* the farmers.

There is a very striking fact that as against the 44 per cent sellers with education upto eighth standard, there were 45 per cent purchasers also with the same educational standard. However, against the 48 percent sellers with high school/intermediate education, there were 55 per cent purchasers with such an educational standard in the market.

This also clearly shows that the persons who received higher education are neither taking active part in selling nor in the purchasing as farmers or traders respectively. Only the people with school education have entered this area. This is also indicative of white collar jobs preferred by the people with education beyond schooling.

Road Linkages and Mode of Transport

The traders, observations show, has collected at the mandi place/yard from the nearby area. All the traders had metalled link roads and they either hired tempo or rickshaw or came by their own vehicles. It was noted that 40 per cent of the traders had their own vehicles while another 40 per cent came by rickshaw. The rest 20 per cent reached the yard by tempo.

This also makes it clear that the traders had collected from a nearby area only. When investigated, it was observed that 35 per cent of the traders came from within 3 km distance only, while another 35 per cent traders had come from a distance more than 5 km. The most of the such traders collected from within 10 km distance from the mandi. However, only 10 per cent of the total traders actually travelled a distance more than 10 km but within 20 km. There were 30 per cent traders who travelled a distance between 2 and 5 km only.

Yard Visits

A very striking feature has come up about the visits by traders to the marketyard. As many as 95 per cent of the traders visited the yard regularly and only 5 per cent traders who were comparatively from distant places did not come to the yard regularly. This is true with both the seasons—Rabi, and Kharif.

There were 40 per cent traders who made upto 10 visits during the Rabi season. Thirty-five per cent of the traders visited the yard for more than 15 times during this season while 25 per cent of the purchasers paid 10-15 visits to the marketyard during the Rabi season.

During the Kharif season as high as 50 per cent of the traders visited the market from more than 15 times. This is quite a striking feature against the Rabi season. Another striking feature is that 25 per cent traders had upto 10 visits while another 25 per cent traders had 10-15 visits to the market during the said period. Thus, as against the Rabi season, the yard visits by the traders during the Kharif season were quite noteworthy.

Market Fee

Every trader has to get himself enrolled with the mandi. After paying a certain fee, he becomes a bona fide trader-purchaser within the mandi yard. It has already been mentioned that during the present days the license fee of traders is Rs. 250 per annum although earlier, it used to be Rs. 50 per annum for the commission agents and Rs. 100 for the traders. The traders who have a lot of business with the market have to pay larger amount as the purchasers pay the fee on the amount transacted. At present, the rate of fee paid on this amount is 2.5 per cent out of which 2 per cent is the mandi fee while 0.5 per cent is the development tax which has been implemented from 15th August,1998. Formerly, the mandi tax used to be 1.5 per cent instead of 2 per cent at present. There were 40 per cent traders who paid Rs. 500 to 1000 as fees for their daily visits to the mandi. However, those who paid upto Rs. 500 as fee and also those who paid more than Rs. 1000 as fee had the same percentage, 30 per cent each. It has been noted that the Daliganj traders pay large amount of fees to the mandi due to their regular multiple visits. Some traders paid even upto Rs. 3000 and many upto Rs. 2000 while so many between Rs. 1000 and Rs. 2000. It is also important to note that mostly the traders belonging to Bania caste (Gupta etc.) make regular and also greater number of visits during both the seasons.

Commodities

As per the availability of crops, the purchasers, make their business. During the Rabi season, while main crops are wheat, barley, gram, arhar, oilseeds (mustard, rye etc.), during the Kharif season, the major crops are paddy, rice, moong, urd, masoor, maize, and potato etc. Paddy, rice and wheat are the crops which have business in large quantities. Many of the traders purchase these crops only.

It is also important to note that the vegetables, and the seasonal fruits are also brought and sold in this market. Amongst the fruits, the major ones are mango, apple, grapes, banana, lichi, coconut etc. Banana is the one which is demanded most. The largest number of fruit purchasers are for this item. Of the total fruit purchasers, 25 per cent come to this mandi for bananas fruit only. Grapes and mango come next, 15 per cent each. Apple purchasers have the third position amongst the fruit purchasers in this market. Their percentage is 10. Other fruits are purchased by traders in smaller numbers.

Suggestions

When contacted, the traders particularly the commission agents expressed that the payments are made to the sellers, generally, immediately after sale but in some cases, the part payments are also made. During the peak season, the payments are, generally, delayed due to heavy business but delays are not taken seriously by the farmers. The commission agents want their commission rates increased as their expenses have gone up on their establishments.

Acknowledgements

The author is grateful to the Indian Council of Social Science Research, New Delhi for awarding Senior Fellowship under which the above research had been completed at the Giri Institute of Development Studies, Lucknow.

REFERENCES

Dixit, R.S. (1998). *Agricultural Markets of Uttar Pradesh—A Spatial Analysis,* Indian Council of Social Science Research, New Delhi, Senior Fellowship Report at Giri Institute of Development Studies, Lucknow, pp. 302-335.

Mandi Samiti (1998). *Lucknow Mandi Samiti Arrivals.*

25

AGRICULTURAL TRANSFORMATION AND RURAL SUSTAINABILITY IN A HIMALAYAN WATERSHED

A Study of Kullu Valley

R.B. SINGH and D.K. MISHRA

INTRODUCTION

The biophysical environment forms a fundamental basis of the existence of human society. It refers to the land, soil, and waterbodies, flora and fauna that surround human habitat and from which human beings derive a living. Basic human subsistence and requirements are met from the different natural resources. Each and every component of the natural environment is significant for our livelihood. Furthermore, the environmental components interlink in a complex regional ecosystem embodying many subsystems whose indiscriminate disruption would upset the delicate ecological balance, with enormous consequences on humankind. It is a matter of great concern that the quality of the natural environment is deteriorating in many parts of the country at a time when the human population is growing at an accelerated rate (Sharma and Minhas, 1993). And also there is a high degree of diversity in land productivity in different regions of the country (Hanumantha Rao, 1994).

STUDY AREA AND METHODOLOGY

The study area consists of 4 villages of upper Kullu valley in the northern most part of Himachal Pradesh. The total area covers about 120 km^2. And the elevation varies between 2000-4000 m from mean sea level. The slope gradient is steep and the region has a thin layer of mountain soil. The river Beas passes through the region in north to south direction. Climatically the region lies in the temperate zone extending to the tundra zone, owing to its high elevation, where snowfall is a common phenomenon. The natural vegetation is dominated by coniferous and mixed forest, with large tundra and alpine meadow areas. The proportion of inhabitable land is low in the region and being a tourist area, the seasonal concentration of tourists is high. The research methodology involves comprehensive studies using a historical perspective and the application of a geographical framework for data representation and understanding both in space and time. Agricultural transformation has been evaluated and the rural watershed sustainability framework has been assessed.

AGRICULTURAL TRANSFORMATION

The unprecedented growth in population is exerting high pressure on the limited land available for cultivation. Agricultural practices have undergone drastic changes since the mid-1960s through the introduction of agro-development programmes and the application of various newly developed techniques. People are giving importance to orchards because of their high profitability. Thus, some agricultural land has been encroached by orchards. In addition, people for orchards and agricultural purposes are now using wastelands. The bulk of agricultural land in the region lies on terraces, along the steep hillsides. The gradient of some agricultural land exceeds 35° (Singh and Mishra, 1995). The cropping intensity, agricultural productivity and additional income generated through agro-horticultural activities, increasing apple orchards are the main factor reducing the cultivated area. Although the orchard plantations were started in the valley during the British period, its commercialization began only during the 1960s and 1970s. The major factors responsible for changing the biophysical environment are deforestation and change in the cropping pattern. The mass felling of trees during the 1950s and 1960s and the clearing of large tracts of forests have made the ecosystems more vulnerable by reducing the biodiversity of the valley. The high increase of consumption of water has considerably influenced the hydrological set-up of the entire valley (Singh, 1998).

INDIGENOUS FARMING SYSTEMS

Nearly all indigenous farming systems in the valley are founded on the principles of organic farming, which a large number of experts consider to be the most efficient way of producing food on an environmentally sustainable basis, *i.e.,* without any major lasting threat to the natural ecosystem. Organic farming exceeds artificial agro-chemicals. It is sustained by natural recycling of natural soil-replenishing nutrients, *e.g.,* biomass, compost, manure and other forms of organic matter. It relies on biophysical methods to control pests and on crop rotations and the inter mixture of species to sustain agroecological stability, food security and balanced utilisation of soil nutrients. Oldest indigenous farming systems were based on the shifting cultivation and land rotation. Under a relatively low population pressure prevailing in the past, this system enjoys the advantages of minimising soil erosion, preserving agro-biodiversity, maintaining ecological stability, optimising utilization of the different soil nutrients, and enhancing food security and a balanced diet, under the influence of the growing pressure on land, which is associated mainly with population growth, the cultivation system has been modified into more settled system (Pandey and Singh, 1984). There is a need for more research in identifying the most productive species, breeding, improved varieties, and testing and management methods. Research into mixed cropping and other traditional farming practices is also yielding interesting results. Increasing number of poor people is also the cause of environmental deterioration, as they are cutting the slopes to produce their crops, feed their livestock and family members (Sarmiento, Monasterio, and Montella, 1993).

CHALLENGE OF SUSTAINABILITY UNDER CHANGING FARMING CONDITIONS

The farming systems are not static. They are changing as a result of changes in the environment, both natural and socio-economic. Some of these changes have rendered the traditional farming unsustainable and somewhat outmoded. A few examples of the changes in the agricultural systems and their impact on sustainability are presented in Table 25.1.

The manner in which changes affect sustainability can be illustrated with two or more of these examples. For instance, introduction of apple orchards can be contributing to increasing stability of production. But it is also true that the introduction of the orchards has often been promoted at the expense of the indigenous

Table 25.1: Major Changes and Effect on Sustainability in Upper Kullu Valley

Sl. No.	*Changes*	*Effect on Sustainability*
1.	Introduction of Apple Orchards	Positive and Negative
2.	Population Explosion	Negative
3.	Commercialisation of agriculture	Largely Negative for less resourceful farmers
4.	Agricultural Chemicals	Largely Negative unless strictly controlled
5.	Fertilizer Use	Negative and Positive
6.	Population Influx	Negative and Positive

Source: Assessed from various observations and field survey.

food crops, some of which are so neglected that they are not much being grown and a considerable degree of bio-diversity has been lost. Population explosion has considerably increased pressures on land, resulting in intensification of farming. Use of agricultural chemicals has different impacts on the environment, where reasonable amount of chemicals is appropriately used. The effects are largely beneficial, where, for example, no fertilizers are used and farming is intensified, the nutrients are depleted and yields drop as soil become degraded. But where excess amounts of farm chemicals are applied, the environment may become polluted and unsustainability is the ultimate result (Berkes and Gardner, 1997).

MICRO ISSUES OF AGRICULTURAL TRANSFORMATION

Till the mid-80s the local people were depending heavily on the agriculture produce in the valley. People were self sufficient in food grains. They never bought grains from the market; in fact buying grains from the market was considered a great shame to the family. The drift from agriculture to horticulture in the mid-80s greatly decreased the agricultural fields. These fields have been converted into apple, plum, pears and cherry orchards. As the local people had land in different far distant villages, they had built temporary houses there, known as *"doghries"*, meaning the second house or a place meant for temporary shelter. People used these places during the sowing and harvesting period. The *"doghries"* which used to have inhabitants at least on four occasions a year in the past, now present a deserted look. The small numbers of agriculturists still around now go by bus, taxi and car to cultivate and harvest the crops and come back home by dusk. Whether these *"doghries"* will ever get an inmate is a big question mark, but the current prosperity has certainly changed the life style of the local population in less than 15 years.

Replacement of Nutritious Traditional Food Crops

Before Independence "Saryara" was an important staple crop in the region, but after Independence it has lost importance because of the innovations of the new high yielding varieties of crops. Though since the beginning of the century the fluctuations in all the 4 villages were a very common feature, but the major decrease has been obvious after 1960-69. "Saryara", is one of the traditional food crops, which are very nutritious, but there has been constant decrease in the sown area under this crop in all the villages. The decrease has been very sharp from 1979-89, with maximum decrease in case of Buruwa village where it has decreased from 35 ha to 7 ha. In Palchan, it has decreased from 18 ha to 3 ha and in Vashisht, it experienced a decrease from 14 ha to 3 ha. It is also reflected in the total area under "Saryara" production where it has decreased from 74 ha to 13 ha. By 1994, the total area under this crop has reduced to only 3 ha with Palchan, Buruwa and Vashisht each having 1 ha (Table 25.2).

"Kodra" is another important crop, which is rapidly losing its importance due to ever-increasing production of rice. The decrease in the net sown area under "Kodra" has not been constant over time in all the villages. In Palchan, the area under this crop has remained around 2-4 ha except during 1959-69, when it has decreased to 2 ha. In Buruwa village the area under "Kodra" till 1959 was quite high, around 60-70 ha, which had a sharp decline in 1969 when it reduced to 6 ha, which further declined to 3 ha in 1979. In case of Vashisht village these decline starts from 1979 onwards *i.e.,* 14 ha in 1969 to 2 ha in 1979 to 1 ha in 1989. In Manali, also the decline in this nutrient crop area has started from 1979 onwards from 19 ha in 1969 to 6 ha in 1979 to 1 ha in 1989. This decrease in net sown area becomes more interesting when we look at the total area under "Kodra" in the valley

Table 25.2: Change in the Sown Area of Saryara Crop in Upper Kullu Valley (1919-94)

(Area in ha)

Year	*Villages*				*Total Area*
	Palchan	*Buruwa*	*Vashisht*	*Manali*	
1919	17.88	41.46	21.54	19.91	100.79
1929	17.07	31.70	17.88	19.51	86.16
1939	15.85	34.55	21.54	24.79	96.73
1949	15.04	31.30	22.76	16.66	85.76
1959	13.41	51.62	18.29	18.69	102.01
1969	20.32	32.00	15.00	16.00	83.32
1979	18.00	35.00	14.00	7.00	74.00
1989	3.00	7.00	3.00	NA	13.00
1994	1.00	1.00	1.00	NA	3.00

Source: *Lal Kitab* (Village Statistical Record), Kullu Tehsil, Kullu, 1919-94.
N.A. = Not Applicable.

where it has decreased from 114.75 ha in 1919 to 59 ha in 1969 to 3 ha in 1994 (Table 25.3).

Table 25.3: Changes in the Sown Area of Kodra Crop in Upper Kullu Valley (1919-94)

(Area in ha)

Year	*Villages*				*Total Area*
	Palchan	*Buruwa*	*Vashisht*	*Manali*	
1919	4.00	74.39	18.69	17.07	114.15
1929	2.00	63.00	16.66	16.26	97.92
1939	2.00	59.75	21.13	14.22	97.1
1949	6.00	70.73	20.32	15.85	112.9
1959	4.00	67.47	15.85	9.75	97.07
1969	2.00	6.00	14.00	19.00	59.00
1979	7.00	3.00	2.00	9.00	21.00
1989	2.00	3.00	1.00	1.00	7.00
1994	N.A.	3.00	N.A.	N.A.	3.00

Source: *Lal Kitab* (Village Statistical Record), Kullu Tehsil, Kullu, 1919-94.
N.A.= Not Applicable.

This general decrease in the land area under traditional and nutritious food crops can be attributed to the following causes:

1. Most of the agricultural land is being converted into orchards.
2. Agricultural land is also being taken up for various construction purposes.
3. Most of the traditional crops are being replaced by cash crops.
4. Growing influx of tourists and increasing commer-cialisation.

Introducing Cash Crops for Economic Upliftment

Introduction of cash crops like maize in the valley has replaced almost all the traditional nutritious food crops like Kodra, Saryara and Kathu. This can be only attributed to the following reasons:

1. Maize as a cash crop fetches a good amount to the farmers during the tourist season.
2. It is also high yielding in nature.

Therefore, we find a general increase in the area under maize cultivation in the upper Kullu valley, with the total area increasing from 41.45 ha in 1919 to 110 ha in 1979 to 268 ha in 1994. At the village level, Buruwa village has the maximum area (153 ha) under maize cultivation (Table 25.4).

Table 25.4: Change in the Area under Maize Cultivation in Upper Kullu Valley (1919-94) *(Area in ha)*

Year	*Village*				*Total Area*
	Palchan	*Buruwa*	*Vashisht*	*Manali*	
1919	N.A.	21.54	11.78	8.13	41.45
1929	4.06	47.15	13.00	16.26	80.47
1939	5.28	51.21	16.26	15.04	87.79
1949	8.94	64.63	18.29	19.10	110.96
1959	15.44	32.11	13.41	20.73	81.69
1969	18.00	37.00	6.00	8.13	69.13
1979	11.00	79.00	18.00	2.00	110.00
1989	37.00	117.00	9.00	50.00	213.00
1994	41.00	153.00	30.00	44.00	268.00

Source: *Lal Kitab* (Village Statistical Record), Kullu Tehsil, Kullu, 1919-94.
N.A.= Not Applicable.

Previously, potato was an important crop in the upper region of the valley, but due to better production in Lahul and Spiti, other crops are replacing the production of potato. Therefore, the general trend of production of potato is showing a negative trend among all the villages. The variation over a last 10 years is maximum in Palchan and Buruwa, where Palchan has shown a negative trend but in Buruwa we find a positive trend (Table 25.5).

Table 25.5 : Change in the Area under Potato and Vegetables Cultivation in Upper Kullu Valley (1919-94)

(Area in ha)

Year	*Total Area under Potato*	*Total Area under Vegetables*
1919	2.00	10.14
1929	4.00	6.07
1939	7.00	6.67
1949	20.34	2.01
1959	65.43	3.38
1969	66.00	25.40
1979	58.00	50.00
1989	36.00	11.00
1994	21.00	3.00

Source: *Lal Kitab* (Village Statistical Record), Kullu Tehsil, Kullu, 1919-94.

Since, the region is a tourist dominated, the production of various types of vegetables should be a common feature of agriculture. But the pattern is not visible. It is because most of the productive fields have been devoted to apple horticulture and on remaining part of the land· farmers produce subsistence crop for their own livelihood. Fluctuations over time have been observed in the production of vegetables, Vashisht has shown a remarkable increase (3 ha) while other villages have stopped the commercial production of vegetables after 1990. The total net sown area under vegetable was 10.14 ha in 1919, which increased to 50 ha in 1979, but due to the orcharding culture in 1994 it declined to only 3 ha in the valley (Table 25.5).

STATUS OF HORTICULTURE

The per unit area production of fruit and income are higher than ordinary cereals grown all over the region. In 1993-94, during late March or early April snowfall caused severe damage to the production of apple. The whole valley suffered from loss of apples worth of million of rupees due to late snowfall. While crossing the area one can observe the growth of apple trees over the cultivated areas. Undoubtedly, increasing number of apple orchards is the main factor in reducing the cultivated area. Therefore, on the basis of intensive fieldwork and analysis (Singh, 1998) of data the following features can be observed:

1. The pattern of agro-orcharding is very common among all the villages. Apple trees are planted in the cultivated areas. Till the trees were small in size; other crops were also grown in the same field. When they become larger and occupied all the space between trees, cultivation of other crops was not possible. In this way cultivated area is decreasing and orchard area is increasing.
2. Though the orchard plantation was started during the British period, commercialisation started during late 1960s and 1970s. Manali was the main centre for apple production and even today it occupies a prestigious place in the Kullu valley.
3. Palchan village, situated in the extreme north of the valley, experienced very low orchard growth due to severe winter and heavy snowfall. Now the apple trees are unable to produce good quality as well as quantity of the apples under such conditions, There was only 1 ha land under apple orchards up to 1985 but between 1985-90 it has increased to about 12 ha.
4. In Buruwa and Vashisht, orchards started during 70s and extensive orcharding was experienced. In Buruwa and Manali, apple orchards increased rapidly during 80s and 90s.

UNSUSTAINABILITY IN AGRICULTURE

The intensification of agricultural production as a result of increasing population pressure, intensification of farming, overgrazing and conversion of land to several uses have resulted in several undesirable changes in the environment with the adverse effect on agricultural production. Changes in the form of intensification of land-use for food production, shortened fallow periods, reduced capacity for soil regeneration, and decline in soil fertility, reduced and declining yields, food shortages, use of marginal sloping lands, overgrazing, loss of vegetative cover, etc. can be seen in the valley resulting in unsustainability in agriculture (Tisdell, 1988).

INTERFACE BETWEEN AGRICULTURE AND ALLIED SECTORS

For appropriate and realistic design of a sustainable farming system it is necessary to ensure that threats to it from other sectors are significantly minimised. For example, poor road construction and maintenance of NH-21 could result in flooding, eutrophication and erosion, all of which can seriously damage agricultural land and stream fisheries. A related example is a policy issue, such as structural adjustment aimed at increasing export earning or reducing debt burdens. This may result in large forest areas being cleared for commercial crops, which will expose the soil to erosion. Similarly, removal of subsidies may result in farmers not using excess pesticides and insecticides, which in turn, will result in minimising environmental degradation. Therefore, sustainable agricultural systems and adhering to the design alone will not ensure sustainability unless policies, strategies, technologies, systems and components of resource management and input use in other sectors are designed to ensure sustainability in development. All these call for not only integrated land-use planning but also adoption of a holistic approach in development planning, in policy formulation and in selecting strategies and the execution of development programmes (Smith, 1993; Singh, 1998).

CONCLUSIONS

In designing sustainable agricultural production system, it is necessary to give due consideration to the characteristics of various resources used in production, the way they are managed in the production process and the indigenous technologies and local practices which render the resultant production system unsustainable. Unsustainability results when impacts of practices and technologies used are economically not feasible and sometimes also culturally unacceptable. In addition to the selection of methods for manipulation of resources and use of practices and technologies, which ensure sustainability, there is a need to ensure that other sectoral activities do not render the farming systems unsustainable. To generate new agricultural practices and technology continuous and systematic research is required incorporating people's response. Agricultural implements and farm inputs needed by farmers should be locally available at reasonable price. Presence of adequate educational, training and support facilities for farmers are urgently needed. Mechanism for planning and directing agricultural programmes should be integral components of overall economic development.

REFERENCES

Berkes, F. and Gardner, J.S. (eds), (1997). *Sustainability of Mountain Environments in India and Canada,* University of Manitoba, Winnipeg.

Hanumantha Rao, C.H. (1994). Agricultural Growth and Rural Poverty in India: Emerging Trends and Perspectives, *Indian Economic Review,* Vol. XXVII, No. 2, pp. 129-140.

Pandey, U. and Singh, J.S. (1984). Energetic of Hill Agro-Eco Systems: A case study from Central Himalaya, *Agricultural Systems,* 13, pp. 83-95.

Sarmiento, L., Monasterio, M. and Montella, M. (1993). Ecological Bases, Sustainability and Current Trends in Traditional Agriculture in Venezuelan High Andes, *Mountain Research and Development,* 13, pp. 167-176 .

Sharma, P.D. and Minhas, R.S. (1993). Land-use and Biophysical Environment of Kinnaur District, Himachal Pradesh, in *Mountain Research and Development,* 13, pp. 41-60,

Singh, R.B. and Mishra, D.K. (1995). Dynamics of Land Use and Cropping Patterns: Historical Prognostication in Upper Beas Basin in R.B. Singh and Martin, J. Haigh (ed.) *"Sustainable Reconstruction of High Land and Head Water Regions"*, Oxford and IBH Pub, New Delhi, pp. 649-661.

Singh, R.B. (1998). Land-use/cover changes, extreme events and eco-hydrological resources in the Himalayan region, *Hydrological Processes,* 12, 2043-2055.

Singh, R.B. (ed), (1998). *Sustainable Development of Mountain Environment in India and Canada (CIDA-SICI Project Experience),* Oxford and IBH Pub., New Delhi.

Singh, R.B. (1998). *Understanding High Mountain Land Use towards Sustainable Environmental Management in Jammu & Kashmir,* Oxford and IBH Pub., New Delhi, pp. 409-422.

Smith, V. Kerry (1993). "Non-Market Valuation of Environmental Resources: An Interpretation Appraisal", *Land Economics,* Vol. 69, No. 1, Feb., pp. 1-26.

Tisdell, C. (1988). "Sustainable Development: Different Perspectives of Ecologists and Economists, and Relevance to LDCs", *World Development,* Vol. 16, No, 3, March, pp. 373-384.

26

TRENDS OF AGRICULTURAL DEVELOPMENT IN U.P. HIMALAYA

SHAMSUL HAQUE SIDDIQUI,
MUMTAJ AHMED and MOHD. TAUFIQUE

INTRODUCTION

U.P. Himalaya occupies an area of 51,125 square kilometres (17.4 per cent) of the State. Here agriculture plays an important role in economy of this Region. U.P. Himalaya consists of eight districts: namely Nainital, Almora, and Pithoragarh of Kumaon division; and Uttarkashi, Pauri, Tehri, and Dehradun of Garhwal division. Here more than 4/5 per cent working population is directly or indirectly engaged in agricultural and forest products. In this region agriculture does not only provide food to 59,26,146 persons (according to Census 1991), but also supplies raw materials to agro-based industries. The other important occupation of this region is Horticulture, sheep-rearing, livestock farming and household cottage industries. There are seventeen important crops grown in U.P. Himalaya and these together account 98.97 per cent (1994-95) of the total cropped area. These crops are cereals, pulses, oilseeds, and cash crops grown in this region.

METHODOLOGY

An attempt has been made to analyse the trend of increase or decrease of major crops, which helps in understanding the direction of agricultural development in the study area during the period from1978-80 to1994-95. The study is based on secondary source of data. The data has been collected from the official record and Bulletin of Agricultural Statistics obtained from the Directorate of Agricultural Statistics and Crop Insurance Krishi Bhawan, Lucknow. The linear

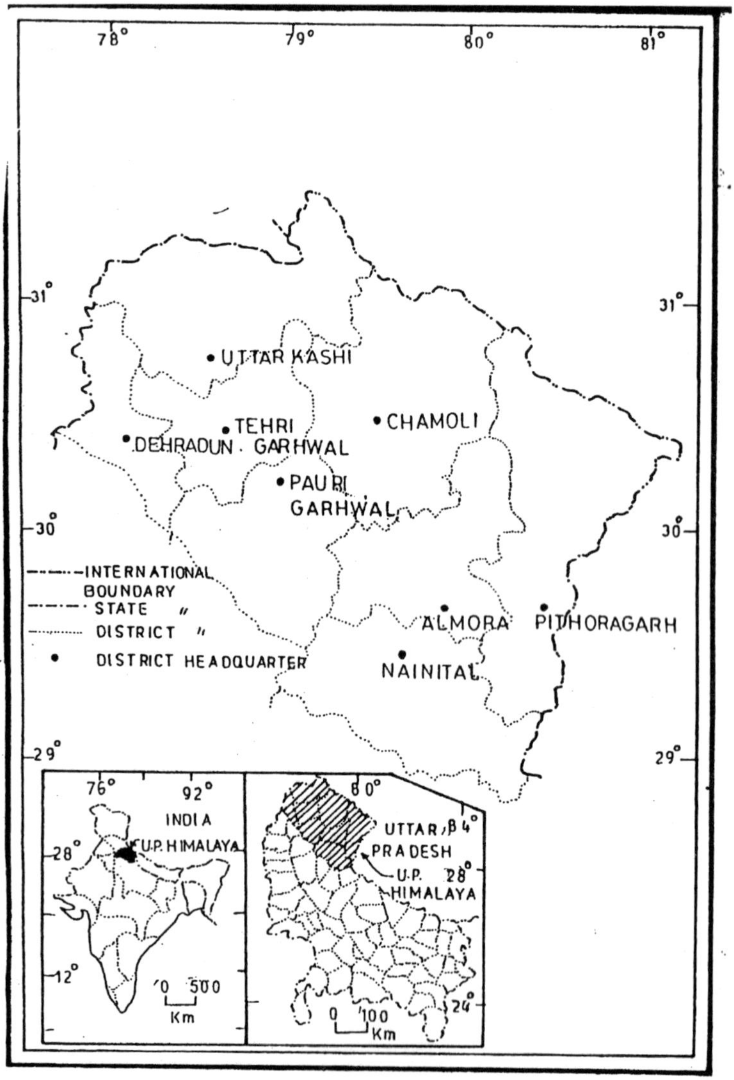

Fig. 26.1 : Uttar Pradesh Himalaya Administrative Division, 1999

growth rate of area and production of major food crops grown in U.P. Himalaya have been calculated using regression equation. It may be expressed as $y = a + bx$, where Y refers to linear growth trend model for area and production separately and X is time variable, a and b are the coefficient. The trend value thus obtained for area and production were plotted on paper and joined to depict the general trend.

LAND-USE PATTERN

Usually the land-use pattern is the manifestation of physical conditions prevailing in an area and it is evident in the district as well. The rise and fall in area available for cultivation as well as net sown area reveals the level of achievement in the field of agricultural development.

In order to follow this progress so that the study is carried on under following broad land-use categories : *(a)* forest, *(b)* land not suitable for cultivation, *(c)* culturable waste, *(d)* pasture land, *(e)* area under miscellaneous trees, *(f)* area available for cultivation, *(g)* fallow land, *(h)* net sown area, *(i)* area sown more than once, and (j) gross cropped area. Figures relating to various categories of U.P. Himalayas are given in Table 26.1 for the years 1979-80,1984-85,1989-90 and 1994-95.

During1994-95 about14 per cent of the total geographical area of U.P. Himalaya was available for cultivation but only13 per cent was devoted to crops and1.0 per cent was left fallow. About 7.5 per cent of the area were cropped more than once so that the gross cultivated area rose up to 21per cent of the total area of the region. The Usar and unculturable lands accounted for about 5.41 per cent while the area under other uses excluding cultivation covered 2.5 per cent of the total area. Forest covered 64.0 per cent while miscellaneous trees and pasture accounted for about 9.0 per cent of the region. About 6.0 per cent land was culturable waste, which can be reclaimed for agriculture in due course of time.

Difficult terrain indicates the fluctuation in the area available for cultivation. It has been seen that the cultivated area increased from14.0 per cent to 21.0 per cent during the session for 1979-80 to1984-85 but again declined to14.0 per cent during the period 1994-95. The net sown area that accounted for13.27per cent in 1979-80, which decreased to13.13 per cent in 1994-95. The area sown more than once also decreased from 9.0 per cent in 1979-80 to1994-95. There are numerous streams "Gad" gadhera in the region but very little use of them is made for irrigation due to rugged topography, difficult undulating landforms features and high cost of lift irrigation scheme. The poor management of land-use, Agricultural implements and machinery and fertilizer distribution caused the decline of net sown area in the region.

Table 26.1 : Land Utilisation in U.P. Himalayas

Area in hectares, Figures given within the brackets show per cent age from the geographical area of the whole region

	1979-80	*1984-85*	*1989-90*	*1994-95*
Reporting Areas	53,23,230	54,18,108	53,72,625	53,69,589
Forests	34,37,056 (64.57)	34,37,405 (63.44)	32,40,767 (60.31)	34,35,513 (63.98)
Usar and not suitable for cultivation	2,89,356 (5.43)	2,98,832 (5.51)	2,84,628 (5.28)	2,90,644 (5.41)
Areas under other uses excluding cultivation	1,18,817 (2.23)	1,23,760 (2.28)	1,31,035 (2.24)	1,36,655 (2.54)
Culturable waste	3,15,726 (5.93)	3,18,832 (5.80)	3,01,781 (5.62)	3,12,158 (5.81)
Permanent pastures and other grazing land	2,17,198 (4.08)	2,72,263 (5.02)	2,11,584 (3.94)	2,56,372 (4.77)
Area under miscellaneous trees	1,85,772 (3.49)	2,07,803 (3.84)	2,04,892 (3.81)	2,17,554 (4.05)
Land under fallows	33,711 (0.63)	4,52,236 (8.34)	68,091 (1,38)	70,745 (1.33)
Net sown areas	7,06,594 (13.27)	7,02,975 (12.97)	7,31,846 (13.62)	7,05,327 (13.13)
Area sown more than once	4,31,103 (9.00)	3,89,138 (7.81)	4,56,552 (8.50)	4,03,690 (7.52)
Cultivated areas	7,59,305 (14.26)	11,55,221 (21.32)	7,99,937 (14.88)	7,76,072 (14.45)
Gross cropped areas	11,37,697 (21.37)	11,52,413 (21.27)	11,07,398 (20.61)	11,26,240 (20.97)

Source : Bulletin of Agricultural Statistics (1979-80 to1994-95), Directorate of Agricultural Statistics and Crop Insurance, U.P., Krishi Bhawan, Lucknow.

MAJOR FOOD CROPS

The trend rate of area, production, yield of individual crops in the region for three annual session *i.e.* 1979-84,1984-89 and 1989-94, and overall growth from1979-80 to1994-95 is shown in Table 26.2. The year-wise trend lines of area and production of cereals, pulses, oilseeds, and cash crops have been depicted through graph. An attempt has been also made to explain the changes in. trends of growth rate in area, production, and yield of each crop in the region.

Wheat

Among individual crops wheat followed by rice, accounted for the highest growth rate in U.P. Himalaya with respect to area, production, and yield per hectares. Wheat is the leading crop and occupied an area of 367.71 thousand hectares in U.P. Himalaya and the area under its cultivation increased to 385.04 thousand hectares in 1984-85. The next alternative period, the area again increased to 391.18 thousand hectares in 1989-90 but it decreased to 371.39 thousand hectares during the period 1994-95. As against this the production of wheat continuously rose from 441.92 thousand tonnes in 1979-80 to 493.72 thousand tonnes in 1984-85 to 572.41 thousand tonnes in 1989-90 and 678.76 thousand tonnes in 1994-95. The average yield rose up from12.0 quintals per hectare in 1979-80 to12.82 quintals per hectare in 1984-85 to14.63 quintals per hectare in 1989-90 and it further increased to18.27 quintals per hectare in 1994-95.

The trend lines of area and production as plotted and depicted in Fig. 26.2 show a rising trend. It may be seen from Table 26.2 that wheat has registered a linear growth in area and production in the region by 53.59 per cent during the period from1979-80 to1994-95, due to the adoption of modern inputs in the region.

Rice

Rice is the second leading crop in this region, it covers an area of 283 thousand hectares in 1979-80, and however the area decreased to 278.18 thousand hectares in 1984-85. But it has further increased to 283.10 thousand hectares in 1989-90 and again declined to 280.41 thousand hectares in 1994-95. Although it recorded a fluctuation in area but the production and yield per hectare achieved a significant increase. The quantum of production rose from 326.38 thousand tonnes in 1979-80 to 490.64 thousand tonnes in 1984-85 to 533.99 thousand tonnes in 1989-90 and reached at the level of 551.19 thousand tonnes in 1994-95. Similarly the yield increased from11.53 quintals per hectare in 1979-80 to17.63 quintals per hectare in 1984-85 to18.86 quintals per hectares in 1989-90 and19.65 quintal per hectare in 1994-95.

The trend lines of area and production of rice shoes an upward movement. It may be evident from Table 26.2 that the area under rice decreased by 0.9 per cent while the production increased by the 68.88 per cent during the session from1979-80 to 1994-95.

Barley

The area occupied by barley was 29.97 thousand hectares in 1979-80 and it decreased to an area of 24.41 thousand hectares in 1984-85. But it further expands to 34.40 thousand hectares in 1989-90 and then declined to 28.40 thousand hectares in 1994-95. It shows a fluctuation in both area and production but the yield shows different, and decreased from11.75 quintals per hectares in 1979-80 to 9.19 quintals per hectare in 1984-85 but it again increased to11.96 quintals per hectare in 1989-90 and12.58 quintals per hectare in 1994-95 respectively.

The trend lines of area and production of barley shows an elevated pattern. The linear growth of area and production are given in Table 26.2. It may be evident from the table that the area under barley decreased by 5.23 per cent while production increased by1.48 per cent from1979-80 to1994-95.

Maize

The area under maize was 38.06 thousand hectares in 1979-80 but it continuously declined to 36.66 thousand hectares in 1984-85 to 34.20 thousand hectares in 1989-90 and 31.69 thousand hectares in 1994-95. As against this the production and yield of maize was 43.65 thousand tonnes and11.47quintals per hectare in 1979-80, which rose to 53.82 thousand tonnes and 14.68 quintals per hectare in 1984-85. But during the subsequent year the production as well as yield declined to 46.86 thousand tonnes and13.70 quintals per hectare in 1989-90 and 37.86 thousand hectares and11.95 quintals per hectare in 1994-95.

The trend lines of area and production of maize depict downward movement. It is obvious from Table 26.2 that the maize has a negative linear growth in area and production in various quinquennial periods. The area as well as production declined by16.75 per cent and 13.27 per cent during the period from1979-80 to1994-95.

Ragi (Madua)

Ragi is a third ranking crop in this region, the area under this crop during1979-80 was168.62 thousand hectares but it declined to 154.11 thousand hectares in 1984-85 and againrose to160.88 thousand hectares in 1989-90 and184.0 thousand hectares in 1994-95. The production and yield shows the same pattern of change. The production and yield of ragi continuously increased from114.36 thousand tonnes and 6.87 quintals per hectare in 1979-80 to169.0 thousand tonnes and10.96 quintals per hectare in 1984-85 to199.0 thousand tonnes and12.35 quintals per hectare in 1989-90 but it decreased to 185.0 thousand tonnes and10.04 quintals per hectare in 1994-95.

The trend lines plotted in Fig. 26.2 shows an upward scoring of area and production, it is oblivious from Table 26.2 that this crop has registered a positive linear growth both in terms of area and production in the region. The aggregate linear growth of area under ragi was 9.11 per cent while production was 61.57 per cent during1979-80 to 1994-95.

Sanwa

In the case of sanwa there have been marginal decrease in area *i.e.* the area decreased from 80.67 thousand hectares in 1979-80 to 79.89 thousand hectares in 1984-85 but it again increased to 84.69 thousand hectares in 1989-90 and 90.75 thousand hectares in 1994-95. The production and yield has improved its position continuously. The total production of sanwa recorded about 48.88 thousand tonnes in 1979-80 in the region which further increased to about 80.76 thousand tonnes in 1984-85 to 96.35 thousand tonnes in 1989-90 and108.60 thousand tonnes in 1994-95. The figures of yield quintal per hectare followed the same trend as the production.

The trend lines in Fig. 26.2 shows an upward movement. The linear growth of the area under sanwa increased by12.48 per cent and production also increased by 84.42 per cent during the study period.

Gram

The area under gram was 4.29 thousand hectares during the session 1979-80 but it decreased to 1.82 thousand hectares in 1984-85 to1.87 thousand hectares in 1989-90 and1.77 thousand hectares in 1994-95. There has also been a gradual decline in the production of gram from1.92 thousand tonnes in 1979-80 to1.48 thousand tonnes in 1984-85 to1.36 thousand tonnes in 1989-90 and further increased to1.56 thousand tonnes in 1994-95 but the yield shows fluctuation during the study period.

The trend lines plotted in Fig. 26.2 shows a downward movement. It may be seen from Table 26.2 that the linear growth rate of area under grain has a negative growth of 58.75 per cent while production has decreased by18.63 per cent during the period from1979-80 to1994-95. Like arhar, gram is also sensitive to frost during flowering time and, therefore, special attention.

Table 26.2 : Growth Rate in Area, Production, and Yield of Major Food Crops in U.P. Himalaya *(Figures in percentage)*

Crop/Year		*1979-84*	*1984-89*	*1989-94*	*1979-80 to1994*
RICE:	A	-1.70	+1.77	-0.95	-0.91
	P	+50.32	+8.83	+3.32	+68.88
	Y	+52.90	+6.97	+4.19	+70.42
WHEAT:	A	+4.71	+1.59	-5.05	+1.00
	P	+11.72	+15.93	18.58	+53.59
	Y	+6.83	+14.11	24.88	+52.25
MAIZE:	A	-3.67	-6.72	-7.34	16.75
	P	+23.29	-12.95	-19.20	-13.27
	Y	+27.99	-6.67	-12.77	+4.18
BARLEY:	A	-18.55	+40.92	-17.43	-5.23
	P	-36.24	+83.35	-13.19	+1.48
	Y	-21.78	+30.14	+5.18	+7.06
RAGI:	A	-8.60	+4.39	+14.36	+9.11
	P	+47.80	+17.54	-7.00	+61.57
	Y	+59.53	+12.68	-18.70	+46.14
SANWA:	A	-0.97	+6.00	+7.15	12.48
	P	+37.16	+19.29	+12.71	+84.48
	Y	+38.35	+12.67	+5.18	+63.97
GRAM:	A	-57.42	+2.40	-5.40	-58.75

Contd. ....

Table 26.2 *(Contd.)*

Crop/Year		*1979-84*	*1984-89*	*1989-94*	*1979-80 to1994*
	P	-22.88	-7.87	+14.53	-18.63
	Y	+81.06	-10.08	+17.65	+91.54
PEA:	A	+57.75	+7.07	+61.36	+172.53
	P	+136.58	+0.82	+33.13	+217.58
	Y	+50.15	-5.85	-17.47	+16.67
ARHAR:	A	+45.46	-4.97	-0.24	+38.15
	P	+178.24	-37.03	-30.07	+22.13
	Y	+91.36	-33.85	-30.07	-11.48
MASOOR:	A	+5.14	-3.00	-3.92	-2.02
	P	+0.71	+10.07	+20.96	+34.10
	Y	-4.12	+13.27	+25.90	+36.74
URAD:	A	+56.76	+21.83	-7.81	+76.08
	P	-1.56	+34.63	+17.02	+55.09
	Y	-37.07	+10.00	+27.27	-11.91
SESAMUM:	A	+55.27	+3.26	-13.90	+38.06
	P	-26.84	+61.83	-25.96	-12.34
	Y	-52.84	+56.89	-14.28	-36.58
MUST-	A	+33.68	-17.46	+2.35	+12.93
ARD:	P	-79.90	-18.38	+24.15	-79.63
	Y	-84.96	-1.18	+21.32	-81.96
SOYA-	A	+30.68	+19.77	+41.24	+121.07
BENE:	P	+28.95	+44.42	+9.02	+103.02
	Y	-1.37	+20.59	-22.77	-8.14
SUGAR-	A	-0.36	+29.82	-3.78	+24.46
CANE:	P.	+22.06	+70.25	-96.44	-92.60
	Y	+22.50	+31.14	-96.30	-94.06
POTATO:	A	+65.31	+14.14	+10.63	+108.95
	P	+113.80	+23.14	+9.59	+188.54
	Y	+29.32	+7.79	-0.90	+38.14
ONION:	A	+13.71	-0.90	+25.35	+41.25
	P	+16.24	-21.42	+32.46	+20.99
	Y	+2.22	-20.70	+5.67	-14.34

Peas

Pea is an important crop as compared to other pulse in U.P.

Himalaya because it uses as pulse and sell in the market as cash crop. The area and production has continuously increased from 0.85 thousand hectares and 0.56 thousand tonnes in 1979-80 to1.34 thousand hectares and1.33 thousand tonnes in 1984-85 to1.43 thousand hectares and 1.34 thousand tonnes in 1989-90 and 2.32 thousand hectares and 1.78 thousand tonnes in 1994-95. As against this the yield increased from 6.60 quintals per hectare in 1979-80 to 9.91 quintals per hectare in 1984-85. But it declined to 9.33 quintals per hectare in 1989-90 and 7.70 quintals per hectare in 1994-95, due to bad weather condition in the region.

The trend lines of area and production as plotted in Fig. 26.2 show a rising trend. It may be seen from Table 26.2 that Pea has registered a gradual linear growth in area and production in the region. The area under pea increased by172.53 per cent while production increased by 217.58 per cent during the study period.

Arhar

Arhar is also most important crop in this region, there have been fluctuations in its area, production as well as yield. The area under arhar was 1.19 thousand hectares in 1979-80 increased to 1.73 thousand hectares in 1984-85 but it declined to1.64 thousand hectares in 1994-95. Similarly the production and yield also followed the same trend as area.

The linear growth rate of arhar is given in Table 26.2 which shows that area under arhar increased by 38.15 per cent while production increased by 22.13 per cent during the period from1979-80 to1994-95.

Masoor

Masoor is a leading crop among the other pulses in this region. The area under the masoor was 15.54 thousand hectares during1979-80 and increased to16.33 thousand hectares in 1984-85. But during the remaining period it declined to 15.84 thousand hectares in 1980-90 to15.22 thousand hectares in '1994-95. As against this the production continuously rose from 8.67 thousand tonnes in 1979-80 to 8.73 thousand tonnes in 1984-85 to 9.61 thousand tonnes in 1989-90 and 11.63 thousand tonnes in 1994-95. But the yield has gradually declined from 5.58 quintals per hectare in 1979-80 to 5.35 quintals per hectare in 1984-85 but it further rose to 6.06 quintals per hectare in 1989-90 to 7.63 quintals per hectare in 1994-95.

The trend lines of area and production in Fig 26.2 show different pattern. The area shows downward while production shows an upward trend. The linear growth rate of both area and production of masoor

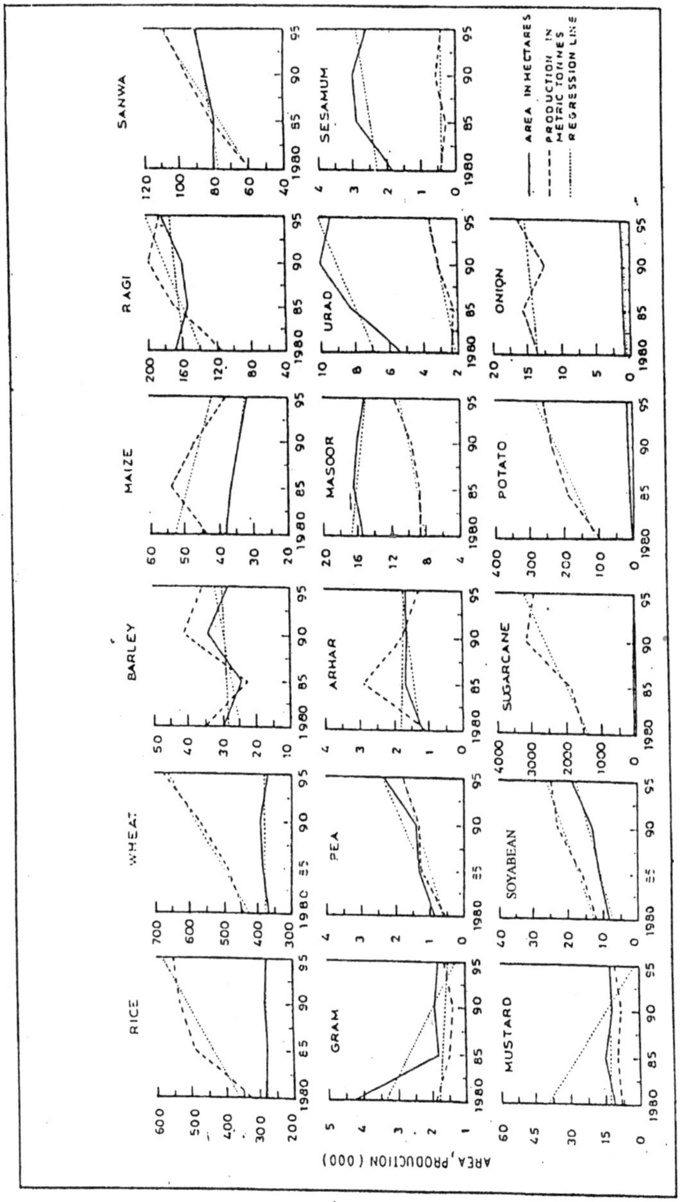

Fig. 26.2 : Uttar Pradesh Himalaya—Area and Production of Major Crops, 1980-95

has been given in Table 26.2, the area under masoor decreased by 2.02 per cent, while the production and yield increased by 34.10 per cent and 36.74 per cent during the period from1979-80 to1994-95.

Urad

Urad is the second ranking crop after masoor. The area under urad was 5.32 thousand hectares during1979-80 but it recorded an insignificant increase of 8.34 thousand hectares in 1984-85 to 10.16 thousand hectare in 1989-90 but declined to 9.37 thousand hectares in 1994-95. But the production and yield followed the same patterns. The production and yield decreased from 2.36 thousand tonnes, and 4.45 quintals per hectare in 1979-80, to 2.33 thousand tonnes, and 2.80 quintals per hectare in 1984-85, but it further expanded to 3.13 thousand tonnes, and 3.08 quintals per hectare in 1989-90, and 3.67 thousand tonnes, and 3.92 quintals per hectare in 1994-95.

The trend lines of area and production as plotted in Fig. 26.2 show an elevated trend. Table 26.2 clears that the area under urad increased by 76.08 per cent while the production increased by 55.09 per cent during the session from1979-80 to1994-95.

Sesamum

The area under sesamum has enhanced from 1.87 thousand hectares in 1979-80 to 2.91 thousand hectares in 1984-85 to 3.0 thousand hectare during1989-90 but it declined to 2.59 thousand hectares in 1994-95. Conversely the fluctuation occurs both in terms of production and yield. The trend lines of area and production as shown in Fig. 26.2, the area shows a rising trend but the trend lines of production depict a straight line. The linear growth rate of area and production of sesamum as given in Table 26.2 shows that area under sesamum increased by 38.06 per cent. As against this the production decreased by 12.34 per cent during the year from1979-80 to1994-95.

Rapeseed and Mustard

Among oilseeds, rapeseeds and mustard have achieved a leading position as they cover greater area than any other crop. The area under mustard was 11.41 thousand hectares in 1979-80 which increased to

15.25 thousand hectares in 1984-85. Later on it declined to12.59 thousand hectares in 1989-90 but it further rose to12.88 thousand hectares in 1994-95. As against this the production and yield has shown a recorded decrease from 51.12 thousand tonnes and yield 44.80 quintals per hectare in 1979-80 to10.27 thousand tonnes and yield 6.74 quintals per hectare in 1984-85 to 8.38 thousand tonnes and yield 6.66 quintals per hectare in 1989-90 and further increased to12.88 thousand tonnes and yield 8.08 quintal per hectare in 1994-95, it is caused by the unfavourable climatic conditions.

The trend lines of area and production plotted in Fig. 26.2 show downward movement. The area under mustard increased by12.93 per cent while the production decreased by 79.63 per cent during the period from 1979-80 to 1994-95.

Soyabean

Soyabean is the main source of protein and hence it occupied the important position in U.P. Himalayas. Soyabean show a significant increase both in terms of area and production, the area under soyabean increased from 8.45 thousand hectares in 1979-80 to11.04 thousand hectares in 1984-85 to13.22 thousand hectares in 1989-90 and18.68 thousand hectares in 1994-95. The production of soyabean increased from12.35 thousand tonnes in 1979-80 to15.93 thousand tonnes in 1984-85 to 23.0 thousand tonnes in 1989-90 and 25.0 thousand tonnes in 1994-95. Conversely the yield has meagre decreased from14.62 quintal per hectare in 1979-80 to14.42 quintals per hectares in 1984-85 but it again increased to17.39 quintals per hectare in 1989-90 but it further declines to13.43 quintals per hectare in 1994-95.

The trend lines of area and production as plotted in Fig. 26.2 which shows an upward movement. It may be seen from the Table 26.2 that soyabean has registered a linear growth in area and production in the region, the area under soyabean increased by103.02 per cent during the session 1979-80 to1994-95.

Sugarcane

U.P. Himalaya is not an important producer of sugarcane as compared to the sugarcane belt of western and eastern U.P. The area and production has increased in the first three subsequent years, later on declined during the last session. The area under sugarcane was increased from 38.49 thousand hectares in 1979-80 to 49.79 thousand hectares in 1989-90 but it gradually declined to 47.91 thousand hectares in 1994-95, but the quantum of production and yield rose from1508.0 thousand tonnes and 391.88 quintals per hectares in 1979-80 to1841.0 thousand tonnes and 480.0 quintals per hectares in 1984-85 to 3135.0 thousand tonnes and 629.0 quintals per hectares in 1989-90. But declined to 2925.0 thousand tonnes and 610.0 quintals per hectares during1994-95.

In Fig. 26.2, the trend lines of area shows a downward pattern but the trend lines of production shows an elevated trend.

Potato

Potato is the most famous and evergreen crops in the world and named as the "KING OF VEGETABLES" In 1979-80 the area under potatoes was 6.90 thousand hectares but in 1984-85 it rose to11.40 thousand hectares in 1989-90 to 13.03 thousand hectares and14.41 thousand hectares in 1994-95. Similarly the production has attained a continuous rise from 90.28 thousand tonnes in 1979-80 to193.02 thousand tonnes in 1984-85 to 237.68 thousand tonnes in 1989-90 and 260.49 thousand tonnes in 1994-95, and the yield also increased from 130.84 quintals per hectare in 1979-80 to169.21 quintals per hectare in 1984-85 to182.39 quintals per hectare in 1989-90 but it gradually declined to 180.75 quintals per hectare during1994-95.

The trend lines of area and production plotted in Fig.26.2 shows an elevated trend. It may be obvious from Table 26.2 that potato has recorded a positive linear growth in area and production, the area under potato expands by108.95 per cent while the production increased by188.54 per cent during the study period. '

Onion

The area occupied by onion was 0.87 thousand hectares in 1979-80 and increased to 0.99 thousand hectares in 1984-85 but it meagrely declined to 0.98 thousand hectares in 1989-90, but it further increased to1.23 thousand hectares in 1994-95. Similarly the production and yields also followed the same trend as area.

The trend lines of area and production show a rising trend. The linear growth rate of area and production of onion as shows in Table 26.2, increased by 41.25 per cent while the production increased by 20.99 per cent during the period from1979-80 to1994-95.

CONCLUSION

The continuous growth of population of U.P. Himalayas forced farmers to engage directly or indirectly in agriculture to fulfil the demands for more and more foodgrains, therefore, the farmers are compelled to pay special attention towards the implementation of their agricultural output. It is quite clear from this study that there has been a gradual increase in various foodgrains, pulses, oilseeds, and vegetables, in spite of slight decline in net sown area and gross cropped area in this region. This increase in food production is due to the advancement of modern agricultural inputs in the study area. The authors have suggested some important improvements of this region.

- Irrigation facilities should be provided in Upland and Lowland areas.
- Total cultivation of the area should be brought under multiple cropping particularly in favourable topographic areas.
- New agricultural technique should be applied in this region, particularly the use of high yielding verieties of seeds, chemical fertilizers, insecticides, pesticides, agricultural implements and machinery, reclamation of wastelands, land consolidation and soil conservation programme must be implemented in the region

REFERENCES

Aggarawal, J.C. and Aggarawal, S.P., (1995). *Uttarakhand Past, Present and Future,* Concept Publishing Company, New Delhi, pp.13-44.

Hussain, M.A. (1978). Education and Agricultural Development—A case study of Western U.P. in Mohd, Ali. *(Ed.) Dynamics of Agricultural Development in India,* Concept Publishing Company New Delhi, pp.133-39.

Mohd, Noor and Bandooni, S.K. (1992). *Monitoring of Land Use and Agriculture in a Hilly Region—A Case Study of Pauri Garhwal Region, The Geographer,* Vol. XXXIX. No. 2, July1992, pp.6-19.

Munir, Abdul. (1992). *Agricultural Productivity and Regional Development,* Manak Publication Pvt. Ltd.

Pant, B.R., Joshi, R.C. and Jalal, D.S. Agricultural Panorama of Uttrakhand, in Mohd, Noor (Ed.): *Spatial Dimension of Agriculture*, Concept Publishing Company, New Delhi, pp.81-94

Shafi, M. (1984). *Agricultural Productivity and Regional Imbalances: A Study of Uttar Pradesh*, Concept Publishing Company, New Delhi, pp. 73-91.

Siddiqui, S.H. (1987). Agricultural Productivity in the North Bihar Plain, *The Indian Geographical Journal,* Madras, Vol. 62, No. 2, pp.130-136.

Siddiqui, S.H. (1990). Trends of Agricultural Development in North Bihar Plain, *The Geographer*, Vol. XXXVII, No. 2, July-1990, pp.46-57.

Singh, O.P. (1983) *The U.P. Himalayas: An Overview*, in Singh O.P. (Ed.) *The Himalaya: Nature, Man, and Culture*, Rajesh Publication New Delhi, pp. 3-29.

27

RELATIVE DEPRIVITY* AMONG VILLAGES

Evolving a Paradigm for Measurement

AGHA M.A. SIDDIQUI

INTRODUCTION

'The objective of social justice as articulated in the Five-Year plans (of India) has two major dimensions. The first dimension is an improvement in the living standards of the poorest groups, and the second is a reduction of inequalities in asset distribution'[1]. Quite often the provision and extension of services, amenities and facilities (SAFs), more importantly roads, to villages has resulted in distinct advantaged and disadvantaged villages mainly due to the approaches followed to this end. The prevalent approaches mostly give priority to villages with a higher level of development *i.e.* villages with more SAFs. This gives an increased opportunity of development to the already developed villages forcing the less developed ones to remain deprived. Relative deprivation is a term that is commonly in use in the field of Economics, and some other disciplines of Arts. The methodology proposed in this chapter is an attempt to explore the possibility of applying this concept on villages for assigning them priorities for their development. The steps followed to this end are : (1) Developing the concept of village level deprivity; (2) Structuring a paradigm for

* The term 'deprivity' has been used as a synonym to 'deprivation'. This term was first used by the author in his paper "Factors for Determining the Levels of Deprivity of Villages—Their Selection and Delimitation", published in the *Spatio-Economic Development Record* (SDR), New Delhi, Vol. 2, No. 1, January-February, 1995.

measuring village level deprivity; and (3) Application of the paradigm on a region.

CONCEPT OF RELATIVE DEPRIVATION

Runciman[2] states that dissatisfaction with the system of privileges and rewards in a society is never felt in an even proportion to the degree of inequality its members are subjected to. Class, status and power are the three dimensions in terms of which Runciman (1966) measures social inequality. These factors in turn give rise to the feeling of relative deprivity.

Relative deprivation is a term borrowed from Social Psychology which when used with another term 'reference group' makes the discussion of social inequality clear and compact. 'The related notions of relative deprivation and reference group both derive from a familiar truism : that peoples' attitudes, aspirations and grievances largely depend on the frame of reference within which they are conceived'.[3]

A person's satisfactions are conditioned by his expectations. The proverbial way to make oneself conscious of one's advantages is to contrast one's situation with that of others worse than oneself (Runciman, 1966).

The term 'relative deprivation' was initially coined by Hyman[4] but was first used by Stouffer *et al.* (1949) in 'The American Soldier'[5]. It was Runciman who gave a detailed explanation in this regard. Let there be two persons *A* and *B*: *A* wants something which he does not have but B has, meaning thereby, *A* is relatively deprived with reference to *B*. Runciman refers to another situation where *A's* expectations are higher than *B's*, or if he was better off than *B* in the past. Under such circumstances, *A* would still have the feeling of being relatively deprived given a chance to be similarly placed as *B*. To quote him, '*A* is relatively deprived of *X* when *(i)* he does not have *X*, *(ii)* he sees some other person or persons, which may include himself at some previous or expected time, as having *X*, *(iii)* he wants *X*, and *(iv)* he sees it feasible that he should have *X*'[6].

THE CONCEPT APPLIED

The deprivation of a village differs from the deprivation of an individual in that an individual may feel deprived of what is dear to him or what he might be longing for; however for a village's deprivation, collective feelings and requirements of people count and not of an individual alone.

The definition of relative deprivation as postulated by Runciman can be modified to define and identify relative deprivation at a village level. A village V_k is relatively deprived when,

(i) the village V_k does not have an SAF, say *m*, located within its precincts;

(ii) the villagers in the village V_k cannot avail *m*, for *m* is located beyond a distance to where the village V_k can be served;

(iii) the villagers in the village V_k know that *m* is located somewhere and is available to villagers in another village;

(iv) the village V_k falls within the service area of an SAF but due to poor or no access, cannot make use of *m*;

(v) the villagers in the village V_k require and aspire for *m*; and

(vi) the villagers in the village V_k find it feasible for *m* to be located within the village V_k or at least be made available to them.

ASSESSMENT OF RELATIVE DEPRIVITY

There are several factors that help in determining the levels of development and/or deprivity. Therefore, such decisions are to be based on multiple criteria. The multiple criteria evaluation methods[7] provide a logical and flexible approach based on quantitative and qualitative criteria that may range from physical to economic, to social, characteristics. The present model is one such model based on the discrete multiple criteria. It ranks villages on their levels of development and subsequently, reverses the order to rank them on their relative deprivity.

There are several criteria on the basis of which two villages can be compared. Using criteria such as availability of drinking water, medical facility etc. only advantages can be measured. There are other criteria such as level of unemployment, pollution in water etc. by which disadvantages can be measured. There may be some decision makers who prefer to cumulate the advantages only (Non-Compensatory Preference). Some others may prefer to cumulate advantages as well as disadvantages for each village, with their respective signs (Compensatory Preference). There may be a third category of decision makers who would prefer to choose the less advantaged villages (Non-Compensatory Non-Preference). Thus there can be three alternative situations solutions to which are given in the sub-models as follows :

Non-Compensatory Preference Indices

Let there be '*K*' number of villages in a region under study which are to be ranked in terms of their levels of development by '*M*' identified criteria.

$$r_{mk} = R_{M*K}$$

such that r_{mk} represents the characteristic value of the *m*th criterion for the village *k* where r_{mk} is an element in the matrix R_{M*K}. To transform the characteristic values of different criteria into a comparable scale, the Village Characteristics matrix is normalised as,

$$r'_{mk} = \frac{r_{mk}}{\left(\sum_{k=1}^{K} r^2_{mk}\right)^{½}}$$

The preference score/weight representing the relative preference attached to the respective characteristic by a class of people or a group of experts is given by,

$$W = [w_1, w_2, w_3, \ldots, w_m],$$

where w_m represents the preference score for the *m*th characteristic as stated by the class of experts. These scores are standardised by the following equation :

$$w'_m = \frac{w_m}{\sum_{m=1}^{M} w_m}$$

Thereafter the Non-Compensatory Preference (NCP) C_{K*K}.

indices indicating the aggregate weightage of those characteristics for which the village k has an edge over village k' are computed as follows,

$$\underset{k \neq k'}{c_{kk'}} = \sum_{m=1}^{M} (w'_m \, c'_m kk') \qquad \text{(for all } k \text{ and all } k')$$

where,

$$c'_m kk' = \frac{[r'_{mk} - r'_m(\min)]}{\sum_{k=1}^{K} [r'_{mk} - r'_m(\min)]}$$

if $\{(r'_{mk} * e_m) \geqslant (r'_{mk'} * e_m)\}$ for all M criteria

otherwise,

$c'_m kk' = 0$
$r'_m(\min)$ = minimum standardised village characteristic in value for criterion m
e_m = —1 when smaller values of village characteristics (*i.e.* unemployment, water pollution etc.) are preferred
e_m = 1 when higher values of the villages characteristics (*i.e.* agricultural yield/ medical facility etc.) are preferred
$c_{kk'}$ always lies between 0 and 1, that is,
$0 \leq c_{kk'} \leq 1$

The NCP indices reflect the relative preference of the village k with respect to k' without compensating for the disadvantages of k with respect to k'.

Finally the aggregate non-compensatory (NCP) differential dominance (ADNCPD) (ar_k) of a village k with respect to all other villages by NCP values, is computed as,

$$ar_k = \sum_{\substack{k=1 \\ k \neq k'}}^{K} c_{kk'} - \sum_{\substack{k=1 \\ k \neq k'}}^{K} c_{k'k}$$

The ADNCPD vector AR_k where $ar_k \in AR_k$, represents the relative levels of development of the'villages. On the basis of ar_k (s) the villages can be ranked in order of their levels of development as,

$$k_1 \geq k_2 \geq k_3 \geq \ldots \geq k_i \geq k_m$$

The ranking / ordering of villages achieved by ADNCPD is in the order of villages' descending levels of development. Reversing this order would give the order of ascending levels of deprivity of villages. That is, the most deprived village falls on the right-most side of the order :

$$k_1 \leq k_2 \leq k_3 \leq \ldots \leq k_i \leq k_m$$

Compensatory Preference Indices

The elements of the Compensatory Preference (CP) matrix CC_{K*K} are called the CP indices $cc_{kk'}$. These indicate the aggregate compensated weightage of the class of villages.

$$\underset{k \neq k'}{cc_{kk'}} = \sum_{m=1}^{K} (w'_m \, cc'_m kk') \qquad \text{(for all } k \text{ and all } k')$$

where,

$$cc'_m kk' = \frac{| r'_{mk} - r'_m (\min) |}{\sum_{m=1}^{K} | r'_{mk} - r'_m (\min) |}$$

if $\{(r'_{mk} * e_m) \geq (r'_{mk'} * e_m)\}$ for all M criteria

otherwise,

$$cc'_m kk' = \frac{| r'_{mk} - r'_m (\min) |}{\sum_{k=1}^{K} | r'_{mk} - r'_m (\min) |}$$

$cc_{kk'}$ always lies between -1 and +1, that is,

$$-1 \leq cc_{kk'} \leq 1$$

The CP indices reflect the relative preference of the village *k* with respect to village *k*' taking all the criteria into consideration and compensating for one another.

Thereafter it is possible to compute the aggregate CP differential dominance (ADCPD) vector RR_k where $rr_k \in RR_k$ as,

$$rr_k = \sum_{\substack{k=1 \\ k \neq k'}}^{K} cc_{k'k} \quad - \quad \sum_{\substack{k=1 \\ k \neq k'}}^{K} cc_{k'k}$$

The ADCPD vector RR_k where $rr_k \in RR_{k'}$^ represents the relative levels of development of the villages. On the basis of rr_k(s) the villages can be ranked in order of their levels of development as,

$$k_1 \geq k_2 \geq k_3 \geq \ldots \geq k_i \geq k_m$$

Reversing this order would give the order of ascending levels of deprivity of villages.

$$k_1 \leq k_2 \leq k_3 \leq \ldots \leq k_i \leq k_m$$

Non-Compensatory Non-Preference Indices

The Non-Compensatory Non-Preference (NCNP) indices ($d_{kk'}$)s measure the degree to which a village *k* is worse than another village *k*′ for some criterion. The crude NCNP index $d_{kk'}$ between villages *k* and *k*′ is obtained as,

$$d_{kk'} = \text{Max} \, (\mid r'_{mk} \, w'_m - r'_{mk'} \, w'_m \mid D'_m kk')$$
$$k \neq k'$$

for all *M* criteria (for all *k* and all *k*')

where,

$$D'_m kk' = 1 \text{ if } \{(r'_{mk} * e_m) \leq (r'_{mk} * e_m)\}$$

for all *M* criteria

otherwise,

$D'_m kk' = 0$

Thereafter the crude NCNP indices are standardised as,

$$d^*_{kk'} = \frac{d_{kk'}}{\Delta}$$

where,

$$\Delta = \text{Max}\ (d_{kk'}),\ k = 1, 2, 3, \ldots, K$$
$$k' = 1, 2, 3, \ldots, K$$

$d_{kk'}$ = crude NCNP index between village k and k'

$d^*_{kk'}$ = standardised non-preference index between k and k'

= maximum crude NCNP index among all pairs of villages

$D'_m kk'$ = constant having values of 0 and 1

$d^*_{kk'}$ always lies between 0 and 1, that is,

$$0 \leq d^*_{kk'} \leq 1$$

Subsequently, the aggregate differential dominance by the NCNP values (ADNCNPD) is computed which represents the level of development of village k with respect to all other villages and is defined as,

$$f_k = \sum_{\substack{k=1 \\ k\neq k'}}^{K} d_{kk'} - \sum_{\substack{k=1 \\ k\neq k'}}^{K} d_{k'k}$$

A negative value of f_k indicates a relative advantage of village k over all other villages.

Banking by Intersection

For each village k with respect to another village k', three sets of aggregate differential dominance relationships (*viz*, ADNCPD, ADCPD, and ADNCNPD) are obtained. The intersection of these sets of rankings takes care of ambiguities/ discrepancies that may remain, and represents a smaller set of relationships for a village as,

$$k_1 \geq k_2 \geq k_3 \geq \ldots \geq k_i \geq k_m$$

Obversing this order, the most deprived village falls on the right-most side of the order.

$$k_1 \le k_2 \le k_3 \le \ldots \le k_i \le k_m$$

DETERMINING THE MULTI-CRITERIA

In this chapter, the Delphi technique[8] has been adopted for the selection of criteria. Views of a chosen group of experts in the field of rural development have been recorded on criteria that should be considered for determining the levels of deprivity of villages so that priorities can be assigned to villages for the provision of SAFs and thereby help promote their development. This survey has been conducted in two rounds.

The experts classify criteria into four categories of importance—extremely important, moderately important, important, and not important. In order to quantify the importance assured by the group of experts to the criteria, a Psychometric scaling technique, called the 'Law of Categorical Judgement'[9] has been used. These quantified values are used as weightages while assigning priorities to villages for road links.

The fifteen criteria that have been finally selected for ranking the villages in terms of their relative deprivity, are : drinking water, electricity, agriculture service centre, market facility, education facility, medical facility, post and telegraph and telecommunication facility, number of marginal workers, agricultural yield, bank, small/ rural/ household industry, veterinary centre, fire service, level of unemployment, and level of accessibility.

The number and weightage to criteria shall vary depending upon the perception of the class of respondents chosen as experts for the interview. The criteria can be determined without going to experts if such criteria have already been tested and accepted.

APPLYING THE PARADIGM

This section deals with the application of the model on Khair tehsil, Aligarh district, Uttar Pradesh, India (Fig. 27.1) that is comprised of 281 villages (Fig. 27.2). The names of villages alongwith their location code numbers are given in Appendix 27.1.

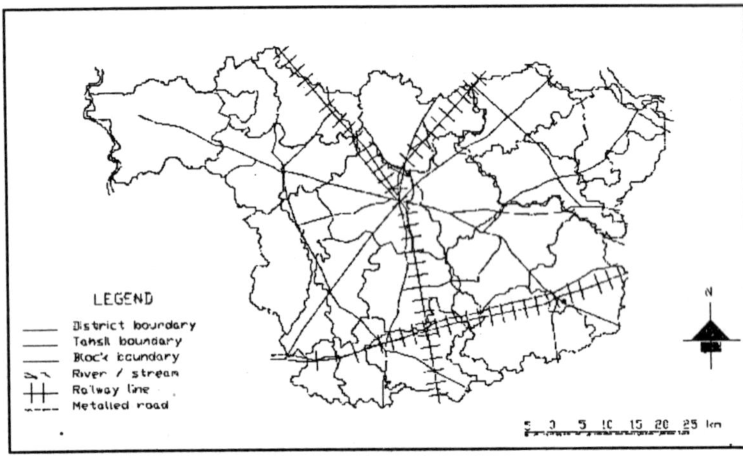

Source : District Census Handbook, Aligarh

Fig. 27.1 : Location Map of District Aligarh

There are norms for determining the location of SAFs as prescribed by the Planning Commission of India.[10] The service area of an SAF and the population served by it depend upon its hierarchy/ order.

A village may fall within the serviceable radius of an SAF but may still remain deprived of it due to the non-availability of an all-weather road (AWR) connection. Therefore, the service area of an SAF is constrained and confined by the non-existence of AWR connections to villages. The hatched portion shown in Fig. 27.3, is the area remaining unserved due to the non-existence of an AWR. Taking this into consideration the service area of SAFs becomes confined.

For the purpose of applying the model, a computer program has been written in 'C' programing language. The ranking of villages on the basis of their relative levels of deprivity arrived at by using the intersection of three sets of ADNCPD, ADCPD, and ADNCNPD rankings has been given in Appendix 27.2 wherein village V_3 appears as the most deprived village. Village V_{35} is the next highest deprived village and so on. Village V_{97} appearing last in the ranking is the most developed or the least deprived village.

CONCLUSION

A modest attempt has been made in this Chapter in developing a

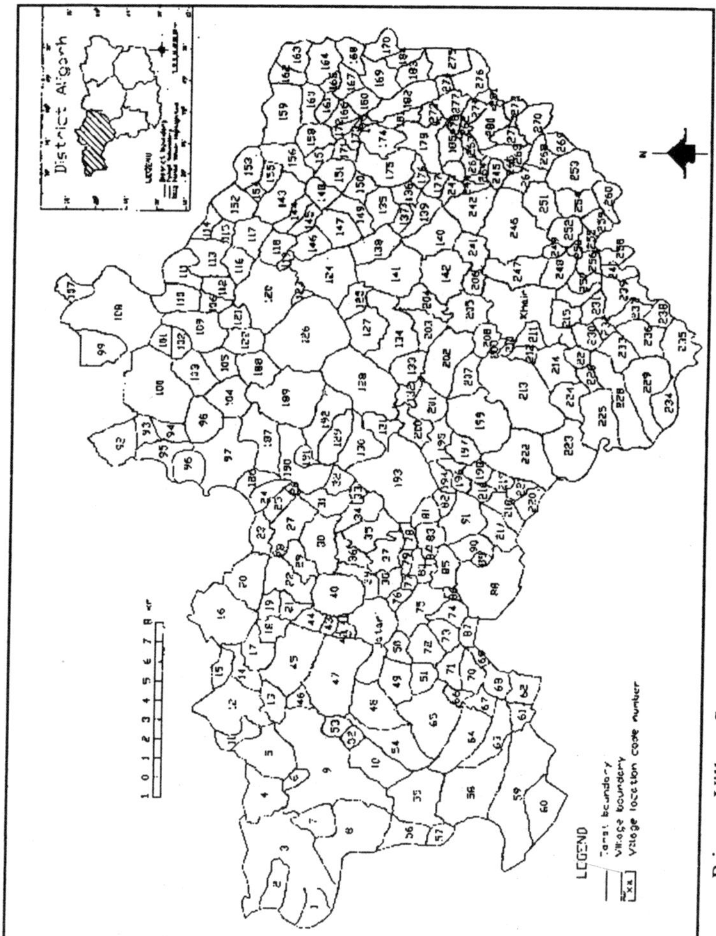

Source : Primary Village Survey Aug.-Oct., 1991 and District Census Handbook, Aligarh

Fig. 27.2 : Tahsil Khair—District Aligarh Region for Application of the Model Village with Location Code Number

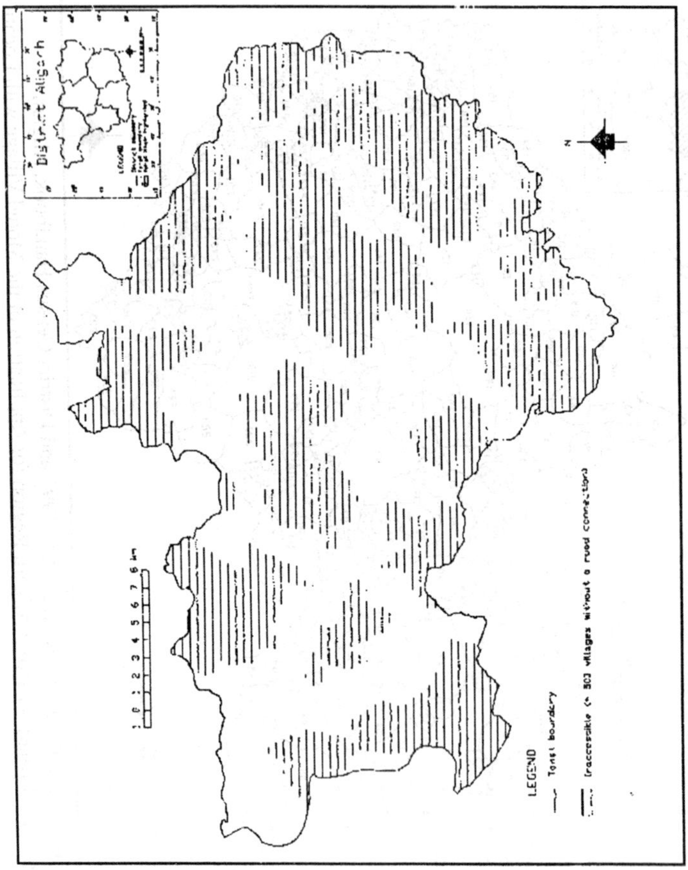

Source : Primary Village Survey Aug.-Oct. 1991, and District Census Handbook, Aligarh.

Fig. 27.3 : Tahsil Khair—District Aligarh Region for Application of the Model Unconnected Villages with + 500 Population

paradigm for determination of village deprivity on the basis of which governments can consider providing development oppportunities to the poor and disadvantaged villages.

The paradigm developed for determining village deprivities may assign rankings to villages that may differ between sub-models (NCP, CP, NCNP, or intersection of these). These inconsistencies are the result of the conditions considered in these indices. In cases where inconsistencies are pronounced/ complementary analyses in the form of differential dominance and the aggregate differential dominance have been suggested. These dominance analyses may not be required when the earlier indices of NCP, CP and NCNP offer consistent rankings. The extent of inconsistencies may vary from problem to problem; however if sufficient justification exists to adopt any of the proposed sub-models the inconsistencies may not appear.

The application of the proposed discrete multi-criteria paradigm would be more in demand in backward and disadvantaged regions where opportunities for socio-economic development/upliftment of poor communities are to be floated.

NOTES

1. Sundaram, I. S. (1984), *Anti-poverty Rural Development in India,* D.K. Publications, Delhi, p 11.
2. Runciman, W.G. (1966), *Relative Deprivation and Social Justice : A Study of Attitudes to Social Inequality in Twentieth Century England,* Routledge and Kegan Paul, London.
3. *Ibid.,* p. 9.
4. Hyman, H. H. (1942), *The Psychology of Status,* Archives of Psychology, No. 269, New York.
5. Stouffer, S. A. *et al.* (1949), *The American Soldier, I : Adjustment During Army Life,* Princeton.
6. Runciman, *Ibid.,* p. 10.
7. Nijkamp, P. and Delft, A. V. (1977), *Multi-Criteria Analysis and Regional Decision-making,* Studies in Applied Regional Science, Netherlands.
8. Linstone, H. A. and Turoff, M. (eds.) (1975), *The Delphi Method : Techniques and Applications,* Addison-Wesley Publishing Company, Massachusetts, p. 3.
9. Hays, W. L. (1969), *Quantification in Psychology,* Prentice-Hall of India Pvt. Ltd., New Delhi.
10. Planning Commission, Sixth Five Year Plan, 1980-1985.

APPENDIX 27.1

VILLAGES AND THEIR LOCATION CODE NUMBER

1.	Shamaspur	2.	Fazilpur Kachchh*
3.	Gharbara	4.	Hamidpur
5.	Siaraul	6.	Dorpuri*
7.	Adampur	8.	Lalpur Raiyatpur
9.	Tappal	10.	Jahangarh
11.	Burhaka	12.	Vaina
13.	Atari	14.	Manpur
15.	Vijaitha*	16.	Maur
17.	Rasulpur	18.	Khera Kishan
19.	Nagla Dada	20.	Basera
21.	Lalgarhi	22.	Palsera
23.	Bhadiyar	24.	Bhojaka
25.	Kadirpur Karah	26.	Pipali Nagla Kadirpur
27.	Shadipur	28.	Nagla Kirpa
29.	Bijna Ki Naglia	30.	Damuaka
31.	Pempur	32.	Balanpur
33.	Hinsel	34.	Bulakipur
35.	Kheria Buzurg	36.	Jalalpur
37.	Bichpuri	38.	Chhajjupur
39.	Itwarpur	40.	Usarah Rasulpur
41.	Sotipura	42.	Nagal Khurd
43.	Nagal Kalan	44.	Niguna Siguna
45.	Bajota	46.	Garhi Surajmal
47.	Khandeha	48.	Kurana
49.	Taharpur	50.	Hasanpur Jarailiya
51.	Hetalpur	52.	Zikarpur
53.	Kirpalpur	54.	Kansera
55.	Narwari	56.	Sherpur
57.	Girdharpur	58.	Malau
59.	Untasani	60.	Pakhodana
61.	Jaidpur	62.	Ghangholi
63.	Utwara	64.	Palar
65.	Simrothi	66.	Maror Garhi

67. Dewaka
68. Gorola
69. Naglia Gorola
70. Fojuaka
71. Bajeedpur
72. Madak
73. Salemapur
74. Shahnagar Sorola
75. Jartoli
76. Hansukh Ki Naglia
77. Bairamganj
78. Kamalpur
79. Bajhera
80. Bharatpur
81. Takipur
82. Lohlari
83. Hajiapur
84. Baluapur
85. Karanpur
86. Kalyanpur
87. Kheria Khurd
88. Salpur
89. Milk Jrana*
90. Jarana
91. Keelpur
92. Dharmpur
93. Rakrana
94. Jalokhri
95. Shahpur
96. Raopur
97. Pisawa
98. Roopnagar
99. Jalakseru
100. Deta Khurd
101. Majupur
102. Balrampur
103. Deta Shedpur
104. Dargwan
105. Sabalpur
106. Badhiyana
107. Dilwarpur
108. Kaseru
109. Sudeshpur
110. Teekri Bhuwapur
111. Baharpur
112. Surajpur
113. Bhojpur Gaiyanpur
114. Bhogpur
115. Bhagatpur
116. Tajpur
117. Jahrana
118. Ramnagar
119. Shakipur*
120. Chandaus
121. Jonhera
122. Allampura
123. Jalakha
124. Rampur Shahpur
125. Anjna
126. Nagla Padam
127. Gangal
128. Umri
129. Bajhera
130. Sahajpur
131. Ganeshpur
132. Nagla Assu
133. Dhariypur
134. Jamaka
135. Hinsel
136. Chandner
137. Aramgarhi*
138. Thanpur Khanpur
139. Udaipur
140. Barka

141. Bisara
142. Bamati
143. Amritput
144. Ogipur
145. Singhpur
146. Cheewanpur
147. Nababpur
148. Imlabara
149. Nagla Sarua
150. Marhki
151. Rasidpur Gorba
152. Kinhua
153. Pahawati
154. Bhagwanpur
155. Nurullahpur
156. Deoroo Chandpur
157. Deopur
158. Somna
159. Madhola
160. Panihawar
161. Nagla Nattha
162. Hursena
163. Arjunpur
164. Morahna
165. Shyampur
166. Rampur
167. Gabhana
168. Lalpur
169. Ogar Nagla Raju
170. Gwalra
171. Khizarpur
172. Khempur
173. Bhimpur
174. Maharajpur
175. Birpura
176. Goroth
177. Karanpur
178. Jakhota
179. Raipur
180. Kanhoi
181. Hasanpur
182. Pairai
183 Chuharpur
184. Lohput
185. Khyamai
186. Gondoli
187. Madha Habibpur
188. Cheeti
189. Mahgora
190. Mirpur Dhonra
191. Raipur
192. Diwa Hamidpur
193. Rajpur
194. Narayanpur
195. Sofa
196. Mohsanpur
197. Khera
198. Palawiran
199. Sujanpur
200. Resari
201. Rooppur
202. Bankner
203. Mau Ka Nagla
204. Kiratpur
205. Lakshmanpur
206. Udaigarhi
207. Bilkhora
208. Raipur
209. Manpur Khurd
210. Chamar Naglia
211. Sangrampur
212. Nagola
213. Gomat
214. Kasison

215. Birol
216. Dhumara
217. Garhi Nagla Shivram
218. Mathna
219. Manpur Kalan
220. Nivsani
221. Madanpur
222. Shiwala
223. Ahiarola
224. Sajna
225. Fatehgarhi
226. Sabhapur
227. Sawai Ragunathpur
228. Aidalpur
229. Kherasattu
230. Nyela
231. Bamni
232. Bhanera
233. Bhagwangarhi
234. Bhareri
235. Arni
236. Palachand
237. Nagla Veerua
238. Nisuja
239. Usaram
240. Bhogpur
241. Ainchna
242. Utwara
243. Bhamrola
244. Neemkhera
245. Sahroi
246. Jarara
247. Rasulpur
248. Lohgarhi
249. Nayabas
250. Bajhera
251. Pipalgaon urf Tarora
252. Arrana
253. Andla
254. Chaudhana
255. Mangola
256. Kunwarpur
257. Gondoli
258. Janera
259. Balipur
260. Tehra
261. Nahrola
262. Fatehpur
263. Basain
264. Ismailpur
265. Bhanoli
266. Kheria Jhallu
267. Biharipur
268. Kakola
269. Haibatpur
270. Akarawat
271. Mukatpur
272. Nandpur Pala
273. Komla
274. Chaumuha
275. Palasallu
276. Korah Rustampur
277. Kadoli
278. Birampur
279. Jagatpur
280. Rait
281. Sujapur

---

* Uninhabited as in 1981.

APPENDIX 27.2

ORDERING OF VILLAGES ON THEIR RELATIVE DEPRIVITY

$V_{3} > V_{35} > V_{202} > V_{178} > V_{223} > V_{158} > V_{190} > V_{88} > V_{87} > V_{191} > V_{129} > V_{130}$
$> V_{203} > V_{134} > V_{205} > V_{128} > V_{15} > V_{2} > V_{6} > V_{137} > V_{119} > V_{89} > V_{31} >$
$V_{105} > V_{265} > V_{264} > V_{255} > V_{250} > V_{279} > V_{220} > V_{53} > V_{14} > V_{43} > V_{42} > V_{8}$
$> V_{7} > V_{51} > V_{44} > V_{55} > V_{280} > V_{50} > V_{278} > V_{60} > V_{18} > V_{226} > V_{92} > V_{56}$
$> V_{32} > V_{138} > V_{181} > V_{272} > V_{36} > V_{204} > V_{157} > V_{173} > V_{124} > V_{260} > V_{135}$
$> V_{74} > V_{39} > V_{37} > V_{33} > V_{224} > V_{212} > V_{161} > V_{269} > V_{95} > V_{136} > V_{115} >$
$V_{216} > V_{281} > V_{259} > V_{94} > V_{63} > V_{242} > V_{225} > V_{234} > V_{165} > V_{96} > V_{21} >$
$V_{227} > V_{208} > V_{241} > V_{270} > V_{214} > V_{221} > V_{237} > V_{69} > V_{155} > V_{5} > V_{28} >$
$V_{13} > V_{12} > V_{45} > V_{1} > V_{114} > V_{174} > V_{131} > V_{271} > V_{117} > V_{116} > V_{162} > V_{151}$
$> V_{267} > V_{164} > {}_{150}C > V_{110} > V_{149} > V_{219} > V_{152} > V_{.206} > V_{133} > V_{93} > V_{266}$
$> V_{261} > V_{198} > V_{148} > V_{192} > V_{58} > V_{52} > V_{17} > V_{23} > V_{163} > V_{3\ 0} > V_{184} >$
${}_{231} > V_{185} > V_{113} > V_{147} > V_{243} > V_{268} > V_{112} > V_{85} > V_{127} > V_{170} > V_{258} >$
$V_{111} > V_{218} > V_{34} > V_{146} > V_{228} > V_{86} > V_{160} > V_{251} > Y_{244} > V_{240} > V_{168} >$
$V_{49} > V_{263} > V_{257} > V_{273} > V_{154} > V_{103} > V_{182} > V_{229} > V_{100} > V_{142} > V_{245}$
$> V_{222} > V_{141} > V_{48} > V_{262} > V_{246} > V_{54} > V_{247} > V_{233} > V_{197} > V_{122} > V_{16}$
$> V_{59} > V_{46} > V_{57} > V_{10} > V_{41} > V_{183} > V_{256} > V_{176} > V_{76} > V_{61} > V_{99} > V_{211}$
$> V_{83} > V_{156} > V_{210} > V_{196} > V_{38} > V_{230} > V_{215} > V_{90} > V_{172} > V_{78} > V_{4} >$
$V_{199} > Vl_{71} > V_{64} > V_{75} > V_{252} > V_{62} > V_{166} > V_{81} > V_{70} > V_{132} > V_{71} > V_{209}$
$> V_{213} > V_{254} > V_{217} > V_{67} > V_{232} > V_{195} > V_{91} > V_{68} > V_{248} > V_{207} > V_{153}$
$> V_{118} > V_{47} > V_{40} > V_{193} > V_{98} > V_{73} > V_{253} > V_{108} > V_{179} > V_{276} > V_{200} >$
$V_{77} > V_{249} > V_{239} > V_{236} > V_{80} > V_{107} > V_{66} > V_{177} > V_{159} > V_{79} > V_{238} >$
$V_{235} > V_{194} > V_{180} > V_{275} > V_{102} > V_{274} > V_{169} > V_{139} > V_{65} > V_{84} > V_{201} >$
$V_{24} > V_{22} > V_{188} > V_{123} > V_{106} > V_{101} > V_{167} > V_{125} > V_{277} > V_{121} > V_{25} >$
$V_{29} > V_{20} > V_{144} > {}_{187} > V_{109} > V_{104} > V_{27} > V_{26} > V_{186} > V_{19} > V_{11} > V_{9} >$
$V_{175} > V_{143} > V_{140} > V_{126} > V_{82} > V_{189} > V_{145} > V_{72} > V_{120} > V_{97}.$

28

REGIONAL DISPARITIES IN AGRICULTURAL PRODUCTIVITY

A Study of Uttar Pradesh, India

ABDUL MUNIR and MOHD. TAUFIQUE

Agricultural productivity is generally used to express the power of agriculture in a particular region to produce crops without regard to that power is due to bounty of nature or to the efforts of man. Measurement and evaluation of agricultural productivity forms a basis for improving productivity at various levels. Agricultural productivity may be defined as the ratio of the index of total agricultural output to the index of total input used in farm production. It is, therefore, a measure of the efficiency with which inputs are utilized in production, other things being equal.

The present study concentrates mainly for a detailed study of variations in agricultural productivity in all the districts of Uttar Pradesh. District-wise data have been taken from the Directorate of Economics and Statistics, Planning and Development Department, Government of Uttar Pradesh. The study has revealed that Western Uttar Pradesh leads in agricultural productivity in respect of all the crops than other regions of the State.

INTRODUCTION

Progress in agriculture also leads to the modernization and social development and better level of living through education and generating propensity to consume other than agricultural goods. The productivity can be increased without increasing production by employing less input for the same production level. But it is commonly agreed that productivity is the ability of a production system to produce

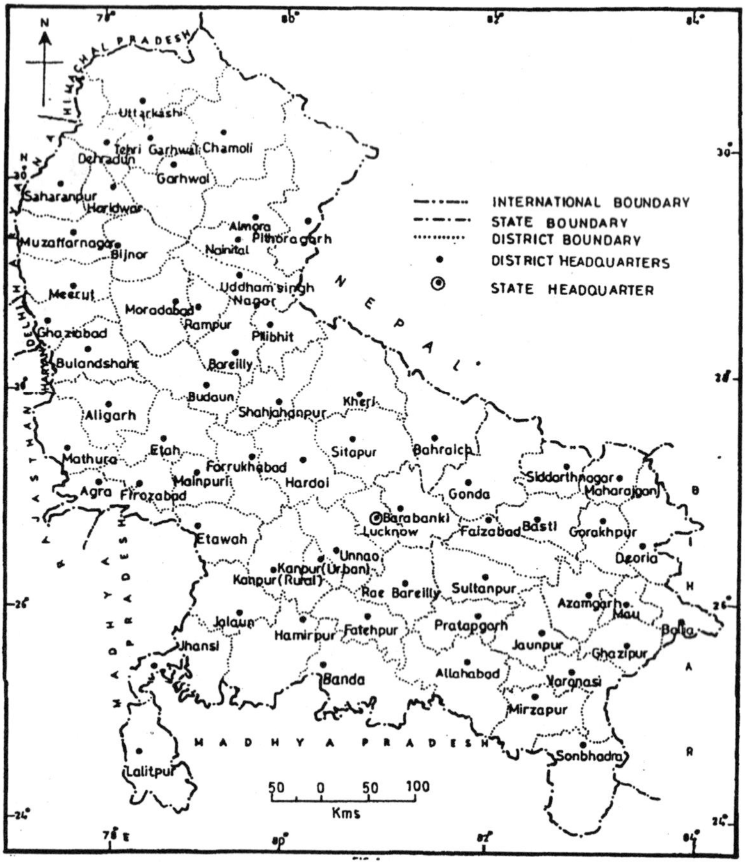

Fig. 28.1 : Uttar Pradesh Administrative Divisions, 1996

more economically and efficiently. Although the modem technology has brought about a great revolution in the levels of production of different regions, yet are areas which are still backward and the farmers are not in position to obtain, two square meals a day.

STUDY AREA

Uttar Pradesh is located in the north-eastern part of India between 23° 52' N to 31° 28' N latitude and 77° 04' E to 84° 38' E longitude, covering an area of 2,94,411 sq. km (9.65 per cent of India's total area) Fig. 28.1. It is the most populous State of India, support a population of 139 million (16.43 per cent of India's population), according to 1991 census. Most of State's area (83.70 per cent) is covered by the fertile alluvial soil deposited by the perennial rivers of Ganga, Yamuna, Gomti, Ghaghra, and its tributaries. The State enjoyed sub-humid climate with three well-marked seasons: summer, rainy, and winter. Agriculture engaging more than three-fourths of the State population directly or indirectly is the main occupation of the people and it provides more than 50 per cent of State's income.

For analysis of agricultural productivity in study area, nineteen major crops, occupying 89.81 per cent of cropped area have been taken. They have been grouped into four categories, *viz.*; Cereals (rice, wheat, maize, jowar, bajra, and barley) accounting for 64.99 per cent of the cropped area; Pulses (urad, lentil, gram, peas, arhar, and moong) with 10.39 per cent; Cash crops (sugarcane, potatoes and onion) with 9.62 per cent; and Oilseeds (mustards, sunflower, sesamum and groundnut) contributes to 4.81 per cent of the total cropped area.

METHODOLOGY

The present study is based on secondary sources of data. The required data for the year 1996-97 has been obtained from the Directorate of Economics and Statistics, Planning and Development Department, Government of Uttar Pradesh. District has been taken as the unit of study. The productivity indices of crops considered for each district were calculated as the statistical technique formulated by Yang and as explained by the author with an example of District Aligarh in Table 28.1 The index represents the yield of all crops in a district compared with the average crop yield of the State.

Before calculating the crop yield index for a particular district the average yield of each of the crops grown is determined. Later on a value in percentage is obtained by dividing the yield per hectare of a crop in the particular district by the average yield of the crop in the entire region. This value gives the index number as shown in col. 5 of Table 28.1. By considering the area devoted to each crop as a weight and multiplying this percentage index, the products are obtained as listed in col. 6 of Table 28.1. By adding the products and dividing the sum of the products by the total cropped area in the district (the sum of col. 4), the average index obtained is the crop index for the particular district, using crop area as weight.

Table 28.1 : Methods for calculating crop productivity index of Cereals for District Aligarh

Name of crop	*Yield in quintal per hectare Average Yield in the State*	*Yield in quintal per hectare Yield in the District*	*Area of crop in the district in hectares*	*Crop Yield in the District as % of the State* $\frac{col.3}{col.2} \times 100$	*Percentage multiplied by area in hectares (col.5×col.4)*
1	2	3	4	5	6
Rice	19.60	13.95	20743	71.17	1476279
Wheat	25.45	32.08	248773	126.05	31357836
Maize	13.25	17.34	53481	130.86	6998523
Jowar	8.52	5.57	166	65.37	10851
Bajra	11.02	16.67	97618	151.27	14766674
Barley	19.73	34.52	32926	174.96	5760732
Total	—	—	454707	—	60370898

Note: Crop Index for the District Aligarh $= \frac{60370898}{454707} = 132.76$

PATTERNS OF AGRICULTURAL PRODUCTIVITY

The State has highly diversified cropping pattern where the selected crops of cereals, pulses, cash crops and oilseeds occupy an important place amongst the crops of the State both in terms of acreage as well as production. On the basis of productivity indices, all the district of Uttar Pradesh have been grouped into three categories (High, Medium, and Low) for cereals, pulses, cash crops and oilseeds

Productivity Region : Cereals

Cereals, the most significant crops acquire very important position in the agriculture of Uttar Pradesh. They occupy 1,66,25,182 hectares of cultivated area which account for about 65 per cent of the total cropped area of the region. The high productivity regions cover about 37 per cent of the total area of the region, while the medium and low productivity regions are shared by about 54 per cent and 9 per cent respectively of the total area of the region. Productivity regions for

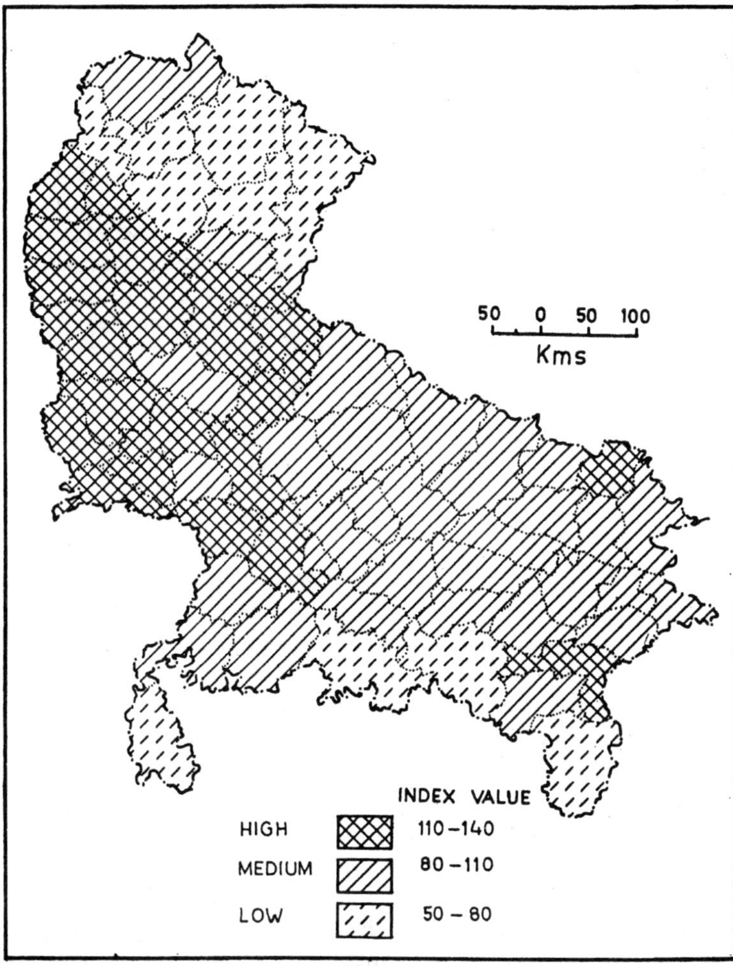

Fig. 28.2 : Uttar Pradesh Productivity Regions, 1996-97, Cereals

cereals have been shown in Fig. 28.2, whereas the number of districts in each category is given in Table 28.2.

The concentration of high productivity lies only in western part of the State, except two districts (Maharajganj and Varanasi—which are located in eastern U.P) with the productivity indices of 110-140. The high productivity of cereals in the region is due the impact of green revolution, availability of irrigation facilities, adequate amount of fertilizer and HYV of seeds. The whole of eastern and central U.P.

Table 28.2 : Levels of Productivity Regions: CEREALS

Category	*Indices*	*No. of Districts*	*Name of the Districts*
High	110-140	23	Haridwar, Saharanpur, Muzaffarnagar, Meerut, Bulandshahr, Ghaziabad, Aligarh, Mathura, Agra, Firozabad, Etah, Bareilly, Shahjahanpur, Pilibhit, Bijnaur, Moradabad, Rampur, Farrukhabad, Etawah, Kanpur (Rural), Varanasi, Maharajganj and Uddhamsingh Nagar.
Medium	80 -110	31	Mainpuri, Budaun, Kanpur (Urban), Fatehpur, Pratapgarh, Jhansi; Jalaun, Hamirpur, Mirzapur, Ghazipur, Gorakhpur, Deoria, Basti, Siddarthnagar, Azamgarh, Mau, Jaunpur, Ballia, Lucknow, Unnao, Rae Bareilly, Sitapur, Hardoi, Kheri, Faizabad, Gonda, Bahraich, Sultanpur, Barabanki, Nainital and Uttarkashi.
Low	50-80	10	Allahabad, Lalitpur, Banda, Sonbhadra, Almorah, Pithoragarh, Chamoli, Tehri Garhwal, Garhwal and Dehradun.

comes under the category of medium productivity region with the indices ranging between 80-110. Only ten districts of the State have low productivity indices (50-80), mostly concentrated in U.P. Himalayas

Productivity Region: Pulses

In our predominantly starchy vegetarian diet, pulses form a very important part as they provide us vegetable protein. Pulse crops include a large number of crops, but here only six major crops (urad, lentil, gram, peas, arhar, and moong) have been taken into consideration. They occupy 26,60,129 hectares of cultivated land. The production as well as productivity of pulses in U.P. constitutes a matter of concern. The low yield of pulses shows that it has not been benefited by the modern inputs. It has also been observed that the area of pulses is replaced for the production of cash crops in the study region. The high productivity region enjoys about 20 per cent of the total area to the region, whereas the medium and low productivity region comprise 62 per cent and 18 per cent respectively of the total area of the region.

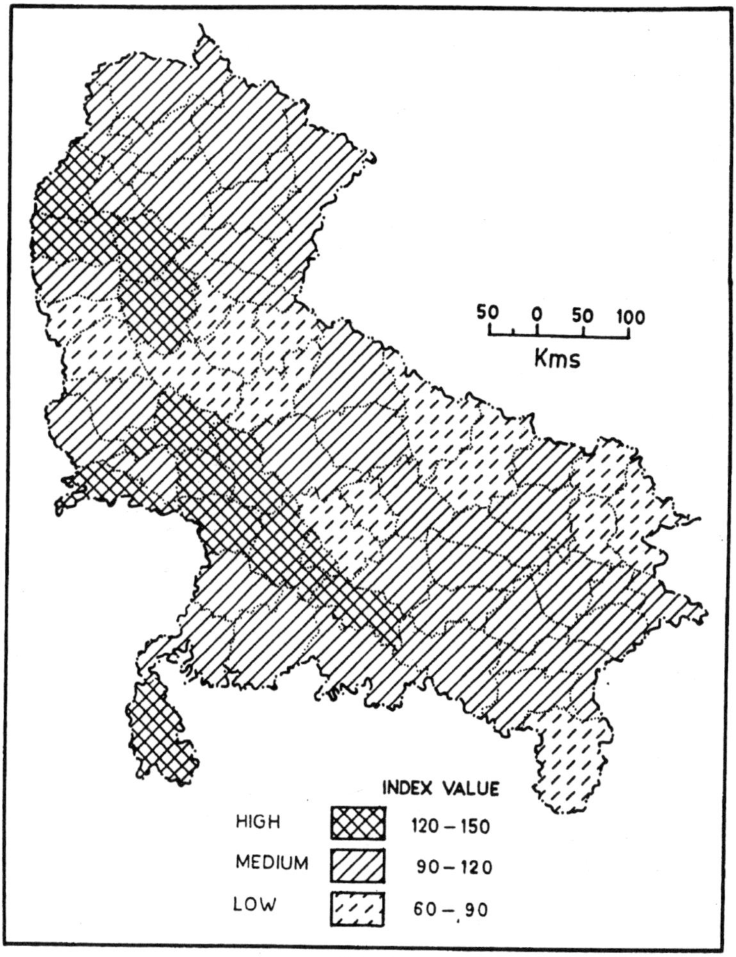

Fig. 28.3 : Uttar Pradesh Productivity Regions, 1996-97—Pulses

Figure 28.3 depicts that the high concentration of pulses is found in western and south-western portion, spreading in 13 districts of the state with the crop indices of 120-150. The high productivity region covers an area of 5,13,822 hectares. As far medium productivity of pulses are concerned, these are scattered in 36 districts with the indices ranging between 90-120. However the low productivity of pulses occupy about 18 per cent of the total cropped area, observed in 15 districts of the study area with the indices of 60-90.

Table 28.3 : Levels of Productivity Region: PULSES

Category	*Indices*	*No. of Districts*	*Name of the Districts*
High	120-150	13	Saharanpur, Muzaffarnagar, Agra, Mainpuri Etah, Bijnaur, Moradabad, Farrukhabad, Etawah, Kanpur(Urban), Kanpur (Rural), Fatehpur and Lalitpur
Medium	90-120	36	Haridwar, Meerut, Aligarh, Mathura, Firozabad, Allahabad, Pratapgarh, Jhansi, Jalaun, Hamirpur, Banda, Varanasi, Mirzapur, Ghazipur, Basti, Siddarthnagar, Azamgarh, Mau, Jaunpur, Ballia, Rae Bareilly, Sitapur, Hardoi, Kheri, Faizabad, Sultanpur, Barabanki, Uddhamsingh Nagar, Nainital, Almorah, Pithoragarh, Chamoli, Uttarkashi, Tehri Garhwal, Garhwal and Dehradun.
Low	60-90	15	Bulandshar, Ghaziabad, Bareilly, Budaun, Shahjahanpur, Pilibhit, Rampur, Sonbhadra, Maharajganj, Gorakhpur, Deoria, Lucknow, Unnao, Gonda and Bahraich.

The reason for low productivity of pulses indicates that it has not been influenced by the modern technology. The use of indigenous seeds, lack of financial resources, non-use of chemical fertilizers and spread of many insects and pests have reduced the overall production of pulses in the region.

Productivity Region: Cash Crops

Cash crops are the most important group of crops grown in the study area. They extend in 24,62,857 hectares of cultivated area. Among cash crops—sugarcanes, potatoes, and onions have been taken into consideration, which contribute to about 10 per cent of the cultivated area. The high productivity region of cash crops acquire 47.83 per cent, whereas the medium and low productivity region capture 45.18 per cent and 6.98 per cent of area respectively (Table 28.4).

According to Fig. 28.4 the western U.P. enjoying the high productivity region with the indices of 105-125, spreading in

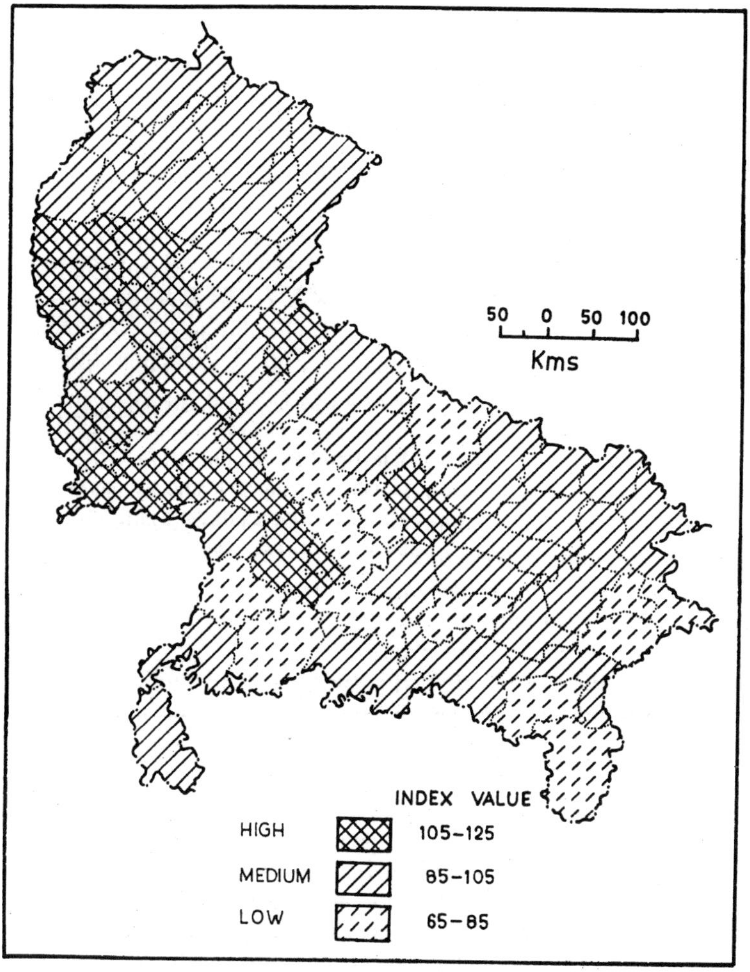

Fig. 28.4 : Uttar Pradesh Productivity Regions, 1996-97—Cash Crops

16 districts of the State. Medium productivity region covers a large area scattered in 35 districts with the indices ranging between 85-105, whereas only 13 districts come under the category of low productivity showing a lower crop indices (65-85), accounts only 7 per cent of the cropped area in cash crops to the total area.

Table 28.4 : Levels of Productivity Region: CASH CROPS

Category	*Indices*	*No. of Districts*	*Name of the Districts*
High	105-125	16	Muzaffarnagar, Meerut, Ghaziabad, Aligarh, Mathura, Agra, Firozabad, Mainpuri, Budaun, Pilibhit, Bijnaur, Moradabad, Farrukhabad, Kanpur (Urban), Kanpur (Rural), Barabanki.
Medium	85-105	35	Haridwar, Saharanpur, Bulandshahr, Etah, Bareilly, Shahjahanpur, Rampur, Etawah, Allahabad, Jhansi, Lalitpur, Banda, Varanasi, Maharajganj, Gorakhpur, Deoria, Basti, Siddarthnagar, Azamgarh, Jaunpur, Rae Bareilly, Sitapur, Kheri, Faizabad, Gonda, Sultanpur, Uddhamsingh Nagar, Nainital, Almorah, Pithoragarh, Chamoli, Uttar Kashi, Tehri Garhwal, Garhwal and Dehradun.
Low	65-85	13	Fatehpur, Pratapgarh, Jalaun, Hamirpur, Mirzapur, Sonbhadra, Ghazipur, Mau, Ballia, Lucknow, Unnao, Hardoi and Bahraich.

Productivity Region: Oilseeds

Oilseeds make the last group of crops grown in the study area. The major crops in oilseeds have been taken as mustard, sunflower, sesamum, sunflower, and groundnut. They occupy 12,32,272 hectares of cropped area. The high productivity area is covered by 46 per cent of the total cropped area, while the medium productivity areas constitute about 34 per cent area and 20 per cent area comes under the category of low productivity region. (Table 28.5)

Figure 28.5 reveals that a comparatively larger area of high productivity is found in western and south-western part of the State with the crop indices of 110-140. The medium productivity region with indices (80-110) is scattered in 39 districts, mostly concentrated in Uttarakhand, central and eastern portion of the State. However, only 12 districts have come out as low productivity region, lie in four pockets of the study region.

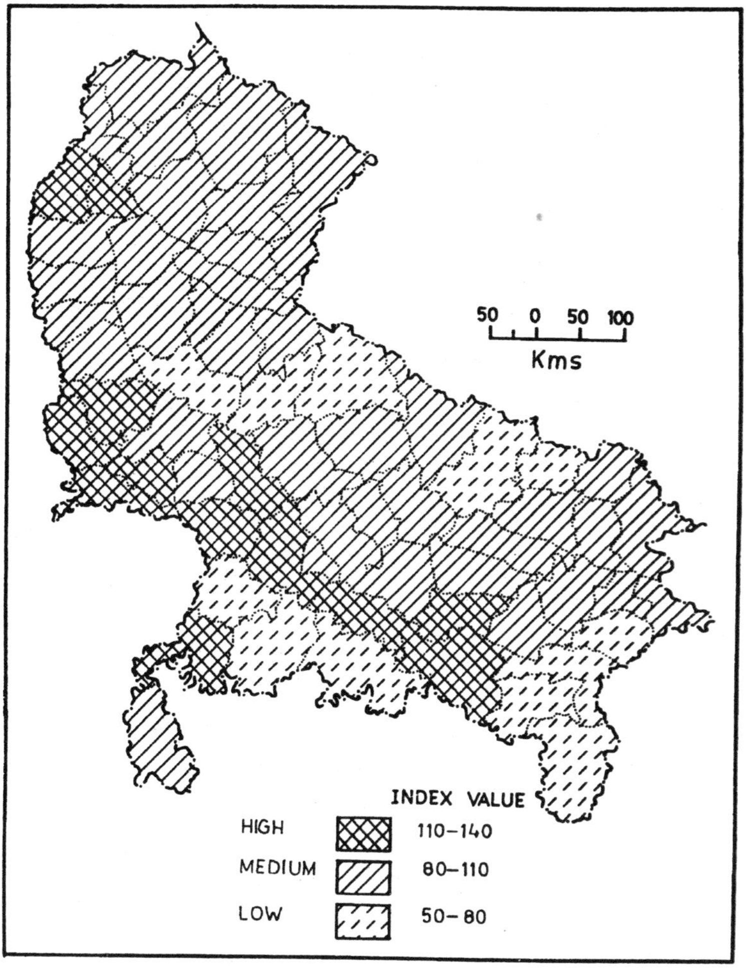

Fig. 28.5: Uttar Pradesh Productivity Regions, 1996-97—Oilseeds

PRODUCTIVITY REGION BASED ON COMPOSITE INDEX

To determine overall pattern of agricultural productivity during 1996-97, a composite index has been prepared after calculating the agricultural productivity for each group *e.g.* Cereals, Pulses, Cash crops, and Oilseeds, in the State of Uttar Pradesh. The values of overall agricultural productivity of each districts are given in Table 28.6, and

Table 28.5 : Levels of Productivity Region : OILSEEDS

Category	*Indices*	*No. of Districts*	*Name of the Districts*
High	110-140	13	Haridwar, Saharanpur, Aligarh, Mathura, Agra, Firozabad, Farrukhabad, Etawah, Kanpur (Rural), Fatehpur, Allahabad, Pratapgarh and Jhansi.
Medium	80-110	39	Muzaffarnagar, Meerut, Bulandshahr, Ghaziabad, Mainpuri, Etah, Bareilly, Pilibhit, Bijnaur, Moradabad, Rampur, Kanpur (Urban), Lalitpur, Maharajganj, Gorakhpur, Deoria, Basti, Azamgarh, Mau, Jaunpur, Ballia, Lucknow, Unnao, Rae Bareilly, Sitapur, Hardoi, Faizabad, Bahraich, Sultanpur, Barabanki, Uddhamsingh Nagar, Nainital, Almorah, Pithoragarh, Chamoli, Uttar Kashi, Tehri Garhwal, Garhwal and Dehradun.
Low	50-80	12	Budaun, Shahjahanpur, Jalaun, Hamirpur, Banda, Varanasi, Mirzapur, Sonbhadra, Ghazipur, Siddharthnagar, Kheri and Gonda.

the resultant levels of agricultural productivity are classified into three groups viz.; high, medium and low and their spatial pattern are shown in Fig. 28.6.

An analysis of agricultural productivity based on composite index reveals the existence of high productivity concentrating in Ganga-Yamuna Doab from the district of Haridwar and Saharanpur to Kanpur, with the productivity indices of 110-130. There are extensive areas moderate productivity, occupying 35 districts in the State with the indices ranging between 90-110 spreading all over study region. However 14 districts of the State come under the category of low productivity with the index value of 70-90, scattering in patches throughout the study region except western U.P. (Fig. 28.6).

CONCLUSION AND SUGGESTIONS

The spatial analysis of agricultural productivity indicates that the high

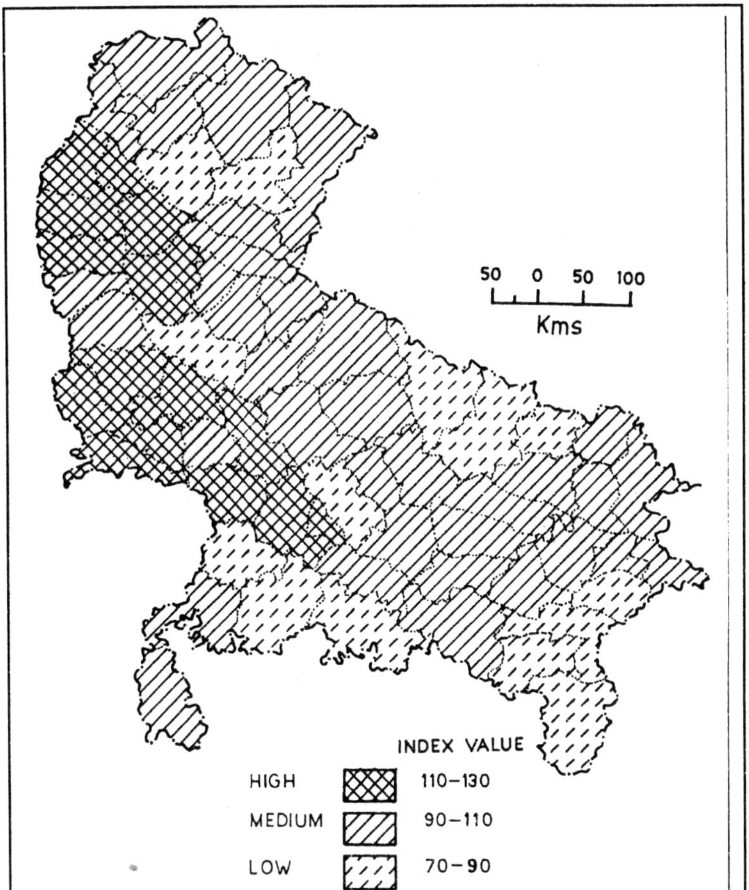

Fig. 28.6 : Uttar Pradesh Productivity Regions, 1996-97—Composite Index

productivity of cereals in Uttar Pradesh is due to the impact of various institutional and technological factors operating in the region which helped in converting low productivity areas to medium and even high productivity areas. However the low productivity of pulses require a serious attention of the farmers and governmental agencies. Efforts should be directed towards the increase of area and production of pulses. It can be achieved through the adoption of new varieties of seeds by safeguarding the interests of the farmers.

Table 28.6 : Levels of Productivity Region: COMPOSITE INDEX

Category	*Indices*	*No. of Districts*	*Name of the Districts*
High	110-130	15	Haridwar, Saharanpur, Muzaffarnagar, Meerut, Aligarh, Mathura, Agra, Firozabad, Etah, Bijnaur, Moradabad, Farrukhabad, Etawah, Kanpur (Urban) and Kanpur (Rural).
Medium	90-110	35	Bulandshahr, Ghaziabad, Mainpuri, Bareilly, Shahjahanpur, Pilibhit, Rampur, Fatehpur, Allahabad, Pratapgarh, Jhansi, Lalitpur, Maharajganj, Gorakhpur, Deoria, Basti, Azamgarh, Mau, Jaunpur, Ballia, Lucknow, Rae Bareilly, Sitapur, Hardoi, Kheri, Faizabad, Sultanpur, Barabanki, Uddhamsingh Nagar, Nainital, Pithoragarh, Chamoli, Uttar Kashi, Tehri Garhwal and Dehradun.
Low	70-90	14	Budaun, Jalaun, Hamirpur, Banda, Varanasi, Mirzapur, Sonbhadra, Ghazipur, Siddarthnagar, Unnao, Gonda, Bahraich, Almorah and Garhwal.

The problem of disparity can be solved through increased output of crops—both in surplus marginal and negative areas of crop production. The increased output of foodgrains can be achieved through: *(a)* extencification of agriculture, *(b)* intensification of agriculture, *(c)* mechanization of agriculture, and *(d)* multiple cropping and cropping pattern. In order to reduce the food deficit, we should emphasize more on irrigation than on other inputs because on irrigated land higher productivity can be achieved through fertilizer only. Moreover educational facilities and extension should be introduced at village level for minimizing the regional imbalances in the State.

The study concludes that the modern farm technology and socio-economic development are entwined in a mutually supportive relationship. The two together form an interesting cycles where each is a means of achieving the other. The message is crystal clear. Advanced farm technology undoubtedly leads to socio-economic well being and it in turn sets the farm technology on wheels towards a greater well-being; and so the cycle remains in motion.

REFERENCES

Annual Season and Crop Report (1996-97), Directorate of Economics and Statistics, Planning and Development Department, Government of Uttar Pradesh; Lucknow.

Bishnoi, O.P. (1992), "The Crop Ecology" in N. Mohammad (Ed.), *New Dimensions in Agricultural Geography,* Vol. 5, Concept Publishing Company, New Delhi.

Kumar, A. (1988), *Foodgrains Production in India*, Criterion Publication, New Delhi.

Mohammad, A. (1992), "Regional Imbalances in the levels of Agricultural Productivity" in N. Mohammad (Ed.), *New Dimension in Agricultural Geography,* Vol.8, Concept Publishing Company, New Delhi.

Munir, A. (1992), *Agricultural Productivity and Regional Development*, Manak Publications, New Delhi.

Shafi , M. (1972), " Measurement of Agricultural Productivity of Great Plains of India", *The Geographer,* Vol. XIX, pp. 4-13, Aligarh.

Shafi, M. (1984), *Agricultural Productivity and Regional Imbalances—A Study of Uttar Pradesh,* pp.148, Concept Publishing Company, New Delhi.

Siddiqui, S.H.; Rahman, H.; and Siddiqui, M.F. (1984), "Regional Analysis of Agricultural Productivity in Bihar ", *The Geographer*, Vol. XXXI, No. 1, pp. 77-85, Aligarh.

Yang, W. Y. (1965), *Methods of Farm Management Investigation*, No. 80, F.A.O., of the United Nations.

ENVIRONMENT, AGRICULTURE
AND
POVERTY

VOLUME THREE

Editors
Ali Mohammad
Abdul Munir
Shamsul Haque Siddiqui
Rocket Ibrahim

CONCEPT PUBLISHING COMPANY, NEW DELHI-110059

ISBN

978-81-8069-270-3 (Set)
978-81-8069-286-4 (Vol. II)
978-81-8069-285-7 (Vol. I)
978-81-8069-287-1 (Vol. III)

Edition 2016

Published and Printed by

Concept Since 1974

Concept Publishing Company Pvt. Ltd.
A/15-16, Commercial Block, Mohan Garden
New Delhi-110059 (India)
T: 25351460, 25351794, *F*: 091-11-25357109
E: publishing@conceptpub.com, W: www.conceptpub.com
Editorial Office: H-13, Bali Nagar, New Delhi-110 015, India.

Cataloging in Publication Data--Courtesy: D.K. Agencies (P) Ltd. <docinfo@dkagencies.com>

Environment, agriculture and poverty / editors, Ali Mohammad ... [et al.].
3 v. cm.
Papers presented at a conference.
Includes bibliographical references.
Includes index.
ISBN 9788180692703

1. Agriculture—Environmental aspects—India—Congresses. 2. Agriculture—Environmental aspects—Asia—Congresses. 3. Poverty—Government policy—India—Congresses. 4. Poverty—Government policy—Asia—Congresses. I. Ali Mohammad, 1946-

DDC 630.954 22

INTRODUCTION

The future of humanity is at critical juncture as the population continues to grow at a high rate in most of the regions surrounding South Asia, which is characterized as world's most deficit region with high level of poverty. The looming shortage of basic needs has compelled the man to irrational exploitation of his immediate surrounding, *i.e.*, natural resources available which has adversely affected the environment. In regional context increasing greenhouse gas emission, ozone depletion and acid rain, loss of bio-diversity, deforestation and loss of top soil and fertility, shortage of water, food and fuel indicate how the natural systems are being pushed even closer to their limits of exploitation.

In order to fulfil the basic needs and to have a hunger free society, we should harness the available resources and the public distribution system in a judicious manner.

In this country and other surrounding countries agriculture is still a main source of survival but highly affected by environmental phenomenon because of the poor infrastructure and poor socio-economic base. With the increase of population and other demographic characteristics we want to raise more food, raw material and for other purpose such as food security and changing demand. Therefore, we have to produce very much more per hectare because we have very little more land, which we can bring under cultivation. Therefore, multiple cropping and intensive input based agriculture is going to be surely a characteristic of the future. This agricultural system require intensive use of chemical fertilizers and irrigation that leads to ecological imbalances. What are going to be the long-term effects of some of these changes which we try to introduce? The high yielding varieties have made a great impact which has aroused the enthusiasm of the farmers but it needs high amount of modern inputs. Probably under these conditions there will be more outbreaks of pests and diseases in the severe form than in the past. What will be the long range effects of some of these practices?

In spite of substantial increase in global food production, FAO has estimated that about 790 million people in the developing countries and about 34 million people in developed and transition economies were suffering from under-nourishment during 1995-1997.

Approximately, 75 per cent of these food deficit population is located in Asia, the Pacific and sub-Sahara region. The largest number of food insecure people is in India (about 200 million).

Agriculture alone cannot solve the problem of hunger and poverty unless and until other supporting factors are not applied. The agricultural production should reach to the weaker sections of the society. Moreover, efforts should be made to generate employment in rural areas. This will improve the purchasing power of the rural masses, particularly the landless labours and artisans.

The regional dimension of environmental degradation is also very dangerous for its inhabitants throughout the world. Change of the environment in one region of country affected others and the problems had to be tackled on a global or regional scale. The issue of environmental diversity with diversified pattern of resource endowments should be understood according to the need of the people living in the region.

An integrated approach is needed to have a sustainable development of all the resources available in the diversified environment, so that it should be viable for man in anchoring the poverty.

About 65 per cent of the livelihood of the people is dependent on agriculture in India by providing employment but poverty has become a way of life for those totally dependent on agriculture because of the small and marginal size of landholding. It is also observed that per hectare productivity has increased two and half times after independence and degradation of environment, especially the soil nutrients depletion has become a major issue for the sustainable agriculture. Input-output ratio is being disturbed in intensive agricultural pockets and it is observed that the model of Punjab agricultural development is not favourable to continue. It shows that environment should be protected in a situation where India is characterised as the home of the largest number of poorest population which is not at all desirable. Therefore, poverty alleviation programme, schemes and planning will affect the environment. But at the cost of development, environment should be protected or not, poverty will prevail at mass level in India.

ALI MOHAMMAD

CONTENTS

VOLUME ONE

ENVIRONMENT

VOLUME TWO

AGRICULTURE

VOLUME THREE

POVERTY

CONTRIBUTORS

A. Taheri Tizro
College of Agriculture,
Razi University,
Kermanshah, Iran.

A.K. Srivastava
Haryana Institute of Public Administration,
Sector 18, Gurgaon,
Haryana.

Abdul Munir
Senior Reader,
Department of Geography,
Aligarh Muslim University,
Aligarh.

Abdul Rauf Shah
Research Associate,
Geography Unit,
Center of Central Asian Studies,
University of Kashmir,
Srinagar (J&K).

Agha M. A. Siddiqui
Department of Architecture and Planning,
University of Roorkee,
Roorkee-247667.

Ahmad Ali Bozorgzad
Islamic Azad University,
Najafabad, Esfahan,
Iran.

Ahmad Mujtaba Siddiqui
Research Scholar,
Department of Geography,
Aligarh Muslim University,
Aligarh.

Ali Mohammad,
Professor, Department of Geography,
Aligarh Muslim University,
Aligarh (U.P.)

Alpana Chandra Scott Lecturer,
I.T. Degree College,
Lucknow.

Archana Gupta
Department of Grography,
Allahabad University,
Allahabad.

B.R. Kareriya
Research Scholar,
Department of Geography,
Banaras Hindu University,
Varanasi.

B.R.K. Sinha
Reader, Department of Geography,
Visva Bharti University,
Shantiniketan-731235.

Balak Ram
Central Arid Zone Research Institute,
Jodhpur-342003.

Beik Mohammadi
Department of Geography,
University of Isfahan,
Iran

D.D. Vishwakarma
Assistant Professor,
Department of Geography,
Government P.G. College,
Chhindwara, (M.P.).

D.K. Mishra
Department of Grography,
Delhi School of Economics,
University of Delhi,
Delhi.

G.M. Mir
Reader, Geography Unit,
Centre of Central Asian Studies,
University of Kashmir,
Srinagar (J&K).

G.S. Chauhan
B.R. Ambedkar College,
(University of Delhi),
Geeta Colony,
Delhi-110 003.

Gheesa Lal
Central Arid Zone Research Institute,
Jodhpur-342003.

H.N. Mishra
Department of Geography,
Allahabad University,
Allahabad.

J.S. Chauhan
Central Arid Zone Research Institute,
Jodhpur-342003.

K.V. Sundaram
Founder Chairman,
Bhoovigyan Vikas Foundation,
641, Sector-A,
Pocket-C, Vasant Kunj,
New Delhi.

Kumar Nikhil
Scientist,
Environmental Management Group,
Central Mining Research Institute,
Dhanbad, Bihar.

M. Ahmad
Scientist,
Environmental Management Group,
Central Mining Research Institute,
Dhanbad (Bihar).

M.P. Tripathi
Department of Geography,
Tribhuwan University,
P.N. Campus, Pokhra,
Nepal

Mehdi Momeni
Najafabad Islamic Azad University,
Najafabad, Iran.

Mohammad Jafar Zomorrodian
Department of Geography,
Ferdowsi University Mashhad,
Mashhad, Iran.

Mohammad Shafi
(Prof. Emeritus) (Padam Shri)
Department of Geography,
Aligarh Muslim University,
Aligarh.

Mohd. Taufique
Research Scholar,
Department of Geography,
Aligarh Muslim University,
Aligarh.

Mumtaj Ahmed
Research Scholar,
Department of Geography,
Aligarh Muslim University,
Aligarh.

Nizamuddin Khan
Senior Lecturer,
Department of Geography,
Aligarh Muslim University,
Aligarh.

R.B. Singh
Department of Geography,
Delhi School of Economics,
University of Delhi,
Delhi-110007.

R.S. Dixit
Reader-Head,
Post-Graduate Department of Geography,
[illegible]abella P.G. College,
Lucknow University,
Lucknow, U.P.

R.S. Sangwan
Department of Geography,
University College (M.D.U.),
Rohtak, Haryana.

Rais Ahmad
Reader,
Department of Agricultural Economics and Business Management,
Aligarh Muslim University,
Aligarh (U.P.).

Rajanish Kumar
G.B. Pant Social Science Institute,
Jhusi-Allahabad.

Rocket Ibrahim
Professor,
Department of Geography,
Aligarh Muslim University,
Aligarh (U.P.)

S. Hasan Ansari
Reader,
Department of Geography,
M.D. University,
Rohtak (Haryana).

S.B. Singh
Head,
Department of Geography,
Banaras Hindu University,
Varanasi.

S.W.A. Ashraf
Professor,
Department of Geography,
Aligarh Muslim University,
Aligarh (U.P.)

Saifuddin
Remote Sensing Application Centre,
7-B, Old Shibli Road,
Aligarh Muslim University,
Aligarh.

Sayed Hassan Hossaini Abari
Department of Geography,
Esfahan University,
Esfahan, Iran.

Shamsul Haque Siddiqui
Senior Reader,
Department of Geography,
Aligarh Muslim University,
Aligarh.

Soorya Lal Amatya
Department of Geography,
University of Kathmandu,
Kathmandu, Nepal.

Sumanta Chatterjee
Research Scholar,
Department of Geography,
Visva Bharti University,
Shantiniketan-731235.

Tapeshwar Singh
Reader,
Department of Geography,
M.M.H. College (C.C.S. University),
Ghaziabad (U.P.).

Yasir Saeed Hanafi
Research Scholar,
Department of Geography,
Aligarh Muslim University, Aligarh.

POVERTY

29

VARIATION IN HEALTH PATTERN OF THE RURAL PEOPLE IN A SEGMENT OF LOWER GANGA PLAIN

B.R.K. SINHA and SUMANTA CHATTERJEE

Health is one of the basic themes of present day study. It contributes directly to welfare by reducing sickness, child mortality, as well as by increasing life expectancy and improving labour's productivity. Health is considered as a pivotal for achieving overall development of a region. In other words, health is regarded as an indispensable human capital which plays a key role in improving the quality of life. The World Bank's Sector Strategy (1997, p. 2) rightly pointed out that good health contributes to the overall quality of life as well as to productivity. Health in this sense is not only an output of society and worthwhile in its own terms, it is also an asset for a community, a resource in the form of energy, ability, talents and other mental and physical capabilities of normal functioning in a given environment. Health is however not a static phenomenon because it varies in space and time, across age, gender, class, education, occupation and income. Such variation is the cumulative effects of unequal distribution and development of social, economic, cultural, political and physical resources of changing nature. African Development Report on Human Capital Development (1998, p. 1), Singh, A.K. (1983, pp. 27-31) and Human Development Report (1998, p. 46) have also laid emphasis on several socio-economic aspects of population particularly on improved food and nutrition, education in improving health of the people as these elements interact positively in many ways in making out a good life at all spatial levels.

OBJECTIVES

In view of the above importance and problems, this Chapter focuses on the variation in the pattern of health of the rural people across age groups and sex, social groups *vis-a-vis* educational and occupational categories. Finally, suggestions are made for improving the health status of the rural people for better life.

METHODOLOGY

Methodology consists of two parts—data base and concept base.

Data Base

The study is exclusively based on the primary data collected through household surveys on various qualitative and quantitative variables of health, and to accomplish the above work a rural Community Development Block (Mangolkote) as segment of Lower Ganga Plain has been selected. Mangolkote is one of the Development Blocks of Burdwan district, West Bengal.

Concept of Measuring Health

Different scholars such as Raza Moonish (1990, p. 47), Trewartha, G.T. (1967, p. 546), Smith, D.M. (1997, p. 47), Basu, S. (1992, pp. 212-226), Pacione, M. (1988, pp. 211-212), Madan, G.R. (1983, pp. 284, 286), and Mukherjee, B.M. (1986, p. 102) and others have used different quantitative and qualitative indicators such as height and weight of the body, body temperature, blood pressure, counts of red blood and white blood corpuscle, infant mortality rate, general mortality, problems of hair, eyes, lips, gums, teeth, skin, gland, life expectancy, number of doctors, hospital and beds per unit population, safe water supply, literacy, education, per capita income, expenditure on health, disease of various types, food and nutrition, calorie intake, capacity to do manual work for standard working hours, clear eye sight and memory condition etc. in measuring health of the people but their quantitative and qualitative indicators are applicable to the common people at regional, State and national levels only.

Investigators have considered only those quantitative and

qualitative parameters that are directly or indirectly related to mental and physical capacities of individuals.

These are the number and types of diseases, condition of appetite, digestive problem, memory capacity, mental tension, blood pressure, eye problem, E.N.T. problems, dizziness, physical and mental problem while working continuously for 6-8 hours standard working period, problems arise during working period in sun, rain and cold, problems in carrying normal head load, state of sleeping, feeling of weakness and tiredness, feeling of laziness, whether enthusiastic to work, whether two times meal and breakfast are taken in time, number of visits to government and private hospitals or clinics, per head total medical cost in the last one year, types of drinking water, number of days of pulse, green and other vegetables taken by the individuals in a week, how many days in a week an individual takes ghee/butter, milk, eggs, meat, fish etc., frequency of taking smoking, tobacco, alcoholic materials, whether the required medicines are purchased timely or not and whether the food is adequately available or not at household level, whether the house is ventilated or unventilated, whether breath is sweet or not, educational achievements including training and skills and income levels etc.

Thus, while preparing master data sheets each variable was given certain positive or negative weightage depending upon its positive or negative conditions of individual.

Therefore, the total composite scores of all variables and also the total positive or negative scores of each individual were separately worked out and serially listed from 1 to 1705.

Finally, for each person the percentage of positive composite score to total composite score (*i.e.* positive composite score + negative composite score = 100 per cent) was calculated. On the basis of variation in the percentages of positive composite scores of individuals five levels of health have been made. These are :

(i) Very good (80 per cent and above positive composite score).
(ii) Good (60 to <80 per cent positive composite score).
(iii) Average (40 to <60 per cent positive composite score).
(iv) Poor (20 to <40 per cent positive composite score).
(v) Very poor (less than 20 per cent positive composite score).

ANALYSIS

Variation in Health Pattern by Age Group and Sex

Table 29.1 and Fig. 29.1 explain the variation in health pattern of the rural people at age and sex level. Variation in health of the infants shows that the percentage of both male and female children is highest in average health category that is followed by good, very good, poor and very poor in descending order in male category and by good, very poor, poor and very good in female category. Variation in health of the school-going age group children shows that percentage of male children is highest in good health category, whereas the percentage of female children is highest in average health category. Next highest percentage of male children is in average health followed by poor, very good and very poor in descending order whereas in female category the next higher percentage is in good health followed by health levels similar to male children. When we look into the health quality of infants and school-going age group children we find that the health of the latter group is relatively better indicated by higher percentage in good health. Variation in health among the younger working age group (15-34 years) persons shows that the percentage to total male or female population of this age group is highest in good health that is followed by the percentage of average, poor, very good and very poor health. Older working age group (35-59 years) persons show their highest percentage in average health, higher in good, high in poor, low in very poor and very low in very good health. Similarly, aged persons of 60 and + years show more variation in their health status as the percentage of males is higher in average and very poor health categories and lower in good and poor health categories, whereas percentage of females decreases from very poor to good health category. Such variation indicates that the health of aged males is normally better than that of its female counterpart.

After looking into health category-wise variation we find that in very good health percentage of lower age group persons particularly below 15 years of age is generally higher and percentage of middle and higher age group persons is lower. Similarly in good health category the percentage of young people of (15-34 years) age group is higher which is followed by children, older persons and old persons in descending order. In average and poor health categories percentage of older persons is generally highest and in very poor and poor

Table 29.1 : Health Status of Surveyed Individuals by Age Groups and Sex, Mangalkote Sample, 1998

Health status	*% range of composite scores*	*Age Groups*										*Total*	
		0-4		*5-14*		*15-34*		*35-59*		*60 & +*			
		M	*F*	*M*	*F*	*M*	*F*	*M*	*F*	*M*	*F*	*M*	*F*
Very Good	>80	6	2	10	3	12	7	9	2	—	—	37	14
		10.17	3.85	8.77	2.38	3.29	2.14	2.77	0.79			(4.03)	(1.78)
Good	60-80	21	13	46	52	175	150	102	69	13	2	357	286
		35.59	25.00	40.35	41.27	47.95	45.87	31.38	27.49	24.07	6.25	(38.93)	(36.29)
Average	40-60	23	28	41	58	139	144	152	130	18	5	373	365
		38.98	53.85	35.96	46.03	38.08	44.03	46.77	51.79	33.33	15.82	(40.69)	(46.32)
Poor	20-40	6	4	14	11	36	24	52	41	7	9	115	89
		10.17	7.69	12.28	8.73	9.86	7.34	16.00	16.33	12.96	28.12	(12.54)	(11.29)
Very Poor	<20	3	5	3	2	3	2	10	9	16	16	35	34
		5.08	9.61	2.63	1.59	0.82	0.61	3.07	3.58	29.63	50.00	(3.82)	(4.31)
Grand Total		59	52	114	126	365	327	325	251	54	32	917	788
		(6.43)	(6.60)	(12.43)	(15.99)	(39.8)	(41.50)	(35.44)	(31.85)	(5.88)	(4.06)	(100.00)	(100.00)

Note : Bracketed lower figures are the percentages to grand total males or females.
Unbracketed lower figures are the percentages to total males or females of respective age groups.
Source : Author's Field Survey, 1998.

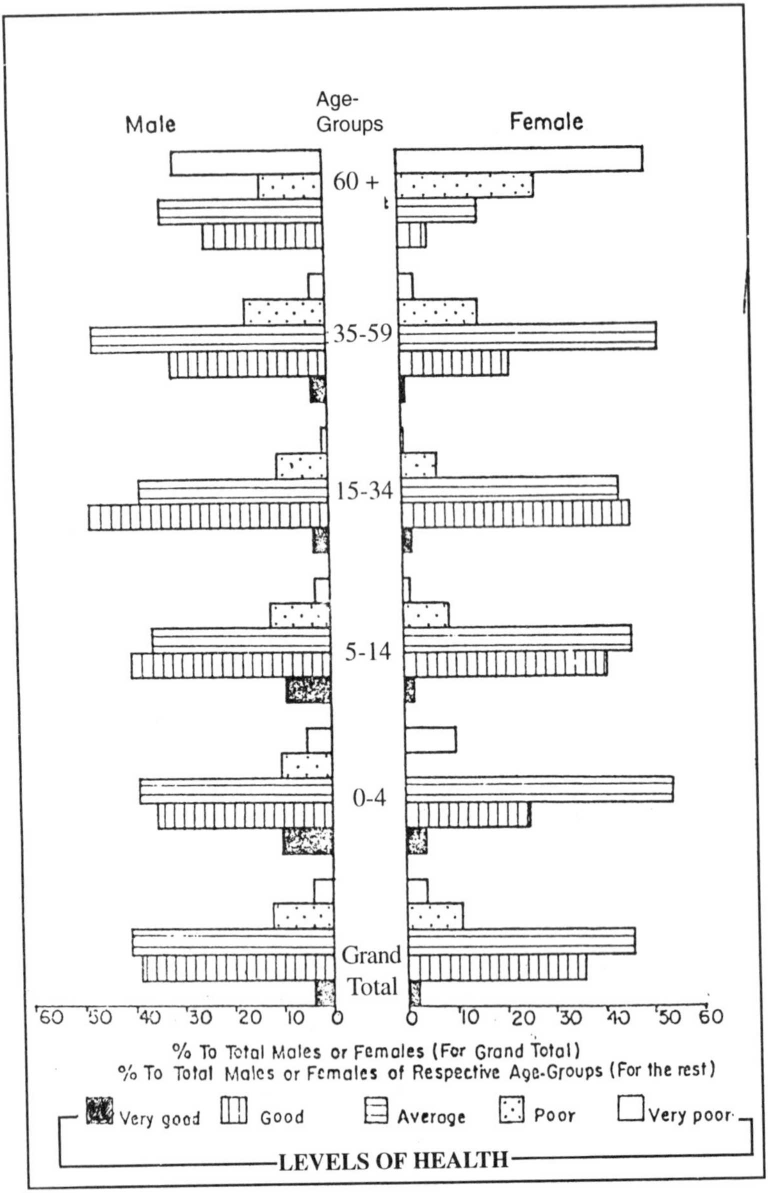

Fig. 29.1 : Mangalkote Block: Health Status of Surveyed Individuals by Age and Sex (1998)

categories the percentage of old persons is higher. As a whole it can be said that generally health of the children and young persons is better than that of older and old persons. Finally, we find that the majority of rural people possess average health followed by good, poor, very poor and very good.

Variation in Health Pattern by Social Group

Social groups and health are the two important aspects of population social group (the results of socio-economic conditions) is considered as a social force in health because each social group has different social pathology in effecting health of its people.

Table 29.2 and Fig. 29.2 deal with the variation in health pattern of the rural people across different social groups. From the table and figure we find that the percentage of persons having very good health decreases from Muslim to O.B.C., higher caste and finally to scheduled caste categories. Percentage of persons having good health also shows a decreasing trend from Muslim category to higher caste, O.B.C., scheduled caste and scheduled tribe. Variation in average health of the rural people is remarkable as the percentage of persons having average health is highest in scheduled caste that is followed by the percentage of scheduled tribe, O.B.C., higher caste and Muslims in descending order. Similarly, the percentage of persons having poor and very poor health is found higher in scheduled tribe, scheduled caste and O.B.C. (in descending order) and lower among the persons of Muslim and higher caste categories.

In fine, we can say that health of the Muslims, higher caste and O.B.C. is comparatively better than that of the scheduled caste and scheduled tribe persons. Such variation in health status may be because of better food and nutrition, education, health awareness and better life style of higher caste, O.B.C. and Muslims and lack of proper food and nutrition's, education, health awareness as well as poor life style, food and drink habits etc. Finally, grand total variations in health status is the same as in the case of age and sex groups.

Variation in Health Pattern by Educational Achievements

Education and health are very important characteristic features of population. Here, education acts as an important factor in determining health level. Education plays a very crucial role in health. According

Table 29.2 : Health Status of Surveyed Individuals by Social Groups, Mangalkote Sample, 1998

Health status	*% range of composite scores*	*Social Groups*										*Total*	
		H.C.		*Muslim*		*O.B.C.*		*S.C.*		*S.T.*			
		M	*F*	*M*	*F*	*M*	*F*	*M*	*F*	*M*	*F*	*M*	*F*
Very Good	>80	15	3	12	4	10	6	—	1	—	—	37	14
		4.95	1.13	11.00	4.30	6.02	4.80		0.59			(4.03)	(1.77)
Good	60-80	168	135	64	60	57	48	37	25	31	18	357	286
		55.45	50.75	58.71	64.51	34.34	38.40	19.68	14.88	20.53	13.23	(38.93)	(36.29)
Average	40-60	103	120	25	22	72	54	104	101	69	68	373	365
		33.99	45.11	22.93	23.65	43.37	43.20	55.32	60.12	45.69	50.00	(40.68)	(46.32)
Poor	20-40	15	7	7	6	24	12	35	37	34	27	115	89
		4.95	2.63	6.42	6.45	14.46	9.60	18.62	22.02	22.52	19.85	(12.54)	(11.29)
Very Poor	<20	2	1	1	1	3	5	12	4	17	23	35	34
		0.66	0.37	0.91	1.07	1.80	4.00	6.38	2.38	11.25	16.91	(3.82)	(4.31)
Grand Total		303	266	109	93	166	125	188	168	151	136	917	788
		(33.04)	(33.76)	(11.89)	(11.80)	(18.10)	(15.86)	(20.50)	(21.32)	(16.47)	(17.25)	(100.00)	(100.00)

Note : Bracketed lower figures are the percentages to grand total males or females.
Unbracketed lower figures are the percentages to total males or females of respective social groups.

Source : Author's Field Survey, 1998.

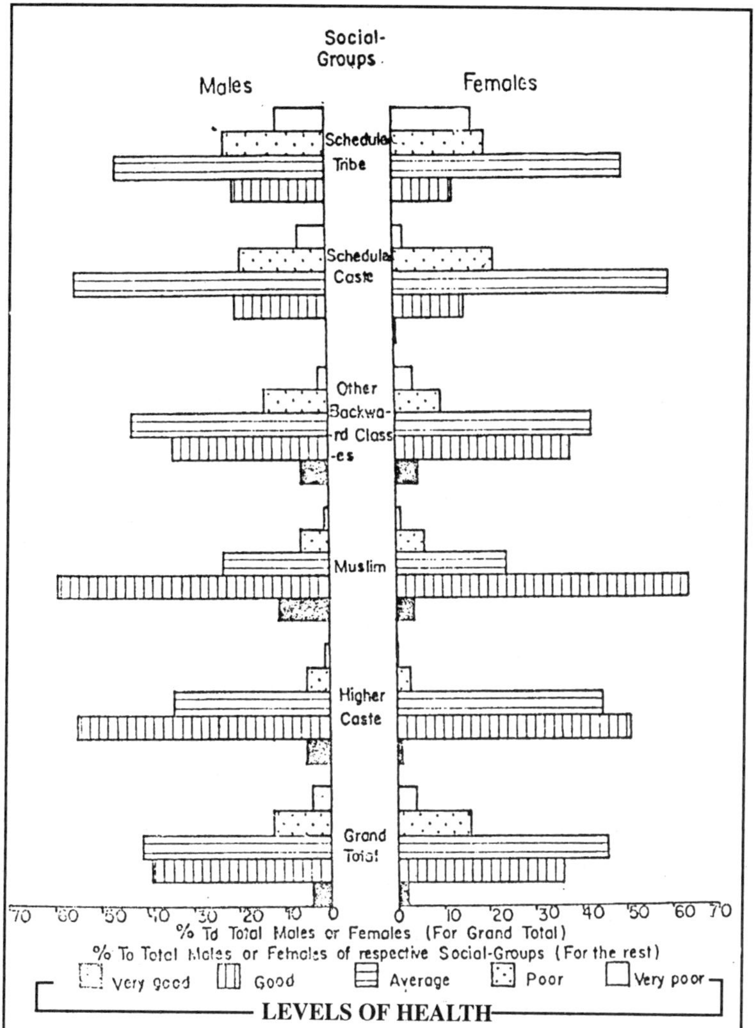

Fig. 29.2 : Mangalkote Block: Health Status of Surveyed Individuals by Social-Groups (1998)

to African Development Report (1988, pp. 110-111) education is considered as an important factor which affects the health and life expectancy of individuals because it equips them with the knowledge and the means to prevent, control and detect diseases. Table 29.3 and

Table 29.3 : Health Status of Surveyed Individuals by Educational Achievement, Mangalkote Sample, 1998

Health status	*% range of composite scores*	*Levels of Education*														*Total*	
		Illiterate		*Primary*		*Jr. Sec.*		*Sec./H.S.*		*U.G.*		*P.G.*		*Prof./Tech.*			
		M	*F*	*M*	*F*	*M*	*F*	*M*	*F*	*M*	*F*	*M*	*F*	*M*	*F*	*M*	*F*
Very Good	>80	1	1	1	1	2	3	5	2	15	3	9	3	4	1	37	14
		0.60	0.48	0.67	0.46	1.01	1.92	2.53	1.23	9.49	10.34	24.32	27.27	50.00	33.33	(4.03)	(1.78)
Good	60-80	38	42	46	87	102	67	95	78	57	8	16	3	3	1	357	286
		22.62	20.19	30.67	39.91	51.52	42.95	47.98	47.85	36.08	27.59	43.24	27.27	37.50	33.33	(38.93)	(36.30)
Average	40-60	68	105	67	108	73	71	81	72	75	5	8	3	1	1	373	365
		40.47	50.48	44.67	49.54	36.87	45.51	40.91	44.17	47.47	17.24	21.62	27.27	12.50	33.33	(40.68)	(46.32)
Poor	20-40	42	37	28	16	16	11	15	10	10	13	4	2	—	—	115	89
		25.00	17.79	18.66	7.34	8.02	7.05	7.57	6.14	6.33	44.83	10.82	18.18			(12.54)	(11.29)
Very Poor	<20	19	23	8	6	5	4	2	1	1	—	—	—	—	—	35	34
		11.31	11.06	5.33	2.75	2.52	2.57	1.01	0.61	0.63						(3.82)	(4.31)
Grand Total		168	208	150	218	198	156	198	163	158	29	37	11	8	3	917	788
		(18.32)	(26.40)	(16.36)	(27.66)	(21.59)	(19.80)	(21.59)	(20.69)	(17.23)	(3.68)	(4.03)	(1.40)	(0.88)	(0.38)	(100.00)	(100.00)

Note : Bracketed lower figures are the percentages to grand total males or females.
Unbracketed lower figures are the percentages to total males or females of respective educational categories.

Source : Author's Field Survey, 1998.

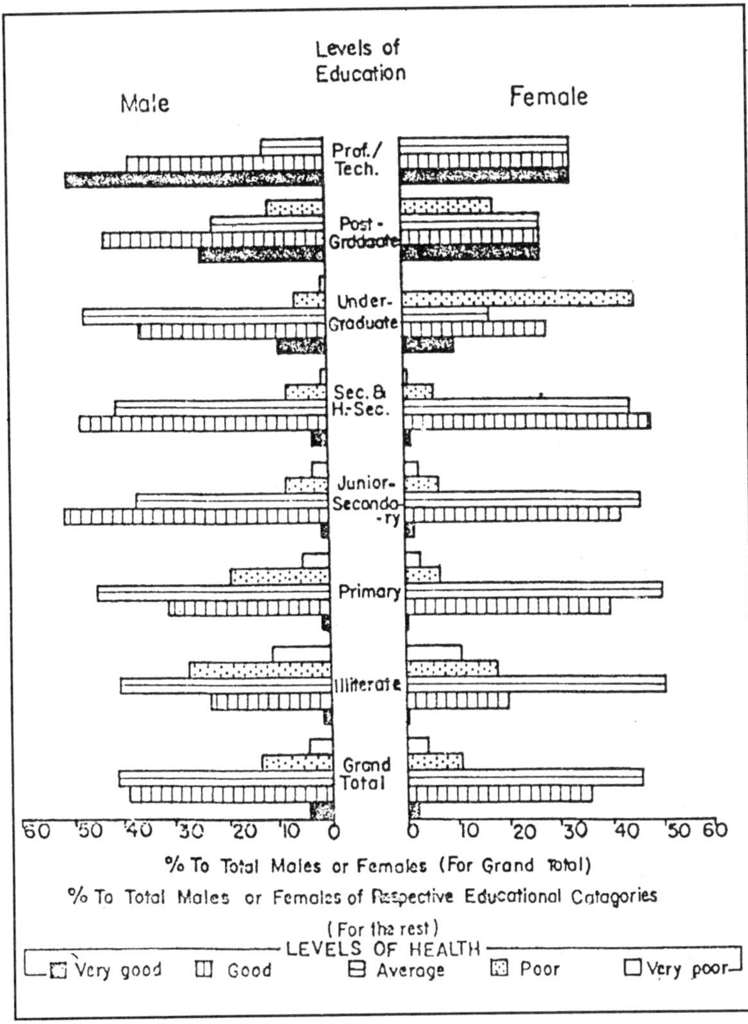

Fig. 29.3 : Mangalkote Block: Health Status of Surveyed Individuals by Sex and Eductional Categories (1998)

Fig. 29.3 analyse the variation in health pattern across different educational categories. It is clear from the table and figure that the percentage of illiterate, primary, and junior secondary females and under graduate males is higher in average health whereas the percentage of good health is higher among junior secondary males, secondary and higher secondary males and females and among post-graduate males. Secondly, percentage of those having professional/technical education and post-graduate females is highest in very good health. In poor health category the percentage of under graduate females is highest which may be because of certain reasons relating to disease.

After looking into the overall picture we find that the percentage of persons having very good health sharply decreases from higher educational levels to lower educational levels. Similarly, the percentage of persons possessing poor and very poor health generally decreases from lower educational level to higher educational levels leaving aside some exceptional cases in under graduate and post-graduate educational categories. Percentage of good health is generally found higher among the persons of medium educational levels. Whereas the percentage of average health is generally found higher towards lower educational levels as well as towards medium levels of education. The above variation thus may be taken into consideration as the result of variation in educational achievements.

Variation in Health Pattern by Different Occupations

Health and occupation are the two important qualitative characteristics of population. Different occupations are found influencing health differently. Different occupation has different problems which in turn affects adversely the health of working persons. Mehta, Rakesh (1998, pp. 263-271) has intensively worked on occupational health in which he has emphasised on environmental, conditions such as work environment, social environment, physical environment, undue physical and mental stress at work place in influencing health of the workers. Marind-Bettolo, Giovanni Battista (1996, pp. 196-199) also discussed about occupational health of workers and pointed out that different types of occupational diseases adversely affect the health of working persons.

Table 29.4 clearly exhibit the variation in health of the rural people across different economic functions. When we look into the variation of health pattern, we find that in very good health the percentage of females engaged in cultivation as cultivators, males engaged in industrial and construction works is relatively higher. In good health category the percentage of persons engaged in commercial and other activities is relatively higher and those belonging to this health category

Table 29.4 : Health Status of Surveyed Individuals Engaged in Different Occupations, Mangolkote Sample, 1998

Health status & % range of composite scores	*Economic Functions*																		*Total*	
	Primary								*Secondary*				*Tertiary*							
	Cultivator		*Agril. Labour*		*L.S. Farming*		*Mining quarry*		*Industrial Activities*		*Const. Work*		*Trans. Comm.*		*Comm. Activities*		*Others*			
	M	*F*	*M*	*F*	*M*	*F*	*M*	*F*	*M*	*F*	*M*	*F*	*M*	*F*	*M*	*F*	*M*	*F*	*M*	*F*
Very Good	4	1	2	1	—	—	1	—	2	—	1	—	—	—	7	—	6	2	23	4
>80	2.21	13.51	1.23	0.77			10.00		16.67		12.50				10.00		9.52	8.69	(4.14)	(1.65)
Good	54	25	43	42	8	4	5	—	7	—	3	—	9	—	36	2	34	10	199	83
60-80	29.03	33.78	26.54	32.31	25.81	50.00	50.00		58.33		37.50		50.00		51.43	40.00	53.97	43.48	(35.85)	(34.16)
Average	68	33	95	53	16	3	4	—	3	—	4	3	2	—	25	—	15	7	232	99
40-60	37.57	44.59	58.64	40.77	51.61	37.50	40.00		25.00		50.00	100.00	11.11		35.71		23.81	30.43	(41.80)	(40.74)
Poor	48	12	11	18	6	1	—	—	—	—	—	—	5	—	1	3	5	2	76	36
20-40	26.52	16.22	6.79	13.85	19.35	12.50							27.78		1.43	60.00	7.94	8.69	(13.69)	(14.81)
Very Poor	7	3	11	16	1	—	—	—	—	—	—	—	2	—	1	—	3	2	25	21
<20	3.88	2.70	6.79	12.31	3.22								11.11		1.43		4.76	8.69	(4.50)	(8.64)
Grand Total	181	74	162	130	31	8	10	—	12	—	8	3	18	—	70	5	63	23	555	243
	(32.61)	(30.45)	(29.19)	(53.49)	(5.58)	(3.29)	(1.80)		(2.16)		(1.44)	(1.23)	(11.61)		(21.61)	(2.05)	(11.35)	(9.46)	(100.00)	(100.00)

Note : Bracketed lower figures are the percentages to grand total males or females.
Unbracketed lower figures are the percentages to total males or females of respective age groups.

Source : Author's Field Survey, 1998.

and are engaged particularly in primary economic functions show relatively lower percentage. Similarly, in average health category the percentage is found comparatively higher among the persons of primary economic functions. In poor and very poor health categories the percentage is fairly higher among the persons found engaged as cultivators, agricultural labourers and livestock farmers.

As a whole we can say that in general health of the persons engaged in secondary and tertiary economic functions is generally better and the health of those engaged in primary activities is relatively poor as indicated by higher percentage in average, poor and in very poor categories. Such variation may be due to better socio-economic conditions of secondary and tertiary workers than that of primary workers. This may also be because of comparatively better education and better health accessibility for those found engaged in the activities other than agricultural.

FINDINGS AND SUGGESTIONS

Findings of the study show that health varies among the persons of different age groups. Children and young persons show comparatively better health than that of older and aged persons. Social group-wise variation shows that health of higher caste persons and Muslims are relatively better than that of other backward class, scheduled caste and scheduled tribe persons. It has also been found that better educated persons possess generally better health. Occupation-wise variation also shows that health of the secondary and tertiary workers is generally better than that of primary workers.

Thus, it is suggested that essential medical and health facilities of different types at different levels in combination with essential educational facilities, food and nutrition, etc. must be made available in all the villages of the block and while distributing or allocating such infrastructures special emphasis should be given on females of different age groups, on older and old people, SC, ST and other backward castes in the block. The children below 15 years should be taken care of their physical and mental health with the provision of compulsory education and health. Once the above measures are adopted and properly utilised, then the male-female, age group-wise, social group-wise and village-wise gaps in health will be considerably reduced.

REFERENCES

African Development Report (1998) on Human Capital Development, African Development Bank, Oxford University Press, New York.

Basu, S. (1992), Nutritional Status, Physical Growth, Trends and Health Problems among the Tribal Population of Bastar District, M.P., in Tewari, P.D. and Tripathi, R.S., *Dimensions of Scheduled Tribes Development in India,* Uppal Publishing House, New Delhi.

Human Development Report (1998) published for the United Nations Development Programme (UNDP), Oxford University Press, Delhi.

Madan, G.R. (1983), *India's Social Problems, Social Disorganisation and Reconstruction,* Vol. 1, Allied Publishers Pvt. Ltd., New Delhi.

Marind-Bettolo, Giovanni Betista (1996), Introduction in Bernardo, Colombo; Paul, Demeny and Max, F. Perutj: Resources and Population: *Natural, Institutional and Demographic Dimensions of Development,* Clarendon Press, Oxford.

Mehta, Rakesh (1998), Occupational Health: An Overview, in Bajpai, P.K., *Social Work Perspective on Health,* Rawat Publications, Jaipur.

Mukherjee, B.M. (1986), Ageing Members and their Health in Changing Techno-Economic Condition, in Chaudhury, B., *Tribal Health—Socio-Cultural Dimensions,* Inter-India Publications, New Delhi.

Pacione, M. (1988), *The Geography of the Third World: Progress and Prospects,* Routledge, London.

Raza, Moonish (1990), *Education, Development and Society,* Vikas Publishing House Pvt. Ltd., New Delhi.

Smith, D.M. (1977), *Human Geography: A Welfare Approach,* Edward Arnold Ltd., London.

Singh, A.K. (1983), Health Modernity Education in India, *Social Change,* Vol. 13, No. 2, *Journal of the Council for Social Development,* 53, Lodi Estate, New Delhi.

The World Bank's Sector Strategy (1997), Health, Nutrition and Population, The Human Development Network, The World Bank's Group, Washington, D.C., U.S.A.

Trewartha, G.T.; Robinson, A.H.; and Hammand, E.H. (1967), *Elements of Geography*: Physical and Cultural, McGraw Hills Book Company, New York.

30

VARIATION IN EDUCATION AND UTILISATION PATTERN OF THE RURAL LABOUR RESOURCE IN LOWER GANGA PLAIN

B.R.K. SINHA

Education and utilisation are the two important aspects of contemporary research. Here, education in the context of labour resource is considered as a consumption or an investment good and utilisation as a process of consuming and producing resources. As an investment good education opens up job opportunities for labour utilisation. When a person is on job or engaged in any activity, use of his/her education and capability becomes inevitable. In this sense, both education and labour have its utility and functionability and in turn become a vital resource upon which the development of any kind largely depends. Srinivasan and Shariff (1997, p. 81) has rightly pointed out that education is a key factor in sustainable development. Education is also taken into consideration as a means and labour as an end. Labour resource is considered as the sum total of knowledge, skill, energy, ability or any other qualitative attributes of working capacity of an individual. Such qualities are consequence upon the education, training, health, food and nutrition, etc. African Development Report (1998, pp. 1 and 110) has rightly emphasised that education directly improves the productivity of people and provides the labour force with the necessary knowledge and skill and hence facilitates their participation in modern economies and societies. Das Gupta (1998, pp. 405 and 406) also stated that appropriate knowledge as the result of education is the only instrument through which one can attain material benefit, exaltation and beatitude.

Education brings material prosperity and makes ethnically strong society, a strong work force and men and women of high character. Kothari (1965, p. 24) and Correa (1963, p. 183) also revealed the importance of education and labour and stated that education is very essential element because educational level of the labour force affects the volume of production and each level of capital intensity requires a labour force with a determined level of education. Wiesbroad and Hughes (1983, p. 150) treated education and labour both as a consumption good and an investment good in the process of production. Hence, education and labour become essential for economic, social, cultural and political development.

Utilisation of labour and education vary by age, gender, social groups and by economic functions. Depending upon the participation of workers in occupations labour has been classified as primary, secondary and tertiary. Primary labour resource is that which is found engaged in cultivation, livestock farmings and mining and quarrying activities. Secondary labour is that which is engaged in construction and industrial economic functions and tertiary labour resource is that which is found engaged in trade and commercial, transport and communication and in other services.

OBJECTIVES

This Chapter aims at examining the variation in education and utilisation pattern of the primary, secondary and tertiary labour resource and to extend appropriate suggestions for improving the quality and optimum utilisation of all types of labour resource.

METHODOLOGY

To accomplish the work 'Burdwan District' an agriculturally dominated part of the Lower Ganga Plain has been selected. The study is entirely based on the primary data collected through questionnaire by conducting survey at household level in 1995. In the study area as a whole only twenty eight Community Development Blocks with 100 per cent rural population as well as with more than 53 per cent rural population were selected. Further, one village from each Community Development Block was randomly selected for the purpose. In order to conduct household survey for primary data collection twenty households comprising equal number (5) from each S.C., S.T., O.B.C.

and higher caste family in each sample village were randomly selected for purposive sample. Thus, in all the twenty eight sample villages the total number of respondents were 560 who furnished the information about 3,002 persons.

To assess the utilisation pattern of labour input in different economic activities person's working hour has been taken into consideration. Therefore, those found working below 6 hours in average throughout the year have been considered as under-utilised, those found working between 6 and 8 hours throughout the year have been considered as adequately utilised and those found engaged in their activities for more than 8 hours throughout the year have been regarded as over-utilised.

ANALYSIS

This part focuses on the variation in education and utilisation pattern of the primary, secondary and tertiary labour resources.

Variation in Education and Utilisation Pattern of the Primary Rural Labour Resource

From the Table 30.1 and Fig. 30.1, we find that child labour with higher illiteracy percentage and primary education were under-utilised. When we compare the variation in education and utilisation pattern of younger working age group (15-34) labour resource, it is found that around 36 per cent males mostly with secondary-cum-higher secondary and slightly more than 22 per cent females mostly with primary and illiteracy educational levels were under-utilised. 50 per cent males mostly with junior secondary, primary and secondary-cum-higher secondary in order of its relative dominance and about 67 per cent females mostly illiterate and some with primary education from 15-34 years of age group were adequately utilised. Only 14.06 per cent males with higher percentage towards lower level of education and only 11.11 per cent females with lower level of education were over-utilised.

So far as the utilisation of 35-59 years of age group labour resource is concerned we find that 34.48 per cent males with higher percentage in secondary-cum-higher secondary and in junior secondary and 26 per cent females mostly illiterate were under-utilised. 38.24 per cent males with decreasing trend from illiteracy to upper levels of

Table 30.1: Variation in Education and Utilization Pattern of the Primary Rural Labour Resource by Age Groups and Sex

		Age Group									
		5-14		*15-34*		*35-59*		*60 & +*		*Total*	
Sex		*M*	*F*	*M*	*F*	*M*	*F*	*M*	*F*	*M*	*F*
Total		22	1	320	126	319	100	41	5	702	232
%		(3.13)	(0.43)	(45.58)	(54.31)	(45.44)	(43.10)	(5.84)	(2.16)	(100.00)	(100.00)
Under-Utilised (Below 6 Hours)	I*	10		13	13	14 +	18	6	1	43	32
		45.45		4.06	10.32	4.39	18.00	14.63	20.00	(6.13)++	(13.79)
	II	10		15	15	14	6	3	2	42	23
		45.45		4.69	11.90	4.39	6.00	7.32	40.00	(5.98)	(9.91)
	III	2	1	23		24		3	—	52	1
		9.04	100.00	7.19		7.52	—	7.32		(7.41)	(0.43)
	IV			49		40	2	18	—	107	2
				15.31		12.54	2.00	43.90		(15.24)	(0.86)
	V			15		17	—	3	—	35	—
				4.69		5.33		7.32		(4.99)	
	VI					1	—	—	—	1	—
						0.31				0.14	
Grand Total		22	1	115	28	110	26	33	3	280	58
		100.00	100.00	35.94	22.22	34.48	26.00	80.49	60.00	39.89	25.00

(Contd.)

Table 30.1 *(Contd.)*

	Sex	Age Group									
		5-14		*15-34*		*35-59*		*60 & +*		*Total*	
		M	*F*	*M*	*F*	*M*	*F*	*M*	*F*	*M*	*F*
Adequately-Utilised (6 - 8 Hours)	I	—	—	37	68	70	49	3	2	110	119
				11.56	53.97	21.94	49.00	7.32	40.00	(15.67)	(51.30)
	II	—	—	39	14	22	—	3	—	64	14
				12.19	11.11	6.90		7.32		(9.12)	(6.03)
	III	—	—	40	2	10	—	—	—	50	2
				12.50	1.59	3.13				(7.12)	(0.86)
	IV	—	—	38	—	12	—	2	—	52	—
				11.86		3.76		4.88		(7.41)	
	V	—	—	6	—	8	—		—	14	—
				1.86		2.51				(1.99)	
	VI	—	—	—	—	—	—	—	—	—	—
Grand Total				160	84	122	49	8	2	290	147
				50.00	66.67	38.24	49.00	19.51	40.00	(41.31)	(63.36)

(Contd.)

Table 30.1 *(Contd.)*

	Sex	*Age Group*									
		5-14		*15-34*		*35-59*		*60 & +*		*Total*	
		M	*F*	*M*	*F*	*M*	*F*	*M*	*F*	*M*	*F*
Over-Utilised (Above 8 Hours)	I	—	—	9	6	30	16	—	—	39	22
				2.81	4.76	9.40	16.00			(5.56)	(9.48)
	II	—	—	13	5	18	5	—	—	31	10
				4.06	3.97	5.64	5.00			(4.42)	(4.31)
	III	—	—	10		19	3	—	—	29	3
				3.12		5.97	3.00			(4.13)	(1.29)
	IV	—	—	8	3	17	1	—	—	25	4
				2.50	2.38	5.34	1.00			(3.56)	(1.72)
	V	—	—	5		3		—	—	8	
				1.56		0.94				(1.14)	
	VI	—	—	—	—	—	—	—	—	—	—
Grand Total		—	—	45	14	87	25	—	—	132	39
				14.06	11.11	27.27	25.00			(18.80)	(16.81)

Source : Field Survey, 1995.

* Levels of Education : I—Illiterate, II—Primary, III—Jounior Secondary, IV—Secondary & H.Sec., V—U.G. and P.G., VI—Professional/Technical.

+ Unbracketed lower figures are the percentages to total males or females of respective age group.

++ Bracketed lower figures are the percentages to total males or females of primary rural labour resource.

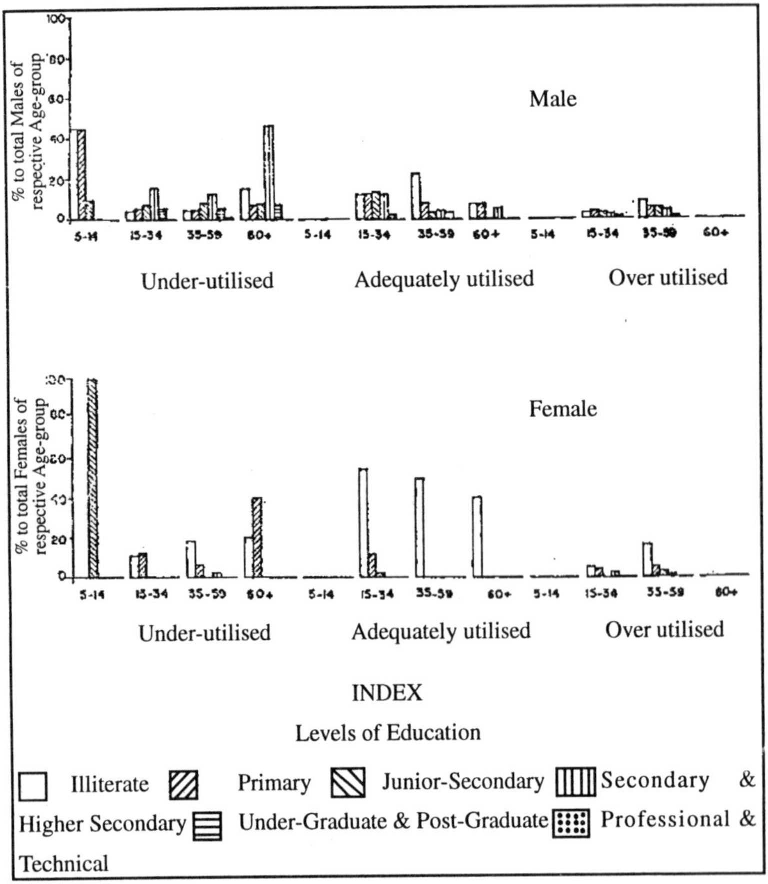

Fig. 30.1 : Variation in Education and Utilisation Pattern of the Primary Rural Labour Resource in Burdwan District by Age-groups and Sex (1995)

education and 49 per cent females with illiteracy were under-utilised. Similarly, 27.27 per cent males and 25 per cent females with education decreasing from illiteracy to upper level of education were over-utilised.

From 60 and above years of age group 80.49 per cent males with higher percentage of secondary-cum-higher secondary and lower education and 60 per cent females having more percentage in primary and illiteracy level were under-utilised. Only 19.51 per cent males with lower education and 40 per cent illiterate females were over-utilised.

From the above we find that around 40 per cent males and 25 per cent females to total primary male or female labour resource were under-utilised in which the percentage of males was higher in medium level education and of females in lower level of education. A closer view of under-utilised males or females shows that the child labour and old age group labour are more under-utilised than the male and female labour resource of 15-34 and 35-59 years of age groups. Similarly, 41.31 per cent males and 63.36 per cent females were adequately utilised. In adequately utilised category the percentage of males and females from 15-34 years age group was higher and it was lower in 35-59 years of age group and lower in 60 and above years age group. In over-utilised category the percentages of males and females were around 19 per cent and 17 per cent respectively. But from among them older working age group (35-59 years) labour shows more over-utilisation, and younger working age group (15-34 years) labour shows less over-utilisation. So far as their education is concerned, it is found that the percentage of both males and females in all utilisation categories gradually decreases from lower to higher educational levels. In under-utilised category the percentage of males is higher in medium level education.

Variation in Education and Utilization Pattern of the Secondary Rural Labour Resource

When we have a close look at the picture of variation in education and utilisation pattern of secondary rural labour resource as mentioned in Table 30.2 and Figure 30.2, it is found that in under-utilisation category there were only males constituted 13.62 per cent to total secondary male labour resource. In this utilisation category the percentage of labour from 35-59 years of age group is more and it is low in 15-34 years of age group. Educationally they were more in junior secondary.

Out of the total secondary rural male or female labour resources the percentage is higher in adequately utilised category. In this category also the percentage of males is higher in junior secondary and of females in primary and illiteracy levels.

In this utilisation category the percentage of labour from 15-34 years of age group and 5-15 years of age group is exceptionally higher than that of 35-59 years age group. Educationally they show the same trend as above.

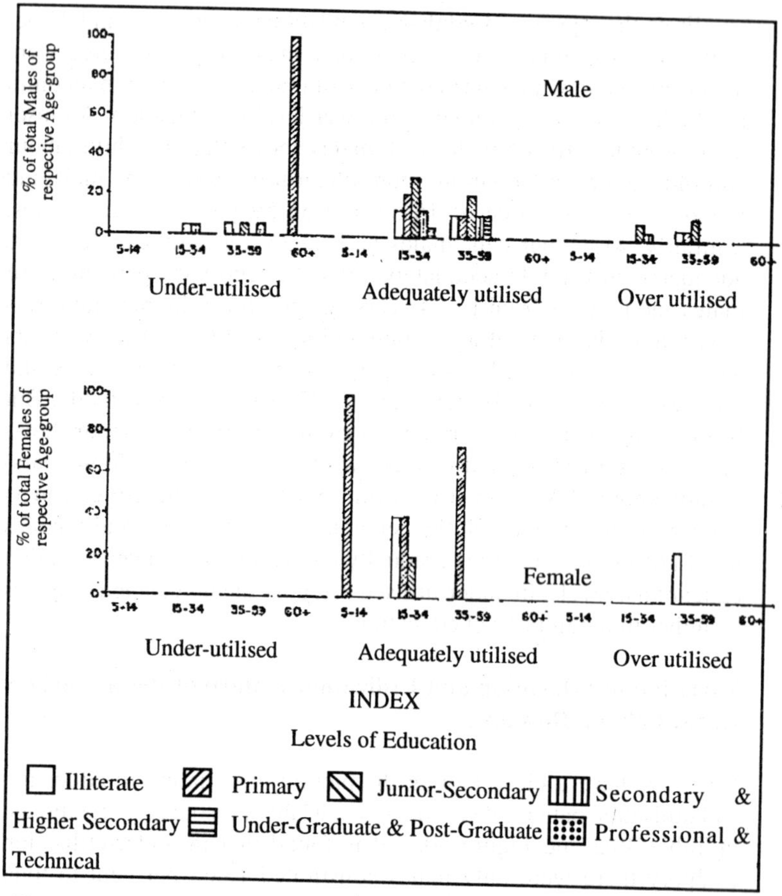

Fig. 30.2 : Variation in Education and Utilisation Pattern of the Secondary Rural Labour Resource in Burdwan District by Age-groups and Sex (1995)

In over-utilised category the percentage of males and females is 15.90 per cent and 10.00 per cent respectively. In this utilisation category also the percentage is higher in junior secondary educational category. Hence older persons are more over-utilised and show similar education.

As a whole we can say that the secondary activities are not so developed that can absorb more labour resource. However, in comparison to primary rural labour resource the education of secondary workers is relatively better. Secondly, the percentage of both males

Table 30.2 : Variation in Education and Utilization Pattern of the Secondary Rural Labour Resource by Age Groups and Sex

		Age Group									
		5-14		*15-34*		*35-59*		*60&+*		*Total*	
Sex		*M*	*F*	*M*	*F*	*M*	*F*	*M*	*F*	*M*	*F*
Total			1	24	5	19	4	1		44	10
%			(10.00)	(54.54)	(50.00)	(43.18)	(40.00)	(2.27)		(100.00)	(100.00)
Under-Utilised (Below 6 Hours)	I*	—	—	—	—	1	—	—	—	1	—
						5.26 +				(2.27)++	
	II	—	—		—	—	—	1	—	1	—
								100.00		(2.27)	
	III	—	—	1	—	1	—	—	—	2	—
				4.16		5.26				(4.54)	
	IV	—	—	1	—	—	—	—	—	1	—
				4.16						(2.27)	
	V	—	—	—	—	1	—	—	—	1	—
						5.26				(2.27)	
	VI										
Grand Total		—	—	2	—	3	—	1	—	6	—
				8.32		15.78		100.00		(13.62)	

(Contd.)

Table 30.2 *(Contd.)*

		Age Group									
		5-14		*15-34*		*35-59*		*60&+*		*Total*	
Sex		*M*	*F*	*M*	*F*	*M*	*F*	*M*	*F*	*M*	*F*
Adequately-Utilised (6 - 8 Hours)	I	—	—	3	2	2	—	—	—	5	2
				12.50	40.00	10.52				(11.36)	(20.00)
	II	—	1	5	2	2	3	—	—	7	6
			100.00	20.83	40.00	10.52	75.00			(15.90)	(60.00)
	III	—	—	7	1	4	—	—	—	11	1
				29.16	20.00	21.05				(25.00)	(10.00)
	IV	—	—	3	—	2	—	—	—	5	—
				12.50		10.52				(11.36)	
	V	—	—	1	—	2	—	—	—	3	—
				4.16		10.52				6.81	
	VI	—	—	—	—	—	—	—	—	—	—
Grand Total		—	1	19	5	12	3	—	—	31	9
			100.00	78.81	100.00	63.13	75.00			(70.43)	(90.00)

(Contd.)

Table 30.2 *(Contd.)*

Sex		5-14 M	5-14 F	15-34 M	15-34 F	35-59 M	35-59 F	60&- M	60&- F	Total M	Total F
Over-Utilised (Above 8 Hours)	I	—	—	—	—	1	1	—	—	1	1
						5.26	25.00			(2.27)	(10.00)
	II	—	—	—	—	1	—	—	—	1	—
						5.26				(2.27)	
	III	—	—	2	—	2	—	—	—	4	—
				8.32		10.52				(9.09)	
	IV	—	—	1	—	—	—	—	—	1	—
				4.16						(2.27)	
	V	—	—	—	—	—	—	—	—	—	—
	VI	—	—	—	—	—	—	—	—	—	—
Grand Total		—	—	3	—	4	1	—	—	7	1
				12.48		21.04	25.00			15.90	(10.00)

* Levels of Education : I—Illiterate, II—Primary, III—Junior Secondary, IV—Secondary & H.Sec., V—U.G. and P.G., VI—Professional/ Technical.

+ Unbracketed lower figures are the percentages to total males or females of respective age group.

++ Bracketed lower figures are the percentages to total male or female secondary rural labour resource.

Source : Field Survey, 1995.

and females is higher in adequately utilised category and lower in under and over-utilised category. Age-wise variation in all utilisation categories remains almost similar to that of primary category.

Variation in Education and Utilisation Pattern of the Tertiary Rural Labour Resource

When we have a closeview on the variation of education and utilisation pattern of tertiary rural labour resource as shown in Table 30.3 and Figure 30.3, we find that children however small in percentage (with primary education) were under-utilised.

45.33 per cent males and 34.78 per cent females to total male or female labour resources engaged in tertiary activities were in 15-34 year of age group and out of them 33.81 per cent males and 62.50 per cent females with medium levels of education in majority were under-utilised; 19.11 per cent males and 37.50 per cent females with majority towards higher educational levels were adequately utilised, and 47.04 per cent males only with different educational levels but show more percentage towards medium and lower levels of education were over-utilised.

Out of the total males or females engaged in tertiary activities 52.66 per cent males and 56.52 per cent females were in 35-59 years of age group and from them about 28 per cent males and 69.21 per cent females showing majority in medium educational levels were under-utilised, 39.22 per cent males only having majority towards higher levels of education were adequately utilised and 32.89 per cent males mainly with lower and medium educational levels and 30.76 per cent illiterate females were over-utilised.

A small percentage of aged males with medium educational levels and illiterate females were under- and over-utilised respectively.

As a whole we find that slightly less than one-third of the total males and about two-third of the total females engaged in tertiary economic functions were under-utilised. So far as the matter of their education is concerned we find that males are mostly in medium and higher levels of education and females are mostly in medium and lower levels of education. When we see age-wise variation then we find that children and aged males were more under-utilised, younger and older working age groups males are less under-utilised and majority of females from all age groups were under-utilised. 29.33 per cent males mainly with higher and medium levels of education

Table 30.3 : Variation in Education and Utilisation Pattern of the Tertiary Rural Labour Resource by Age Groups and Sex

		Age Group									
		5-14		*15-34*		*35-59*		*60&+*		*Total*	
	Sex	*M*	*F*	*M*	*F*	*M*	*F*	*M*	*F*	*M*	*F*
	Total	1	1	68	8	79	13	2	1	150	23
	%	(0.66)	(4.34)	(45.33)	(34.78)	(52.66)	(56.52)	(1.33)	(4.34)	(100.00)	(100.00)
Under-Utilised (Below 6 Hours)	I*	—	—	—	—	1	2	—	—	1	2
						1.28+	15.38			(0.66)++	(8.70)
	II	1	1	—	1	1		—	—	2	2
		100.00	100.00		12.50	1.28				(1.33)	(8.70)
	III	—	—	4	3	3	4	1	—	8	7
				5.88	37.50	3.79	30.76	50.00		(5.33)	(30.43)
	IV	—	—	9	1	10	2	1	—	20	3
				13.23	12.50	12.65	15.38	50.00		(13.33)	(13.04)
	V	—	—	10	—	6	1	—	—	16	1
				14.70		7.59	7.69			(10.67)	(4.35)
	VI	—	—		—	1	—	—	1		
						1.28				(0.66)	—
Grand Total		1	1	23	5	22	9	2	—	48	15
		100.00	100.00	33.81	62.50	27.87	69.21	100.00		(32.00)	(65.22)

(Contd.)

Table 30.3 *(Contd.)*

Sex		Age Group									
		5-14		*15-34*		*35-59*		*60&+*		*Total*	
		M	*F*	*M*	*F*	*M*	*F*	*M*	*F*	*M*	*F*
Adequately-Utilised (6 - 8 Hours)	I	—	—	—	—	1	—	—	—	1	—
						1.28				(0.66)	
	II	—	—	1	—	3	—	—	—	4	—
				1.47		3.79				(2.67)	
	III	—	—		1	4	—	—	—	4	1
					12.50	5.06				(2.67)	(4.35)
	IV	—	—	5	—	5	—	—	—	10	—
				7.35		6.32				(6.67)	
	V	—	—	2	—	6	—	—	—	8	—
				2.94		7.59				(5.33)	
	VI	—	—	5	2	12	—	—	—	17	2
				7.35	25.50	15.19				11.33	8.70
Grand Total		—	—	13	3	31	—	—	—	44	3
				19.11	37.50	39.22				(29.33)	(13.04)

(Contd.)

Table 30.3 *(Contd.)*

Age Group		*Age Group*									
		5-14		*15-34*		*35-59*		*60&+*		*Total*	
Sex		*M*	*F*	*M*	*F*	*M*	*F*	*M*	*F*	*M*	*F*
Over-Utilised (Above 8 Hours)	I	—	—	4	—	7	4	—	1	11	5
				5.88		8.86	30.76		100.00	(7.33)	(21.74)
	II	—	—	6	—	7	—	—	—	13	—
				8.82		8.86				(8.67)	
	III	—	—	6	—	5	—	—	—	11	—
				8.82		6.32				(7.33)	
	IV	—	—	9	—	4	—	—	—	13	—
				13.23		5.06				(8.67)	
	V	—	—	5	—	3	—	—	—	8	—
				7.35		3.79				(5.33)	
	VI	—	—	2	—		—	—	—	2	—
				2.94						(1.33)	
Grand Total		—	—	32	—	26	4	—	1	58	5
				47.04		32.89	30.76		100.00	(38.67)	(21.74)

* Levels of Education : I—Illiterate, II—Primary, III—Junior Secondary, IV—Secondary & H.Sec., V—U.G. and P.G., VI—Professional/Technical.

+ Unbracketed lower figures are the percentages to total males or females of respective age group.

++ Bracketed lower figures are the percentages to total male or female tertiary rural labour resource.

Source : Field Survey, 1995.

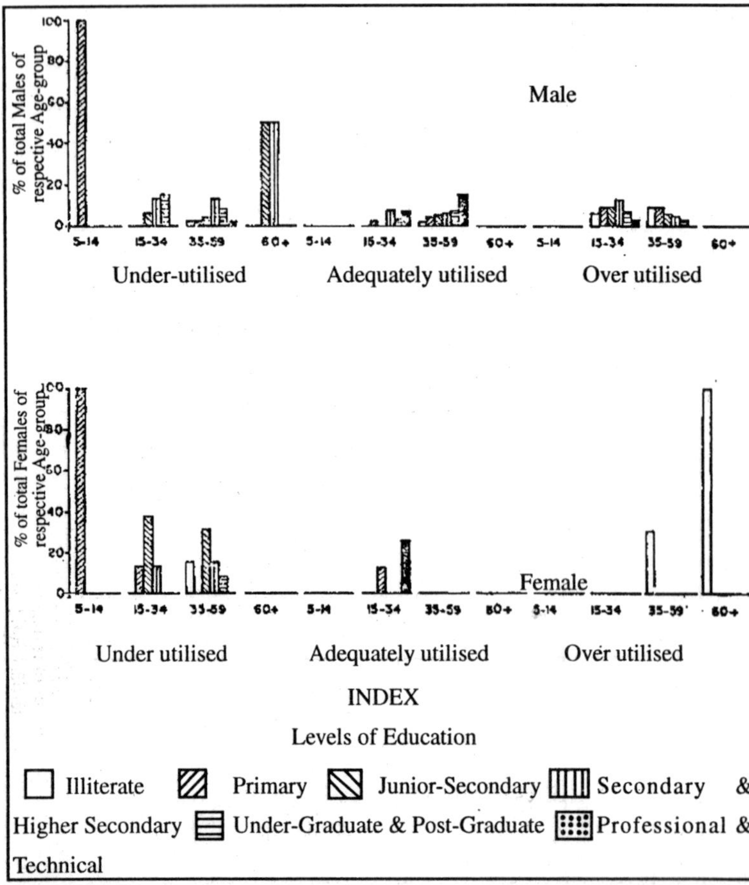

Fig. 30.3 : Variation in Education and Utilisation Pattern of the Tertiary Rural Labour Resource in Burdwan District by Age-groups and Sex (1995)

and 13.04 per cent females mainly with higher educational levels were adequately utilised. In this utilisation category older age group (35-59 years) males are more than the younger age group (15-34 years) persons. About 39 per cent of males showing majority in medium and lower levels of education and about 22 per cent illiterate-females were over-utilised. But when we see age-wise variation we find that younger age group males are more over-utilised and older age group males and females less over-utilised.

Relatively more over-utilisation of younger working age group of males is observed due to their more participation in terms of working hours in commercial and transport activities rather than other services mainly governmental. Similarly, more under-utilisation of females from working age groups may be because of their less participation in tertiary activities. More percentage of males than that of females in adequate utilisation is due to the fact that males particularly from government service age group (35-59 years) are more employed than females.

CONCLUSIONS AND SUGGESTIONS

After having gone through the above it has been found that the primary rural labour resource comprises 78.00 per cent males and 87.00 per cent females to total rural male or female labour resources of the study area. Secondary rural labour resource comprises about 5 per cent males and 4 per cent females whereas, the tertiary rural labour resource constitutes around 17 per cent males and 9 per cent females to total rural male or female labour resources of the study area.

Most of the child labour from 5-14 years age group (particularly with illiteracy and primary education) were engaged in cultivation and more under-utilised. Similarly, most of the aged workers from over-working age group 60 and above with the majority in medium and lower level education (in male category) and in lower level education (in female category) were found working in cultivation as cultivators or agricultural labourers and most of such workers were under-utilised and some with similar level of education were adequately utilised in primary economic function and some others were under-utilised in secondary or tertiary economic functions.

Age, education and utilisation-wise variation shows that in primary economic functions the percentage of males or females to total primary rural male or female labour resources were higher in adequate utilisation category. In this utilisation category younger working age group males and females were more than the older working age group males or females. Education of former category labour is seen better than that of the latter age. About 40 per cent males and 25 per cent females were in under-utilisation category in which males education (mostly being in medium and lower level education) is far better than that of females. Likewise, about 19 per cent males and 17 per cent females were over-utilised in which the percentage of older working persons is found more than that of the younger working persons. In all utilisation categories the percentage of education generally decreases from lower to higher levels of education except few cases in exception. Most of the primary female labourers were found working as agricultural labourers.

Secondary rural labour resource is small in percentage. Out of the total secondary male or female labour resources 70 per cent males and 90 per cent females were adequately utilised in which younger working persons were more than that of older working persons. Those who were under-utilised comprised about 14 per cent males only. Similarly, about 16 per cent males and only 10 per cent females were over-utilised. In all utilisation categories the percentage of junior secondary persons was more and the percentage gradually decreases towards primary and illiteracy side.

In tertiary group around one-third males and around two-third females were adequately utilised. In this utilisation category the percentage of younger working males and older working females were more than that of the younger working females and older working males. Only 7.33 per cent males and around 22 per cent females with more percentage from older working age group were adequately utilised and about 39 per cent males and 22 per cent females were over-utilised. In this utilisation category the percentage of younger working males are more than that of older working persons. In all three utilisation categories the percentage of tertiary rural labour resource was higher in upper and medium levels of education.

Lastly, we can say that in terms of quality measured by educational achievement tertiary labour resource is better. Secondary labour resource is of moderate quality and primary labour resource is of low quality.

Keeping in mind the above situation it is suggested that primary rural labour resource particularly females should be imparted vocational education/training, extension education in agriculture through formal and non-formal educational and training centres. Functional literacy should also be provided to both male and female rural labour resources through governmental and non-governmental agencies. It is imperative to suggest that different types of household industrial and commercial activities should be developed in the District to absorb more males or females for their efficient utilisation and for quality development they should be linked with skill upgrading educational systems. With a view to absorb more labour resource particularly females in tertiary economic function efforts should be made for creating additional job opportunities. Skill upgrading formal and non-formal educational systems should also be introduced. Thus, for proper generation of labour resource potential and its efficient utilisation the benefit of good training, education, health facilities, job opportunities and the gain of material advancement should be made available to the persons of all age brackets and all segments of the society.

REFERENCES

African Development Report (1998), *Human Capital Development*, African Development Bank, Oxford University Press, New York.

Correa, H.C. (1963), *The Economics of Human Resources*, North Holland Publishing Company, Amsterdam.

Das Gupta, S.P. (1998), Resource Development in the Light of Indian Tradition, in *Geographical Review of India,* Vol. 6, No. 4, December 1998, The Geographical Society of India, Calcutta.

Kothari, V.N. (1965), *Investment in Human Capital*, Popular Prakashan, Bombay.

Srinivasan, K. and Shariff, A. (1997), Towards Population and Development Goals, UNFPA for United Nations System in India, Oxford University Press, Delhi.

Wiesbroad, B. and Hughes, H. (1983), *Human Resources, Employment and Development,* Vol. 3, The McMillan Press Ltd., London.

31

RURAL POVERTY ALLEVIATION IN BASHAGARD, IRAN

SAYED HASSAN HOSSAINI ABARI

Description

Bashagard is a physical socio-economic and cultural region in south-east Iran, facing the Oman Sea in south and Hormoz strait in its south-west. The most isolated and inhospitable and sparsely populated region, but its mountain ramparts has not been the scene of activity and development. Its area of 16,000 km, which is in the most north of the Bashagard Mountains, extends down to the Oman coast. Most of the region is mountainous with deep valley and steep slopes. According to the (I.H.) 1,375 (1996 A.M.) sensous its population was about 75,000, living in about 900 scattered villages.[1]

In the long history of Iran upto the Islamic revolution (1979), the region of Bashagard, though toward above the Oman sea and Hormoz strait remained in isolation, so that, all in the geographical and historical records of Iran, are not referred to the region. Up to those times even the central government had not a full authority in Bashagard. In the times of English colonization on Indian subcontinent even, the colonial agents could not penetrate all through the region (Gabriel, 192).[2]

According to these circumstances the region remained totally isolated and frustrated.

At the time of re-exploration, 1361 (1982) the common features of Bashagard were as follows :[3]

1. Not any road had been made all through the region, but totally animal paths.
2. About 75,000 population lived in about 900 hamlets, without

any city or town (the average households of each palm leaf strawed 12 ones).

3. The illiteracy was common (only 2 per cent of men were educated).
4. The disease was in full number and common place (*i.e.* malaria, tuberculosis, typhoid, small pox, leprosy, and a number of skin pests).[4]
5. No electric light, and water supply and not any path, the scarce watertables and weak springs were both for drinking and washing.
6. The Khan was governing the region in full authority from the early times.
7. The caste system was main social system. The society was classified in nine layers, three of them were higher and 6 were lower (the slaves).
8. Being mainly the polygyny wives was the common way of life, 2, 3, or sometimes 4 wives.[5] The men in any condition of oldness could marry the young girls over 9 years old. As the result there were fatherless children in a large number everywhere in the area.
9. From the economic point of view, the lower caste had no right to the economic resources. They could live only as slaves.
10. The Bashagardy people mostly lives in poverty, illness, social pressure and so on.
11. The majority of population had no direct contact with outside area.

It should be mentioned that there was a shortage of irrigation water, cultivable land, and the pastures were scarce, so that, the river valleys do not tend to be flat floored owing to the deposition of material eroded from the adjacent slopes.

Soils

The alluvial fans are not wide separated, but along the river basins, the soil which is one of the most vital material resources of any region, is not possible, to take place.

Among the factors that are of controlling soil formation, are climate, parent material, vegetation, slope hydrology, micro-organisms man and time. In Bashagard but the parent material (sand looms-clay

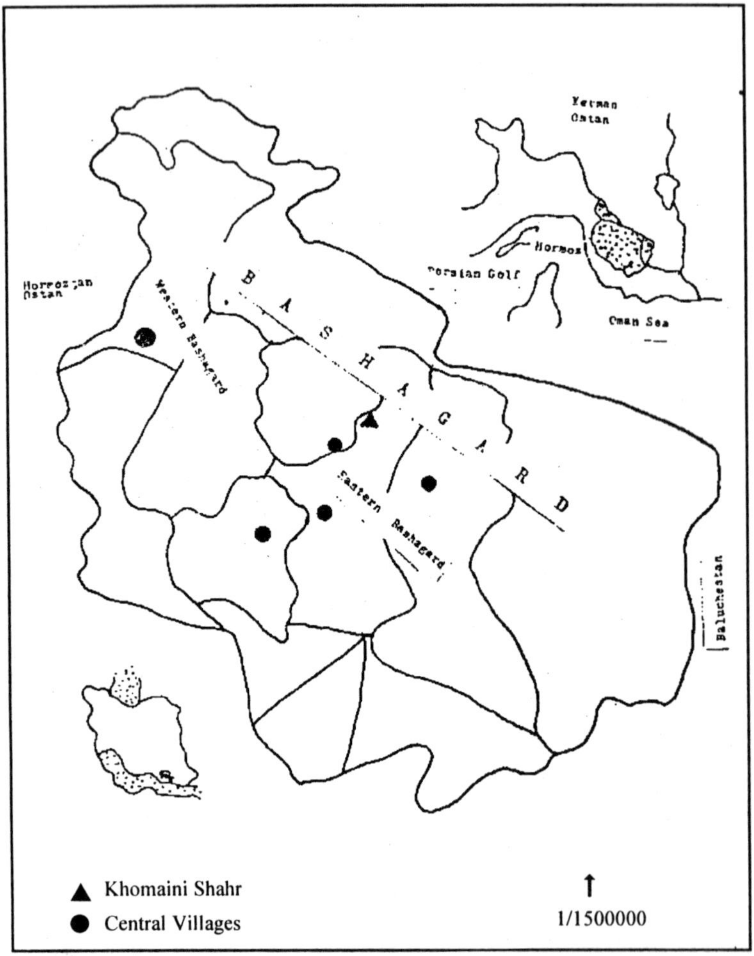

Fig. 31.1 : Bashagard

and claylooms) and climatic conditions (hot summer and heavy rainfall) and shortage of vegetation are the most significant (Fig. 31.1).

Generally speaking, the most widely occurring soils in the region are without doubt, the red or reddish gray soils found throughout the north region.[6] Owing to the dryness of the region in which these soils are found animal and plant the growth within them is minimal and as a consequence, the humus content is very low. As a result in most parts of the region, two major problems exist with regard to the utilization of the soil resources. These are soil unfertility, and soil erosion, which occurs when water flows and when slopes are steep. Specially when the heavy rains occur in summer and autumn.[7]

Control measures against soil erosion has been the main problem for the local people all through the history of the region, but their tools and construction materials is weak and unstable, therefore, after each flood, they are to be reconstructed.

Relief, Geology, Geomorphology and Soils

In Bashagard a very complex pattern of mountain ranges described as the Bashagard mountains are covered the entire area (about 900 per cent) attaining only 2000 m in altitude, sloping to the dunes in the coast of Oman sea in the south. Here in the coast some of the dunes are more than 200 m in height.[8]

In the region catastorphic geomorphic events play an important role in the evolution of land forms and at the same time greatly influence human activity. Most of those events occur during the season of maximum precipitation, and are a direct result of water action. In upland areas, mud flows are common occurrences on unstable slopes with less vegetation cover. More generally, floods can devastate the river valleys and lowland areas.[9]

Despite the humidity, which prevails over the region, fluvial action tends to be the most important geomorphic process though it might operate only on a few days per year in the most parts. There in this region the effects of the winds and salt are very rare.

The out crop as usually resistant rock types and very steep slopes characterizes the uplands. The amplitude of relief may be hundreds and even more than thousands of metres.

Animal and agriculture, which are the backbone of the region, are the basic feature of Bashagard, combining internal influence and indigenous element. It means that it brings an annual cultivated land only about 3 square metres per head. Goat (the main milch animal) is only 1 and the palm (date tree) one plant.

The Nature in Bashagard

The region of Bashagard as a slopy plateau in south-east Iran lies at an altitude of 200 m in the north to the sea level on the border of

Oman sea. The area to the north where the head streams of the great floody rivers Jagin and Gabrik rise. Several closely spaced mountain ranges cross it, mostly cutting the bordering heights. To pass beyond the main rivers and their tributaries one has to cut across these heavily dissected region, having passed the high mountain ranges and river gorges among the rocks and stones which plunges into the steep depths of the next river valley.

The climate in the region is mostly hot average annual 25°C. But it differs in four seasons (autumn, winter, a short spring, and a long summer).

Climate

The weather patterns of Bashagard can best be expained by reference to the succession of cyclones, which pass over the region. Two types of rainy disturbance have been recognized in the region. The first are shallow waves making rapidly in the upper troposphere, which cross the area. Rain associated with these disturbances is a maximum during the winter months.

The cold troughs in the upper atmosphere creates almost stationary in the region from the second type of the group. They draw in moist air and cause precipitation in the region. The maximum precipitation associated with these systems occurs during autumn and winter.[10]

Another is monsoon airmass, which brings rain in the late summer. Temperatures are almost high. The highest mean daily temperatures at the 6 months of summer passes more than 45°C. Even on the low lands in south day time maximum can often pass 48°C. In winter the mean daily temperature in January comes down to 2°C.

Table 31.1 : Mean Monthly Temperature in Angohran (Bashagard 197-193)

D	*N*	*O*	*S*	*A*	*J*	*J*	*M*	*A*	*M*	*F*	*J*	*MON.*
0.6	15.2	20	24	26	23	18	23	26	13	10	8.8	Average min.
12.5	26	33	38	38.6	39.2	12.8	37	28	24.3	20	18.2	Average max.
10.9	18.8	13	12	8.6	13.4	14	14	8.6	11.3	10	3.4	The range

In short in Bashagard, actually two long seasons occurs during the year, hot season about 6 months and moderate 6 months.

Table 31.2 : Monthly Potential Evaporation in Bashagard

J	*F*	*M*	*A*	*M*	*J*	*J*	*Au.*	*S*	*O*	*N*	*D*	*Local Month*
85	102	175	230	325	42	330	220	220	220	195	120	2632

Table 31.3 : Seasonal Average of Precipitation in Sardasht

Winter	*Autumn*	*Summer*	*Spring*	*Amount*
220.9	50.3	13.7	23.4	mm
71.7	16.3	4.45	7.55	

Water Resources

Despite the great size of the Bashagard, there are no constant water resources but only seasonal flows, in Jagin, Gabric and some unimportant waterways into the Oman sea.

Maximum precipitation occurs during autumn, winter and some in the Indian monsoons, almost all the rivers of the region are characterized by region hydrographs with maximum charge in all the seasons' flows in to the sea. Not any certain records is known or available for the amounts of the runoff in Bashagard. But some estimations show that about 45 per cent of the precipitation of the area directs to the sea. The remainders penetrates in the rocks and crashes making the very weak springs, effecting vital cultivation in the region. Now groundwater and Qanats cannot be fond there in Bashagard.

The first and most common method of erosion control all through the region is the construction of physical barriers which would reduce the speed of surface runoff, as well as penetrate it in the ground, charging the sources of the weak streams.

Another, which is most widely used as the soil conservation technique, is terracing. All the methods of constructing are done by hand labour and have been a feature of rural agriculture in the region. Since 1361 (1982) modern conservation methods have been increasingly introduced. These include Gradoni Terracing, contour walling, check-dam, and storage dams contruction.[11]

Agriculture

Though pasturalism (mainly goat rearing) and agriculture are the only types of land-use in Bashagard's and their role is of most importance in the life of the region. But physical conditions, especially scarcity of water and soil are the most important constraints on rural land-use, there.

About the land tenure in the region, there is not any State owned and any big land owner in Bashagard, but only private system, not trial, the total land belongs to the three upper sects of the society, and the lower sects absolutely has not any right to the land, they only work as the sharing slaves.

Traditional pattern of agriculture in Bashagard survived right up to the early times of its history preponderance of food crops. For the long period the indigenous varieties of cooling (a small-sided variety of sorghum), might come from East Asia.[12] The maize moved of west from the west Asia, barley broad beans form inner Iran, several varieties of date palm from the sorrounding areas and citrus fruits as sour lemon, for dried fruits, with the lower average yield. The above cereals became the main crops in the region, and was grown wherever possible in the region. In many areas of the region soil is sandy and infertile and rainfall slight, and also drought are frequent, and no water is available for irrigation especially in between the long periods of rainy seasons. Some indigenous vegetables are also cropped. But not as the cash crop, only for local use.

Settlement

Settlements in Bashagard are a special pattern in all around Iran. It means that the 75,000 population of the region live in 900 small houses with averagely 10 household each, and the largest 200 as the traditional headquarter of the area. It is mainly because of scarcity of irrigating water. Moreover the houses almost all are straw made. The water comes from the springs, along the tributaries of the rivers, and geostructure of the region is made of the limestones. The basins constantly are washing away and the source of the spring demolishes. Therefore, after some distance another new spring starts in that continuation. The oasis should move after the spring. So that it may be that during one generation, an oasis once or twice is moved following the spring. Actually in Bashagard, the springs always change their location and the houses following them afterwards.[13]

Bashagard through the Alleviation Times 1361 (1982)

After the re-exploration of Bashagard (1361), several groups of government staffs and discussion experts visited the region. And as a consequent, two main decisions were discussed about its future.

The first decision was to move the population to the areas with broad and rich economic resources. The second was to leave them so that they choose themselves what and wherever they liked.

The writer and the other geographers in the group, found the area capable for development as a source of about 8,000 million cubic m of water running out to the Oman sea, a moderate climate in compared with the coast, as well as its political position in south-east of the country.

After about four years of discussions, it was accepted that the people should remain there in the region and start its poverty alleviation due to a sustainable development programme.

To make an attempt to start poverty alleviation in Bashagard, the planning group faced the region and area of full range of absenties, *i.e.* (1) The absence of irrigation water, (2) The scarce cultivable land, (3) Lack of social services, (4) A strong caste system within the majority of landless people, (5) No road and accesses, (6) No trade system as well as the lack of capital, and so on.[14] First, an attempt was made in the strategy for development, not for an agrarian community, nor for rural development. But an approach to rebuild a totally ruined area, with the 900 small poor peasant places.

The beginning of the discussion was made on the basis of comprehensive regional development with several various objectives. It then identified certain pay element of a model on which lead to a set of economic, social and cultural actions for the efficacy of a particular poverty alleviation strategy.

The second was to organize, develop and utilize the potential resources of water, land and manpower in such a manner that the entire region depended on these resources had a constant step for development to meet at first a minimum their basic needs of food, clothing and shelter with the reasonable facilities for education, health and employment.[15]

As already mentioned, the Bashagard social system based on castes and social categorizing in 9 categories totally up to down. The social and economic tradition was against the developing affairs. So, the planning should be mentioned indigenous traditions carefully. Therefore, at first step all of those were as they were with some special care. But gradually should be changed. However decided that the poverty alleviation in Bashagard should start on the three basic main points.

1. Started the watershed management and water and soil control, everywhere.
2. In the field of cultivation the introduction of new cereals and new plants as well as developing the indigenous cereals was taken in hand.
3. Storage and marketing for the local productions and new cash crops was of main attempt.

Handicrafts, (indigenous and new ones) as the means of the ladies income also were put in hand. Due to the social needs also a long range of attempts started. Among them, the schools of three levels (elementary, middle and high schools as well as teacher training and religious schools) both for the boys and girls, suitable to the number and sexes of students were built, and the educational needs and food were supported for all the students in the schools far away from the student's own villages. Diseases and pest control started rapidly and health services and medical aids were made in suitable accounts.[16]

The third attempt which made to develop the accessibility both economic and social, first of which a widespread road making, postal, telephone, trade system, mini-bus services, co-operative stores, marketing for cultivated crops repairing shops, petrol shop, road tea shops, public and private transportation, finance systems as Islamic and common banks were made.

The system is totally non-governmental, but a human Islamic task. In this way it made a long range of helps in food, medicines, roads, clothes, and a kind of special aids mentioned as Shahid Rajai's project. That is to give some monthly salaries to the olders. Another aids were to make a social charity to the poor, parentless children, sicks, and so on, under the Imam Khomaini's committee. Moreover the finance came from the three more resources, as :

1. From the personal aids as the vowes, indocument, qarz-olhassaneh (a kind of loan without interest).
2. Voluntary manpower from outside the region, as the scientist and experts, for researchers and consultants.
3. The local labour in place of their share in constructions (road, water and soil control, etc.).

Due to those aids some special belief took shape among the local population. So that whatever advices came from the development groups they without any hesitation obeyed it. And it was the main source for the group.

The second approach was planning for regional development as the main attempt for poverty alleviation all through the region.

The strategy emphasised on the three dimensions of human need.

1. Livelihood.
2. Social needs.
3. Accessibility.

Achievements

According to a comprehensive group study on the achievements of poverty alleviation in Bashagard, after 15 years of its start time (1982-1996) the socio-economic features of the region was more or less changed. Respecting to the above mentioned, three dimensions of poverty alleviation it can be approached as follows :

1. The family livelihoods[17].
2. The social services.
3. The accessibility.

Amongst all of the needs for everyone, is to meet his family needs as his occupation, food, shelter, health, clothes and so on in Bashagard where the people almost are the most frustrated portion of Iran's nation, the first attention seemed to start with growing agriculture through the water and soil control. Either as the food or the cash crops. But it is basically depended on the operations for water and soil control, due to it may rise the occupations, such as gathering the local structural materials, levelling the land, digging the waterways, transport work, cultivation, driving the machineries, marketing, the tools, working on the roads and a large number of relevants. To bring about a good believe and hope to the natives. Therefore, 15 years of experience there showed that:[18]

(*a*) Each of 10 kms of the area gives at least an amount of one million cubic fresh water of which, an area of 50 hectares of irrigated crops can be made.

(*b*) About 100 hectares of level land is beyond the soil control dam ready for plantation of date palms, as well as being a good pasture for sheep and goat rearing.
(*c*) Increasing the discharge of springs along with the riversides downwards.
(*d*) A total of about 60 persons occupation in cultivation and its relevants.

Since in Bashagard for each of the farmers about one hectare of irrigated land satisfies to crop his foodstuffs, and cash crops, water and soil control was accepted as the main way of household poverty alleviation. The operations followed under three types of schemes of construction.

1. The construction of dams.
2. The building of diverginal dams to direct the floods in irrigation.
3. The construction of water reservoirs as the main water resources.

The Managing Systems of Water and Soil Control in Bashagard

The system for watershed management and flood control in Bashagard was not adopted from any other countries. But entirely based on traditional system which long has been operated in Islamic Iran. The model is fluctuated and opened. So that, with rare change in it, can be operated to all the socio-economic actions in which the benefits come to the privates, such as: The construction of the irrigation dams, the levelling for cultivable lands. The house makings and so on. In the construction the large and public projects, such as the roads, schools and mosques and also the large water reservoirs, the Imdad Committee of Imam Khomaini, the Charitants and the respond government officers make some sort of participation. Figure 31.1 shows the watershed management system in Bashagard.

The Social Services

When speaking about the social needs in Bashagard, it is not the special ones, as in the big cities but it refers on the general and critical possibilities. It is because the needs in each place or region differs widely.[19]

In the region of Bashagard, there are very primary services on which the people can start their future irrigational development. Therefore, during the mentioned period, the following services were made:

- *(a)* More than 135 schools all through the region of which 117 was elementary schools, 10 middle schools, for the boys and girls, 6 high schools for the both sexes, one teachers training and one religious school in the centres of the region, due to it at present. About 40 Bashagardians are learning different branches in the universities.
- *(b)* A number of health centres and its relevants are working in the region.
- *(c)* In each of the central villages a public bath is constructed and is working.
- *(d)* All the government regional offices now are located in the region.
- *(e)* The 900 scarcely populated oasis of the region collected, to a number of first 450 and at last will be reduced to 150 villages.[20]

The Accessibility

The term accessibility means simply "The ability of people to make about and this can be measured by *(a)* the movement and nature of trips actually travelled. (Whether on foot or using a vehicle) and *(b)* the ease of movement experienced by the individual according to personal and social economic criteria and transport criteria, such as availability of and ability to use private or public transport.

In Bashagard due to the dense mountain ranges, deep valleys and the fearfulness of the area, up to 1361 all the road accesses was foot paths and animal routes. So that a trip to outside the region took more than 15 days (to Minab or Jask) because not any automobile could cross the region. Consequently, socio-economic relations between the local people and outsiders was very poor, and also the information technology was not possible because not any government office was established, thus not any member could stay in Bashagard, due to fear of disease and lack of insurance.

Regarding the accessibility and mobility in Bashagard, the following resources is made during the mentioned period.

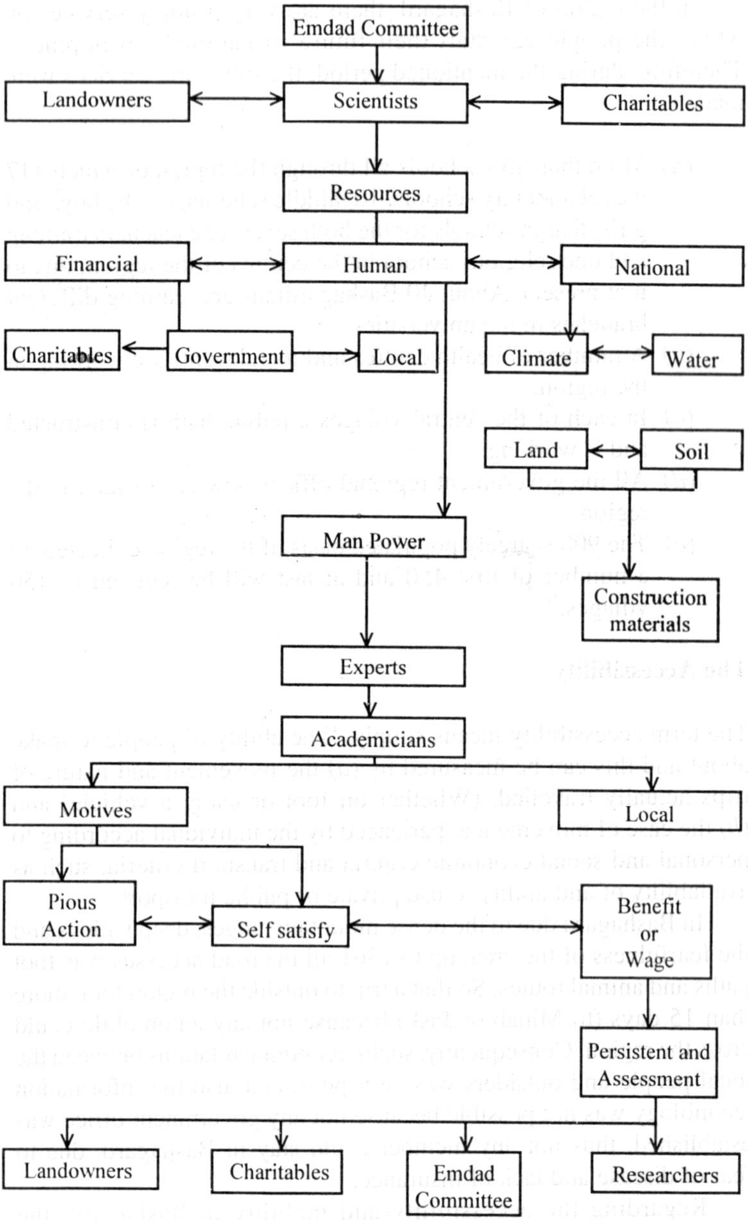

Fig. 31.2 : The Watershed and Development Management System in Bashagard

(*a*) Some about 1500 km of roads made. 400 km hand made and 100 km mechanized. Out of which only 150 km is concrete and the rests are dusty.[21]

(*b*) More than 250 km of electric lines which due to it about 20 per cent of the total population are benefited.[22]

(*c*) A number of about 5 telephone lines now are operating in the central villages.

(*d*) The marketing system of the region developed with some 30 cooperative stores and a lot number of private stores all through the area.

(*e*) Though in each of the axis of Bashagard the private tracks now are in the transportation of human and loads, to facilitate the mobility in the region a four government mini buses move the people in a cooperative cost to the region. But the out road people should still walk a long distance to reach the stations.

(*f*) The post offices started working all through the region.

NOTES

1. Hossaini Abari, S.H. (1990), Bashagard, *A Look on the Dimensions of Poverty*, Esfahan University, p. 250.
2. Gabriel, Alfons, *The Geographical Studies on Iran*, translated in Farsi by Khaje, Nuri, 1954, p. 415.
3. Hossain Abari, S.H., Bashagard, The Poverty, "Environmental or Political", *Geographical Research* (Seasonal, Mashad, 1993, Vol. 33, p. 51).
4. Hossaini Abari, S.H., *ibid.*, p. 27.
5. Hossaini Abari, S.H., *ibid.*, p. 45.
6. Atai, Hooshmand, 1993, The Hydro Climate of Angohran River, M.A. Thesis, Department of Geography, Esfahan University, p. 125.
7. Beaumont, Peter, and others, 1976, The Middle East, J.W. Arousemith, Bristol, p. 25.
8. Beaumont, Peter, *ibid.*, p. 25.
9. Beaumont Peter, *ibid.*, p. 27.
10. Angohran, The Climatologícal Station, the data.
11. Angohran and Sardasht, Climatological Stations, the data.
12. Hossaini Abari, S.H., 1993, *ibid.*. p. 52.
13. Hossaini Abari, S.H. and others, The Socio-Economic and Cultural Effects of Imdad Committee of Iman Khomaini in Bashagard, Central Office, Tehran, 1996, Vol. 1, p. 215.
14. Kolb, Albert, East Asia, Methuen, London, 1971, p. 105.

15. Paul, J. Cloke and Chris, C. Park, 1986, *Rural Resource Management*, Croamhelm, London and Sydney, p. 315.
16. Natural Resources and Development, 1992, Institute for Cooperations, Helm Tubingen, Vol. 51, p. 85.
17. Hossaini Abari and others, *ibid.*, Vol., p. 215.
18. Hossaini Abari, S.H., 2001, Bashagard, Poverty and Poverty Alleviation, University Press, Esfahan University, p. 129.
19. Hossaini Abari, S.H., 1995, The Experiences of Watershed Management in Bashagard, Geographical Researches, Mashad, Vol. 37, p. 98.
20. Paul, J. Clock and Chris C. Park, *ibid.*, p. 309.
21. *Ibid.*, p. 303.
22. According to the Field Investigations, Jan. 2001.

32

POPULATION, POVERTY AND ENVIRONMENT IN SISTAN AND BALUCHESTAN

BEIK MOHAMMADI

ENVIRONMENTAL AND GEOGRAPHICAL CONDITIONS

Geographical and Situation Bounds

Sistan and Baluchestan is located south-east of Iran and its borders with Pakistan, Afghanistan and the Oman sea have made it very important in the region.

The province is 1,81,000 km^2 vast, and has the third rank after Khorassan and Kerman regarding extent.

The population of this province in 1996 was equal to 17,22,000. Sistan and Baluchestan consist of two sections : Sistan in the north and Baluchestan in the south. These two sections are considered two separate areas in geographical, environmental economic and cultural aspects, but they have joined and shared the fate in their geographical isolation, environmental: poverty and political divisions.

Sistan is a desert and semi-arid area with an altitude of 500 m. Its agricultural economy and people's sustenance are dependent on Hirmand River.

Baluchestan is a vast, high and semi-arid mountainous area which, in the past, extended to the Indus and later in the time of Qajar some of its parts separated from Iran as part of Pakistan. (Afshar- e Sistani, 1987, p, 201)

So, there is a close relationship between the two parts.

Relief, Geomorphology, Soil and Natural Vegetation

The region of Sistan and Baluchestan is located at the intersection area of eastern mountain ranges of Iran and the continuation of Bashagard heights.

The heights of this area are relatively young belonging to the second and third geological periods, and geomorphic events play an important role in the evolution of landforms and formations, and at the same time, it has been greatly influenced by human activities. Most of these events occur during the season of maximum precipitation, and are a direct result of water action.

Among these heights, there are various pits and basin, the most notable of which are Sistan and Mashkel pits, Jazmoorian plain, and Oman coastal region.

The general outlook and geomorphological perspectives of the region consist of erosional valleys, alluvial plains, deep and steep valleys, high bare rocks, and sand hills, which can be seen in different places of the region.

Because the region is located in a warm and dry climate, most of the soil face limitation due to more salt, lack of drainage, marshy nature of the land, unevenness, low depth, gypseous and loamy compositions and coarse structure.

Therefore, the expansion of alluvial and fertile soil in the province is very unlikely and can only be seen in the limited part of the region such as, east of Sistan Bampoor plain and around Simish valley in Saravan.

On the whole, only 1 per cent of the province land is fertile which is about 106/599 hectares (Statistics Note Sistan and Baluchestan Province 1996, p. 245).

Climate and Water Resources

The climate of the province, in general, is warm, dry and of a desert type and locally, in the northern half is of a moderate desert type. Also in the southern half, it is of dry-steppe with arid winters . In general, the more we go south and the more we move from the mountainous area to the valley and pits in the lowlands, the moderation of weather lessens and the heat will increase.

Throughout summer, temperatures are high almost everywhere. The highest mean daily temperatures in this period, of more than 34°C are recorded in Iranshahr and Sistan and some part of the southern region of Baluchestan. In the coastal region such as Chah-Bahar, diurnal temperature ranges tend to be smaller than in the interior parts. During the summer months, very few places, however, are free from relentless heat during midday. Even on the high-lands of the region, day-time maxima can often surpass 40ºC (Table 32.1).

The average annual rainfall is 65 mm that ranges from 24 mm in Hormak Valley to 213 mm in Ghasre Ghand (near Chah-Bahar).

The wind, too, blowing from the north and north-west, is extremely hot, and owing to its high velocities, is heavily dust laden.

Maximum temperature of the province in almost all the towns is +40ºC. The warmest city is Iranshahr with 51ºC and the coolest is Zehedan with 25ºC.

Table 32.1 : Comparative between July temperatures at stations with Mean Daily Temperatures greater than 34°C

Area	*Average daily*		*Mean*	*Absolute*	
	max.	*min.*	*daily range°C*	*max.*	*min.*
Bushehr	35	29	6	44	23
Chah-Bahar	35	28	10	37.5	24
Tehran	37	22	15	43	15
Sistan	40	26	14	48	17

Source : Afrakhtah, H. (1996) *The role of Chah-Bahar in south-east coastal of Iran,* Vaghefi Press, Mashhad, Iran, p 19.

On the whole, considerable heat, scarcity of rain, relatively low humidity, high evaporation, high velocity winds accompanied by sand movement and dust can be considered is the main characteristics of the climate of the region. These are, of course, relative environmental factors affecting all different aspects of the economic, social, and cultural life of the people in the region and, as a consequence, causing general poverty and scarcity of population.

The needed water supply of the region is obtained from rivers, wells, springs, and under-ground canals. These resources, due to the climate conditions are not stable. Therefore except Hirmand river in Sistan, which is permanently flowing, other rivers in the region are seasonal and have little water and sometimes face floods and periodical drying up.

As the consequence, the severity water level fluctuations and the weakness of technology and capital investment have caused unstability and vulnerability in the agriculture.

Vegetative Cover

The region's vegetative cover is generally poor and consists of halophilic species resistant to the province climate. The pastures are of type 3, and due to the climate, destruction by man, abnormal grazing, they weaken gradually.

Only in part of the province southern region, and along Sarbaz river, there are considerable arborial species of palm, citrus fruits, banana, trees of arborial water melon, olive and tamarind (Beik Mohammadi Hassan, 1993, p. 6).

POPULATION AND POVERTY

Population and Growth Rate

The population of this province has approximately doubled during 1976-1996, and its growth rate has been more than 4 per cent. Yet, in the past decade, due to family planning all over Iran, its population rate has lowered, and the population has increased from 11,19,000 persons in 1986 to 17,22,000 persons in 1996 which implies a 3.6 per cent population growth rate (Beik Mohammdi Hassan, 1992, p. 141). High level of the rural population percentage, tribal texture and dominance of religious and ethnic culture, geographical isolation, cultural severe poverty, and the low level of marriage age have been more effective on that high growth rate than other factors.

Population Distribution and Density

Due to geographical conditions and specially shortage of precipitation, the population distribution is not balanced and there are lots of differences in the province in different regions.

Apart from the province north-western part which is nearly vacant, population distribution in its other regions is very dispersed and seen as non-harmonious nuclei.

Though occupying 11 per cent of Iran area, this province contains less than 3 per cent of our country's population. Its population density is 9 individuals which is very low in comparison with Iran's total relative density (37 individuals in 1996).

In this connection, the quantity and state of water resources, fertile lands and precipitation rate have been more influential than any other factors.

In Sistan, man's life and plant's survival merely depend on Hirmand river water flow. The southern coasts of the province, due to severe shortage of water (drinking and farming) and soil salinity, have very low population. Yet, on the south-eastern corner, the gulf of Chah-Bahar, trade and fishing activities have been very effective in the development of Chah-Bahar port and its +50,000 individual population.

Literacy and Education

The problem of literacy and education is an important limitations and indicators of Sistan and Baluchestan poverty. This problem has intertwined with the issue of employment and development so that objectives may ameliorate. Illiteracy has been much more acute in the past, yet it has now-a-days mitigated.

Statistics shows that in 1956, only 6 per cent of the population of the province have been literate. This value has increased to 36 per cent in 1986 and to 57 per cent in 1996 (total value in Iran, 79.5 per cent).

These figures though show a positive growth, yet, it is not so satisfactory compared with other regions of Iran.

This shows that Sistan and Baluchestan is one of the most deprived provinces of Iran.

Literacy rate is 70 per cent in urban regions and 45 per cent in rural parts. This rate for 6-14 year old individuals is only 72 per cent.

In the educational year 1997-1998, there have totally been 3,671 schools in the province—35 boy units, 12 girl units and 578 mixed units of which locating in rural district have been huts (lodges made with tree leaves, straws and branches). Some classes too, are attended in open space and in tree's shadows, while 6.5 per cent of which use non-standard and ruined buildings. In some villages only primary and almost guidance courses are attended.

Age Structure, Activity and Employment

Based on the last official census statistics of Iran (1996), this province has had 17,72,000 people that is 46 per cent of which have been urban and 54 per cent of which have been rural habitance.

The population sex ratio is 10:3. This ratio is 10:6 in urban region and amounts to 10 :1 in rural districts. Sex ratio for children (-1 year) is 10:4, yet it is 14:7 for adults (+65 years). In 1996, 29:8 per cent of the population having 10 years old and over have been employed and unemployed, and 12 per cent of the economically active population are jobless.

The province economically active whole population consists of 93 per cent men and 7 per cent women. Of the total employed population, 33.1 per cent were active in agriculture (agriculture, forestry and fishing), 22.7 per cent in manufacturing industries (mining and quarrying, construction and electricity, gas and water) and 44.2 per cent in services (trade, transportation, financing, community and other services). In urban areas, these figures were 5.50 per cent, 29.77 per cent and 64.73 per cent, and in rural areas 55.9 per cent, 16.9 per cent and 27.2 per cent respectively.

Table 32.2 : Employed Population in Major Occupation Groups for Urban-Rural Areas

Major occupation groups	*Total (%)*	*Urban areas (%)*	*Rural areas (%)*
Total employed	100	100	100
Agriculture forestry and fishing	33.1	5.50	55.9
Industry	22.7	29.77	16.9
Services	44.2	64.73	27.2

Source : National Census of Population and Housing, Nov. 1996. Statistical Centre of Iran. Sistan and Baluchestan province, p. 18.

In this province, due to household size (5.9 individuals per family) the father has to provide for 5 other persons apart from himself.

The comparison of the province major employed groups in the year 1966 and 1996 shows that the employed population gradually decreases in the agricultural sector and increases in the service sector.

Table 32.3 : The Trend of Economic Structure in Sistan and Baluchestan in 1966-1996

Activities	*1966*	*1996*
Agriculture	66 per cent	33 per cent
Industry	14 per cent	23 per cent
Services	20 per cent	44 per cent

Source : National Census of Population and Housing, 1996 Sistan and Baluchestan, Iran, Statistical Centre of Iran, p. 25.

The statistics of 1996 shows that approx. half of the population of the province is under 15 years. So, based on Verners test or formula (Germain, Veyret - Verner 1959, p. 62). and based in Werthiem's test (Amani, M. 1973, Principal of Demography, p. 13), the population structure of the province is young.

While total fertility rate is 6.4 babies in Iran (Farid, Y. 1995, p. 44), this rate increases to 8.8 infants in Sistan and Baluchestan. So it is clear that mortality incidence is more probable in pregnant women (Zanjani, H. 1990, p. 34). What preceded shows that population characteristics, employment and activity conditions in this province lack desirability, yet, there is a lot of depreciation. In this connection its environmental limitations, cultural severe poverty, tribal texture, traditional agricultural economy, high level of rural population rate, and low level of public literacy, and as a result, abnormal population growth and young structure have been influential.

Table 32.4 : Population Structure for Major Age Groups in Sistan and Baluchestan—1996

Age groups	>60	15-60	<15
Per cent	4.9	46.1	49

Source: National Census of Population and Housing, 1996, Sistan and Baluchestan, Iran, Statistical Centre of Iran, p. 18.

Biological Density

Based on the 1996-1997 agricultural statistics (province agricultural headquarter, 1997), total cultivated lands of the province extend to 106.599 hectares, That is 1 per cent of which is fertile (Province Statistics, 1996, p. 245). Regarding population, it can be said that the province biological density is 16 individuals. This value varies from 2.7 in Zabol to 50 in Zahedan. The province high biological density is indicative of farming land shortage, and the high difference of 2.7-50 individuals shows that there is no equilibrium in the province farming lands distribution.

In other words, while 2.7 individuals live in one hectare of agricultural land in Zabol city, this value is 50 individuals in Zahedan city.

Regarding agricultural lands limitation and population increasing trend, generally, more attention to technology in the direction of produces increase per area unit is inevitable. Otherwise, we shall encounter, every year, increased volume of agricultural products imports to the province.

Hygienic Indicators

Sistan and Baluchestan province is yet far from actualizing ideals and attaining development standards despite the authorities efforts and struggles, due to situations, social, climatic and geographical.

In 1986 there has been one physician for 4,400 individuals. This ratio is now one for 6,650 individuals!

It is interesting to know that, while the province population is about 2 per cent that of Iran, 37 per cent of those afflicted with trachoma belong to this province.

Hygienic information poverty, economic indigence and as a result, nutritives per capita consumption deficit, cultural privation, villages dispersion, open borders of Pakistan and Afghanistan and inappropriate climate are among factors reversely influencing the province health growth trend.

Table 32.5 : Comparison of some province development indicators with those of Iran (1996) totally 26 provinces

Indicator type	*Literacy (%)*	*Employed % Iran*	*Employed active Population (%)*
Province rank in Iran	24	22	26
Indicator type	Unemployment	Dependency rate	% of house owner families
Province rank in Iran	16	4.5	24
Indicator type	Ratio of physician	% of families having piped water	Families having electricity
Province rank in Iran	22	21	22

Source : Urban development indicators—Iran, Ministry of Housing and Urban Planning, 1998, p. 35.

It is worthy to note that besides manpower deficit, health and therapeutic facilities inappropriate distribution across the province, non-access to sound drinking water non-stability of management and manpower are other problems of this province. Vicinity with Pakistan and Afghanistan which by themselves have a lot of hygienic difficulties has also caused us to always witness the incidence of the type of dangerous epidemics in the borders of the province.

CONCLUSION AND SUGGESTIONS

The province of Sistan and Baluchestan is located south east of Iran in a strategic situation. Due to geographical isolation, non-compliance of climatic and natural conditions, topographical situation on the one hand, and ethno tribal cultural structure, cultural severe poverty, production activities indigence, open borders, non-official activities development, and finally lack of sufficient attention by federal governments in planning and investing in the region over history on the other hand, have undeniably influenced all the province socio-cultural and economic structures and have imposed lots of deprivation on the region people. So, it is the geographers task to participate in the development planning optimization by recognizing man and environment, understanding the relevant inter-relation and recognizing differences. On the other hand, It is their duty to guide the programmers at national, region and district levels by discussing every region's deprivations reasons problems and difficulties.

From what preceded it can be concluded that Sistan and Baluchestan province is one of the most deprived regions in Iran. Environmental limitations, cultural structure, authorities and planners and programmers have been influential in this connection. So the most important bottlenecks and deprivations of this province can be summarized in 2 environmental-geographical and human - economic. groups as below :

Environmental Limitations

- — Lack of precipitation, severe dryness of the weather and desert climate.
- — Geographical remoteness, distance from Tehran, and neighbouring with Afghanistan and Pakistan.

— Rivers seasonally, droughts, and sometimes periodic floods.
— Severe shortage of agricultural soils, and soil salinity specially in Baluchestan.
— Severe winds blowing and running sands.
— Hot and dry 120 days winds blowing in Sistan plain.
— Poverty of vegetative cover including pastural and forest species.
— Marshy and malarial regions in the province southern parts.
— Lack of sound drinking water in all the province.

Human Deprivations

— Rapid population growth, and population young structure.
— Economic poverty and the unemployed numbers increase specially in tribal and rural communities.
— Cultural severe pauperism and low level of public literacy.
— Tribal cultural structure.
— Low level of active population ratio, and woman reduced share in employment.
— Per capita income decrease and family size increase.
— Opium prevalence and the unemployed inclination to false and unofficial jobs.
— Presence of over 4,00,000 Afghani refugees and occupation of the province few job opportunities by them.
— Incidence of dangerous epidemics types in border regions.
— Poverty of planning programming, investing and technology.

Suggestions

The following are proposed in the direction of Sistan and Baluchestan deprivation minimization:

1. Decrease of the province population natural growth increasing rate through education and enhancement of literacy level specially for sisters.
2. Recognition and utilization of the province natural positive capabilities specially of the province 300 km water border.
3. Development of more job opportunities through investment and trade activities expansion.
4. Development and extension of border markets, and provision of border region residents living.
5. More recognition of mineral resources and their optimized utilization.
6. Increase coastal band fishing ships of the province.
7. Management of the floods and conservation of water of the region.
8. Participation of the people in the region of the province in planning and programming.
9. Management of the eastern border of the province and prohibition of drugs traffic and distribution.
10. Enforcement of vegetative cover and prohibition of severe soil erosion.
11. Fighting cultural severe pauperism through public literacy level enhancement.
12. Recognition of environmental potentials capabilities.
13. Programming and investing in region having production facilities.
14. Preparation of the required sub-structures for making use of planning models.

REFERENCES

Afrakhteh. H. (1996). *The Role of Chahbahar in Coastal Region of South East Iran,* first printing, Vagefy published.

Afshare Sistani, I. (1984). A look to Sistan and Baluchestan. Tehran, *Amine Khazrai,* issues.

Amani, M. (1973). *Principals of Demography*, University of Tehran.

Beik Mohammadi. H. (1993). *Locality Analytics of Population in Sistan and Baluchestan,* Ph.D. thesis. Islamic Free University of Tehran.

Beik Mohammadi, H. (1992). *The Patterns of Migration,* in Sistan and Baluchestan Eighth Geographical Congress in Iran, University of Esfahan.

Farid, Yadollah, (1995), Third edition, University of Tabriz, Iran.

Germaine Veyret-Verner, (1959), *Population, movements, Repartition*, ed., Arthaud. Paris.

John Coward and John Pollard, (1983), *Worked Exercises in Human Geography,* third printing, Cambridge University Press.

National Census of Population and Housing, (1996), Iran, Plan and Budget Organization. Statistical Centre of Irán.

National Census of Population and Housing Sistan and Baluchestan Province, (1996), Iran, Plan and Budget Organization, Statistical Centre of Iran.

Urban Development Indicators, (1998), Iran Ministry of Housing and Urban Planning.

Zanjani, H. (1990), Issues and Methods of Urban Planning, Population, Iran, Ministry of Housing and Urban Development.

33

ROLE OF CROPPING PATTERN IN FOOD SECURITY IN IRAN

AHMAD ALI BOZORGZAD

Introduction

The analysis of cropping pattern and the trends in the production of various cropping pattern and the trends in the production of various crops so far has made us enable to understand that the cultivators of these crops have not properly been guided by and contorolled through a suitable policy during last two decades. It seems that the farm producers produced different commodities in order to maximize their total revenue guided by market forces of supply and demand. What they did is true to every rational producer. If a certain policy is made to guide and help them to maximize their returns it would certainly be conducive to raise the production of each and every crop and thus the total agriculture production in the country. With regard to the cropping pattern we find that the farmers are producing even such crops where they have not only but very low productivity. Our main task is not to make an alternative cropping pattern which would raise the total agricultural production in the country. Moreover, under the proposed cropping pattern it is possible to utilize more area of land for double cropping also. It would result in the increase of more cultivation of area under various crops with the result that higher productivity level can be achieved.

The proposed cropping patern is based upon the following assumptions :

The first assumption regarding the quality of soil which is used for the production of particular crop is an essential factor governing the cropping pattern. The quality of soil, therefore, constitutes an important variable in determining the production of any crop. Since we are interested in extensive cultivation, in which more emphasis is laid on the more area under cultivation, the above mentioned factor do play here also a major role in the determination of an overall production.

The second assumption is regarding the adequate supply of chemical and mechanical inputs required to produce a particular crop in a given province. The regular supply of required inputs would help to undertake not only larger area under cultivation, but also raise the productivity.

The third assumption is about the government policy supporting such cropping pattern for the farmers to maximize their return on farm income. It should also be given due priority.

The above analysis is based on three years average period *i.e.* 1993-95. Therefore the deletion and addition of area under cultivation of particular crop is mainly done on national average productivity level. But in few cases we have included even areas of those provinces where productivity is below national average simply because their productivity in general is very low.

Most of the wheat producing provinces have recorded an increase in alternative area under cultivation.

These provinces include Esfahan, Semnan, Mazandaran, Kohgiluyeh-va-Boyer Ahmad and Fars. They have recorded an increase in area under cultivation by 9.4, 8.5, 49, 62 and 5.1 per cent respectively under our proposed pattern. On the other hand provinces namely Markazi, Bakhtaran East Azarbaijan, Khuzestan and Zanjan have been considered to reduce their area under cultivation of wheat by 36.4, 13.5, 6.3, and 28 per cent respectively.

Apart from these provinces the other have been assumed to have the same area under the cultivation of this crop. Under this pattern Gilan has been completely deleted to produce wheat.

This alternative pattern of land-use for the production of wheat registers a decrease of 5 per cent in the use of area without any change in the level of total production. It shows that a decrease of 5 per cent in area will not affect production with a proportional increase in its productivity.

The area of 300,000 hectares under cultivation of wheat may be saved and used for any other suitable crop in the given provinces.

In case of barley there was increasing in area in three provinces. These provinces inculding Tehran, Markazi and west Azarbaijan have recorded an increase in area under cultivation by 6.6, 92.8 and 16.5 per cent respectively. On the other hand provinces namely Kohgiluyeh-va-Boyer Ahmad and Mazandaran have been considered for complete of area under cultivation. The remaining provinces have been assumed to have the same area under cultivation. This proposed pattern of land-use production of barley registers a decrease of 4.6 per cent in the use of area without any change in the level of total production. It alsc shows that productivity of barley has gone upto 4.6 per cent.

The area of 96000 hectares under cultivation of barley may b(saved and used for any suitable crops in the given provinces.

Table 33.1 : Wheat: Actual and Proposed Land Area under Cultivation

	Actual (1993-95)			*Alternative*		
Province	*Area (000 hect)*	*Yield kg/hect*	*Production 000 tonnes*	*Area (000 hect)*	*Yield kg/hect*	*Productior 000 tonnes*
Esfahan	145.7	2790	406.0	159.35	2790	444.60 ++
Yazd	16.7	2110	35.3	16.7	2110	35.33
Semnan	39.3	2060	81.0	42.64	2060	87.85++
Kerman	93.0	2010	187.0	93.0	2010	187.0
Theran	53.0	1950	103.2	53.0	1950	103.2
Mazandaran	306.3	1840	306.3	456.9	1840	840.7++
Kohgiluyeh-va-Boyer Ahmed	60.3	1460	88.0	97.7	1460	142.64++
Fars	446.7	1350	604.0	469.6	1350	634.0++
Charmahal Bakhteyari	91	1310	119.7	91.0	1310	119.70
Lorestan	243.7	1270	310.0	243.7	1270	310.0
Hormozgan	14.7	1110	16.3	14.7	1110	16.33
Markazi	288.0	1060	304.3	183.0	1060	194.00+
Seistan-va-Baluchestan	110.0	1000	110.0	110.0	1000	110.00
West Azarbaijan	288.0	978	279.0	288.0	970	297.0
Bakhtran	296.6	968	285.0	256.6	960	264.34+
East Azarbaijan	638.6	920	591.6	598.60	920	558.71+
Ilam	87.7	910	79.6	87.7	910	79.60
Khorasan	867.3	880	767.7	867.3	880	767.78

(Contd.)

Table 33.1 *(Contd.)*

Province	*Area (000 hect)*	*Yield kg/hect*	*Production 000 tonnes*	*Area (000 hect)*	*Yield kg/hect*	*Production 000 tonnes*
Hamedan	375.3	820	309.0	375.3	828	309.0
Gilan	32.3	740	24.0	-	-	+++
Khuzestan	566.6	710	405.0	389.4	710	276.47+
Kordestan	428.3	710	303.2	428.3	710	303.28
Zanjan	464.0	580	267.3	334.0	580	193.72+
Bushehr	111.7	210	24.0	111.0	210	24.0
Total	6865.0	1030	6264.0	5767.49	1084	6255.12

\+ Decrease in area.
++ Increase in area.
+++ Complate deletion in area.

Source : (i) Computed on the basis of data available in Ministry of Agriculture.
(ii) Annual Agricultural Reports of 1993, 1994 and 1995 Rural Research Centre and Agricultural Economy, Tehran, Iran (Persian).

Table 33.2 : Barley: Actual and Proposed Land Area under Cultivation of Barley

	Actual (1993-95)		*Alternative*			
Province	*Area (000 hect)*	*Yield kg/hect*	*Production 000 tonnes*	*Area (000hect)*	*Yield kg/hect*	*Production 000 tonnes*
Esfahan	48.5	2740	132.7	48.5	2740	132.70
Tehran	30.3	2520	76.3	32.3	2520	81.48++
Yazd	2.4	2420	5.8	2.4	2420	5.80
Kerman	26.3	1940	51.0	26.3	1940	51.00
Semnan	23.6	1920	45.2	23.6	1920	45.24
Markazi	115.0	1590	182.8	221.7	1590	352.50++
Lorestan	78.7	1240	97.3	78.7	1248	97.33
Fars	203.3	1210	246.7	203.3	1210	246.70
Charmahal Bakhteyari	25.3	1200	30.0	25.3	1200	30.34
West Azarbaijan	105.0	1130	118.3	113.3	1130	128.10++
Bakhtaran	65.2	1090	71.3	65.2	1090	71.34

(Contd.)

Table 33.2 *(Contd.)*

Province	Actual (1993-95) *Area (000 hect)*	*Yield Kg/hect*	*Production 000 tonnes*	Alternative *Area (000hect)*	*Yield kg/hect*	*Production 000 tonnes*
Seistan-va-Baluchestan	20.5	1040	21.4	20.5	1040	21.40
Khorasan	339.6	1040	355.2	339.6	1040	355.20
Hormozgan	4.6	980	4.5	4.6	988	4.53
Ilam	40.2	948	38.0	40.2	948	38.0
Mazandaran	200.6	930	187.0	-	-	+++
Hamedan	75.0	900	67.4	75.0	900	67.40
Gilan	10.3	900	9.3	32.6	900	29.34++
East Azarbaijan	191.6	890	71.7	191.6	890	171.70
Zanjan	141.3	710	100.5	141.3	710	100.50
Kordestan	44.0	650	28.7	44.0	650	28.70
Khuzestan	210.0	640	135.3	210.0	640	135.30
Kohgiluyeh-va-Boyer Ahmad	34.3	410	20.0	-	-	+++
Bushehr	48.3	230	11.0	48.3	230	11.00
Total	2084	1060	2208	1988.3	1109	2206.0

++ Increase in area.
+++ Complete deletion in area.
Source : (i) Computed on the basis of data available in the Ministry of Agriculture.
(ii) Annual Agricultural Reports of 1993, 1994 and 1995 Rural Research Centre and Agricultural Economy, Tehran, Iran (Persian).

The position of rice is altogether different from barley and wheat. The overall proposed area has increased by 30 per cent to that of actual area under cultivation. It has shown more increase in production than that of area. It has gone up by 34.7 per cent than the actual production, while alternative productivity has also followed the same trend, with an increase of 3.3 per cent in our proposed productivity of crop. It is clear from Table 33.3 that five provinces, out of thirteen have undergone an increase in alternative area as compared to that of actual one. The provinces including Mazandaran, Zanjan, East Azarbaijan, Chahrmahal Bakhteyari and Gilan have show an increase in alternative area by 25.3, 1875, 833.3, 75 and 5.5 per cent respectively. On the other hand three provinces, namely Khuzestan,

Kohgiluyeh-va-Boyer Ahmad and West Azarbaijan have been completely deleted to produce rice. It is thus obvious from foregoing table that the crop has recorded only two alternative positions—an increase in area and deletion of area, whereas not a single province has shown any decrease in area.

In cases of pulses the overall alternative area has recorded a decrease of 10.5 per cent, but production and productivity levels have increased by 5 and 17 per cent respectively. Six provinces, namely Tehran, Esfahan, Kerman, Khuzestan, Semnan and West Azarbaijan have assumed an increase in alternative area. Out of these provinces high increase in alternative area was recorded in Esfahan and West Azarbaijan, while remaining provinces have not reached up to this level.

Table 33.3 : Rice: Actual and Proposed Land Area under Cultivation of Rice

	Actual (1993-95)			*Alternative*		
Province	*Area (000 hect)*	*Yield kg/hect*	*Production 000 tonnes*	*Area (000 hect)*	*Yield kg/hect*	*Production 000 tonnes*
Esfahan	4.8	4110	49.6	4.8	4110	49.60
Mazandaran	198.0	3910	775.5	248.0	3910	969.70++
Fars	23.3	3450	80.5	23.3	3450	80.50
Zanjan	3.2	3440	11.0	63.2	3440	217.41++
East Azarbaijan	4.8	3380	15.9	44.8	3380	151.42++
Charmahal Bakhteyari	2.6	3200	8.4	4.5	3200	14.56
Ilam	0.7	3050	2.2	0.7	3050	2.28
Gilan	181.9	2860	520.5	191.9	2860	845.83++
Khuzestan	22.8	2050	46.8	-	-	+++
Kohgiluyeh-va-Boyer Ahmad	3.1	1960	6.1	-	-	+++
Seistan-va-Baluchestan	2.0	1550	3.1	2.0	1550	3.10
West Azarbaijan	0.5	1550	0.8	-	-	+++
Khorasan	0.5	1350	0.7	0.5	1350	0.78
Total	448.3	1350	1491.0	583.7	3448	2008.00

++ Increase in area.

+++ Complete deletion in area.

Sources : (i) Computed on the basis of data available in the Ministry of Agriculture.
(ii) Annual Agricultural Reports of 1993, 1994 and 1995 Rural Research Centre and Agricultural Economy, Tehran, Iran (Persian).

Table 33.4: Percentage Change in Actual and Proposed Cropping Pattern

Crop	*Actual (1993 -95)*			*Alternative*			*Percentage change (Alternative to Actual)*		
	Area (1000 hect)	*Yield kg/hect*	*Production 1000 tonnes*	*Area (1000 hect)*	*Yield kg/hect*	*Production 1000 tonnes*	*Area (1000 hect)*	*Yield kg/hect*	*Production 1000 tonnes*
Pulses	436.8	710.0	311.0	391.7	810.0	325.3	–10.0	14.0	4.6
Cereals	8597.3	1159.0	9963.0	8339.5	1255.0	10469.1	–3.0	5.0	8.3
Foodgrains	9033.3	1137.0	102.7	8731.2	1236.0	10794.4	–3.3	5.1	8.7

Source : (i) Computed on the basis of data available in the Ministry of Agriculture.
(ii) Annual Agricultural Reports of 1995, Rural Research Centre and Agricultural Economy, Tehran, Iran (Persian).

On the other hand provinces, namely Gilan and Bakhtaran have been considered to completely delete production of pulses. The remaining provinces like Seistan-va-Baluchestan, Fars, Lorestan, Kohgiluyeh-va-Boyer Ahmad, Charmahal Bakhteyari, Khorasan, Mazandaran, East Azarbaijan, Markazi, Hamadan, Zanjan Kordestan and Ilam have been assumed to have the same area under cultivation of this crop.

However we can sum up from the proposed plan that in case of pulses, cereals, and foodgrains the foodgrains alone have recorded a decrease in actual area by 3.3 per cent while the production and productivity have also shown an increasing trend. The production level of foodgrains has shown an increase of 8.7 per cent while the productivity has increased by 5 per cent. Table 33.4 shows that through this alternative land-use pattern of production, pulses registered a decrease of 10 per cent in area but production and productivity level have increased by 14 and 15 per cent respectively. This increase in production as well as productivity level is more than of cereals of 5 per cent in the production level and 8.3 per cent in productivity level. While the proposed area has shown a decrease of 3 per cent.

However this proposed cropping pattern required full support of the government for providing necessary facilities as also mentioned in the assumptions.

In order to arrive at optimum solution we require a study of Iran's local demand of vital agricultural commodities expected in the long-run, which depends upon prediction operation on agricultural commodities. Ultimately it depends upon several variables such as changes in population and growth rate of domestic demand.

By taking 1996 as a base year in both population and income in 2000 year assuming that population growth rate is 1.58 per cent and income growth rate is 6 per cent per annum after 1996, the population which stood at 60,000 million in 1996 will touch a mark of 63,800 million persons in 2000 year showing an increase of 3,800 million persons representing more than the total population estimated in 1996. While GNI of Iran estimated at 3,306 billion Rials will touch a mark of 7,471 billion Rials in 2000 year, showing an increase of 4,165 billion Rials and representing 126 per cent more than that of the 1996.

The personal income which in 1996 was 68,746 Rials will increase to 95,300 Rials in 2000, showing an increase of 26,584 Rials and representing about 39 per cent more than the year of 1996, which is shown in Table 33.5.

Table 33.5 : Expected Population, National Income, and Per Capita Income during 1996-2000

Year	*Population ('000)*	*National Income (Billion Rials)*	*Per Capita Income (Rials)*
1996	60000	5918	98633
1997	61000	6273	102836
1998	61900	6649	107415
1999	62800	7048	112229
2000	63800	7471	1171003

Note : Computed according to population's growth rate equal to 1.58 per cent per annum and for GNI, growth rate equal to 6 per cent per annum.

By applying the Geometric Progression formula* Table 33.6 shows the expected individual demand for basic agricultural commodities in kilogram during 1996-2000. Table 33.7, present the total demand for basic commodities estimated on the basis of estimated data given in Table 33.5 and the estimate of individual demand given in Table 33.6. It is shown as to how this increasing demand can be met by the end of twentieth century. Under the first estimate it is assumed that the average physical productivity is taken as constant what it was during the average of 1993-95 level as base year as shown in Table 33.9, while the second alternative assumes that the avarage productivity must increase annually by 2 per cent after 1996, as shown in Table 33.6.

Conclusions and Suggestions

In the first alternative the area under cultivation of wheat has gone up

* The Geometric Progression Formula

$$Y_t = Y_{t0} \left(\frac{d_t - a_{t0}}{a_t} \right) 2$$

Where

Y_t = Expected personal demand in the year concern.
y_{t0} = Actual personal demand in the base year (t_0).
a_t = Expected personal income in the base year concern.
a_{t0} = Actual personal income in the base year (t_0).
e = Income elasticity of demand.

Table 33.6 : Expected Individual Demand for Basic Commodities during 1996-2000

*Commodities**	*Wheat*	*Rice*	*Sugar*	*Fruits*	*Vegetable*	*Onion*	*Pulses*	*Cotton*	*Potato*
Income Elasticity of Demand**	0.20	0.30	0.60	0.64	0.60	0.60	0.70	0.35	0.85
Year									
1996	153.76	35.37	36.08	53.06	42.59	12.27	6.46	7.59	23.43
1997	154.69	35.64	36.74	54.09	43.36	12.49	6.59	7.68	24.02
1998	155.59	35.90	37.39	55.11	44.12	12.71	6.73	7.77	24.60
1999	156.47	36.16	38.03	56.11	44.88	12.93	6.99	7.58	25.18
2000	157.33	36.41	38.66	57.11	45.62	13.15	7.12	7.93	25.94

* Normally as per the standard of FAO, a person requires about 2728 calories per day and of this total calories about 425 calories are obtained from animal products, while the rest 2307 calories are obtained from following agricultural products.

Consumption of 397 grams of wheat gives 1414 calories per day.
Consumption of 90.4 grams of rice gives 356 calories per day.
Consumption of 83.6 grams of sugar gives 334 calories per day.
Consumption of 123.3 grams of fruits gives 61.6 calories per day.
Consumption of 49.3 grams of potato gives 42.7 calories per day.
Consumption of 28.2 grams of onion gives 7.6 calories per day
Consumption of 98.6 grams of vegetable gives 33.5 calories per day.
Consumption of 15.1 grams of pulses gives 53.5 calories per day.

For wheat, rice, sugar, fruits, vegetable, onion, pulses and potato, the year 1986 is taken as base year, while estimated individual consumption was 145, 33, 30.5, 45, 10.3, 5.3, 6.6 and 18 kg, respectively.

** Income elasticity of demand is estimated. Income elasticity of demand in case of wheat, sugar are low, because these commodities are under government control and granted subsidy to keep their prices at low rate.

Table 33.7: Total Local Demand of Basic Commodities during 1996-2000

Commodities	*Wheat*	*Rice*	*Sugar*	*Fruits*	*Vegetable*	*Onion*	*Pulses*	*Cotton*	*Potato*
Year									
1996	10480	2411	2459	3617	2903	836	446	517	1597
1997	10918	2515	2593	3818	3860	881	465	542	1695
1998	11372	2624	2733	4028	3225	29	492	568	1798
1999	11842	2737	2878	4246	3397	8	529	594	1906
2000	12330	2853	3030	4476	3575	1070	550	621	2033

Source: Computed on the basis of Tables 33.5 and 33.6.

Table 33.8 : Total Area Necessary to Produce Important Crops according to Local Demand, assuming Stability of Average Yield as it was in the Average Period of 1993-95 during 1996-2000

Unit : 1000 hectare

Commodities	*Wheat*	*Rice*	*Sugar**	*Fruits*	*Vegetable*	*Onion*	*Pulses*	*Cotton***	*Potato*	
Av. Yield 1993-95 kg/hect	1030	3300	4172	4350	11700	18670	710	559	15350	Total Area
1996	10174	731	589	831	248	45	620	925	104	14267
1997	10600	762	621	878	261	47	655	969	110	14903
1998	11041	795	655	926	276	56	693	1016	117	15575
1999	11497	839	690	976	290	52	745	1063	124	16266
2000	11971	865	726	1029	305	55	786	1111	132	16980

* For sugarbeet, ten tonnes crude sugarbeet give only 1.7 tonnes of sugar, the average yield of crude sugarbeet is 24540 kg per hect which gives 4172 kg of refined sugar.

** For cotton one tonne crude cotton gives only one-third of a tonne of ginning cotton. The average yield was 1677 kg/ hectare divided by 3 = 559 kg of ginning cotton.

Source : Computed on the basis of Tables 33.6 and 33.7.

Table 33.9 : Total Area Necessary to Produce Important Crops to Local Demand, assuming an Increased of Output per hectare after 1996 as Two per cent during 1996-2000

Unit : 1000 hectare

Crops	*Year*									
	1996		*1997*		*1998*		*1999*		*2000*	
	Average Yield Kg/hect	*Cultivable Area*	*Average Yield Kg/hect*	*Cultivable Area*	*Average Yield Kg/hect*	*Cultivable Area*	*Average Yield Kg/hect*	*Cultivable Area*	*Average Yield Kg/hect*	*Cultivable Area*
Wheat	1256	8344	1281	8523	1386	8707	1332	8890	1360	9866
Rice	4022	58	4103	613	4185	627	4569	641	4350	656
Fruits	5086	484	5187	500	5291	517	5397	533	5505	550
Vegetables	5302	682	5409	706	5517	730	5627	754	5740	780
Onion	14261	203	14547	218	14838	217	15134	224	15437	231
Pulses	22758	37	23214	38	23678	39	24152	40	24634	42
Cotton	865	509	883	527	900	547	918	576	937	595
Potato	18711	85	19885	89	19467	92	19857	96	20254	100
	11702		11986		12277		12575		12861	

Source : Computed on the basis of Tables 33.7 and 33.8.

just double of area in the base year. It was 6.19 million hectares in 1996 and it is likely to be 11.97 million hectares in the year 2000. Almost the situtation may arise in case of rice and non-foodgrain production. The area was recorded 0.47 hectares in 1996 and it is likely to reach the level of 0.86 million hectares in the year 2000. But in the case of onion and potato, the area records no substantial increase as it was 0.039 million hectares in case of onion and 0.11 million hectares in case of potato, and it is likely to reach 0.055 and 0.132 million hectares respectively in the year 2000. The total area of 8.3 million hectares which was under cultivation of crops (total cropped area) in 1986 is expected to reach a level of 17 million hectares in the year 2000.

In the second alternative, in case of wheat (an increase of productivity by 2 per cent annually) the area is expected to reach only 9 million hectares in the year 2000. According to this estimate there would be less utilization of land by about 3 million hectares. Same situation prevails in the remaining crops like rice, sugarbeet, fruits, vegetable, onion, pulses, cotton and potato. Among these crops more area has been saved in case of rice, pulses and cotton. It is seen that more than 10 million hectares of area can be saved in the year 2000. While there is just an increase of only 2 per cent in productivity level.

On the basis of the alternative methods of estimating food requirements of Iranian population by the turn of century we can say that there is need for an increase in investment in the agriculture. But this increase in investment may be different in size, types and forms. In order to raise the agricultural productions to double in the year 2000 at level in 1996 and assuming yield level as constant. We require to make the total land under cultivation double the present level. There is no problem in raising the cultivated area to extent because at present only one-third of the cultivable land is under cultivation. Thus the total cultivated area would be about two-third as estimated. In this case the country would need to double the capital and the labour to cultivate in this land assumed at present ratio of these inputs. Here the gross investment represent the capital consumption and net investment in agriculture would be relatively low without much financial application of modern technology on the land already under cultivation would help us to raise the productivity at 2 per cent per annum reaching double amount at the end of the century. The adoption of modern teachnology is only possible when heavy investment is made in agriculture. It would not be applied only to the area already under cultivation, but also the new area would be taken under cultivation due to more possibilities and opportunities for the farmers. From the economic development point of view the fulfilments of the increasing requirements of food is not the only goal but the shift from the traditional agriculture to the modern agriculture is the basic goal. Unless and until a country takes the nation out of that situation she can never think of economic development.

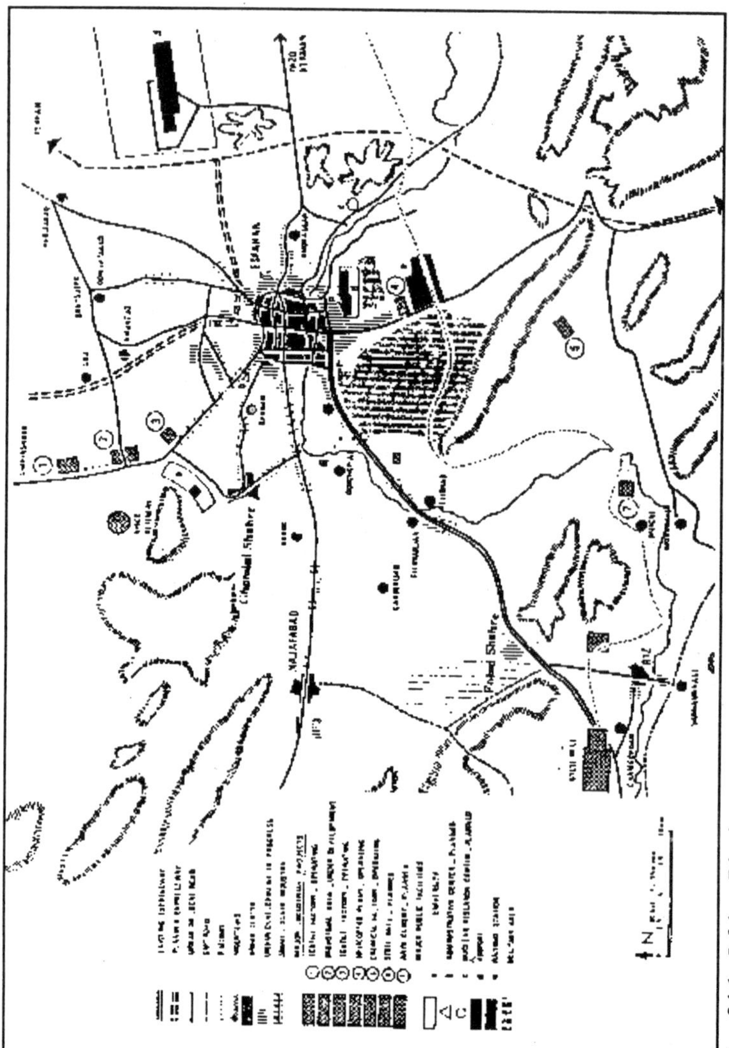

Fig. 34.1 : Isfahan District

The district is characterised by the Zayandehrude river, the sole water source of the region which originates in Zardukh in Bakhtiary district, eastern Zagross, flowing a 360 kilometres west-east course and emptying in the Gavkhoni Marches.

In 1971, a large dam was constructed across the river to secure water for Isfahan Steel Mill and other projects in the area, to irrigate Isfahan plain, to secure electric power for the project to the city of Isfahan, especially its historical bridge (Ministry of Water and Powers, 1973).

Izard Model Based Analysis on Economic Improvements of the District

To determine a correlation between the district's economic improvements and immigration, a linear comparison has been made betwen employees in different economic sectors in the countries of the region (Isfahan, Komeini Shahr, Lenjan, and Flaverjan) and those in all of the province in three periods of data gathering (Tables 34.1 and 34.2) then, the obtained percentages have been marked on X and Y coordinates. After drawing the related Izard graph, (Izard, 1979) an analysis has been made of the economic, quality of the district, its comparison with other periods and the immigration trend as well.

The first period coincides with the laying of the foundation of Isfahan Steel Mill. In the Table 34.1, the employment rates in different economic sectors are given:

Table 34.1: The Number of Employees in Economic Sectors in the Object of Study and that of the Province in the year 1967-1977

Economic sectors	*Employees in Isfahan district*			*Province Employees*		
	1967	*1977*	*%*	*1967*	*1977*	*%*
Industry & Mine	59551	66319	111	170674	198792	116
Agriculture	81200	93270	114	191745	136727	71
Private and State Service	26105	60990	233	107004	138632	129
Business and Trade	39515	40232	101	37673	89204	236
Total	153607	176868	115	507096	562756	110

Source: Plan and Budget Organisation, Isfahan Province Statistics, 1967-1977.

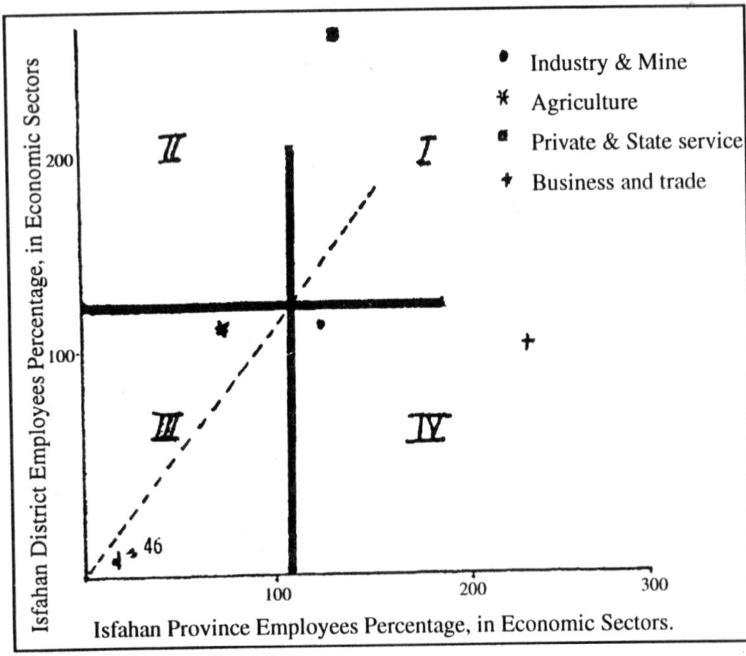

Graph 34.1 : Isfahan District Economic Conditions in 1967-1977

I. All sectors lying in the first quarter have developed more than the mean in the district and in the province.
II. All economic sectors lying in the second quarter have developed more than the mean in the district and less than that in the province.
III. All sectors lying in the third quarter, district or province, have developed less than the mean.
IV. All sectors laying in the fourth quarter have developed less than the average in the district and more than that in the province.

Comment

Graph 34.1 reveals that in one economic sector only—private and State service—in the 1967-1977s, growth has been above the mean in the object of our study and the province, while in other sectors — Industry and Mine and Business, the growth in the province exceeds that in the district. In agriculture sector the growth in both district and province is below the mean. In other words, in this decade one notices still no development in the economy of the district.

The second period (1977-1988) indicates the socio-economic impacts resulting from the industrialisation of the district and also the employment rate in different economic sectors. See Table 34.2.

Table 34.2 : Economic Sectors Employees in Isfahan District and Province in the 1977-1987

Economic sectors	*Employees in Isfahan district*			*Province Employees*		
	1977	*1987*	*%*	*1977*	*1987*	*%*
Industry & Mine	66319	135873	189	198193	194768	98
Agriculture	93270	64581	69	136727	157791	115
Private and State Service	60990	164019	267	138632	136109	98
Business and Trade	40232	118450	294	89204	62795	70
Total	260818	472923	267	562756	557460	97

As Graph 34.2 indicates, growth in private and State services, and business in the district exceeds that at province level, whereas in agriculture sector the growth of the economy of the province overtakes that of the district, indicating inconsistencies in agriculture compared with the uncontrolled development in the industry.

If we base our comments on the angle, we notice extensive socio-economic changes resulting from the industrialisation of the district.

In the previous decade the angle is 45° and in the 1977-1987s it reaches 70°. So is the situation in the next decade too. Here, inspite of some sought changes in the province, the graph is similar to that of the 1977-1987.[1]

The economic changes resulting from the development of industrial sectors have entirely impressed immigration trend, so that until 1960, due to the shortages in socio-economic facilities the people emigrated from Isfahan plain. But from 1963 on, thanks to these changes, it joined the areas where they receive immigrants. Isfahan population in 1960 amounted to 5,48,711, people, 50 per cent of whom lived in urban community and the rest in rustic one. Supposing, the

1. α - 45° economic coordination between District and province.
   $\alpha > 45°$ economic development in district compared with province.
   $\alpha < 45°$ economic development in province compared with district.

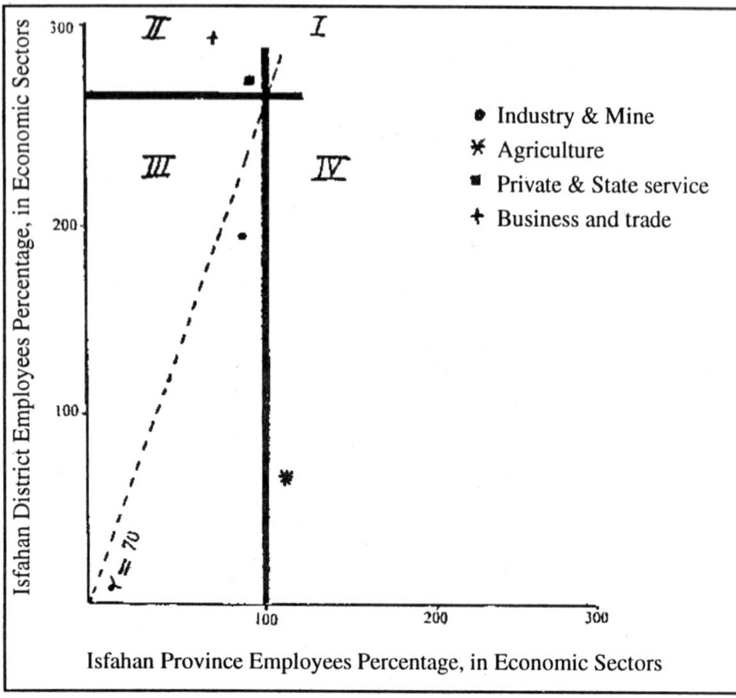

Graph 34.2: Isfahan District Economic Conditions in 1977-1987

natural population growth in Isfahan community, in all periods, as constant 2.4 a year, we expected a population increase of 59,935 people in the decade ending in 1967. But the actual increase, at that date, was 2,18,881 people or 1,28,998 people more than the expected natural growth. In other words, the urban community had accepted 33.48 per cent immigrants in the period. This implies an annual growth of 3.34 per cent in immigration or annual increase in Isfahan district population.

The population of the object of our study, based on the available statistics reached 7,82,281 by 1967, 49,874 of whom lived in urban areas and the rest in villages. The natural growth in population in the urban community by 1977 was 1,44,720 people, almost 1,13,675 people less than the actual increase rate in the same year. This implies an annual increase of 1.82 per cent in Isfahan district immigrant urban community.

Table 34.3 : Isfahan Urban District Immigration Status from 1960-1997

Survey period	*Population at the outset*	*Natural growth percentage*	*Natural increase*	*Population end of period*	*Actual population increase*	*Net immigrants*	*Period's immigrant*	*Annual immigration percentage*
1957-1967	275693	2.6	89935	494574	218881	128949	33.48	3.34
1967-1977	494574	2.6	144720	752969	258395	113675	18.22	1.82
1977-1987	752969	2.6	220335	1179250	926281	205946	21.3	2.13
1987-1997	1179250	2.6	345074	1670015	+30765	85691	14.8	1.4

Source : Iran's Statistics Centre. Statistical calendars from 1960 to 97.

The statistics reported reveal that Isfahan population reached 9,44,362 people by 1977, 7,52,969 of whom lived in urban community and the remainder resided in the rural one.

As statistics show, the natural population growth by 1987 was 2,20,335 people, 2,05,948 people less than the actual population growth. In other words, the urban community has reached 21.3 per cent immigrants in the period — an annual 2.1 per cent increase through immigration.

The 1997 census shows that Isfahan district population reached 1610015 people by that same year. This implies an annual 1.4 per cent increase in immigration or 85691 people in all despite a drop in the number of immigrant (Table 34.3).

Table 34.3 shows immigration rate from 1960 to 1967 remarkably exceeds those of other coming decades. The reason for this outstanding discrepancy is the socio-economic developments which Isfahan Steel plant and land reforms had brought about in this period, causing villagers, who had lost their jobs in the implementation at the first phase of land reforms, to rush to Isfahan to hunt a job.

A glance at the changes in urban and rural population of Isfahan province clarifies the situation (Table 34.4).

Irrespective of the motives affecting the immgrant—sending rural areas, the industrial regions *per se* are so attractive to the villagers that they absorb them even when the rusitc areas are prospering (Zanjani 1998, p. 23). These changes, the immediate results of the new industrial project with new promising prospectives in wages and employment opportunities for the villagers are quite noticeable in Flaverjan and Lenjan. Industrial activities too, expand the third class urban social businesses such as : services, trade and administrative affairs (Soltani 1997, p.135) (Table 34.5).

Table 34.4 : Change in Isfahan Province Urban and Rural Poplation from 1960 to 1967

Population	*1960*	*1967*	*Annual growth percentage*
Total province population	1141244	1422543	2.1
Urban province population	327886	731187	7.6
Rural province population	813357	691355	–1.62

Source: Iran's Statistics Centre, 1960-1967.

Table 34.5 : Employment Rate in Agriculture, Industry and Service

Activity type	*1964*	*1977*	*1987*	*1997*
Agriculture or affiliated activities	191745	136737	136737	157791
Mine	3357	3672	1157	1218
Industry	167317	195581	193611	193611
Construction	44586	73580	75971	80286
Power	2980	3703	8712	9491
Total	218240	276476	279451	290743
Trade	37673	39339	62795	71243
Transportation	12853	35813	44332	48195
Service	46585	84501	7091	8421
Total	97111	159653	144218	127859

Source: Plan and Budget Organization, Isfahan Province Statistical Report 1967 to 1997.

As Table 34.5 reveals, agricultural emloyment power has lost 55,000 people of its manpower at province level and simultaneously 58,000 people have been employed in industrial sector and about 62,542 people have bean added to business and service sector.

Therefore, however much a district's economic and industrial conditions, stimulated by historical, technical and geographical considerations, speed up, and industries, thanks to specific mechanisms, progress; transportation and roads, due to a general economic boost expands, the attraction of the industrial centres increases as much and absorbs multitudes of the rustic people from distant places.

The attraction of the industries in Isfahan steel plant area, for instance, as shown in Fig. 34.2 is a good example of industrial attraction.

This affair is confirmed by a report dealing with an urgent need of Isfahan steel plant to skilled workers in 1970 which was met by the sending of over 2000 workers to the area from the provinces of Azerbaijan, Sistan, Baluchestan and Khuzstan.

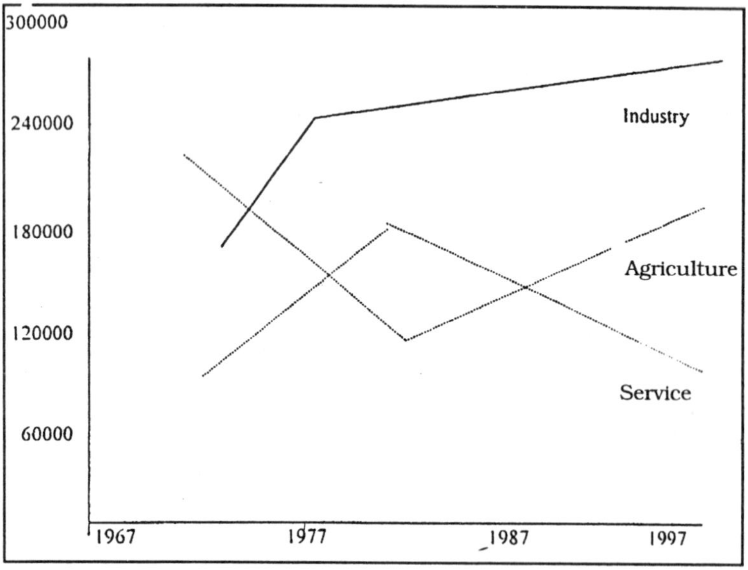

Graph 34.3 : The Trend of Changes

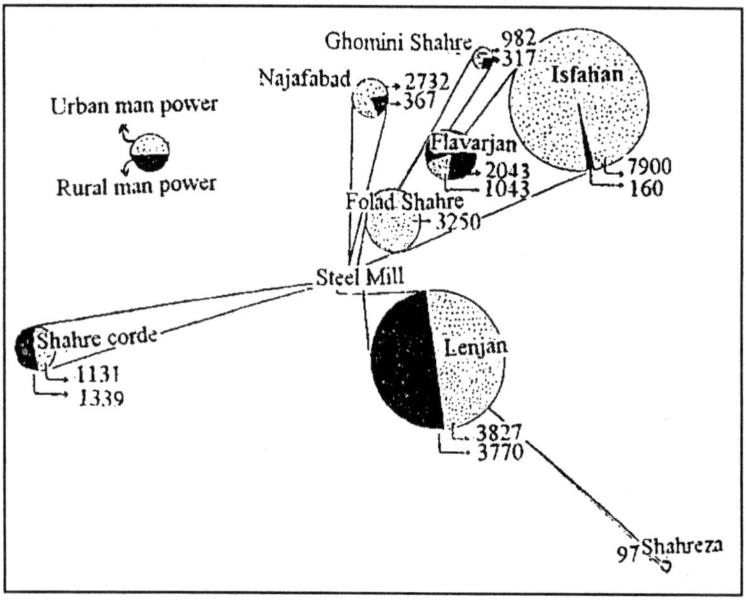

Fig. 34.2: Isfahan Steel Plant Population Attraction

Conclusions and Suggestions

Centralised planning approaches in propitious districts lead to district unevenness on the one hand and imbalance and heterogeneity of provincial and national level on the other hand. This results in hasty emigration from the undeveloped district to the developed ones.

The reason why the governments of poorly developed countries execute plan particularly in developed areas could be their backwardness and the fact that they are in a hurry to raise the economic position of these places to an acceptable local and international level. They invite huge capitals — foregin or domestic — hoping to achieve an early development, but unaware of its socio-economic outcomes. They also execute the expansion of the projects in economical places having infrastructural facilities and an accessible market. This affair exactly happened in Isfahan plain. The remote and deprived areas received no projects or investments. In other words, the plans were executed in the appropriate but economically developed places rather than the underdeveloped ones.

This trend usually leads to obvious discriminations among the districts and their underprivileged people, contrasting cities, centralized and industrial areas and their inevitable problems.

It seems imperative for the Islamic Iranian government, utilising special planning, to carry out an evenness in the distribution of national forces (administrative, government, social, political and economic) in the third socio-economic plan, which will lead to the establishment of some sort of socio-economic equilibrium among all places and groups of people in the province. To achieve these goals the following problems should be taken into consideration.

1. Decentralisation and District Expansion.
2. Control of the foreseen job — seeking populace of the district.
3. Establishment of socio-economic installation, fitting the deprived places, to attract manpower by suitable investment.

REFERENCES

Charles, N. (1968), *Population and Society*, Boston, Houghton Miffin Company.

Farid Yadollah, Favid (1975), *The Geography of Population*, Tabriz Publication

Hauser, Philip (1959), *The Study of Population*, The University of Chicago.

Iran's Statistics Bureau, *Statistical Yearbook*, 1960-1997.
Izard Walter (1979), *Location and Space Economy*, Translator : Dr. Daryosh Kazemzade, National University of Iran.
Mahdavi, Massood (1996), *The Tenets and Bases of Geography of Population* Tehva.
Ministry of Water and Power (1973), *Developmental Plan of the Zayandehroud River Valley.*
Momeni, Ahmad, *Employment and Development of Man Sources,* A Quarterly of Scientific Research in Population.
Momeni, Medhi (2000), *Principles and Approaches in District Planning,* Goya Publication. Isfahan.
Pediology Bureau and Soil Fertility (1974), *A Survey on Isfahan Eastern Lands.*
Pierre, George (1993), *The Geography of Population*, translator; Siroos Sahami, Ph.D., Nika Publications, Mashad.
Plan and Budget Organisation, *Isfahan Province Statistics*: 1967-1977.
Plan and Budget Organisation, *Isfahan Province Statistics*: 1977-1978.
Plan and Budget Organisation, *Isfahan Province Economic Fund Functions Costs*: 1977.
Shafaghi, Siroos (1972), *Familiarize Yourself with Iran's Steel Mill.*
Zanjani, Habibollah (1998), *Demography Analysis,* Samt Publications.

35

CHALLENGES OF POVERTY ALLEVIATION IN NEPAL

Challenges Ahead

SOORYA LAL AMATYA

Introduction

Nepal is one of the least developing countries in the world with an estimated population of 23 million in 1998. The country covers an area of 1,47,181 sq.kms with average density of 144 persons per sq. km.The annual rate of population growth is about 2.4 per cent. It is still an agricultural country with about 80 per cent of the total population depending on agriculture for their livelihood and about 88 per cent of the total population live in rural areas. Poverty has been the main problem in the country.

Poverty Syndrome

On the basis of conventional methods of per capita income of the people and minimum requirement of calories, the report of Independent South Asian Commission on Poverty Alleviation had estimated that there were approximately between 330 to 440 million people below poverty line in SAARC region in 1991. The report had categorically stated that the number of people under the poverty line will substantially increase in future if the present trends in population and economic growth will continue.[1] In the countries of SAARC region relatively large percentages of the total population below poverty line live in rural areas and depend on agriculture for their livelihood.

Quite a large proportion of the total population live below the poverty line in Nepal.The government agencies have estimated the percentages of the total population below the poverty line in different years.The first estimation of the percentage of the people below poverty line was made in 1977 by the National Planning Commission (NPC) which was about 36.2 per cent. According to a survey conducted by Nepal Rastra Bank in 1985, the percentage of the total population below poverty line was approximately 42.6 per cent.On the basis of the findings of national survey jointly conducted by Central Bureau of Statistics and World Bank. NPC estimated in 1998 that about 45 per cent of the total population were below the poverty line. Some studies have, of course, indicated relatively higher percentages of the total population below the poverty line. According to an estimation, the number of poors has almost doubled in Nepal from 4.9 million in 1992 to 9.5 million in 1997.There is heavy concentration of poors in 25 districts of Mid-Western and Far Western Development regions. Poverty is also an ethnic issue in Nepal. Quite a number of ethnic groups are characterized by absolute poverty like Chepangs, Raute, Majhe, Satiar and others. However, two facts are relevant in the Nepalese context. Firstly, the number of people living below the poverty line has increased which has been substantiated by all the surveys and studies. Secondly, the development efforts on the part of the government has miserably failed to address the poverty problem in the country.

Agriculture Perspective Plan (APP)

APP was prepared in 1995 as a perspective plan of 20 years (1996/97 to 2014/15) for the development of agriculture sector in Nepal. The APP was endorsed by the government and the Ninth Five Year Plan has incorporated the major policies and programmes as recommended in the APP. The top priority was accorded to development and management of irrigation, roads, technology and chemical fertilizers in APP. The plan visualized the acceleration of agricultural growth rate from about average 3 per cent to 5 per cent per annum.[2] It has also assumed that the population growth rate will decline from 2.5 per cent to 2.0 per cent per annum.The additional increment of 2 per cent growth in agricultural production per annum will obviously boost up food production; create employment opportunities in the rural areas; and help in poverty alleviation in the country.

Poverty Alleviation Planning

The National Planning Commission has estimated that approximately

45 per cent of the total population were below poverty line at the end of the Eighth Five Year Plan and has drafted an ambitious plan to alleviate the poverty percentage from 45 per cent to 10 per cent in 20 years plan period.[3] To achieve an ambitious target of poverty alleviation, NPC has proposed to accelerate the GDP at an annual rate of 7.2 per cent growth during 20 years plan period.

Poverty alleviation has been the main objective of the Ninth Plan and it aims to reduce the percentage of the total population below poverty line from 45 per cent to 32 per cent. Special emphasis has been given to accelerating high economic growth rate in all the sectors. It has been assumed that agriculture sector will play a lead role in poverty alleviation. As such the implementation of Agriculture Perspective Plan has been emphasised in the Ninth Plan. Along with creating extensive employment opportunities, it has been aimed to uplift the socio-economic status of the people of backward areas, remote regions and weaker sections of the society by providing the access to primary health, primary education and drinking water. For poverty alleviation, priorities will be given to development programmes relating to skill development, rural credit, social mobilization, population planning, women's empowerment, rural electrification and others.

A number of programmes had been implemented in rural areas with the main objective of uplifting the life of the people by providing credit for development activities.These are Small Farmers Development Programme (SFDP); Production Credit for Rural Women (PCRW); Micro Credit Related NGO Programmes; Grass Root Movement of Saving and Credit groups; Intensive Banking Programme (IBP); Rural Development Banks; Bishweswar Programme and others.

SFDP has been in operation for more than two decades and has extended its programme in all the districts of the country.It covers 652 Village Development Committees (VDCs) and has provided credit on group basis and other services to approximately two hundred thousand households[4]. In terms of coverage of the VDCs, it accounts only for about 16.6 per cent of the total VDCs in the country.

PCRW has been implemented in different parts of rural areas for the last seventeen years with the main aim of improving the socio-economic condition of poor women through institutional credit, skill training and community development activities. It has extended its programme to 67 districts.With Asian Development Bank's support micro credit to women projects has so far been extended to 12 districts.

Rural Development Banks have also been playing an active role in poverty alleviation by providing credit to rural poors for self-employment and income generating activities in all the five development regions. Under the Intensive Banking Programme, Nepal Bank Ltd., Rastriya Banijya Bank and other banks have been providing credit at relatively low interest rate to poor groups in all the 75 districts. In addition, there are Remote Area Development Programme and specified local development programmes which are obviously concerned to the welfare of the people of remote and backward communities. All the formal credit programmes had a record of reaching a cumulative outreach to 23 per cent of the rural population in 1997.[5]

Quite recently, of course, the government has initiated an ambitious Bishweswar Poverty Alleviation Programme. The main aim of this programme is to alleviate poverty in the country. Under this poverty alleviation programme100 poor households under the poverty line will be identified each year in each of the 205 electoral constituencies. Each of the identified poor household will be provided micro credit for income generating as well as self-employing activities.This year the government has earmarked about 10 million rupees for this programme. The detailed working procedures of this programme has yet to be finalized. The government has already constituted district level committees for 30 districts. It is yet to be seen how this programme will help in poverty alleviation.

Concluding Remarks

Keeping in view of the sluggish upward trends of general economic development of the country and poverty alleviation related programmes in particular it can be predicted that the poverty alleviation target envisaged in the Ninth Plan will be very difficult to be achieved. The institutional credits provided by government, semi-government agencies and NGOs altogether accounted only for about 30 per cent of the total amount of rural credit in 1991-1992.[6]

A number of constraints can be pointed out in this context. The government does not seem to be seriously committed in implementing poverty alleviation programmes. There is a lack of co-ordination in execution of the package programmes of credit disbursement and development of agricultural activities. The entire design of poverty alleviation programme has been visualized from the centre and local self-governing institutions like District Development Committees

(DDCs), municipalities and VDCs are not at all involved in framing the poverty alleviation programmes. As such, these self-governing institutions are not much concerned in implementation of poverty alleviation programmes. Local leaders can play a vital role in poverty alleviation. In Nepal there are 75 DDCs, 58 municipalities and 3,912 VDCs. In total there are approximately 1,88,000 elected representatives to local self-governing institutions. They can contribute efficiently in local development and poverty alleviation programmes if required authority is provided to them. However, their performances need to be properly monitored, supervised and evaluated by district level government agencies.

Specified poverty alleviation related programmes alone cannot contribute significantly in poverty alleviation. There has to be effective implementation of the development programmes in agriculture, social service, industry and other sectors of the economy. It is only with general economic development in the country, the specified poverty alleviation related programmes can be more effective in poverty alleviation. So high priority has to be given in effective implementation of development programmes. Regional aspect, remote and backward areas need to be given special priority while planning the development programmes and project allocation.

Local self-governing institutions should be vitalized and given authority in planning, co-ordinating, supervision of the local level development projects and activities. Development funds need to be created in each of the DDCs, VDCs and municipalities which should be utilised for the development of school education, primary health service, drinking water and social welfare activities. There seems to be hesitancy on the part of the government in decentralizing power to the local self-governing institutions in resource mobilization, planning, monitoring and executing the local level development programmes.

The women, depressed and backward communities are to be given adequate representation to the local self-governing institutions. Relatively there is higher rate of population growth among the depressed and backward communities in Nepal. The high growth rate of population is particularly true of the people living below the poverty line. So concerted efforts should be made in launching awareness campaign, literacy movement and family planning drive.

The informal credit flow in rural areas is substantial and need to be properly regulated. The money lenders and traders usually charge at relatively higher rates ranging from 36 to 43 per cent per annum. It is reported that approximately 39 per cent of the total rural households are under debt. Large percentages (86 %) of these households have borrowed from informal sources for non- productive social activities.

Local level development programmes initiated on the basis of social groups have been remarkably successful in Nepal. We can cite examples of forest user groups; irrigation user groups; small farmer credit groups; poor women groups; and micro credit groups. A success story of Annapurna Conservation Area Project (ACAP) in mid-hill district of Western Development Region of Nepal can be mentioned in this context. ACAP has successfully helped in the establishment of Conservation and Development Committees (CDCs) at the village level and various groups notably mother groups at grass root level.[7] CDCs have been operating very successfully in conservation and management of forest and other natural resources. The mother groups in particular has contributed substantially in road paving, regular cleaning, management of community forests, and fund raising for local development and income generating activities. ACAP has been regularly providing technical and financial support as well as monitoring the conservation and development activities. The participation of women, depressed and backward communities should be promoted through such social grouping. All the poverty alleviation programmes should be implemented through the active participation of the targetted communities by forming social groups.

NOTES

1. SAARC, 1992, *Meeting the Challenge*, Report of the Independent South Asian Commission on Poverty Alleviation, November, 1992.
2. APROSC and John Mellor Associates, INC, 1995, *Nepal Agriculture Perspective Plan* (APP), Kathmandu.
3. National Planning Commission, HMG, Nepal, 2054 b.s., *Baseline Paper of the Ninth Five Year Plan 2054-2059* (1997 to 2002), Kathmandu, p. 4.
4. National Planning Commission, HMG, Nepal, 1998, *The Ninth Plan (1997-2002)*, Kathmandu, Nepal, July, 1998, p. 199.
5. Nara Hari Dhakal,1999,"Expansion of Micro-finance Services : The Development of Community Based Saving and Credit Co-operatives in Nepal", *Agricultural Credit*, Bi-Annual Journal, Vol. 31, July, 1999, p. 31.
6. Kishor Kumar Guru Gharana, 1997, "Issues in Accesss to Rural Credit in Nepal", *Agricultural Credit*, Bi-Annual Journal, ADB/N, Vol. 29, Kathmandu.
7. ACAP/Nepal & SNV/Nepal, 1994, *Final Evaluation Report of Agro Forestry and Community Development Project,* July, 1994, Kathmandu.

36

POVERTY ALLEVIATION AND INTEGRATED RURAL DEVELOPMENT PROGRAMME

A Study

RAIS AHMAD

Indian economy has been an agricultural economy in nature as well as in character. Being the largest sector of the country's economy, income level of the people has been at the low ebb and as a result the purchasing power of the people is very low resulting into higher degree of poverty, unemployment and backwardness. They were poor, they are poor and they will remain poor unless and until some concerted efforts are not made to raise the level of the income of the rural masses, as it has been an undisputed fact that 'rural India is the real India'. Hence the need of the day is to bring a quality improvement in the life of the common men in general and the rural poor in particular. In this direction, a scheme, programme and policy was the need. Integrated Rural Development Programme (IRDP) has been one of the unique schemes in this direction *i.e.* to bring the desired level of improvement in the quality of life of rural people.

Prior to introduction of IRDP a large number of programmes for all-round rural development throughout the country were implemented by the Government such as Community Development Programme, National Extension Service, Intensive Agricultural District Programme, Hill Area Development Programme, Tribal Area Development Programme etc. But it was found that benefts from these programmes were reaped by the rich farmers/persons and the rural poor did not get the benefits from these programmes who were in majority and poverty was increasing in rural areas. Therefore, the

Government of India designed a new scheme known as IRDP, only for those rural masses who were living below the poverty line. The IRDP was a major poverty alleviation programme and it was started as back as in 1978 in 2300 selected blocks throughout the country and the scheme was found an effective scheme for eradication of poverty from rural areas. Therefore, from 1st October, 1980, it was implemented in all the 5,011 blocks of the country. The very basic objective of IRDP was to enable identified rural poor families to raise above the poverty line. For this purpose productive assets and inputs were given to target groups. Assets may be needed for farming business, non-farming business, were given through financial assistance by way of Government subsidy and term credit from financial institutions comprising, Commercial Banks, Regional Rural Banks and Primary Agricultural Cooperative Credit Societies. The IRDP was a centrally sponsored scheme funded on a 50:50 ratio by the Centre and State Governments.

The target group includes or covers small farmers and marginal farmers, agricultural labourers and rural artisans. The IRDP focus is on the poorest of the poor. The poverty line was fixed during the (1980-85) at Rs. 3500 per annum *i.e.* a family whose annual income from all sources was less than Rs. 3500, considered living below the poverty line. This limit was raised to Rs. 6400 during Seventh Plan (1985-90) and it was raised to Rs.11,000 during Eight Plan (1992-97) and financial assistance is targetted primarily to families with annual income below Rs. 8,500 with utmost priority to families with annual income upto Rs.6,000. Amongst the target groups priority is given to scheduled castes and scheduled tribes, women, liberated bonded labourers and physically handicapped persons. The IRDP is implemented through District Rural Development Agency (DRDA) at district level and at grass root level, the block staff is responsible for programme implementation. At State level Coordination Committee monitors the programme and at Central level the Ministry of Rural Development is responsible for policy formation, overall guidance, monitoring and evaluation of the programme. At the time of introduction of this scheme in 1978, about 50 per cent of the population of the country had been living below the poverty line which has come down to 37 per cent in 1993-94.

Now to make the poverty alleviation programmes more effective, the various programmes have been merged in a new scheme namely 'Swarnjayanti Gram Swarozgar Yojna' (SGSY) which is the single self-employment programme for the rural poor, launched in 1st April,1999, the programme replaces the earlier self-employment and allied programmes — IRDP, TRYSEM, DWCRA, SITRA, GKY and MWS, which are no longer in practice. The main objective of IRDP and now SGSY is the eradication of poverty from rural areas by providing income generating assets to the rural poor. For this purpose credit-cum- subsidy is disbursed to the selected poor families.

RESEARCH METHODOLOGY

Eighty families who received financial assistance under IRDP scheme 3-5 years before the date when this survey was conducted (1990), were selected from eight villages of Lodha and Dhanipur Blocks of Aligarh District (U.P.) comprising 10 families from each village. The beneficiaries were interviewed with the help of a well designed questionnaire. Personal contacts and enquiries were also made with the administrative staff of the banks and Government agencies involved in the implementation of the IRDP and their remarks also were sought for an effective implementation of IRDP scheme.

In this Chapter an attempt has been made to find out as to how many of the surveyed families were able to cross the poverty line after passing of 3-5 years of receiving the financial assistance under IRDP and to analyse whether the amount granted to the beneficiaries was sufficient or not. The schemes/projects given to beneficiaries were successful or not and finally evaluation to identify and analyse the main problems which are retarding an effective implementation of the IRDP scheme.

Performance Appraisal

In order to raise the families living below the poverty line *i.e.* having an annual income less than Rs. 6400 (during the time of survey conducted), financial assistance is provided to them under IRDP scheme. Subsidy is sanctioned by the Government and loan is distributed by the commercial banks and cooperative banks.

It is clear from the data given in Table 36.1 that a maximum of 31.25 per cent of the sampled beneficiaries received financial assistance in the range of Rs. 4500-5000, 23.75 per cent in the range of Rs. 3500-4000, and 11.25 per cent in the range of Rs. 5000-5500.

The average amount of net loan and subsidy given to a sampled beneficiary family comes out to Rs. 4424. The average amount of subsidy alone stood at Rs. 1684 per family. Table 36.2 shows trends in assets given to different people under IRDP.

Table 36.1 : Financial Assistance given to the Sampled Families under IRDP

Total amount of assistance (subsidy + Credit) Rs.	*Number of families*	*Percentage to total*
500 - 1000	3	3.75
1000 - 1500	1	1.25
1500 - 2000	1	1.25
2000 - 2500	1	1.25
2500 - 3000	3	3.75
3000 - 3500	2	2.50
3500 - 4000	19	23.75
4000 - 4500	7	8.75
4500 - 5000	25	31.25
5000 - 5500	9	11.25
5500 - 6000	6	7.50
6000 and above	3	3.75

Source : Data is collected and compiled by the author from the sampled families with the help of a questionnaire.

Purpose-wise utilization data given in Table 36.2 indicates that a maximum of 33.75 per cent of the sampled benefited families have been given assistance for the purchase of the Buffalo. Assistance provided for purchase of she-buffalo comes to 31.92 per cent of the total assistance given to 80 benefited families. At second place, 16.25 per cent of the sampled benefited families have been assisted for starting the business of provision store; and financial assistance given for this purpose comes to 18.22 per cent of the total assistance given to all the sampled beneficiaries. Next come assistance provided to 10 per cent of sampled families for the blacksmithy. Assistance given for this purpose comes to 10.44 per cent of the total assistance given to the beneficiaries.

It was explained by the surveyed families that the amount granted for purchasing of milch cattle was less than the market prices and they had to arrange some more funds from their own sources/private

Table 36.2 : Showing the Assets given to Sampled Families

Assets given	*No. of beneficiaries*	*Percentage to total*	*Total assistance (Rs.)*	*Percentage to total assistance*
She Buffalo	27	33.75	113000	31.92
Buffalo Cart	6	7.50	32075	9.06
Horse Cart	5	6.25	25500	7.20
Bullock Cart	2	2.50	12900	3.64
Bullock pair	2	2.50	5500	1.55
Provision store	13	16.25	64500	18.22
Diesel Pump Set	2	2.50	13950	3.94
Cycle Repairing shop	2	2.50	8000	2.26
Sewing Machine	3	3.75	2200	0.62
Blacksmithy	8	10.00	37000	10.44
Cloth selling	2	2.50	11000	3.10
Rassi Making	1	1.25	1200	0.33
Rickshaw	4	5.00	9400	2.65
Radio Repairing shop	1	1.25	6700	1.89
Tea stall	1	01.25	6000	1.69
Goat rearing	1	01.25	5000	1.41

Source : Same as for Table 36.1

money lenders on a very high rate of interest. The cattle also were not of good quality because they were bound to purchase through the staff of the block headquarters.

During the course of discussions with the beneficiaries it was found that the loan given for sewing machines, provision stores, cloth selling, tea stall etc., were not useful for increasing the income of the beneficiaries. Rather these loan became burden over them. It was so because the staff of the concerned block was interested only in granting loans to beneficiaries or fulfilling the target without considering that a loan granted under a particular scheme may increase the income of the beneficiary or not.

On the other hand assistance provided under various schemes such as she-buffalo, rikshaw, horse cart and bullock cart, diesel pump set etc., was very much successful in increasing the income of the beneficiaries. Hence under IRDP scheme loans should be granted only under those schemes in which the beneficiary is having sufficient experience and knowledge so that his income may be increased. If the beneficiaries are not trained in any particular scheme which is going to be sanctioned by the Block Development Officer, a proper training should be provided by the Block Headquarters so that beneficiary may achieve the benefits of the scheme and he may be able to increase his income with assistance provided under IRDP scheme.

From the above analysis, it has brought out that the IRDP has started making breakthrough in eradication of poverty. But much has been left to make the scheme as a model. There are many factors affecting performance of IRDP which are being discussed in the following pages.

Problems

On the basis of personal observations during the field work, the following main problems have been identified which are coming in the way of an effective implementation of the IRDP.

1. The main objective of the IRDP is to raise the income of the poorest of the poor families in the rural areas. However, the identification of the poorest among the poor is a difficult task. They are selected by the Gram Sabhas and a final list is prepared by the Village Level Workers (VLWs). During the course of discussions and interviews with the IRDP beneficiaries and non-beneficiaries, a number of irregularities and corrupt practices have been identified. It was reported that the VLWs recommend the names of only those persons from whom they receive some payment. Moreover, the selection procedure is not strictly in accordance with the guidelines of the Government of India. To ensure that the assistance under the IRDP is sanctioned to the most deserving poor families, the Government has incorporated in the official guidelines that the list of identified families prepared by the staff at the block level, should be presented in a Gram Sabha meeting and should be ratified by it. However, it is learnt that in actual practice the procedure of the villagers' participation and an open selection did not take place.
2. It was further observed that the persons responsible for the programme implementation are much interested in the target

fulfilment. They do not assign due to weightage to the viability of the projects. Similarly, no consideration is given to the assets creation possibility for the families covered under the IRDP benefits. Due to this reason, the assets become a liability for the poor families and they fail to cross the poverty line.

3. The value of assets is very small to generate additional income for the targetted families to cross the poverty line. For example, a cow or a buffalo is limited upto two units for a family and in reality most of the families have received only one unit and the second unit has not been made available to them. Therefore, to provide an opportunity for the generation of additional income, the number of units should be increased.
4. Some beneficiaries expressed doubt by saying that the full payment of the subsidy is not given to them. Rather, it is less than the amount sanctioned under the progra-mme. It is observed that there is a lack of coordination between the bank manager and the administrative officers of the block. As a result the beneficiaries have to wander here and there for getting loan.
5. In certain cases we were informed that the beneficiaries have sold their assets obtained under the scheme after getting the total assistance which is against the objective of the programme.
6. Some of the beneficiaries have pointed out that there is undue delay in the sanction of the loan by the banks. As per the RBI directives, the banks should dispose of the IRDP proposals within 15 days. But the banks do not take prompt action in processing the loan applications. The Block Development Officers do not care whether the scheme given to the beneficiaries is income generating or not. It was found, in some cases, that 5-10 women beneficiaries were given sewing machines who were living in the same area and they were trained at the block headquarter only in sewing 3 very ordinary items which can be prepared by every women. Therefore, they do not get sewing work and earn nothing and no change has taken place in their economic life by the IRDP assistance.
7. In some cases, a number of neighbouring women were given

financial assistance for organizing provision stores or (*Kirana* shops). The goods given to them from the approved shops were found to be of higher prices than the open market. Thus the articles sold by these provision stores are costlier than in the other shops of the area and the beneficiaries were forced to utilize the items in their houses and their business was over. There are two reasons for the failure of these provisions stores. One is the higher prices than the open market rates, nobody purchases from the IRDP beneficiaries as their selling rates are higher than the other shopkeepers; and the other is sanctioning of the same scheme to a number of beneficiaries living in the same locality. This means lack of diversification in the operations under IRDP.

8. We were also informed that the beneficiaries had got their animals insured but in the event of mortality, they do not get the insurance payment. This is because the doctor of their block does not give the death certificate of the animal for the payment of insurance claim without getting some payment. In many cases, the poor beneficiaries tell that they are unable to give to the doctor the amount demanded by him so they do not get the insurance payment.
9. It has been reported by the various beneficiaries that the purchasing of raw material for manufacturing goods is completed in a number of instalments and it takes several months resulting no economic benefit to them. Moreover, there is no generation of additional income to cross the poverty line.
10. Most of the beneficiaries complained for inadequate assistance under the programme, which is the reason for the ineffective performance of the programme. The size of the beneficiary families has not been taken into consideration while sanctioning the loans.

SUGGESTIONS

In spite of a number of problems discussed above which are coming in the way of the effective implementation of the IRDP, the programme is successfully attracting the attention of the rural people and is quite successful in removing poverty from rural areas. It may be more productive and purposeful if the following measures are implemented by the concerned authorities:

1. For the correct identification of the poorest of the poors, it is suggested that the Gram Sabhas should be organized regularly and the beneficiaries should be selected in the Gram Sabhas so as to get rid of the illegal payment which is a great hurdle in the selection of genuinely poor families. The selection procedure of the beneficiaries should strictly be in accordance with the guidelines of the Government. The concerned authorities should take stringent action against those who infringe these established rules and regulations.
2. Regarding the selection of the viable economic projects, the present limit of the loans should be increased to cover the price rising effect. Sufficient financial assistance should be made available to the beneficiaries so that they are able to purchase viable assets of good quality. In this way, the beneficiaries would be saved from the problem of borrowing from the private money lenders. However, the present ceiling of the subsidy should sufficiently be increased for motivating the rural poor to come forward for their socio-economic development. In other words, the poorest of the poors will be encouraged to help themselves by their own efforts.
3. Proper care is needed in the selection of the schemes for the beneficiaries. Only such schemes should be granted to the beneficiaries in which they have sufficient experience and knowledge. At the time of consideration of the applications for financial assistance under the IRDP schemes, the interest and choice of the beneficiaries should be given top priority. In this way they will be able to make proper utilization of the assets.
4. For milk generating cattle, it is suggested that the financial assistance to the IRDP beneficiaries should be sanctioned in one instalment and at least four units should be provided so that a reasonable income could be earned by the benefited families to cross the poverty line.
5. Necessary arrangements should be made for regular monitoring of the assets given under the IRDP scheme. Whenever it is found that any of the beneficiary has sold his assets, immediate and stringent action should be taken against such a beneficiary, and he should be punished for violating the norms.
6. Sufficient amount of loan should be granted to the

beneficiaries so that they may be able to purchase assets of good quality and may earn sufficient additional income to cross the poverty line. Before giving any assistance under the IRDP, the family size of the beneficiary should be taken into consideration. If the family size is taken into consideration, the small family would get more benefit than the big family. This will be an incentive to limit the family size. It is also suggested that the banks should take urgent action on the loan applications of the IRDP beneficiaries and the loans should be granted within 15 days after the receipt of the applications which is as per the directives of the Reserve Bank of India.

7. Non-income generating schemes should not be given to the beneficiaries which defeat the goals of the programme. The low income generating assets should also be avoided and to create interest among the beneficiaries only high income generating scheme should be implemented.
8. In order to eliminate corrupt elements in the process of purchasing, the amount of assistance should be given in cash and the beneficiaries should be left free to purchase his assets from any of the markets. In the case of misutilisation of loans, stringent legal action should be taken against such beneficiaries so that no one could dare to misutilise the loans.
9. To prevent fake transactions, encouragement should be given for making the purchases outside the district.
10. The banks may be directed to make some changes in their repayment schedule. The recovery instalments should be reduced and it should spread over a longer period. Thus, a part of the increased income will be left with the beneficiaries.
11. In some genuine cases, the District Rural Development Agency and the banks should be authorized to write off the loan.
12. Since independence Government of India has made very sincere efforts for eradication of poverty from rural areas and on average every year one programme has been introduced for this purpose and allocated a lot of funds. But it is found that implementation machinery of the programmes is ineffective or corrupt. It was accepted by the then Prime Minister (late) Mr. Rajiv Gandhi that only 15 paise reaches to the rural poor out of Re. one released by the Government for poverty alleviation schemes.

Thus it is clear to the Government of India that in addition to other lacuna in the poverty alleviation programme the required funds/ financial assistance is not reaching to the selected families for bringing up them above the poverty line. Hence the poverty is not being removed at a desired rate. Therefore the Government should take it as a mission to remove the roots of the corruption from the implementation machinery by taking stringent actions against those who are found guilty so that the action against one may become a threat to others and the poverty alleviation programme may be implemented in more effective manner.

Hence, if above mentioned suggestions are carried out then there is no doubt that Sawarnjayanti Gram Swarozgar Yojna (SGSY) could bring the desired level of increase in income of the poorest among poors and would become a scheme of purposeful and meaningful nature to eradicate the poverty from the rural masses which is a distant dream of rural poors.

37

PROBLEM OF RURAL POVERTY AND MICRO-FINANCING THROUGH SHGs

RAIS AHMAD

INTRODUCTION

It has universally been accepted that the micro-financing is a very successful instrument for eradication of poverty on the model of Grameen Bank of Bangladesh. Most of the Governments in developing countries are encouraging micro-financing for alleviation of poverty and empowerment of women in their countries. In India also NABARD encouraged lending through Self-Help Groups (SHGs) after achieving successful results from a pilot project of 500 SHG Bank linkage programme in 1992. Now it is the largest micro-financing programme anywhere in the world as till March 2003 according to NABARD 7,17,360 SHGs that are linked to the banking system. Over 12 million poor households have gained access to the formal banking system through the SHG Bank Linkage Programme as on March 2003. The recovery of loans so far is remarkable 96 per cent. In India now various international donor agencies are working for extending micro-credit to the rural poors for eradication of poverty. World Bank also has sanctioned various micro-financing projects for people living below poverty line in India.

In this Chapter an attempt has been made to review as to why the formal financial institutions were not accessible for the poorest of the poors to cater to their micro credit needs and to evaluate the progress of micro- financing programme in India and its impact on rural poverty. It is found that the micro-financing through SHG Bank Linkage Programme of NABARD is working very effectively and the credit needs of the rural poor is catered by the rural financial institutions through SHGs. As a result of this the socio-economic status of rural folk specially of rural women is improving and poverty is reducing.

WHAT IS MICRO-FINANCE ?

Micro-finance is an enabling, empowering, bottoms-up tool to poverty alleviation that has provided considerable economic and non-economic externalities to low-income households in developing countries. Micro-finance is being hailed as a sustainable tool to combat poverty, combining a for-profit approach that is self-sustaining, and a poverty alleviation focus that empowers low-income households. Micro-finance is increasingly becoming a tool to exercise developmental priorities for governments in developing countries. (Srinivas, H.)

Micro-finance as an alternative source of credit for the rural poor has received wide attention in recent years. It is a "programme that provides credit for self-employment and other financial and business services (including saving and technical assistance to very poor persons" (World Bank, 1997). In many countries in the world micro credit programmes have succeeded in generating self-employment by providing access to small capital to people living in poverty and has proved to be an effective tool in freeing people from poverty (United Nations, 1998). According to Task Force (NABARD) a working definition of micro-finance is "provision of thrift, credit and other financial services to the unserved poor enabling them to raise their income levels and improve living standards".

Micro-finance—International Experience

Over the past few decades several informal and innovative approaches in financing the poors in a sustainable manner have been experimented in many developing countries. Grameen Bank and Bangladesh Rural Advancement Committee (BRAC) of Bangladesh, Efforts of Bank Rakyat Indonesia (BRI) in Indonesia, Bank for Agriculture and Agriculture Cooperatives (BAAC) in Thailand, Amanah Ikhtiar Malaysia (AIM) of Malaysia and Agriculture Development Bank of Nepal (ADBN) are some of the examples which have yielded encouraging results in empowering the poors through credit in South-East and South Asia (Satish P., 2001).Thus we find that micro-financing is a very effective tool for combating poverty.

Formal Financial Institutions in Rural India

After more than five decades of independence, the formal financial institutions have not been able to ensure better and even distribution of rural credit. Out of 63 million operation holdings below 1 ha of land, less than 25 per cent hand access to formal agricultural credit in 1990-91 (Desai and Namboodiri 2001). In India the majority of operational holdings are small and marginal, which are not economically viable. The condition of other rural poors, who largely depend on off and non-farm activities, is also precarious as their resource and income status is even below that of tiny landholders. Perpetual poverty and lack of adequate credit have remained the major constraints in the economic upliftment of rural households. Credit promotes capital investment and adoption of new technology, leading ultimately to better standards of life due to increased production and incomes (Kamal Vatta, 2003). Rural areas dominated by the marginal and small farmers and poorer sections of population, if neglected, will lead to imbalances. Rural urban divide may deepen, inter-regional disparities may widen and income inequalities may intensify [Agrawal *et al.* (1997)]. This is mainly attributed to the failure of financial intermediaries in fulfilling their basic functions, *viz.*, production credit to finance income generating activities, consumption credit to maintain and expand human productive capacity, and quality saving schemes for increasing risk-bearing capacity of the rural households. Moreover, these institutions have failed to promote any of its social objectives (Desai and Namboodiri, 1996). Despite such a large presence, an overwhelming majority of the rural poors belonging to the lower income group find the Rural Financial Institutes (RFIs) out of their reach and resort to borrowing from informal sources. Also, despite their access to enormous funds in the shape of deposits, refinance and equity support from central and State governments, these RFIs fell too short of their expectations (Tomar J.S., 1995).

The formal financial institutions performance particularly in their lending to the poors in India has been unsatisfactory. They face a number of constraints in broadening their services to the poors. A large number of rural households are with limited land resources and small economic activity accompanied with poor technology. But their demand for credit has been rising due to growing family size, increased consumption requirements, social obligations and so on. But the institutional agencies not only lacked the required mechanism to assess their credit needs but often overlooked their demand for credit on the ground that their needs are for non-productive purposes. Besides, perceived high risks, transaction costs in small-scale rural lending and absence of collateral securities kept the poors away from the fold of formal financial intermediaries (Namboodiri and Shiyani, 2001). In such environment weaker section in the society depended on money sharks. Micro- finance institutions have good potential to reach the rural poor and to address the basic issues of rural development where the formal financial institutions have not been able to make significant headway (Kamal Vatta, 2003).

NABARD's Initiative for linking Formal and Informal System

Micro-financing by 'non-formal' financial organizations had already started. SEWA (Self-Employed Women's Association) owned by women of petty trade groups was established on cooperative principle in 1974 in Ahmedabad. Working Women's Forum (WWF) started promoting working women's cooperative societies in Tamil Nadu since 1980. Shreyas in Kerala actively got involved in micro-finance operations since 1988 with the objective of promoting people's cooperatives (Dasgupta R., 2001).

Some research studies conducted by NABARD in early eighties which revealed that the country has a wide network of rural bank branches, consisting of 33,000 rural branches of Commercial Banks, 14,500 RRB branches and 92,000 Primary Agricultural Cooperative Credit Societies, these branches were implementing poverty alleviation programmes for creation of self-employment opportunities through bank credit for almost two decades, a large number of rural poors were not benefited from the formal banking system. These studies further pointed out that the present banking system, procedures and policies, and deposit and loan products were perhaps not most suited to meet the immediate needs of the very poors. In cases where credit was made available to the poor through special programmes, absence of an integrated savings component and something to fall back upon in case of any adversity was leading to poor repayment performance. The problem was further compounded as the beneficiaries refused to distinguish between State subsidy support and bank credit. The political expediency for removing poverty at a stroke was putting resources for running micro enterprises in the hands of the poors without nurturing them to handle such resources. The high cost of appraisal and monitoring meant compromise on appraisal and supervision, leading to poor repayments of loans. In the absence of institutional support, the poor continued to approach intermediaries in the informal credit markets like moneylenders, traders, and wholesalers for meeting their consumption and production needs (Nanda Y.C., 2003).

Thus in spite of the huge expansion by the banking sector in terms of rural branches, there were still a very large number of people they bypassed. In 1992, therefore, NABARD began an experiment. It was looking at innovative ways of reaching the banking system to these people. The organization thus started a project on a pilot basis involving the creation of informal structures of Self-Help Groups (SHGs) of people and trying to link them to the banking system. NABARD started with a pilot project of 500 such Self-Help Groups (SHGs) (Majumdar, 2003) This experiment of NABARD achieved great success to cater to the microcredit need of the rural poor through SHGs. Under the SHGs NABARD ensured that very poor people come together, start saving, start lending to each other, and once they are able to reach a stage that they can enforce the will of the group on the individual, the banks can start coming into the picture and lending them funds. Today the SHG concept is a resounding success in India and is one of the most successful programme run by NABARD (Majumdar, 2003).

SELF-HELP GROUPS

Self-Help Groups (SHGs) are necessary to overcome exploitation, create confidence for the economic self reliance of rural poors, particularly among women who are mostly invisible in the social structure. These groups enable them to come together for a common objective and gain stength from each other to deal with exploitation which they are facing in several forms. A group becomes the basis for action and change. It also helps building of relationship for mutual trust between the promoting organization and the rural poors to constant contact and genuine efforts (Singh Om Raj, 2003). SHG is a small voluntary association of poor people, preferably from the same socio-economic background. The maximum number of members in a SHG should not exceed 20.

Group members usually create a common fund by contributing their small savings on a regular basis. Groups evolve flexible systems of working sometimes with the help of (NGOs) and manage pooled resources in a democratic way. Loan requests are considered by groups in periodic meetings and competing claims on limited resources are settled by consensus. The loan amounts are small, frequent for short duration and are mainly for conventional purposes. The rates of interest vary from group to group and the purpose of loan. It is higher than that of banks but lower than that of moneylenders. At periodic meetings, besides collecting money, social and economic issues are also discussed. Defaults are rare due to group pressure and intimate knowledge of the end use of credit (Singh Om Raj, 2003).

Impact of Self-Help Groups

The economic impact of SHGs was relatively more pronounced on the social aspects than the economic aspects (Puhazendhi & Satyasai, 2001). The SHG is really a boon in the rural areas, which gives financial autonomy to the rural women and make them economically independent (Laxman, 2001). The SHGs have become a platform for exchange of experiences and ideas (Tilekar *et al.,* 2001). The women have tremendous energies to start their own enterprises given the right opportunities. They have developed abundant self-confidence and self-esteem through SHG movement. Not only economic poverty but also social and gender issues can be tackled effectively through this process. (Chiranjeevulu, 2003). Although the loan amount is small, it helps in meeting the requirement of the poor. Small amounts of loans coupled with financial discipline ensure that loans are given more frequently and hence credit needs for a variety of purposes and at shorter time intervals can be met. This is a better mechanism to reduce poverty gradually as against one time loan for productive asset which may or may not lead to sustained increases in income (Madheswaran and Dharmadhikary, 2001). The micro-financing to women through SHGs has helped the groups to achieve a measure of economic and social empowerment. It has developed a sense of leadership, organizational skill, management of various activities of a business, right from acquiring finance, identifying raw material, market and suitable diversification and modernization (Manimekali and Rajeswari, 2001). Hence micro-financing through SHGs is contributing to the development of rural people in a meaningful manner. It is seen that significant changes in the living standards of SHG members have taken place in terms of increase in income levels, assets, savings, borrowing capacity and income generating activities (Sharma, 2001).

It has been observed that the overall impact of micro-finance through SHGs is very effective in combating poverty, unemployment and empowerment of women.

Progress of Self-Help Group Bank Linkage Programme of NABARD

The highlights of overall progress of SHG Bank Linkage Programme are shown in Table 37.1 from 1997 to March 2002. The progress shows that the achievement of NABARD under micro-financing programme through SHGs is remarkable. This is the largest programme anywhere in the world for combating poverty, unemployment and empowerment of women. It is seen from the table that number of SHGs linked with banks during the year 1997 were only 3,841 which increased to 5,719 till March 1998 registering a growth of 49 per cent. In the successive year 1999 the growth has grown as high as to 227 per cent. It is revealed that during the span of six years from 1997 to 2002 the highest growth has been registered during the year 2000 *i.e.* 338 per cent and the total number of SHGs which were only 3841 in 1997 have mounted up to 1,97,653 during the year 2002.

From the begging of this programme the total number of SHGs linked to banks till March 1997 were only 8,598 which has gone up as high as 4,61,478 till March 2002 and according to NABARD this achievement has further gone up to 7,17,360 till March 2003. This is the indication that the SHG Bank linkage programme is steadily moving towards right direction. NABARD should ensure that this pace is maintained on continuous basis without any obstacle in its way. The involvement of women is these groups is 90 per cent. Till March 2003, 500 banks are participating in this programme with their 30,000 branches. On time payment of bank loan was above 95 per cent from SHG members. About 14,000 participants from lesser experienced NGOs in micro-finance have been trained by NABARD in association with experienced NGOs.

CONCLUSION

The micro-financing programme is working very effectively, but a major challenge for this programme is the viability of non-farm economic activities. Two major problems are, firstly, to find an economic activity that will yield a rate of profit necessary to cover the

Table 37.1 : Progress of Self-Help Group Bank Linkage Programme — Highlights, March 1997-2002

(Rs. in millions)

Sl. No.	*Items*	*March 1997*	*March 1998*	*March 1999*	*March 2000*	*March 2001*	*March 2002*
1.	No. of SHGs linked during the year	3841	5719	18678	81780	182045	197653
	%growth over previous year		(49)	(227)	(338)	(123)	(8.57)
2.	No. of SHGs linked (cumulative)	8598	14317	32995	114775	263825	461478
3.	% of women groups	76	78	84	85	85	90
4.	No. of participating banks	120	150	202	266	314	444
	(i) Commercial Banks	29	31	38	40	41	44
	(ii) Regional Rural Banks (RRBs)	79	102	129	165	166	191
	(iii) Co-op. Banks	12	17	35	61	107	209
5.	No. of Provinces	20	21	24	24	27	All Provinces
6.	No. of districts covered		221	280	362	382	488
7.	No. of NGOs	220	291	550	718	750	2155
8.	Bank Loan (cumulative) Rs. Millions	118.36	237.6	570.7	1929.82	4808.7	10263.38
	US $ Million	2.75	5.53	13.27	44.88	104.54	209.46
9.	Refinance (cumulative)	106.5	213.8	520.6	1501.26	4007.2	7964.81
	US $ Million	2.48	4.97	12.11	34.91	87.11	162.55
10.	No. of families assisted	146000	243000	560000	1900000	4485000	7845126
11.	Average Loan/SHG Rs.	13766	16596	17297	16814	18227	22240
	US $	320.14	385.95	402.26	391.02	396.24	483.48
12.	Average Loan / Family Rs.	810	978	1019	1016	1072	1308
	US $	18.84	22.74	23.70	23.63	23.30	28.43
13.	Model-wise linkage (cumulative) :						
	i Directly to SHGs without intervention/ facilitation of any NGO	13%	18%	17%	14%	14%	16%
	ii Directly to SHGs with facilitation by NGOs and other formal agencies	45%	46%	56%	70%	70%	75%
	iii Through NGO as facilitator and financing agency	42%	36%	27%	16%	16%	9%

Source : www.nabard.org.

interest rate on the loan. Secondly, marketing of the produce is a problem. The main market for non-farm activities is in the urban areas, hence, when these activities are taken up by rural women the produced goods cannot meet the standards of the urban market. Moreover, the distance also imposes a cost of marketing, which these women cannot bear (Madheswaran and Dharmadhikary). It is also seen that at macro level, there is the problem of sustaining the business that could engage much larger number of people, "unless there is a proper marketing system, merely producing the goods would not help" (Sharma, 2003). If the goods produced by the rural poors are sold at a right time with profitable price, then the poors will be able to improve their savings potential, credit handling capacity and access to financial institutions, inculcate entrepreneurial skill, develop an urge for investment and also risk taking attitude through SHGs. Thus micro-financing can do wonders to the rural poor living even in remote villages in improving their economic condition and protecting them from the clutches of the village moneylenders (Singh *et al.*).

In nutshell, it can be said that the overall performance of SHG-Bank Linkage Programme of NABARD is remarkable as it is the largest micro-financing programme in the world and it is moving towards the right direction in combating poverty, unemployment and empowerment of women. Need of the hour is that this pace of growth and development of micro-financing must be continued through SHGs without any obstacle for achieving the long-cherished objective of poverty alleviation and rural development.

REFERENCES

Agrawal, K.P.; Puhazhendhi, Styasai K.J.S. (1997), Gearing Rural Credit for the Twenty-First Century, *Economic and Political Weekly*, 2717-28, October, 18.

Chiranjeevulu, T. (2003), Empowering Women Through Self-Help Groups; Experiences in Experiment, *Kurukshetra*, Vol. 51, No. 5, March, 16-19.

Das Gupta, R. (2001): Working and Impact of Rural Self-Help Groups and other Forms of Micro Financing, *Indian Journal of Agricultural Economics*, Vol. 56 (3).

Desai, B.M. and N.V. Namboodiri (2001), Organisation and Management of Rural Financial Sector: Text, Cases and Exercises, Oxford and IBH Publishing Co., New Delhi.

Desai, B.M. and Namboodiri N.V. (1996), Whither Rural Financial Institutions, *Economic and Political Weekly*, Vol. 31 (31).

Dadhich, C.L. (2001), Micro Finance—A Panacea for Poverty Alleviation : A Case Study of Oriental Grameen Project in India, *Indian Journal of Agricultural Economics*, Vol. 56 (3), 419-43.

Kamal Vatta, (2003), Microfinance and Poverty Alleviation, *Economic and Political Weekly,* Vol. 35 (5), 432-433.

Lakshmanan, S. (2001) Working of Self-Help Group with Particular Reference to Mallipalayam Self-Help Group, Gobichettipalayam Block, Erode District, Tamil Nadu, *Indian Journal of Agricultural Economics*, Vol. 56 (3), 457.

Majumdar Aloka (2003): Breaking new ground - Nabard is now depending more on the market for finances, *Business India*, 61-66, June 9-22.

Majumdar N.A. (1997) : Overhauling the Somnolent Rural Credit System, *Economic and Political Weekly*, October 18.

Madheswaran, S. and Dharmadhikary, A. (2001), Empowering Rural Women through Self-Help Groups: Lessons from Maharashtra Rural Credit Project, *Indian Journal of Agricultural Economics*, Vol. 56 (3), 427-43.

Manimekalai, M. and Rajeswari, G. (2001), Nature and Performance of Informal Self-Help Groups—A case from Tamil Nadu, *Indian Journal of Agricultural Economics*, Vol. 56 (3), 453-54.

Nanda, Y.C. (2003), Financing the Informal Sector—NABARD's Catalytic Role, *The Hindu Survey of Indian Industry,* 49-51.

Namboodiri, N.V. and Sjhiyani, R.L. (2001), Potential Role of Self-Help Groups in Rural Financial Deepening, *Indian Journal of Agricultural Economics,* Vol. 56(3), 401-9.

Puhazhendhi V. and Satyasai K.J.S., (2001), Economic and Social Empowerment of Rural Poor through Self-Help Groups, *Indian Journal of Agricultural Economics,* Vol. 56 (3), 455-61.

Pear Daniel and Philips Michael M. (2001), Gramin Bank Popularised Loans for the Poor, Now It's got a debt problem, *Economic Times,* December, 11.

Satish, P. (2001), Some Issues in the formation of SHGs, *Indian Journal of Agricultural Economics,* Vol. 56 (3), 410.

Singh, Om Raj (2003), Role of NGOs in Fostering SHGs: A Case Study of Myrada, Kurukshetra, Vol. 51 (4), 33.35.

Srinivas, H., www.gdrc.org

Sharma, K.C. (2001), Micro Financing through Self-Help Groups, *Indian Journal of Agricultural Economics,* Vol. 56(3), 460-61.

Sharma, Anil (2003), Micro-credit plan pays big time to self-help groups, *The Economic Times,* June, 28.

Tomar J.S. (1995), Rural Financial Institutions—A Revamp is Overdue, *Economic Times,* May 4, New Delhi.

Tilekar, S.N., Naikade, N.J.; Deshmukh, B.J.; and Patil A.A. (2001), Woman SHGs Affiliated to Chaitanya : A Case Study in Pune District, *Indian Journal of Agricultural Economics,* Vol. 56 (3), 458.

United Nations (1998), "United Nations General Assembly Resolution 52/194 passed in December 1997 (1998)" in *Countdown* 2005, Newsletter of the micro-credit summit campaign, Vol. 1, No. 3, February/March.

World Bank (1997), "Introducing Savings in Micro Credit Institutions : When and How"? CGAP Focus Note, Consultative Group to Assist the Poorest of the Poor, Washington, D.C., USA.

38

LAND DEGRADATION AND FOOD SECURITY IN INDO-GANGETIC PLAIN OF INDIA

A Basic Issue

ALI MOHAMMAD, S.W.A ASHRAF and ROCKET IBRAHIM

As compared to all the relief features and regions of Indian sub-continent, the Indo-Gangetic plain has been most overused historically for various ways to satisfy the needs of the highly dense population. As a consequence a decline in natural soil fertility seems to have occurred over all civilizations. Overgrazing and deforestation are the two most important factors causing the soil degradation. Nutrient loss from agricultural systems are broadly recorded through leaching, various forms of erosion, losses through product removal and volatilization. A characteristics of faster declines in soil fertility is also related to the large demand of the expanding population, nutrient mining of agricultural areas with concomitant shifts of produce to cities, and intensification of agricultural activities without proper regard for long-term maintenance of fertility by application of fertilizers, recycling of organic wastes, fallowing, rotation of crops and prevention of large scale soil erosion. Agricultural policy has often encouraged soil fertility decline and soil degradation. Lastly, the changing cropping pattern in the Indo-Gangetic plain is a major threat to food availability and food security that is going in an undesirable direction rather in an unplanned manner especially in the western region of the great plain. All these rapidly emerging characteristics must be thoroughly studied to save the future of food problem in the region and in future agricultural scientists must implement different policies and programmes.

SOIL FERTILITY STATUS OF INDO-GANGETIC PLAIN

As compared to all India district level fertility status, which ranges between 45 to 85 per cent, Indo-Gangetic plain variations is ranging between 60 to 85 per cent, it is because of the meso-level homogeneity throughout all the 122 districts identified in Indo-Gangetic plain. Soil fertility is often equated with soil health which is defined as "soil's ability to supply essential nutrients in adequate amounts and proportions for plant growth throughout the plants growing season (DeLuca and Jacobson 1988)". It is measured generally in relation to yield variations subject to the proper availability of water etc. These two resources (fertility and water) are the main factors of 'land resource' and yield factor is largely dependent on them. But the yield result is so complicated that if any fertility aspect is lacking then water and fertility alone may not be held responsible for the yield gap. Therefore, every aspect of the land resource is equally important. For example the district Ludhiana of Punjab and Bastar district is having the same level of soil rating Index (fertility index) but there is a very high level of yield gap because of the management of water gap between these two locations. Similar example can be traced from the district of Bahraich of (U.P.) where fertility (soil rating Index) is the highest, i.e., equal to Ludhiana but yield gap is well marked and till 1973 Bahraich was showing the lowest yield in U.P. But the danger of soil degradation is being observed at a very faster rate as discussed above. The dominance of small and marginal farm holdings are largely facing the highest level of loss of soil fertility because of the intensive soil use. A horizontal and vertical variations in soil fertility should be analysed and soil-rating index should be measured at micro level as it is going to emerge as a major problem in agricultural policy.

SOIL RATINGS INDEX (FERTILITY) AND PRODUCTIVITY IN INDO-GANGETIC PLAIN OF INDIA

There are numerous studies related to the fertility and productivity and ecologically suited crops but not often comparable because of the human and social factors operating with the various levels of inputs and disturbing the comparability. For example the districts of Ludhiana of Punjab and Lakhimpur Kheri of Uttar Pradesh belonging to the same soil rating index but variations in productivity is well marked because of the great variations in input levels. The districts of Aligarh,

Etah, Badaun, Moradabad, Pilibhit etc. belongs to the fourth order of soil rating index but productivity level is of second order in Indo-Gangetic plain. Similar type of uncomparable contrasts are available throughout the area under study but many comparable pockets are also available such as Bihar is identified as lowest order both in terms of fertility index and productivity, i.e., the lowest level in the study area.

Soil Ratings are available at district level[1] by calculating the factors such as profile of the area, permeability, degree of weathering, texture, structure, stoniness, salinity, natural fertility, soil moisture retaining capacity and others.

Mostly the characteristics of Indo-gangetic plain is non-comparable between fertility level and existing productivity because of the innovation and diffusion of high yielding variety of seeds, input variations of fertilizers, rate of irrigation, man-land relations, pesticides and other micro factors. Wherever, human interference is of lesser degree, the impact of natural fertility and soil ratings index is positive in Indo-Gangetic plain.

The major reason of non-comparability is also related to some micro factors exercising their impact. Relationship between fertility and productivity is disturbed especially in the low fertility districts of western Uttar Pradesh where productivity is of second order while fertility index is under fourth order. On the whole wherever human and social factors are involved, the relationship is disturbed and comparability is missing. The degree of non-comparability depends on the level and intensity of the human factors.

TRENDS IN AGRICULTURAL DEVELOPMENT IN INDO-GANGETIC PLAIN OF INDIA

Indo-Gangetic Plain of India consisting of 122 districts located in Punjab, Haryana, Uttar Pradesh, Bihar and West Bengal is a homogeneous geographical unit with little heterogeneity affecting the characteristics of agriculture. Only the rainfall variation is exercising its impact in changing the cropping pattern only.

Out of the gross cultivated area of the region about 36 million hectares are under foodgrains producing about 75 million metric tonnes of foodgrains, *i.e.* about 37 per cent of the total foodgrain production of India while the share in cultivated area is only 30 per cent of the total of the country. But the region is holding about 38 per cent of the total population of India.

Soil degradation is a major aspect which is posing a very serious threat and productivity without artificial nutrients is declining to the extent that the farmers are not 'feeling good' to continue it for a longer time. Declining man-land ratio is another factor for the heavy application of inputs for more production. However, as compared to other physical zones of India the Indo-Gangetic plain is going to face a major threat to the survival of the people living on land resource in this region.

TRENDS IN FOODGRAIN PRODUCTION IN INDO-GANGETIC PLAIN

About 36 million hectares land is devoted for foodgrain production which is producing about 77 million metric tonnes out of the total national production of about 200 million metric tonnes.

There exists a major regional inequality in the productivity from Upper Indo-Gangetic plain to the lower plain that is ranging between 1000 kg/hect to more than 4000 kg/hect.

On the basis of the existing variations the great plain is divided into five major productivity regions, i.e., very low, low, medium, high and very high. The regional picture is more or less compact except few locations. The highest productivity districts are located in Punjab and Haryana.

The highest productivity region is occupying about 16 per cent of the GCA and contributing about 26 per cent in the total production while the lowest productivity region is occupying about 20 per cent of the GCA but contributing only 12 per cent of the total production. The regions below the medium level of productivity is occupying about 50 per cent of GCA but contributing only 30 per cent in the total production while the region below the average occupying about 30 per cent but contributing about 45 per cent in the total production. On the whole, backwardness in agriculture in the plain is characterized by 70 per cent area which is contributing only 55 per cent in the total production.

If the agriculturally backward regions of the plain are developed at par with the developed pockets, the Indo-Gangetic plain alone can produce 70 per cent in the total foodgrain production of India. In other words, if the input level of Punjab is introduced to all the regions of the plain, the region alone can feed two-third of the total population of the country.

Cereals Productivity Variations

In Indo-Gangetic plain cereals occupy a major and highly significant position because of the absence of the pulse crops as these occupy only 6 per cent of the gross cropped area while cereals occupy about 94 per cent. The regional variation of cereals is more or less of the same characteristics of foodgrains because of the lesser percentage share of pulses and very low per hectare productivity.

Because of the higher rate of irrigation in the Indo-Gangetic plain the pulses area was shifted to either wheat or rice crops. Pulses crisis in the plain is very serious and unless a HYV variety is not introduced the situation may not be tackled properly with this vital source of protein among the under-age population of the region. A separate plan to tackle the situation is needed keeping in view the ecology and economic value of other crops.

Pulses Productivity Regions

In the Indo-Gangetic plain pulses are the most neglected crop because of the much lower productivity as compared to the wheat and rice. For instance wheat and rice productivity is ranging between 1500 kg/hect to 4000 kg/hect while pulses productivity is ranging between 500 to 1500 kg/hect. A fast decline in this crop is also recorded because of the substitutes available for consumption are much cheaper such as leafy vegetables and potatoes.

Pulses in the Indo-Gangetic plain is occupying about 2 million hectares land and contributing less than 2 million metric tonnes in the total of 76 million metric tonnes of foodgrains. The per head share of pulses in the dietary system is far below than the standard requirement causing various types of malnutritional abnormalities especially among the vegetarian food consumers.

Regional variations in pulses productivity is ranging between 500 kg/hect to less than 1500 kg/hect. It is being cultivated in all the 122 districts of the great plain. A very low level of productivity is recorded only in three districts of the plain, i.e., Sitapur, Midnapore East and Cooch-Behar where productivity per hectare is lesser than 500 kg/hect. A very high level of productivity is recorded in only one district of the great plain, i.e., the district of Champaran West of Bihar where productivity is more than 1500 kg/hect. (Fig. 38.1 and 38.2)

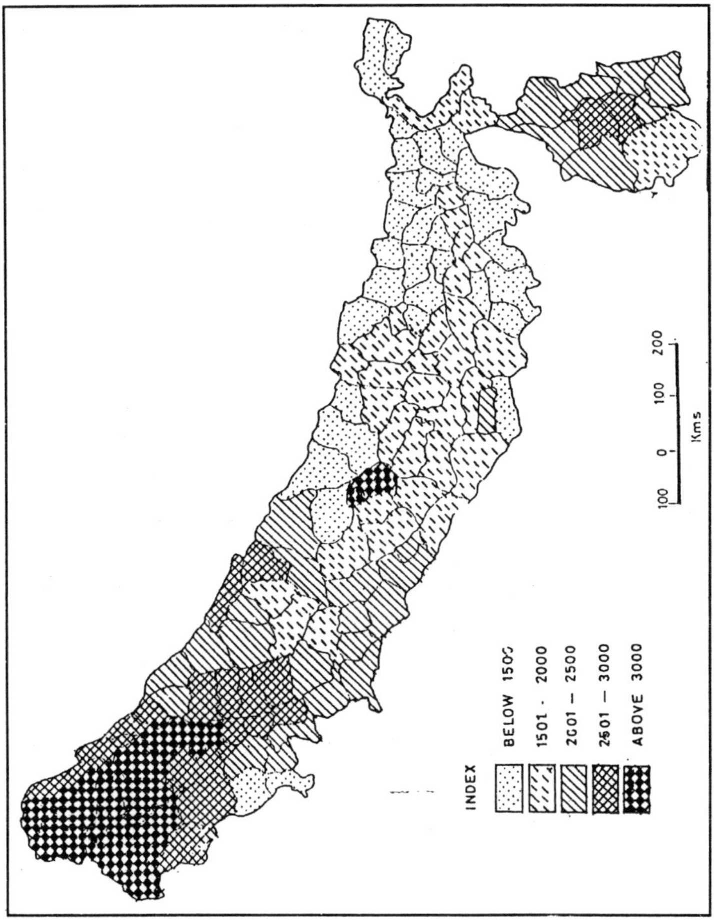

Fig. 38.1 : Indo-Gangetic Plain—Levels of Agricultural Development of Foodgrains

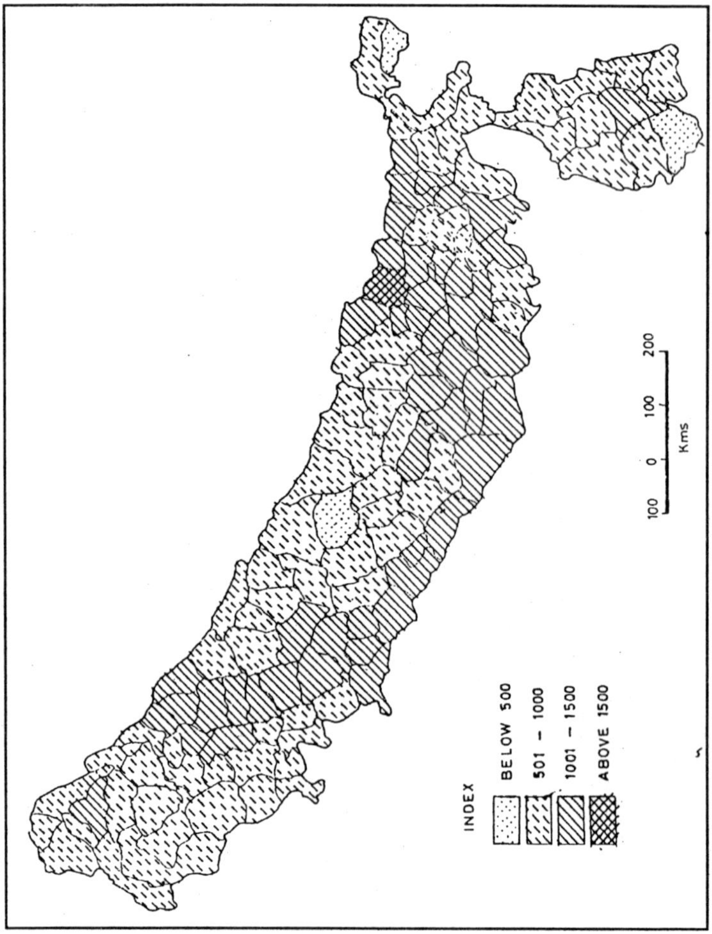

Fig. 38.2 : Indo-Gangetic Plain—Levels of Agricultural Development of Pulses

CROPWISE REGIONAL CHARACTERISTICS

Cropping pattern in Indo-Gangetic plain is dominated by rice and wheat as they occupy more than 30 million hectares out of 37 million hectares of gross cultivated area. Other crops such as maize and bajra which are occupying less than 2 million hectares each while barley is occupying only 0.5 million hectare. After wheat and rice, jowar is occupying the third position as it occupies about 4 million hectares. Maize is a fourth ranking crop occupying about 1.7 million hectares. Other crops such as grain, ragi and tur are occupying less than one million hectare.

Rice Production and Productivity and Regional Variations

Rice is a dominant crop in terms of area as it occupies 15.11 million hectares and producing about 30 million metric tonnes while wheat is a dominant crop in terms of production as it occupies 14.89 million hectares land but producing 42 million metric tonnes.

Regional variations in rice productivity is very high from western part of the great plain to the eastern. Productivity of rice in non-traditional pockets are very high because of the application of high level of inputs such as HYV, fertilizer, irrigation and pesticides. Variations in productivity is ranging between 1500 kg/hect to 4500 kg. About 55 per cent area is under low level of productivity contributing about 46 per cent in the total production. About 20 per cent area is identified under high level of productivity contributing about 30 per cent in the total production.

Wheat Productivity and Production and Regional Variations

Wheat is the most significant fooodgrain crop in the diet of the people in Indo-Gangetic plain. During 1970s due to the increase in area as well as in production its share has increased in almost all the districts of Indo-Gangetic plain. It is being harvested in all the 122 districts of the great plain whereas grown in 120 districts.

It occupies about 40 per cent of the gross cultivated area of the plain while rice is being grown in 42 per cent of the GCA. As compared to rice it is a higher remunerative crop. It is obvious from Table 38.1 'rice and wheat production and productivity level' that rice is contributing only about 29 million metric tonnes and wheat is

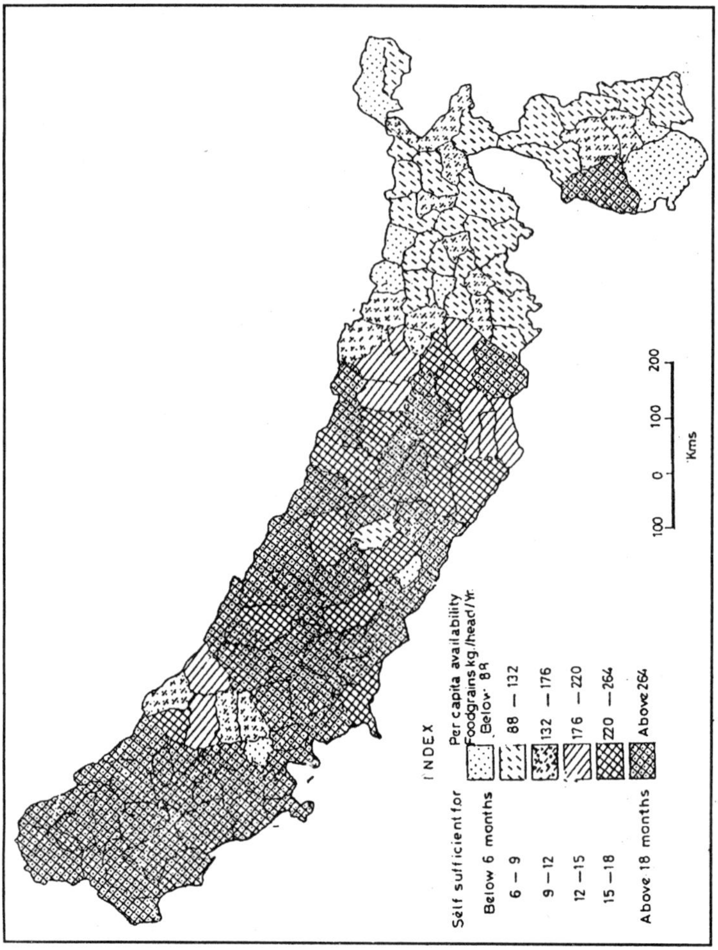

Fig. 38.3 : Indo-Gangetic Plain—Foodgrain Availability (Kg./Head/Yr.)

1. *Cereals* –They are primarily energy and protein sources. Cereals are also considered as basic food items and as an essential food items needed to measure the food security in true sense, as non-basic food items and its production is not needed for food security such as the items like animal products and poultry items. Therefore, a region should be self-sufficient in terms of basic food items. The situation of Indo-Gangetic plain seems to be satisfactory after 1985 and till 2004 the situation has improved dramatically because of the continuous favourable weather after 1989 a rare climatic (rainfall) phenomenon in the past century.

Foodgrain fluctuations is another characteristics of the great plain. This should be analysed as even after good weather the fluctuations in foodgrain production was observed because of the micro-weather conditions in many of the pockets of the plain.

2. Pulses availability is characterized as a serious issue throughout the great plain. Both the area and production declined at a much faster rate than the average of India. It is because of the introduction of HYV seeds in two major high productivity crops (rice and wheat) and introduction of high rate of irrigation which replaced the pulses with these two major crops. Only the marginal land remained under this crop but such type of land is not available at large scale as compared to other regions of the country. For example such type of land is not more than the range of 5 to 13 per cent.
3. Oilseeds and root corps in the great plain of India poses a unique characteristics. The cultivation of root crops has increased phenomenally but the oilseeds has declined to such an extent that the importation of this crop has increased the prices manifold which is not affordable as it was during the period just after independence. It is a major source of protein and energy.
4. Vegetables are energy sources of food but very low in protein. These crops are grown in a very negligible area in the Indo-Gangetic plain.

A balanced cropping pattern is needed for the balanced development of the overall economy and balanced dietary system, otherwise the health condition of some of the major age group population may not suffer (Table 38.1).

Table 38.1 : Per Head Production of Each Crop

(kg/Head)

Regions	*Rice*	*Wheat*	*Jowar*	*Bajra*	*Maize*	*Barley*	*Gram*	*Tur*
Punjab plain	140.50	180.70	1.14	1.70	2.10	0.50	15.07	10.50
Western U.P.	105.70	130.50	1.20	1.35	2.70	0.70	19.05	11.50
Central U.P.	10.70	130.50	2.70	3.70	3.77	3.30	15.07	18.70
Eastern U.P.	90.00	102.70	2.28	2.30	3.60	—	16.07	10.70
Bihar	170.00	30.70	3.78	5.80	8.40	10.50	10.09	10.40
Bengal	200.0	5.70	3.70	—	—	—	—	—

A unique phenomenon is also emerging for analysis, i.e., most of western U.P. districts along the border of Haryana State has rejected the surplus production of foodgrains because of the market problems and governmental ineffective policies regarding the procurement of foodgrains. Therefore, this pocket presenting a picture of food deficit regions because the farmers have devoted 70 per cent land to the commercial crops and remaining 30 per cent to the foodgrains. As such a new model of agricultural development needs a separate micro-level analysis and its application should be analysed in rest of the Indo-Gangetic plain of India.

SUSTAINABLE LIVELIHOOD IN SELECTED VILLAGES OF LOWER GANGA PLAIN

A balanced livelihood and self-sufficiency can be achieved only by increasing the agricultural productivity through the application of HYV, irrigation intensity and high doses of fertilizers. But soil degradation threat is losing the sustainability and decreasing input-output ratio. In such a situation only a micro-level survey and analysis can give the authentic result for a sustainable agriculture.

A large number of villages were selected from the Lower Ganga plain to find out the reliable result because the population pressure in these villages as compared to other villages of the Upper and Middle Ganga plain is much higher. The declining man-land ratio, per head share in production is only on the margin (not surplus nor deficit), therefore, sustainability is always in danger. The level of input is still at lower side, hence, only the increase of input at par with the developed pocket can solve the problem and a better livelihood can be attained.

In such a situation the village households have been divided into three economic units, i.e., poor household, medium and rich

Table 38.2 : Number of Households under Marginal and Small Farmers

S.No.	*Villages*	*Total Number of Households*	*Small farmers (% of total households)*	*Marginal Farmers (% of total households)*
1.	Hainsi	170	30 (17.64)	25 (14.70)
2.	Farid Chak	220	30 (13.63)	35 (15.90)
3.	Para	600	102 (17.00)	95 (15.83)
4.	Noonhara	580	95 (16.37)	68 (11.72)
5.	Nizampur	120	30 (25.00)	25 (20.83)
6.	Susundi	101	20 (19.80)	15 (14.85)
7.	Arzanipur	150	10 (20.00)	20 (13.33)
8.	Hussainpur	40	10 (25.00)	5 (12.50)
9.	Nasirpur	70	15 (21.42)	10 (14.28)
10.	Fatehpur	180	26 (14.44)	25 (13.88)
11.	Khalispur	670	107 (15.97)	112 (16.71)
12.	Chaurahin	45	15 (33.33)	10 (22.22)
13.	Chhaterpur	75	18 (24.00)	17 (22.66)
14.	Bainsagar	112	25 (22.32)	5 (4.46)

households. The poor households include the households having deficit availability of foodgrains through their small landholdings. These households include the small and marginal landholders as well as landless and agricultural labourers etc. Their own production is not sufficient because of the small size of landholdings. The intensity of cultivation has already reached to the saturation point because of the small size of holdings and household labour. The cropping pattern on these landholdings is most diversified and according to the need of the households. The population size of the marginal and small farmers is most important to be analyzed to measure the households below or above poverty line. No doubt outside income is more important to measure the poverty level.

It is obvious from the marginal and small farmers that still one-third of the total number of households are under the small and marginal farmers, who are unable to produce the sufficient amount of foodgrains for the whole year (Table 38.2).

The sustainability is in danger and going to face a critical situation after one decade or after the land is sub-divided among the family members. And if a concerted plan is not started the surplus labour force in the Indo-Gangetic plain may become a serious problem and politically the situation may be disturbed.

Concluding Remarks

1. It was observed that fluctuating situation in foodgrain production is dangerous for a sustainable livelihood even after a favourable weather for agriculture during the period 1989 to 2002. Therefore, if assured base of irrigation is not provided the rate of fluctuations will remain as a major threat to food security system.
2. Land degradation is going to be a major environmental hazards for a sustainable agricultural development as among all the environmental aspects, agricultural land is facing the major threat. The land degradation and soil fertility may be regarded as a major issue for the planning purposes. Generally the application of artificial supplement of fertility and nutrients to the crops may become uneconomic after some years for the sustainable agriculture.
3. It has been observed that a generalized picture of fertility is not very useful unless and until the horizontal variations in fertility and vertical variations should be analysed. Horizontal fertility is related to the variations in fertility from one cropping fields to another, while vertical is related to the depth of soil and at every depth fertility varies as different crops have different rooting system to exploit the fertility of the soils.
4. Diversification is another issue to be tackled at different parts of the Indo-Gangetic plain according to the environment and ecological balance available for different agricultural pursuits such as horticulture, food crops, apiculture, aquaculture, dairy farming, etc.

NOTE

1. O.H.K. Spate and Learmouth, A.T.A. prepared a map of Soil Rating Index by calculating the data available from S.P. Raychaudhuri and K.B. Shome in 'Proceeding of National Institute of Science of India (1960), pp. 260-69.

REFERENCES

Alagh, Y.K. and P.S. Sharma. (1980) Growth of Crop Protection : 1960-61 to 1978-79 – Is Decelerating? *Indian Journal of Agricultural Economics*, Vol. 35, No. 2, pp. 104-118.

Alexandratos, N. (Ed.) (1980). *World Agriculture Towards 2000*. New York University Press, New York.

Ali Mohammad (1980). *Situation of Agriculture, Food & Nutrition in Rural India*, Concept Pub. Co., New Delhi.

Ali Mohammad (1990). *Food Production, Food Problem in India*, Concept Pub. Co., New Delhi.

Ali Mohammad (2002). *Agricultural Development in India – Issues and Challenges*, Sursawera, New Delhi.

Joshi, P.K. and D. Jha. (1990). Environmental Externalities in Surface Irrigation Systems in India. In *Environmental Aspects of Agricultural Development*, IFPRI, Washington, D.C.

Moore, R. and Jan Ryan (Ed.) (1995). *Sustainable Development : Policy and Practice*. New Age International (Publisher), New Delhi.

Moorti, T.V.; K.D. Sharma; and D.R. Thakur. (1991). Trends in the Production of Pulses and Oilseeds in Himachal Pradesh. *Agricultural Situation in India*. Vol. 46, No. 5, pp. 303-308.

Pierce, John T. (1990). *The Food Resource*. Longman Scientific and Technical, New York.

Roy, R.N. (1987). Trees : Appropriate Tools for Soil and Water Management. In Bernhard Glaeser (Ed.). *The Green Revolution Revisited*, Alleh & Unwin, London, pp. 56-57.

Sekhon, M.K. and Manjeet Kaur, (1996). Green Revolution in Punjab : Other Side of the Coin. Vol., LII, No. 11, *Agricultural Situation in India*, 11, pp. 213-19.

Shah, Deepak, (1997). Foodgrain production in India : A Drive Towards Self-Sufficiency. *Artha Vijnana*, Vol. XXXIC, No. 2, pp. 219-239.

Sharma, T.C. and O. Coutinho, (1998). *Green Revolution Gap*. Rawat, Jaipur.

Singh, G.B., (1998). Natural Resource Management for Sustainable Agriculture in 21st Century. *Indian Farming*, Vol. 44, No. 5, pp. 7-12.

Singh, Tapeshwar (1978). *Drought Prone Areas in India*, People's Publishing House (PPH), New Delhi.

Singh, Tapeshwar (1995). *Drought Disaster and Agricultural Development in India*, PPH, New Delhi.

Spitz, Pierre, (1987). The Green Revolution Re-examined in India. In B. Glaeser (Ed.) *The Green Revolution Revisited*. Allen & Unwin, London, pp. 56-57.

Swaminathan, M.S. (1991). Towards Sustainable Agriculture. *Update Agriculture*, July 1991.

Sud, Surinder. (1997). Crop Productivity and Sustainability. *Yojana*, Feb. 1997, pp. 5-8.

Tandon, H.L.S. (1995). Fertilizer and Integrated Nutrient Recommendations for Balance and Efficiency. *Fertilizer Development and Consultation Organisation* (FDCO), New Delhi, p. 4.

TERI. (1997). *Looking Back to Thin Ahead.* GREEN India 2047, p. 2.

Vandana, Shiva. (1989). *The Violence of the Green Revolution.* Research Foundation for Science and Ecology (RFSE), Dehradun.

INDEX